Cardiac and Vascular Biology

Volume 9

This book series publishes review articles discussing important aspects in research and in clinics in the field of cardiac and vascular biology. Basic research aspects of medically relevant topics are covered and the latest advances and methods from diverse disciplines such as genetics, epigenetics, mechanobiology, platelet research or stem cell biology are featured. The book series is intended for researchers, experts and graduates, both basic and clinically oriented, that look for a carefully selected collection of high quality review articles on their respective field of expertise.

Markus Hecker • Dirk J. Duncker
Editors

Cardiac Mechanobiology in Physiology and Disease

Springer

Editors
Markus Hecker
Department of Cardiovascular Physiology
Heidelberg University
Heidelberg, Germany

Dirk J. Duncker
Department of Cardiology, Thoraxcenter
Erasmus MC, University Medical Center Rotterdam
Rotterdam, The Netherlands

ISSN 2509-7830 ISSN 2509-7849 (electronic)
Cardiac and Vascular Biology
ISBN 978-3-031-23964-9 ISBN 978-3-031-23965-6 (eBook)
https://doi.org/10.1007/978-3-031-23965-6

This Springer imprint is published by the registered company Springer Nature Switzerland AG
The registered company address is: Gewerbestrasse 11, 6330 Cham, Switzerland

Foreword

Biomechanical forces play a major role in organ development, shape and function. When exceeding the physiological range, however, they can become detrimental to organ structure and function. This is probably best exemplified by the cardiovascular system, with both the heart and blood vessels being continuously exposed to the biomechanical forces exerted by the pressure and flow of blood, which not only produce acute changes in cardiovascular function but also result in structural changes in the cardiovascular system.

The heart is a unique organ that, in concert with the resistance to blood flow in the pulmonary and systemic arterial beds, generates arterio-venous pressure gradients that provide the energy for the forward flow of blood through the pulmonary and systemic vascular beds. The constant beating of the heart together with the pressures it generates has resulted in a highly specialised muscle of which the shape and function are under the constant influence of the forces generated by and exerted on the cardiac muscle in a close interaction with the vasculature to which it is coupled.

Throughout life, the mechanical loading conditions of the heart change, starting during embryonic development, through birth and subsequently during physiological growth of the body, and these changes stimulate the heart to ultimately reach its adult size and shape. In the healthy heart, the sudden variations in mechanical loading conditions, e.g. during postural changes, physical activity or emotional stress, can be accommodated for by acute changes in cardiac function. However, when the heart is exposed to either repeated or sustained increases in loading conditions, it can adapt with either physiological remodeling (e.g. in response to exercise training) or pathological remodeling as occurs, for example, in response to chronic pressure or volume overload (e.g. due to valvular disease or hypertension) or in response to loss of viable myocardial tissue (e.g. in acute or chronic myocardial ischemia).

The mechanisms by which all these changes occur as well as the contributions of the various cell types in the heart remain incompletely understood. All these aspects of cardiac mechanobiology along with many more facets of this fascinating, timely and highly clinically relevant field of research are addressed by the various chapters in this book. They have been written by several leading experts on this subject from around the globe and cover a wide range of topics from basic research to disease-relevant mechanisms and therapeutic options.

We wish you an interesting and enjoyable read of this ninth volume in the book series, which is the sequel to the Vascular Mechanobiology volume in the Cardiac and Vascular Biology book series.

Department of Cardiology,
Thoraxcenter, Erasmus MC, University
Medical Center Rotterdam, Rotterdam,
The Netherlands
Dirk J. Duncker

Department of Cardiovascular
Physiology, Heidelberg University,
Heidelberg, Germany
Markus Hecker

November 2022

Contents

Cardiac Mechanoperception and Mechanotransduction: Mechanisms of Stretch Sensing in Cardiomyocytes and Implications for Cardiomyopathy 1
Matthias Eden, Lucia Kilian, Derk Frank, and Norbert Frey

Mechanotransduction in Heart Development 37
Alexandra E. Giovou and Vincent M. Christoffels

Mechanobiology of Cardiac Growth in Health and Disease 51
Gloria Garoffolo and Maurizio Pesce

The Role of Mechanosensitive Signaling Cascades in Repair and Fibrotic Remodeling of the Infarcted Heart 61
Claudio Humeres, Harikrishnan Venugopal, and Nikolaos G. Frangogiannis

Mechanobiology of Cardiac Fibroblasts in Cardiac Remodeling 101
Luke R. Perreault and Lauren D. Black III

Mechanobiology of Cardiac Remodelling in Cardiomyopathy 121
Thomas Randall and Elisabeth Ehler

Biophysical Stretch Induced Differentiation and Maturation of Induced Pluripotent Stem Cell-Derived Cardiomyocytes 141
Jiabin Qin, Alain van Mil, and Joost P. G. Sluijter

Mechanical Considerations of Myocardial Tissue and Cardiac Regeneration 181
Ignasi Jorba, Milica Nikolic, and Carlijn V. C. Bouten

Mechanobiology of Exercise-Induced Cardiac Remodeling in Health and Disease 211
William K. Cornwell III and Peter Buttrick

Cardiac Microvascular Endothelial Cells and Pressure Overload-Induced Cardiac Fibrosis 229
Jaafar Al-Hasani and Markus Hecker

Cellular and Subcellular Mechanisms of Ventricular Mechano-Arrhythmogenesis 265
Breanne A. Cameron, Peter Kohl, and T. Alexander Quinn

Computational Biomechanics of Ventricular Dyssynchrony and Resynchronization Therapy 299
Angela W. C. Lee, Marina Strocchi, Cristobal Rodero, Caroline Mendonca Costa, and Steven A. Niederer

Therapeutic Innovations for Heart Failure 337
Anca Remes, Derk Frank, and Oliver J. Müller

Cardiac Mechanoperception and Mechanotransduction: Mechanisms of Stretch Sensing in Cardiomyocytes and Implications for Cardiomyopathy

Matthias Eden, Lucia Kilian, Derk Frank, and Norbert Frey

1 Introduction

Increased myocardial distension, as a result of acute or chronic hemodynamic pressure overload, as it occurs in settings of persistent arterial hypertension and valvular or ischemic heart disease, is one of the most frequent challenges for the heart [1]. From a hemodynamic point of view, the short-term adaptation to increased cardiac load as an initial increase in systolic force generation has been described more than 100 years ago and is still known as "Frank-Starling mechanism" [2].

But since, the ability of cardiac tissue and in particular the cardiomyocyte itself, to respond and chronically adapt to various loading conditions throughout development and disease, crucial to maintain hemodynamic stability and prevent circulatory congestion or to compensate for increased wall stress, has also been extensively studied.

In particular, it has been analyzed how cardiomyocytes that are coupled chemically, electrically, and mechanically sense active contractile forces and passive tissue stiffness and how this affects signal transduction and gene regulation [2–5]. The rapid development of new biochemical, genetic, and imaging techniques has allowed us to gain even more insight in the innumerable and diverse molecular components and signaling processes that are involved herein. These intracellular and extracellular

M. Eden (✉) · N. Frey
Department of Internal Medicine III, University of Heidelberg, Heidelberg, Germany

German Centre for Cardiovascular Research, Heidelberg/Mannheim, Germany
e-mail: matthias.eden@med.uni-heidelberg.de

L. Kilian · D. Frank
Department of Internal Medicine III (Cardiology and Angiology), University Hospital Schleswig-Holstein, Campus Kiel, Kiel, Germany

German Centre for Cardiovascular Research, Hamburg/Kiel/Lübeck, Germany

M. Hecker, D. J. Duncker (eds.), *Cardiac Mechanobiology in Physiology and Disease*, Cardiac and Vascular Biology 9,
https://doi.org/10.1007/978-3-031-23965-6_1

processes mostly include conformational protein alterations as initial events resulting in altered protein-protein interactions, differential protein expression, post-translational modifications (i.e., phosphorylation), and changes in subcellular protein localization, as well as altered ion channel functions and generation of small volatile molecules like nitric oxide (NO) and reactive oxygen species (ROS) [3, 6–12]. Of note, the latter are mediators that are involved in almost all mechanotransduction processes and provide ultra-short-term mechano-electrical feedback and also enable mechanochemical transmission [6, 13].

Beyond these structural and signaling-based mechanisms, cardiomyocytes feature a complex network of calcium entry mechanisms and calcium storages, as well as a large number of mitochondria, responsible for synthesizing and delivering ATP molecules ensuring constant and proper sarcomeric contraction [2, 14–16]. Structural and regulatory components involved in these metabolic mechanisms are closely aligned near the cardiac sarcomere, forming important microdomains that are also able to rapidly adapt to increased energy/metabolic and calcium demands under pressure overload, hypertrophy, and heart failure and thus participate in muscular mechanoperception [11, 17].

Scientific evidence is constantly growing that there are distinct cellular pathways that mediate cardiomyocyte mechanotransduction under either pathological or physiological (i.e., exercise) conditions, involving mostly different intracellular signaling cascades [18]. To add another layer of complexity, these signaling pathways and involved components are believed to also change upon developmental and aging processes [4, 5, 19, 20]. In general cardiomyocyte hypertrophy, denoting initially an adaptive cardiac response to compensate for increased wall stress, results in a pathological maladaptive condition upon persistent activation, so there have to be common mechanisms (and potential future therapeutic targets) dividing physiological or adaptive cardiac hypertrophy signaling cascades from those driving intracellular and extracellular processes leading to increased cardiac morbidity and mortality [1, 5]. To precisely dissect these highly variable and complex mechanoperception and mechanotransmission mechanisms, signaling cascades will help to design future gene therapy-based or pharmacological strategies to improve patient outcomes.

2 The Z-Disc and Mechanotransduction

Multiple studies have identified the cardiac sarcomere with its complex assembly of myofilament proteins as a key structure for mechanotransduction [9, 21, 22]. In particular, the cardiac Z-disc as the lateral boundary within each sarcomere not only consists of multiple layers of α-actinin aligned in antiparallel organization [17] but also forms a complex protein network for sensing sarcomeric strain. The view of the cardiac Z-disc as a mere structural sarcomeric component has extensively changed over the years and is now believed to be the most important signaling nodal point for strain-related downstream signal transduction [23]. Beyond carrying myofilaments, the sarcomere connects to cardiac sarcolemma by cytoskeletal protein complexes that link in particular the Z-disc to the integrin and dystroglycan complexes at costameres. Indirectly, this interface can connect the Z-disc to components of the cardiac extracellular matrix (ECM) through components containing transmembrane domains [5, 8, 17, 24]. Among others, the intermediate filament desmin has also been shown to connect the Z-disc to the cardiomyocyte nucleus [25, 26], while other proteins like mink (minimal potassium channel subunit) or telethonin/TCAP are believed to tether cardiac t-tubules to the sarcomeric Z-disc [27, 28]. Taken together, this places the sarcomeric Z-disc within the center of multiple cardiomyocyte signaling microdomains.

Titin, muscle LIM protein (MLP), calsarcin 1, and filamin C are likely critical players in force transmission and sensing within the sarcomeric Z-disc [23]. But a constantly increasing number of Z-disc proteins (TCAP, LMCD1, ALP, ABLIM, Cypher/ZASP/Oracle, PICOT, FHL1/2, CEFIP, muscle-specific ankyrin repeat protein (MARP), nexilin, myopalladin, myotilin) have been recently discovered and shown to participate in complex signaling networks implicated in mechanotransduction, hypertrophy signaling, and cardiomyocyte contractile (dys)-function as discussed below [5, 22, 23, 28].

2.1 Strain Sensing and Mechanotransduction at the Sarcomeric Z-Disc/I-Band

Titin is a giant protein that pervades the sarcomere from the sarcomeric Z-disc to the center of the sarcomere at the M-band [29, 30]. Titin contains multiple distinct protein domains that have been shown to serve as anchoring hubs for a variety of additional proteins that are all involved in myocardial stretch responses during cardiac hypertrophy and failure [29–31]. The overall cellular location along the sarcomeric filaments, as well as the unique structural properties of each titin molecule itself, makes titin an ideal candidate for sensing systolic or diastolic biomechanical forces acting on the sarcomere and those being generated by the sarcomere itself during active contraction [29, 32]. A large variety of functional roles for titin have been revealed, including the control of correct sarcomere assembly, sarcomere length and stability, the generation of stretch-adapted passive sarcomeric stiffness, and as signal transducer in response to pressure overload [16, 29–31, 33–36]. I-band titin contains unique domains and interdomain linkers that are believed to follow a "stepwise extension" model. Initially, upon moderate strain application, at the sarcomeric I-band, N2BA titin is elongated by straightening their Ig interdomain linkers [29, 37]. When this capacity is worn out upon further increase of strain forces, the PEVK-domain and finally the N-Bus domain further extend and unfold to expose phosphorylation and binding sites [30, 31, 34]. Taken together, titin and its unique structural and regulatory features are thus believed to represent the potential molecular fundament for the aforementioned Frank-Starling mechanism.

2.2 *The Z-Disc/I-Band Titin/MLP/TCAP Complex: Stretch Sensing Within Sarcomeric Borders*

The muscle LIM domain protein (MLP) is specifically expressed in all striated muscle cells and has been implicated in a broad range of different cellular functions [5, 21, 28]. It contains two LIM domains and thereby interacts with multiple proteins, including α-actinin, titin, and TCAP at the Z-disc and other subcellular compartments including intercalated discs and even the nucleus [21]. Of note, MLP protein was also found at sarcomeric I-bands. At Z-discs, MLP-associated TCAP is hereby able to weave two titin filaments at the Z-disc [5]. This protein complex has been shown to be exceptionally resistant to biomechanical forces, due to multiple hydrogen bonds between TCAP and titin [5, 30]. Moreover MLP is believed to shuttle into the nucleus to promote nuclear signaling in muscle cells (possibly via GATA4 and SRF) [21]. MLP knockout mice display a severe dilated cardiomyopathy phenotype resulting in heart failure and premature death, whereas in vitro, MLP deficiency resulted in abrogation of hypertrophic signaling upon neurohumoral or strain-related activation [21, 23, 28]. The W4R-MLP polymorphism, found in human DCM, results in an altered MLP structure, incapable to interact with TCAP [38]. Knock-in mice with homozygous expression of the W4R-MLP show spontaneous cardiac hypertrophy without pressure overload possibly via alteration of elastic intrinsic properties of Z-disc titin and/or shuttling of mutant MLP to the nucleus [28, 38]. Yet, cardiac-specific overexpression of MLP does not protect against increased biomechanical stress due to pressure overload [39]. In addition to the aforementioned MLP functions, it is believed to recruit the phosphatase calcineurin to the Z-disc, a major signaling enzyme linked to pathological cardiac hypertrophy as discussed below [21] (Fig. 1).

Myopodin, also known as Synaptopodin2, was initially described as actin-bundling protein enriched in the heart and skeletal muscle and is believed to be responsive to cellular stress induced by heating of myoblasts to 43 °C [40]. The fact that myopodin is also part of a Z-disc signaling complex containing alpha-actinin, calcineurin, Ca^{2+}/calmodulin-dependent kinase II (CaMKII), muscle-specific A-kinase anchoring protein, and myomegalin suggests an important function in Z-disc's associated signaling [41]. This is further supported by the finding that protein kinase A (PKA) or CaMKII-dependent myopodin phosphorylation mediates its binding to the chaperone 14-3-3 and subsequent nuclear import, whereas dephosphorylation of myopodin by calcineurin counteracts this process. Another relation of myopodin function to cellular tension was reported in mammalian A7r5 cells, where tension-induced unfolding of filamin disrupted a complex of myopodin, BAG3, and filamin which led to autophagosome formation and subsequent filamin degradation [42].

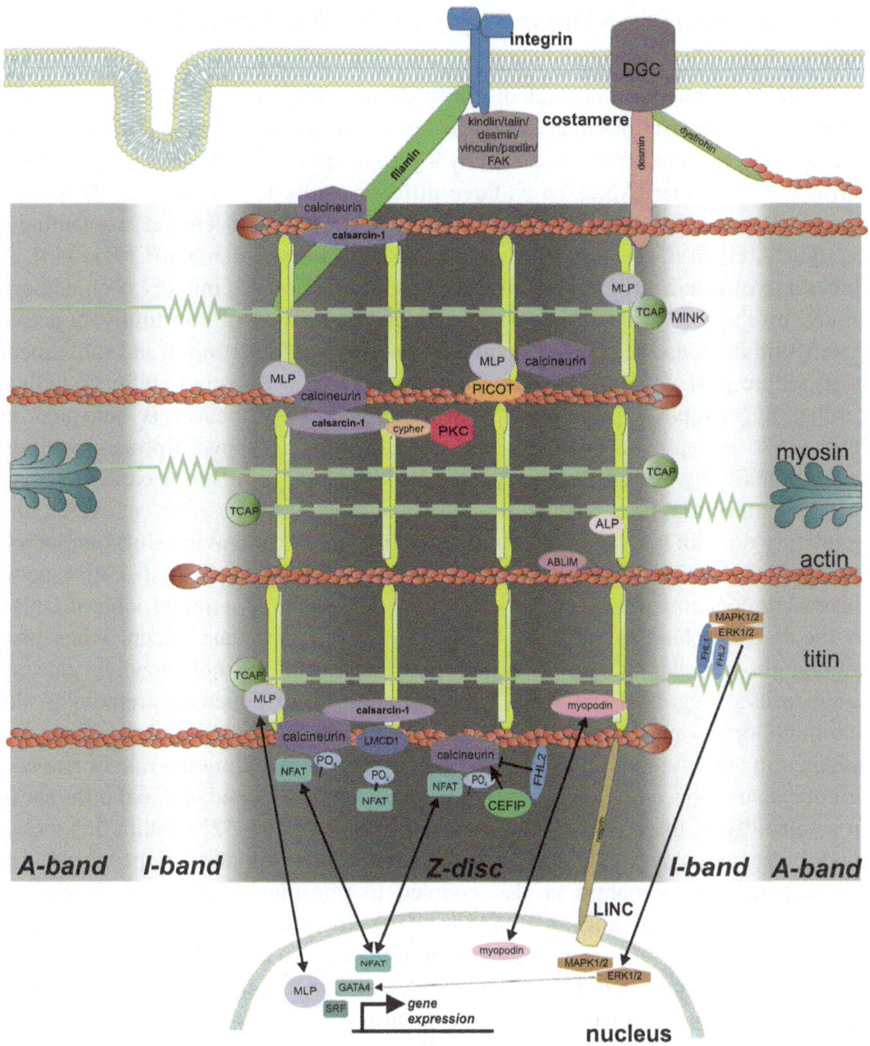

Fig. 1 Signaling molecules embedded in the sarcomeric Z-disc in context to the nuclear and cellular lamina. Schematic illustration of typical multiprotein complexes involved in cardiomyocyte mechanotransduction. At the sacromeric Z-disc, the calcineurin-NFAT pathway tethered via calsarcin 1 onto the sarcomere is shown. NFAT as well as sarcomeric MLP is believed to shuttle between the Z-disc and the nucleus. At the sarcomeric I-band, a titin-associated complex involving FHL1 and 1 and MAP-Kinases Erk1/2 is displayed. ERK1/2 is also believed to shuttle in the nucleus to promote hypertrophic gene expression via transcription factors like GATA4. *GPCR* G-protein-coupled receptor, *At1R* angiotensin II type 1 receptor, *SR* sarcoplasmic reticulum, *LINC* linker of nucleoskeleton and cytoskeleton

2.3 Calcineurin as Major Player in Cardiac Stress Signaling

The calcium/calmodulin-dependent phosphatase calcineurin (CaN) and its downstream targets, transcription factors of the nuclear factor of activated T cells (NFAT) family, play essential roles in cardiomyocyte signaling [10]. Calcineurin typically forms a heterodimer expressed as three different isoforms, namely, αCaN, βCaN, and γCaN, while the structure of each isoform shows similar features, containing a catalytic A chain (CnA) and calcium-binding regulatory B domain (CnB1) [43]. A direct link of calcineurin activity to mechanotransduction has initially been demonstrated by the fact that in vivo, transgenic overexpression of constitutively active calcineurin in mouse hearts leads to massive ventricular hypertrophy and subsequent heart failure [44]. In contrast, mice lacking βCaN show impaired ability to induce cardiac hypertrophy upon pressure overload [45]. Calcineurin is activated in response to increases in local calcium concentrations as stepwise process with an inactive, partially active, and fully activated state [43]. At basal cardiomyocyte calcium concentrations, calcineurin remains in its inactive state, in which the regulatory domain is folded onto the B chain binding helix, leaving CnB1 unbound. The partially active state occurs when loose calcium ions bind to CnB1 upon increasing concentrations, resulting in interaction of the N-terminal lobe of CnB1 to the B chain binding helix of CaN. The fully active state occurs when also calmodulin (a calcium-sensing protein) associates to the holoenzyme, causing calcineurin's regulatory domain to correctly fold, thereby removing its autoinhibitory domain from the catalytic site [43]. To ensure tightly controlled calcineurin activity, many factors are likely involved, including the rate of rise and fall of calcium concentrations, association with calmodulin, and regional differences in availability of these factors in certain microdomains [46–52]. Within this global model, cardiomyocyte calcium entry is linked to dephosphorylation, whereas in contrast, calcium signaling is also coupled to phosphorylation through various calmodulin-modulated kinases (CAMKs) [10]. In this interplay, these processes offer a distinct coordination between calcium-induced calcium release and excitation contraction coupling and phosphorylation/dephosphorylation processes affecting signaling and gene regulation. While calcineurin is primarily located in the cytosol, various protein binding partners, like MLP, calsarcins, and A-kinase anchoring proteins, are believed to tether calcineurin to different subcellular signaling microdomains, like the t-tubule-Z-disc interface as well as the nuclear envelope. This localization in close proximity to voltage-activated L-type calcium channels (LTCC) and transient receptor potential canonical (TRPC) channels at t-tubule membranes and ryanodine receptors (RyR2) in the sarcoplasmic reticulum (SR) place calcineurin in an ideal location for sensing calcium fluxes upon short-term increases in local intracellular concentration [43–49]. The most studied substrates of calcineurin are the family of nuclear factors of activated T cells (NFATs). Desphosphorylation of these transcription factors by calcineurin in the cytosol results in exposure of a nuclear localization signal, resulting in NFAT translocation to the nucleus, thereby activating prohypertrophic gene expression programs [10, 43]. Interestingly, NFATs function as heterodimers in cooperation with other prohypertrophic transcription factors like MEF2 and GATA4, which can also be directly activated by calcineurin-mediated dephosphorylation [10]. Because of its crucial functions in maintaining cardiac homeostasis, calcineurin activity is tightly regulated by multiple endogenous mechanisms. The calcineurin inhibitor CABIN-1/CAIN interacts with calcineurin as well as acting directly as a co-repressor of the SRF-associated transcription factor MEF2. Since CABIN-1 is a large, multidomain protein, it is assumed to function as a calcineurin scaffold, facilitating interactions with other regulatory proteins [10]. Carabin is another endogenous calcineurin inhibiting protein showing reduced abundance in models of pressure overload and in human heart failure. Knockout of carabin in vivo results in exaggerated pressure overload-associated cardiac hypertrophy and heart failure, whereas cardiomyocyte-specific overexpression of carabin has shown to be protective [10]. The family of regulators of calcineurin (Rcan1, 2, and 3) consists of small proteins that can potently inhibit calcineurin activity through a unique C-terminal calcineurin binding domain. Of note, the expression of the cardiac Rcan1.4 isoform is regulated under almost exclusive calcineurin/NFAT transcriptional control and thus provides a direct feedback loop for calcineurin activity [10]. Other negative regulators of calcineurin activity at the sarcomeric Z-disc include calsarcin 1 and the protein PICOT (protein kinase C-interacting cousin of thioredoxin) [48, 53]. PICOT directly interacts and colocalizes with MLP at the Z-disc and thereby disturbs the MLP-calcineurin interaction, resulting in a concentration-dependent displacement of calcineurin from the Z-disc [5, 10, 23]. The calsarcin family of proteins includes three known members, all localizing specifically to the Z-disc. In this regard, calsarcin-1, representing the only isoform expressed in the adult heart, has been shown to be a negative regulator of calcineurin activity [5, 10, 23, 47].

2.4 N2B Titin Controlled Mechanosensing Hubs

An I-band localized subdomain of N2B (N2Bus) titin can act as a molecular spring element, able to unravel upon heavily increased sarcomeric strain and subsequent diastolic distension, thereby functioning as a diastolic length sensor (see above) [32, 36, 37]. At the N2Bus, titin interacts with 4½-LIM domain proteins, FHL1 and FHL2, that are believed to form a unique multiprotein complex with the kinases MAPK/ERK1/2, directly linking mechanical strain to hypertrophic signaling and growth. Load-dependent unfolding of the N2Bus spring-like subdomain releases activated MAPK/ERK2 and enables their active shuttling to the nucleus to promote hypertrophy-associated gene expression [5, 23, 34]. FHL2, which is known to show additional interactions in cardiac muscle, also connects the N2B-domain to integrins, thereby connecting titin to the integrin-associated mechanotransduction pathway at costameres. FHL protein-associated signalosomes gained more attention, since recently a previously uncharacterized protein named cardiac-enriched FHL2-interacting protein (CEFIP) was discovered. CEFIP is upregulated in cardiomyopathy and shows a heart- and skeletal muscle-specific expression profile. CEFIP also localizes to sarcomeric Z-discs where it interacts with FHL2, thereby enhancing calcineurin activity [46]. Beyond FHL2 regulatory processes, titin N2Bus can be phosphorylated by multiple cardiac kinases, including ERK1/2, PKA, and CamKII, possibly in order to provide local adaptation of this regulatory pathway by reducing titin/sarcomeric passive tension [5, 23, 34].

The muscle-specific intermediate filament component desmin not only connects nuclei and mitochondria but also intercalated discs to the Z-disc [54]. Mutations in desmin cause a variety of cardiac diseases like dilated cardiomyopathy, arrhythmogenic right ventricular cardiomyopathy (ARVC), and familial restrictive cardiomyopathy (often sharing the common feature of variable associated conduction system defects) as well as skeletal myopathies [54–56]. Due to its assumed unique elasticity, desmin is believed to sense cellular and protein complex deformations, which may trigger certain conformational changes in response to strain-related mechanical alterations, resulting in diverse signaling events as well as structural alterations (further discussed below) [5, 9, 23, 57].

The Z-disc protein filamin C (FLNC) is also recognized as a potential molecular strain sensor in cardiomyocytes. Moreover, filamin C can act as a structural crosslinker of actin rods at the sarcomeric Z-disc [22]. For all three filamins (filamin A, B, and C), a subcellular localization at either the sarcomeric Z-disc, the intercalated discs, cell membranes, and myotendinous junctions has been described. It is speculated that due to its unique structural characteristics, in particular filamin A and filamin C can also serve as a molecular interface for mechanotransduction [22]. As it has been assumed for other proteins, filamin molecules are able to straighten upon tension, thus changing affinities for protein-protein interactions and subsequent signaling [22]. This seems of special interest, since at Z-discs, filamin C interacts with various proteins implicated in mechanotransduction like calsarcins, myotilin, myopodin, and others [23], and binds to sarcolemma via integrin-1β and sarcoglycan-delta. Mutations in the filamin C gene have also been linked to a variety of cardiac diseases like dilated cardiomyopathy, arrhythmogenic right-ventricular cardiomyopathy (ARVC), familial restrictive cardiomyopathy, and hypertrophic cardiomyopathy (sharing some important features of desmin-related diseases like increased rates of associated cardiac conduction defects) [22].

3 Sensing of Mechanical Strain and Signal Transduction at the Center of the Sarcomere, the M-Band

Beyond its function as a structural anchor for thick and thin filaments within the center of the sarcomere, the M-band is a precisely organized nodal point in signaling and has been recognized as biomechanical strain sensor and regulator of sarcomeric force imbalances during active muscle contraction [58]. Additionally, it is involved in transfer of these biomechanical signals into altered signal transduction and has its role in hypertrophic signaling, cardiac protein turnover, and cardiomyocyte calcium handling. Like the Z-disc, the sarcomeric M-band is composed of a multiprotein structural and signaling hub with numerous interactions and associated pathways [29–31, 33, 34].

From a structural point of view, it has been largely accepted that regular M-band assembly is needed for proper packaging of the thick filaments with appropriate distances and overlap to the thin filaments at the onset of sarcomeric contraction (termed "M-bridges") [58]. In this context, the M-band domain of titin plays a pivotal role besides the proteins myomesin 1 and myomesin 2 that have been shown to directly interact with M-band titin and also function as "M-bridges." Myomesins are organized into antiparallel dimers to the central zone of myosins and to the C-terminal region of titin and thereby crosslink thick filaments [58].

As outlined above, due to its elastic I-band domains and its unique filament-like structure, titin is an ideal stretch sensing molecule and adaptable molecular spring that can integrate various biophysical signals about acute and chronically altered myofilament tension [30, 34]. The M-band domain of titin shows multiple interaction hubs linking the M-band to the protein turnover machinery (via Nbr1/p62/Murf2) [59]. Hence, upon inactivation in the skeletal muscle, titin kinase (TK)-mediated conformational change results in disruption of the Nbr1/p62/Murf2 complex and subsequent inhibition of SRF signaling [58]. Single-molecule analysis experiments by atomic force spectroscopy revealed that exposure of recombinant TK domain to stretch forces causes a conformational change required for ATP binding and access of the autoinhibitory tyrosine, which fully activates the kinase by removal of the carboxy-terminal autoinhibitory tail by yet unknown protein cofactors [29, 58].

Beyond that, M-band titin interacts with myospryn, which is known to also bind the cardiac phosphatase calcineurin, thereby indicating an SRF-independent sarcomeric hub at the M-band for prohypertrophic or anti-hypertrophic signaling [49]. Analyses in titin knockout mouse models revealed that specific depletion of M-band titin leads to atrophy and preserved cardiac function, whereas depletion of full-length titin leads to dilated cardiomyopathy and heart failure, while both phenotypes are associated with premature cardiac death before day 40 [35]. In line with these results, a mouse model with a conditional depletion of the M-band-associated titin kinase region developed dilated cardiomyopathy with altered sarcomeric architecture and dissociation of MURF1 [60]. A mouse model mimicking the human DCM related c.43628insAT mutation, as well as a mouse model expressing the A178D missense variant, develops a dilated cardiomyopathy phenotype that can be aggravated under neurohumoral stimulation via prohypertrophic factors [61].

But beyond titin's functions, the overall M-band composition varies upon different developmental stages and under pathological conditions. Differential expression levels of mature myomesin and an alternatively spliced embryonic ("EH"-) myomesin isoform are thought to contribute to specific functional and structural features at different cardiac developmental stages. Alterations of M-band assembly with re-expression of EH-myomesins and "fuzzy"—less stiff—sarcomeres are believed to accompany and contribute to the progression of cardiomyopathy

[58]. A similar phenomenon was described for titin itself that exists in two main co-expressed isoforms in the mammalian heart, which are differentially expressed in the developed and diseased heart ("titin switch" with exchange to N2BA (more compliant) from N2B (stiffer) or vice versa) [8, 16, 30]. Another novel component of the sarcomeric M-band is the leucine-rich repeat containing protein myomasp/LRRC39, which binds to the rod domain of MYH7 at the M-band, and that is downregulated upon pressure overload. Loss of myomasp in vivo results in M-band alterations and reduced contractility associated with impaired SRF signaling, supporting the notion that it is required for M-band integrity and associated stretch signaling [62].

4 Stretch Sensing Between Cardiomyocytes and the Extracellular Matrix via the Intercalated Discs and Costameres

The cardiac cytoskeleton can be broadly subdivided into intermediate filaments, microtubules, myofibrils (with the sarcomere as main contractile unit), and the intercalated discs that all are connected via multiple mechanisms involving multiprotein complexes [5, 8, 13, 17]. Myofibrils are also attached to the plasma membrane at adherens junctions (zonula, fascia, punctum adherens) as zonal subtype of intercalated discs (ID), whereas in turn the sarcomeric Z-disc is anchored to the plasma membrane at the costameres [63]. Additional specialized subtypes of cell-cell contacts at the ID are desmosomes and gap junctions as well as adherens junctions, all of which have been proposed to form a specific joint structure in the heart termed "area composita" [64, 65]. All these highly specified substructures are involved in cell-cell adhesion, cell connection to the extracellular matrix, bidirectional force transmission, and electrical coupling of cardiac muscle cells [13, 63, 66]. Importantly, the extracellular matrix is not only regarded as an environmental milieu but also organizes, connects to, and communicates with neighboring cardiomyocytes and fibroblasts, while absorbing and transmitting biomechanical forces through interaction with cytoskeletal networks [66, 67]. Their main components are laminin proteins and various collagen subtypes with different levels of molecular stiffness (collagen I stiffer than III) [68]. The main connection of the ECM to the cytoskeleton is achieved by transmembrane receptors from the integrin family that link the ECM with extracellular domains and also connect to macromolecular complexes containing kindling/paxilin, talin, vinculin, focal adhesion kinase, desmin, and VASP (vasodilator-stimulated phosphoprotein) with cytosolic domains [8]. Integrin signaling complexes hereby transduce mechanical forces *outside-in* and *inside-out*, a process that is tightly regulated by multiple signaling mechanisms [63, 67]. Beta1D integrin was found to be expressed at intercalated discs as well as costameres and binds to alpha 7B integrin, which forms a heterodimeric complex that can phosphorylate focal adhesion kinase (FAK) at Tyr 397. This signaling complex also includes scr/PI3K/Grb2, and has also been associated with mitogen-activated kinase (MAP-Kinase) signaling (e.g., MEK1/2 and ERK1/2-), directly linking ID-based mechanotransduction to hypertrophic signaling and cardiomyocyte growth [69–72]. In addition, the protein melusin has also been shown to interact with FAK at costameres to promote strain-related hypertrophic signaling. Melusin knock-out mice fail to execute a hypertrophic response upon pressure overload and directly undergo cardiac failure, emphasizing melusin's role as additional strain sensor [63, 67, 73, 74].

4.1 The Extracellular Matrix: More Than Just Collagens

As it has already been outlined, the ECM is not only a static environmental milieu interacting with cytoskeletal networks. Acute and chronic tissue biomechanical strain in the myocardium is associated with profound and dynamic changes in the composition of the ECM [75, 76]. Under basal conditions, quiescent cardiac fibroblasts are responsible for constant renewal of ECM proteins, but upon biomechanical stress, fibroblasts are able to proliferate and convert into myofibroblasts, contractile and matrix remodeling cells. This is one of the earliest effects upon cardiac pressure overload, thus leading to deposition of additional collagenous matrix and expansion of the interstitium. Together with ECM crosslinking, matrix deposition results in a profound increase in overall myocardial stiffness [75].

Within the extracellular matrix, cardiac pressure overload is associated with a local induction and activation of various collagenases (MMP-1, MMP-8, MMP-13), gelatinases (MMP-2, MMP-9), stromelysins/matrilysins, and also membrane-type MMPs. Within this context MMP induction may be mediated through activation of proinflammatory signaling pathways like TNF-α and IL-1β. Members of the MMPs family have been implicated in TGF-β activation and may cleave transmembrane receptors, such as integrins or syndecans in order to modify proinflammatory or fibrogenic cascades. MMPs may also act as intracellular and intercellular mediators, promoting degradation of contractile proteins in cardiomyocytes or modulating signal transduction responses in interstitial cells. In contrast to MMP activation, rather decreasing myocardial stiffness, matrix-preserving mediators like members of tissue inhibitors of metalloproteinases (TIMPs) control the deposition of structural ECM proteins, thus increasing myocardial stiffness as a counter regulatory process. Imbalances between those two regulatory systems might accompany the transition of compensated cardiac hypertrophy to decompensated heart failure during long-term cardiac remodeling [75–77].

Stress-induced fibroblast activation in the pressure-overloaded myocardium may also be indirect, involving paracrine factors and interaction via cardiomyocytes and various immune cells involving the release of fibrogenic growth factors. The most prominent cascade is activation of the renin-angiotensin-aldosterone system (RAAS) triggering broad inflammatory signaling and leading to downstream stimulation of TGF-β and other associated pathways [75, 78].

Besides those characterized local and systemic pathways, matricellular proteins play a pivotal role in the paracrine regulation of the pressure overload-associated stress response. Members of the matricellular protein family thereby serve as dynamic integrators of microenvironmental changes between ECM and any other myocardial cell type. Fibronectin deposition is believed to be crucially involved in myofibroblast transdifferentiation and has been implicated as an important mediator in subsequent cardiomyocyte hypertrophy. Increased myocardial load or strain triggers marked upregulation of various splice variants of fibronectin which has in turn been linked to TGF-β pathways promoting fibroblast to myofibroblast conversion [75].

Further members of the matricellular family are tenascin-C; tenascin-X; SPARC; thrombospondin-1 (TSP-1), TSP-2, and TSP-4; osteopontin; periostin; and the members of the CCN family. All protein members of this family share the common function that they are able to modulate fibroblast proliferation, survival, and activation and in myofibroblast conversion that they control certain subpopulations of resident and infiltrating immune cells, directly interfere with cardiomyocytes to control cell survival and apoptosis, and are involved in ECM crosslinking processes [75]. This inducible paracrine repertoire offers a tissue wide adaptive control connecting various cell types and the ECM, and also offers systemic regulatory functions. Whereas tenascin-C can act as modulator of macrophage phenotypes in pressure-overloaded hearts, osteopontin is present at very low levels under basal conditions but becomes rapidly induced upon pressure overload and may act by stimulating a fibrogenic program in cardiac fibroblasts. Beyond that, ECM fragments (so-called matrikines) generated through protease activation are able to induce proinflammatory cascades or activate pro-apoptotic pathways in cardiomyocytes upon pressure overload [76, 78, 79].

But not only structural and paracrine factors are involved in ECM-based mechanotransduction and mechanoperception. The Hippo pathway-associated transcriptional complex YAP/TAZ is believed to be involved in ECM-based mechanosensing since it plays a crucial role in adult cardiac fibroblast migration, proliferation, and differentiation. In adult murine heart's post myocardial infarction, cells with nuclear YAP/TAZ localize at the infarct border zone, suggesting a prompt response of resident stromal cells to ischemia or increased biomechanical load.

4.2 Stretch Transmission at Costameric Integrin-Talin-Vinculin Clutches

The integrin interacting protein talin also seems to be of special interest as potential local molecular stretch sensor, since it can undergo conformational extension upon increased strain, unfolding a binding site for vinculin and thereby easing subsequent actin binding [70, 80, 81]. Talin has been mainly found at costamers but is also highly associated with intercalated discs. This integrin-talin-vinculin-actin-ECM-cell adhesion complex hereby re-forms within seconds, representing a dynamic transmission-like interface with constant engagement and disengagement reinforced by cyclic vinculin binding [80, 81]. Comparable to other mechanotransduction initial events, stepwise unfolding of talin domains represents a key feature, enabling stabilized vinculin actin adhesions under conditions of increased tissue tension [81, 82]. Beyond that, binding of talin to integrin per se activates integrin signaling through open conformation stabilization [69]. Of note, talin exists in two isoforms (talin 1 and 2), of which talin 2 represents the most prominent isoform in adult hearts. Upon pressure overload and in human cardiomyopathy, a re-expression of fetal-like talin 1 occurs, which is considered as rather maladaptive response [81]. Beyond that, talins also connect costameres to the sarcomeric Z-disc via interaction with gamma-actin and alpha-actinin [81, 82]. The intermediate filament protein desmin has also been shown to connect costameres to sarcomeric Z-discs and serves as strain-sensing molecule (as discussed above) and additional shock absorber in force transduction [54].

4.3 Stretch Signaling at Costameric Dystrophin Glycoprotein Complexes

Besides integrin-talin signaling and desmin, the cardiomyocyte dystrophin glycoprotein complex (DGC) is another important signaling hub for costameric mechanotransduction, also regarded as a "shock absorber" [24]. Known members of the core DGC complex are dystrophin, dystroglycan, and sarcoglycan-sarcospan subcomplexes as well as dystrobrevin and syntrophin [83]. DGC also associates with ILK/PINCH/Parvin and MLP [67]. Besides its role in absorption mechanical forces, it has been assumed that DGC protects costameres from fragmentation during strong contraction events as they can occur in skeletal but also cardiac muscle cells [67, 84]. DGC-associated hypertrophic signaling mainly involves MAPK and Rac1 [51]; additional signaling molecules include nitric oxide as highly volatile second messengers (see below) [67]. Dystrophinopathies are a group of specific clinical entities due to mutations in the dystrophin gene and include the Duchenne muscular dystrophy (DMD), Becker muscular dystrophy (BMD), and X-linked dilated cardiomyopathy (XLDCM) [85]. In all three subforms, the heart muscle can be affected to various degrees, depending on the precise genotype and the stage of the disease. Mutations in the dystrophin gene thereby result in altered protein structures, possibly affecting the function of the overall cardiac DCG, ultimately leading to regional replacement of myocardium by fibrotic tissue or fat [86, 87]. During cardiac progression of dystrophinopathies, left ventricular dysfunction and ventricular arrhythmias due to increased fibrosis can occur, ultimately leading to heart failure and sudden cardiac death in final stages [88]. Of note, in particular XLDCM represents a rapidly progressive myocardial disorder, starting in young male variant carriers as dilated cardiomyopathy, leading to death from refractory heart failure within 1–2 years after diagnosis [86]. Of further note, gene therapy for

DMD has recently been successfully applied in large animals, paving the way for future treatment of patients affected with this severe disorder [89] (Fig. 2).

5 Force Sensing and Transmission at Cell-Cell Contacts in the Heart

Intercalated discs (ID) are also believed to be involved in cardiac mechanotransduction through conformational changes in associated proteins and in the sensing of stiffness of the ECM by various mechanisms [63]. While costameric stretch sensing mainly involves talin proteins, ID strain sensing is highly linked to the cadherin family of proteins on a molecular level [63, 90]. Cadherins bind through their cytoplasmic domain to β-catenin which in turn tethers cadherin to the actin cytoskeleton via α-catenin. Like in single talin molecules, catenins can undergo conformational straightening upon increased biomechanical load, exposing a vinculin binding domain and thereby facilitating vinculin binding [63]. Recruitment of vinculin and metavinculin to intercalated discs is believed to rather stiffen the whole ID complex, whereas loss of vinculin or metavinculin, as well as missense mutations in vinculin, have been linked to development and progression of dilated cardiomyopathy [63, 91]. Phosphorylation of vinculin protein at residue Tyr 822 occurs upon stretch and is recognized as another fine-tuning principle in ID-based vinculin signaling [63]. At intercalated discs, the protein N-RAP (nebulin-related-anchoring protein) is able to form another N-cadherin/integrin signaling complex by its interaction with talin, vinculin, MLP, and alpha-actinin and also crosslinks cadherin and integrin signaling. N-RAP is upregulated in MLP knockout mice and dilated cardiomyopathy [63, 92], while transgenic N-RAP overexpression in mice results in right ventricular cardiomyopathy [63], emphasizing its role in proper heart muscle function.

5.1 Stretch Signaling at Cardiac Desmosomes

The desmosome, another specific cell-cell contact structure, represents a multiprotein complex composed of transmembrane cadherin family members desmoglein (DSG2, which has been proposed to sense mechanical forces) or desmocollin, whose extracellular domain interacts with opposite facing cadherin molecules at neighboring cardiomyocytes [13, 93]. Their corresponding intracellular domain thereby interacts with intermediate filaments via the desmosomal proteins plakoglobin, plakophilin and desmoplakin [63]. As outlined, desmoglein is believed to be directly involved in desmosomal strain sensing, since targeted deletion of the extracellular domain of desmoglein in mice leads to a biventricular form of ARVC including biventricular dilatation and dysfunction as well as sudden cardiac

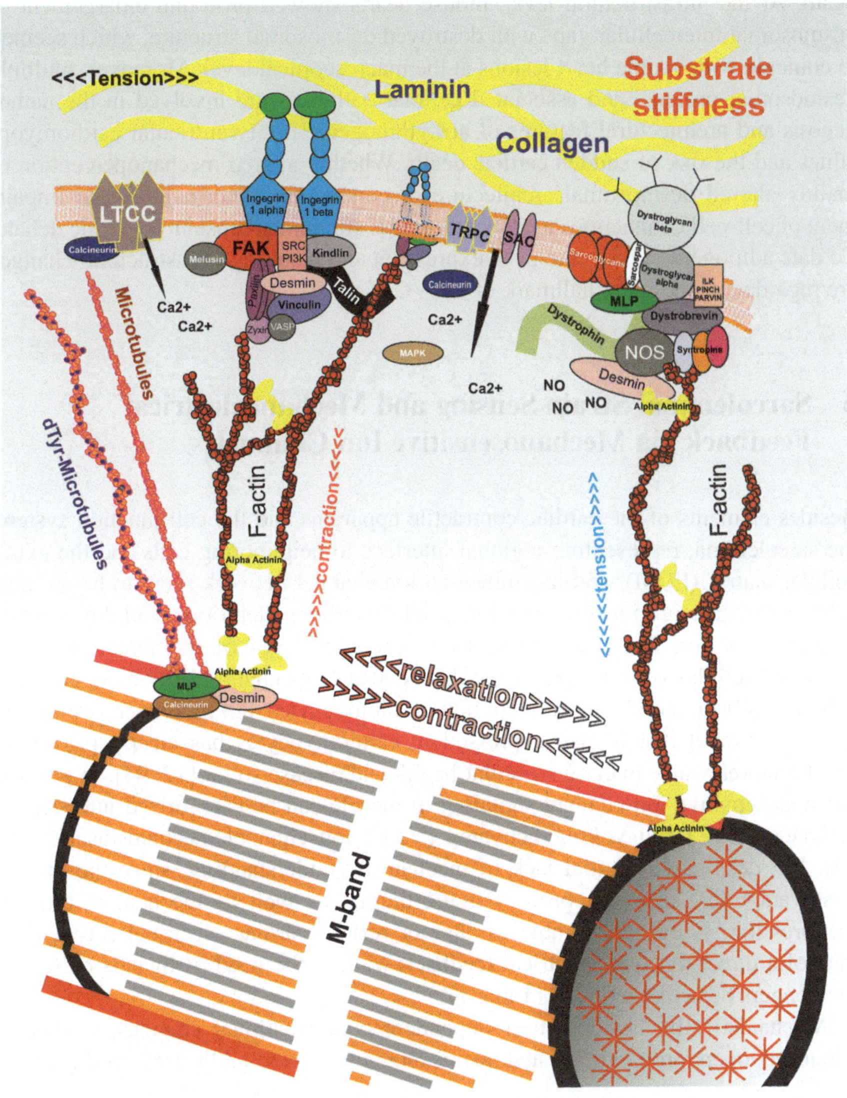

Fig. 2 Signaling molecules and associated pathways at cardiac costameres. Schematic overview of multiple signaling principles connecting cardiac sarcomeres to the cardiomyocyte's sarcolemma via costameres involving the integrin signaling complex and the dystrophin glycoprotein complex (DGC). Both signaling and sensing hubs directly connect intracellular stretch perception mechanisms to extracellular matrix components that are embedded in and also able to adapt to various loading conditions. Via sensing of ECM substrate tension and via adaptation of ECM stiffness mechanoperception, as well as via integration of the contraction-relaxation status of the sarcomere, cardiac mechanosensing is organized by inside-out and outside-in directed mechanisms. *LTCC* L-type calcium channel, *TRPC* transient receptor potential channels, *SAC* stretch-activated channels, *NO* nitric oxide

death. At the ultrastructural level, mouse hearts showed profound enlargement of desmosomal intercellular gaps with destroyed desmosomal structure, which seemed to coincide with visible heart lesions at the macroscopical level. Moreover, multiple desmosomal proteins and associated cellular pathways are involved in the pathogenesis and architectural features of arrhythmogenic right ventricular cardiomyopathies and the risk of sudden cardiac death. Whether altered mechanoperception or mainly altered desmosomal architecture or conduction abnormalities and impairment of cell-cell connections are key features is still under constant scientific debate. To date adipogenic/fibrogenic gene expression and resulting ultrastructural changes are regarded as the main hallmark of ARVC.

6 Sarcolemmal Strain Sensing and Mechanoelectrical Feedback via Mechanosensitive Ion Channels

Besides elements of the cardiac contractile apparatus and the cell junction system, the sarcolemma, representing a global interface to neighboring cells and the extracellular matrix (ECM), and its embedded ion channel network seem to be an ideal subcellular structure for force sensing, conformational alterations, and downstream regulation of ion fluxes according to differential stages of membrane strain or increased cellular volumes [12, 15, 94]. The idea of mechanosensitive ion channels (MCS) has been largely analyzed in non-cardiomyocytes, but moreover, even every specific channel that is also expressed in cardiomyocytes has clearly proven its functional relevance in cardiovascular health and disease so far [12, 94]. In general, mechanosensitive ion channel signaling in muscle can be categorized into stretch-activated (SAC) and volume-activated (VAC) ion channel mechanisms [12, 15, 94]. But despite a potential lack of mechanistic understanding, since the clinical observation that a strong precordial fist thump, as well as mechanical thoracic compression, is eventually able to alter or restore cardiac electrical activity, the idea of a direct mechanoelectrical feedback without the need of ligands or second messengers has emerged [12] (Table 1).

Mechanistically, besides strain-dependent conformational changes, leading to enhanced open probabilities and increasing ion fluxes, additional modifications

Table 1 List of known mechanosensitive cardiac ion channels, encoding genes, and conducted ions

Channel	Encoding gene	Conducted ion
Polycystin 1	PKD1	Calcium, potassium
Polycystin 2	PKD2	Calcium, potassium
TRPC1	TRPC1	Calcium, potassium
TRPC3	TRPC3	Calcium, potassium
TRPC6	TRPC6	Calcium, potassium
TREK-1/TRAAK	KCNK2/4	Potassium
Piezo1	PIEZO1	Cations nonselective
Piezo2	PIEZO2	Cations nonselective

like phosphoinosite binding, phosphorylation, or altered protein-protein interactions have been described [94]. In terms of phosphoinosite-dependent sarcolemmal stretch signaling, the phosphatidylinositol 3-kinase (PI3K) pathway plays a pivotal role in cardiomyocytes [95]. In response to increased cardiomyocyte mechanical strain, the phosphoinosite-converting enzyme PI3K translocates to the plasma membrane to locally convert phosphatidylinositol (4,5)-bisphosphate (PIP_2) into phosphatidylinositol (3,4,5)-trisphosphate (PIP_3) [95]. In turn, PIP_3 activates the phosphoinositide-dependent protein kinase-1 (PDK1) through its pleckstrin- homology (PH-) domain, and PDK1 subsequently phosphorylates and activates the serine/ threonine-specific protein kinase Akt at threonine residue 308, which is further stabilized by additional phosphorylation of serine 473 by mammalian target or rapamycin (mTOR). Akt activation results in a variety of prohypertrophic and prosurvival signals, and is also believed to directly regulate the strain-dependent titin isoform switch [95]. PIP_3 accumulation also subsequently recruits gelsolin to the plasma membrane, thereby negatively regulating its activity and pathological actin remodeling [95]. Finally, it has been recently shown that the protein kinase Akt together with the strain-sensitive channel polycystin-1 can stabilize sarcolemmal LTCC levels by phosphorylation-dependent prevention of proteosomal degradation [96]. This finding represents another important principle, directly and indirectly linking cardiomyocyte sarcolemmal stretch sensing to calcium-induced calcium release and excitation-contraction coupling.

6.1 Mechanosensitive Transient Receptor Potential Channels (Polycystins) in the Heart

Various transient receptor potential channels (TRPs), though initially and sometimes extensively studied in non-cardiac cells, are expressed in cardiomyocytes and are believed to play a pivotal role in mechanoelectrical feedback processes [94]. The subfamily of transient receptor potential polycystins (TRPPs), polycystin 1 and 2 (encoded by PKD1 and PKD2 genes), are involved in sensing mechanical forces and fluid shear stress, triggering multiple intracellular signaling pathways [97, 98]. The notion that cardiac specific polycystin-1 knockout mice develop spontaneous cardiomyopathy supports the concept that they are required to maintain normal cardiac function [99, 100]. Moreover, polycystin-$1^{-/-}$ mice displayed shorter action potential and shortening of QT intervals in surface ECGs, a phenotype that has also been linked to loss of function mutations in the LTCC coding CACNA1C gene or other mouse models of compromised LTCC function-associated cardiomyopathies [101]. TRPP2/polycystin-2 is expressed in cardiomyocyte endoplasmatic reticulum and has also been linked to sarcoplasmatic ryanodine receptor 2 (RYR2) function [97, 98].

6.2 Other Mechanosensitive Calcium Channels in the Heart

Besides TRPP channels, there are over 33 genes that encode TRP subunits, which form a superfamily of mammalian channels, further subclassified as TRP-C, TRP-V, TRP-M, TRP-P, TRP-ML, TRP-N, and TRP-A subfamilies [94]. Several members of the TRPC ("C" for canonical) family channels, especially TRPC1, TRPC3, and TRPC6, are considered to be mechanosensitive and important for accurate cardiomyocyte function and signaling [102, 103]. TRPC1 expression has been shown to increase in an induced rat heart hypertrophy model, while TRPC1$^{-/-}$ mice seem protected from cardiac hypertrophy, which underlines its crucial function in pathological hypertrophic signaling [104]. Most surprisingly, the functions of these multiple TRPC channels do not seem to be redundant in cardiomyocytes, potentially because they act through different regulatory mechanisms [105]. Mice with heart-specific overexpression of TRPC6 develop spontaneous cardiac hypertrophy and pathological remodeling, whereas TRPC6 deletion or inhibition has been shown to be cardioprotective [106, 107]. Of note, TRCPC6 channels are also partially regulated by PI_3K-dependent exocytosis.

Since it has been shown that expression of TRP cation channel, subfamily C, member 3 (TRPC3) is increased in response to calcineurin signaling [52, 102], a potential link from TRPC6 function to calcineurin has also been analyzed. Interestingly, TRPC6 abundance was differentially upregulated in mouse hearts in response to pressure overload and, most strikingly, in failing human hearts. Two conserved NFAT consensus sites in the promoter of the TRPC6 gene have been discovered, representing a potential reciprocal activation circuit in which calcineurin activation results in increased TRPC6 expression [50, 108]. In turn, in vivo pathological cardiac overexpression of TRPC6 was associated with enhanced calcineurin signaling leading to pathologic cardiac growth and heart failure [50].

6.3 Mechanosensitive Potassium Channels in the Heart

Besides TRP channels, multiple types of mechanosensitive K^+ channels are expressed in different parts of the heart [12, 94]. TREK-1 [(*T*WIK)-*re*lated *K*$^+$ channel], also known as potassium channel subfamily K member 2, is expressed in both the atria and ventricles, and extensive research has shown that TREK-1 and its human homologue TRAAK (*T*WIK-*r*elated *a*rachidonic *a*cid stimulated *K*$^+$ channel) are mechanosensitive [109–112]. TREK-1 channels are aligned in longitudinal stripes on the cardiomyocyte surface, which potentially facilitates bidirectional strain sensing [112]. TREK-1 channels that can easily be opened by either pipet suction or pressure in vitro have thereby been linked to arrhythmogenesis and the onset of atrial fibrillation as well as heart failure [12, 94]. The role for TREK-1 in arrhythmogenesis is further strengthened by the fact that several anti-arrhythmic drugs, including lidocaine, mexiletine, propafenone, dronedarone, and vernakalant, inhibit TREK-1 channel function. Compared to TREK-1, the human relevance of TRAAK is less well understood [12, 94].

Recently two newly discovered proteins, Piezo1 and Piezo2, which assemble as transmembrane trimers, were also proposed as bona fide stretch-activated channels with striking characteristics in sarcolemmal force sensing and potentially regulating calcium influx [94]. Piezo1 channels are upregulated in rodent heart failure but are downregulated and potentially delocalized in in vitro cardiomyocyte stretch models [113]. Being largely characterized in endothelial cells and shear stress scenarios, more research in hearts and cardiomyocytes is needed to fully clarify the role of Piezo channels in mechanotransduction in cardiac muscle health and disease.

7 The Microtubule Network in Cardiomyocyte Stretch Sensing: *More Than Just a Skeleton*

The view of the microtubule's role in cardiac function and mechanotransduction has experienced a fundamental change along with the availability of more sophisticated observational and manipulative molecular techniques [114, 115]. Like with other molecules involved in cardiac strain sensing, direct observation of dynamic microtubule networks in beating myocytes also suggests a complex behavior with spring-like functions that is additionally fine-tuned by post-translational modifications [115]. Maybe most relevant, targeted detyrosination (dTyr) of microtubules is a result of enzymatic cleavage of a C-terminal tyrosine residue and is thereby able to facilitate specific interactions with intermediate filament proteins like desmin that link microtubules to the sarcomeric Z-disc or components of the nuclear envelope [26, 116, 117]. Beyond that, recent progress in elegant live imaging techniques has helped to visualize growing microtubule networks in cardiomyocytes. These experiments revealed that microtubules likely grow from Z-disc to Z-disc and even tend to pause, drift, or deviate on this central sarcomeric signaling hub. Strikingly, this growth behavior implicates local protein-microtubule interactions and possible modifications at the sarcomeric Z-disc in affecting microtubule architecture [115]. Modified microtubules are known to alter cardiomyocyte stiffness, resulting in overall increased myocyte viscoelasticity conditioned by elevated detyrosination [114]. Microtubules are also believed to modify cardiac contractility and cytoskeletal mechanosignaling through a variety of processes involving cellular organelles like mitochondria and the nucleus [7, 14, 25, 115]. Moreover, microtubules have been shown to form interactions with structural components and ion channels within the t-tubule system and sarcoplasmic reticulum, controlling ion channel membrane trafficking and indirectly EC coupling [114, 115]. Elevated levels of dTyr microtubules can be a result of elevated levels of reactive oxygen species (ROS) and levels of nitric oxide (NO) as they occur during increased mechanical stretch in cardiomyocytes [115]. dTyr microtubule abundance is also increased in human cardiomyopathy, and increasing detyrosinated micotubules herein correlated with declining systolic function [116]. Moreover, dTyr microtubules are believed to have in particular an increased affinity to bind the intermediate filament and strain-sensing molecule desmin [115]. In this setting, also proper positioning of organelles like mitochondria and the cardiomyocyte nucleus involves the cardiac actin network, desmin, and the microtubule system [26]. Alterations in all these structures (a hallmark of heart failure remodeling) also result in impaired cardiomyocyte mitochondrial metabolism and Ca^{2+} cycling as well as impaired nuclear function, integrity, and even possible DNA damage [25, 115, 116].

7.1 Mechanotransduction at the Cardiomyocyte's Nuclear Lamina

The cardiomyocyte nucleus, representing the command center of cardiomyocyte gene regulation, is another important organelle involved in mechanoperception [25, 26, 118]. Lamins are the main protein components of the inner nuclear lamina and form stable filament webs inside the whole nucleus [119, 120]. Lamins are intermediate filament proteins divided into A-type lamins, derived from alternative splicing of the LMNA gene, and B-type lamins that are encoded by distinct LMNB1 and LMNB2 genes, respectively [121, 122]. Experimental evidence has demonstrated that lamin structures play pivotal roles in the maintenance of normal nuclear mechanics and cardiomyocyte mechanotransduction [119, 123]. Besides its nuclear functions, lamins affect the mechanical properties of the cytoplasm and the organization of cytoskeletal elements as well as regulation of the overall cell shape [119, 122, 124]. Hence, lamins are believed to interact with multiple proteins and thereby can regulate gene expression, chromatin homeostasis, and even nuclear positioning [124, 125]. Cells from lamin knockout mice show decreased association of desmin at the nuclear surface and severe alterations of actin-, vimentin-, and tubulin-based filament structures. A-type lamins, represented by lamins A and C, are developmentally regulated proteins found in high abundance in the skeletal and cardiac muscle. The crucial function of these proteins is emphasized by the facts that mice deficient in lamin A and C develop severe muscular dystrophy and die prematurely at the age of 6–8 weeks [124]. Moreover, human relevance is underlined by a group of diseases caused by mutation variants in the LMNA gene sequence [126, 127]. Laminopathies are usually classified into four groups, according to the number and the types of the affected tissues. The first group represents lamin myopathies affecting both the skeletal and the cardiac muscles [128]. Mutations in the lamin A gene are generally considered one of the most common mutations associated with this disease. LMNA gene defects are believed to account for almost one third of dilated cardiomyopathy cases accompanied by atrioventricular block [128]. Given the knowledge that cells expressing mutated A-type lamins display histological lobulations in the nuclear envelope, loss of peripheral heterochromatin, and anomalous nuclear pore complex distribution, two main models were hypothesized to explain the onset of laminopathies [25, 124]: according to the "structural model," mutations in A-type lamins impair the nuclear resistance to mechanical stimuli, resulting in fragility, increased stress sensitivity, and possibly premature senescence [25, 124]. This model would explain why in particular striated muscle tissues, which are frequently exposed to mechanical strain, are mainly affected by "laminopathies." The mechanistic role of lamins in cardiac mechanotransduction is also related to their unique filamental structure and biophysical behavior. Assembly of the nuclear lamin mesh starts with lamin dimer assembly and subsequent dimer aggregation head to tail, finally forming small polymer filaments [121].

Under increasing cardiomyocyte tension, integrins transmit a strain-dependent external impulse to the cytoskeleton, which then transfers the stimulus to the nuclear lamina structure. In turn, lamins are rearranged at the molecular level, leading to unfolding of lamin A immunoglobulin domains and/or alterations of flexible linkers L1, L12, and L3 in the protein structure, as well as alterations in the dimer head-tail interaction. Within these proposed "altered protein complex" models, mechanical strain straightens the lamin proteins as well as the overall lamin-nuclear-mesh potentially exposing protein-binding regions [128]. The global lamin-mesh rearrangement has been correlated with a higher nuclear localization of cardiac transcription factors (cFOS, RARG, JNK, SRF, MRTF, Yap1, and ERK), and their specific affinity to chromatin by exposing binding sites for these transcription factors to increase transcription of cytoskeletal components [119, 125, 128, 129]. Moreover, cardiomyocytes can modulate the nuclear biophysical properties by changing the phosphorylation level of the lamins, thereby affecting both the structural lamina conformation and stiffness. In this case increased force transmission results in hiding of the sites for phosphorylation. The inhibited phosphorylation activity increases the amount of lamins at the nuclear envelope with a consequent increase in lamina stiffness [130, 131].

7.2 *Mechanotransduction by Protein Complexes Within the Nuclear Envelope*

But besides the lamin web, there are other integral parts of the nuclear envelope that are involved in mechanotransduction like LINC complexes (*LI*nker of *N*ucleoskeleton and *C*ytoskeleton) that contain different nesprins, emerin, and SUNs (named for SUN domain family members (*S*ad1p, *Un*c-84). LINC complexes can sense and regulate myocyte-wide strain transfer to the nucleus itself [25, 26, 132]. LINC complex disruptions have been analyzed and described in cells that typically are prone to experience high mechanical strain, such as myocytes and cardiomyocytes. Nesprins share the common capabilities of stretch sensing molecules, as they have been shown to unfold upon envelope strain, exposing binding sites that promote either dimerization or recruitment of additional binding proteins, facilitating complex stability and rigidity [25, 26]. In mice, deletion of nesprin 1 and 2 gene function results in cardiomyopathy and altered gene expression in response to myocyte pressure overload [133, 134]. Linking the nucleus to the cytoskeleton, microtubules not only can provide compressive forces on nuclei but also show multiple interactions to nesprin 1 and 2 [135, 136]. It is speculated that loss of desmin-nesprin-microtubule interactions or disruption of LINC in nuclear envelopes results in nuclear membrane infolding, driven by external forces applied through external microtubule-network compression [26]. Since correct assembly of LINC, desmin, and the microtubule network is required for accurate nuclear shape, the integrity, positioning, and homeostasis, nesprin-dependent regulation of LINC stability might represent another important cardiac mechanotransduction principle [25, 26, 134–138].

8 Mechanotransduction and Mechanosensitive Gene Induction

Extensive gene expression profiling in cardiomyocytes using cDNA microarrays or RNA sequencing has helped to gain more insight into strain-associated gene regulation and led to the discovery of cardiac mechanosensitive gene programs [139]. In vitro models of cardiomyocyte stretch as well as animal models of experimental pressure overload by transverse aortic constriction (TAC) have consistently demonstrated increased expression of natriuretic peptide genes and re-expression of a so-called fetal gene program. Beyond that Mt1 (encoding metallothionein 1) and various other genes encoding proteins involved in mitochondrial metabolism or the cytoskeleton are believed to be expressed or re-expressed in a strain-dependent manner in cardiomyocytes [140]. More cardiomyocyte genes with potentially specific mechanosensitive expression have been discovered like suppression of tumorigenicity-2 (ST2) and the TGF-beta superfamily member growth differentiation factor-15 (GDF15) [141], which subsequently have been developed as a clinical biomarker for myocardial infarction, hemodynamic cardiac load, or heart failure [142–145]. Myocardial gene expression can be regulated on the posttranscriptional level by a variety of mechanisms like certain microRNAs that are also believed to be differentially expressed in various loading conditions or that are involved in the regulation of mechanotransduction. MicroRNAs are short, non-coding RNAs that bind complementary mRNAs to control mRNA degradation or subsequent protein translation. In an effort to identify specific microRNAs regulated through myocyte stretch, we and others used microarrays under different experimental loading conditions in cardiomyocytes in vitro. These experiments identified miR-20a as being responsive to both stretch and simulated ischemia reperfusion and that overexpression of miR-20a was sufficient to protect cardiomyocytes from apoptosis [146].

9 Translational Perspectives

The increasing insight into molecular mechanisms of mechanoperception and transmission (mechanotransduction) is the potential basis for the development of new therapeutic approaches. Biomechanical stress sensors and downstream signaling pathways involving the above described sarcomere-associated molecules are essential for physiological cardiac function [2, 19]. Impairment or dysregulation of this complex network is directly associated with a variety of congenital and acquired cardiac diseases such as laminopathies, dystrophinopathies, desminopathies, arrhythmogenic right ventricular cardiomyopathy, familial restrictive cardiomyopathy (RCM), valvular and ischemic heart diseases, and dilated and/or hypertrophic cardiomyopathy (DCM, HCM) [27, 54, 56, 58, 63].

The causes for laminopathies, dystrophinopathies, and desminopathy are known mutations in specific genes encoding for proteins associated with sarcomeric proteins, namely, lamins (LMNA), dystrophin (DMD), and desmin (DES) [55, 56, 147–149]. In addition, dozens of mutations in genes encoding for different sarcomeric proteins involved in mechanoperception and translation of stretch-induced signals have been identified to contribute to different cardiomyopathies [38, 98, 150, 151]. Basic research using knockdown approaches in in vitro and in vivo models as well as OMICs studies analyzing patient data can help to identify even more mutations and polymorphisms as risks factors for cardiomyopathies (Fig. 3).

Besides congenital mutations, chronically increased biomechanical stress on the ventricular myocardium, which occurs as a result of pressure overload in hypertension and valvular but also after remodeling processes in ischemic heart diseases and dilated/hypertrophic cardiomyopathies, leads to alterations in signaling cascades triggered by mechanotransduction and subsequently accelerates disease progression to heart failure. A better understanding of the structural and regulatory role of sarcomere-associated proteins in mechanotransduction has already led to tests of new pharmaceutical approaches. In recent years especially ion channels/transporters have come into focus as targets to block detrimental signaling induced by biomechanical stress in the myocardium leading to cardiac remodeling and heart failure. Examples are cariporide treatment of patients with ischemic heart disease and in in vivo models of cardiac hypertrophy [152], rimeporide treatment of children with dystrophinopathy [153], and inhibition of K^+ channels in models of atrial fibrillation [154, 155]. However, further research is essential, as the inhibition often lacks specificity and may increase the risk for side effects (e.g., strokes, as reported for the cariporide treatment [152]), and the molecular mechanisms are still not completely understood (e.g., association of anti-arrhythmic drugs with inhibition of TREK-1[13] on the one hand but correlation of impairment of TREK-1 with arrhythmogenesis on the other hand).

As described above, many other members of the network of cardiac mechanoperception and mechanotransduction are currently studied in in vitro and in vivo experiments with regard to their exact function, and further studies might

CARDIOMYOPATHY	MUTATED GENE/PROTEIN
HYPERTROPHIC CARDIOMYOPATHY	Myosin binding protein C
	ß-Myosin heavy chain
	α-Myosin heavy chain
	α-Tropomyosin
	Troponin T
	α-Cardiac actin
	Troponin I
	Titin
	Myosin light chains
	Junctophilin 2
	Caveolin 3
	RYR2
	PDLIM3
	Troponin C
	Vinculin
	Muscle LIM protein
	α-Actinin 2
	FHL1
	Telothonin
	Calsarcin 1
	Vinculin
	Myomesin 1
	Filamin C
DILATED CARDIOMYOPATHY	β-Myosin heavy chain
	Desmin
	N-cadherin
	α-Cardiac actin
	α-Tropomyosin
	Muscle LIM protein
	δ-sarcoglycan
	Lamin A/C
	TAZ
	Titin
	Phospholamban
	Vinculin
	Troponin I
	Troponin T
	SCN5A
	Presenilin 1 & 2
	Troponin C
	α-Cardiac actinin
	Plakoglobin
	Nexilin
	Nebulette
	Cypher, ZASP
	ILK
	FHL1
	FHL2

Fig. 3 List of mostly accepted cardiomyopathy-associated genes in relation to their disease phenotypes. Genes/proteins that are believed to be directly or indirectly involved in cardiac stretch sensing and mechanotransduction are highlighted in red. Web access to comprehensive lists of cardiomyopathy-associated genes and gene variants is possible via https://www.ncbi.nlm.nih.gov/clinvar/?term=cardiomyopathy, https://www.cardiodb.org/acgv/, https://seidman.hms.harvard.edu/?page_id=1476

	α-Actinin 2
	Calsarcin 1
	Crystallin Alpha B
	Myopalladin
	Myotilin
	RBM20
	Filamin C
RESTRICTIVE CARDIOMYOPATHY	Troponin I
	Desmin
	α-Cardiac actin
	Troponin T
	Tropomyosin α
	Bcl2-Associated Athanogene 3
	Myopalladin
	Calsarcin 1
	ß-Myosin heavy chain
	Filamin C
ARRHYTHMOGENIC RIGHT VENTRICULAR CARDIOMYOPATHY	Plakophilin-2
	Junctional plakoglobin
	Desmocollin-2
	Desmoglein-2
	Desmoplakin
	Filamin C
	Catenin A3
	Lamin A/C
	Titin
	TGF-β B3
	RYR2
	Transmembrane Protein 43
	Lim Domain-Binding 3
	Arrhythmogenic Right Ventricular Dysplasia, Familial 3, 4, 6
	Phospholamban
	Desmin
MUSCULAR DYSTROPHY–ASSOCIATED CARDIOMYOPATHY	Dystrophin
	Sarcoglycan
	Dystroglycan
	Dystrobrevin
LEFT VENTRICULAR NONCOMPACTION	TAZ
	α-Dystrobrevin
	FKBP12
	ß-Myosin heavy chain
	α-Cardiac actin
	Troponin T
	Calsequestrin 2
	E3 Ubiquitin Protein Ligase 1
	Lim Domain-Binding 3
	T-Box 20
	Pr Domain-Containing Protein 16

Fig. 3 (continued)

demonstrate their potential in therapeutic approaches (e.g., titin, MLP, LMCD1, myomasp/LRRC39, melusin, polycystin, talin, TRPC1 and TRPC6, and nesprin).

In this context, not only the development of conventional pharmaceutics is of interest, but also the constantly growing field of gene therapies amplifies the potential targets and therapeutic approaches including a "personalized medicine" approach for genetic disease.

The most promising methods for gene therapy are CRISPR/Cas-based and vector, mostly adeno-associated virus (AAV), mediated approaches. In recent years more and more successful gene editing in vitro and in vivo has been performed, e.g., for removal of HIV proviral DNA sequences or host receptors to treat HIV infections, targeting the convertase PCSK9 which is associated with high blood lipid levels to treat cardiovascular diseases or to correct mutations causing sickle cell disease in stem cells [156]. Although gene editing in humans is ethically controversial, first human cells with induced disruption of genes associated with inhibition of anti-tumor responses and introduced tumor recognition genes have been used to treat cancer patients [157]. With regard to heart diseases, AAV9-mediated CRISPR/Cas9-based gene editing has been used to target specific genes in cardiomyocytes in animal models, too [156, 158]. Along these lines, a number of new animal models for, e.g., dystrophies, cardiovascular diseases, and dilated cardiomyopathies have been created for further research [156]. Furthermore, first in vivo corrections of mutations which cause heart diseases have been described [156, 159], e.g., an AAV9-Cas9-based approach to restore functional dystrophin levels in a canine model for Duchenne muscular dystrophy [89, 156]. The potential of gene therapies in general and for genetic and acquired cardiomyopathies in particular is immense, as beyond correcting mutations, possible targets for CRISPR/Cas approaches to treat cardiac diseases are discussed [159, 160]. In this context, proteins involved in mechanotransduction are promising targets and await further research. But beyond these potential promising possibilities of gene therapeutic approaches in genetic cardiomyopathies, there are numerous scientific, ethic, and economic limitations with respect to a broad bedside use in humans.

10 Conclusions

Cardiac and skeletal muscle cells bidirectionally sense and transmit (outside-in and inside-out) mechanical forces between the extracellular matrix, the contractile apparatus, and various organelles, and they consecutively respond via structural changes and altered signal transduction, a process known as cardiac mechanotransduction. Increases in mechanical load of cardiac myocytes lead to biochemical signals and induce cellular hypertrophy, an initially adaptive response that, through persistent strain exposure, ultimately leads to pathological hypertrophy, increased tissue fibrosis and stiffness, predisposition to various arrhythmias, and subsequent terminal heart failure. Over the past decade, increasing scientific evidence has emerged that sensing cardiac load involves not *the* "cardiac stretch sensor" but rather divergent, different, and multiple mechanisms involving a large variety of structural and regulatory proteins at distinct subcellular localizations. These recent scientific advances provide better mechanistic insight into the earliest manifestations of pathologic cardiac hypertrophy and heart failure. And beyond this, such progress also promises future possibilities of better therapeutic interventions for diseases still leading mortality and death rates in the Western world.

Declarations

Funding: This work was supported by the German Research Foundation (DFG) and the German Center for Cardiovascular Research (DZHK).

Availability of data and materials: Not applicable.

Authors Contributions: ME conceived and wrote the manuscript, NF supervised and reviewed the manuscript. DF and LK added the chapter "translational perspective and Fig. 2."

Ethical approval: This article does not contain any studies with human participants by any of the authors. In all experiments involving animals and that have been conducted by the authors, ethical permit was granted by local committees of the University of Heidelberg (BW) and Kiel (SH).

Disclosure of interests: The authors declare that they have no conflicts of interest.

Consent for publication: All authors declare that they permit publication of this manuscript in its present form.

References

1. Gjesdal O, Bluemke DA, Lima JA (2011) Cardiac remodeling at the population level–risk factors, screening, and outcomes. Nat Rev Cardiol 8:673–685. https://doi.org/10.1038/nrcardio.2011.154
2. Neves JS et al (2015) Acute myocardial response to stretch: what we (don't) know. Front Physiol 6:408. https://doi.org/10.3389/fphys.2015.00408
3. Garoffolo G, Pesce M (2019) Mechanotransduction in the cardiovascular system: from developmental origins to homeostasis and pathology. Cells 8. https://doi.org/10.3390/cells8121607
4. Kitmitto A, Baudoin F, Cartwright EJ (2019) Cardiomyocyte damage control in heart failure and the role of the sarcolemma. J Muscle Res Cell Motil 40:319–333. https://doi.org/10.1007/s10974-019-09539-5
5. Lyon RC, Zanella F, Omens JH, Sheikh F (2015) Mechanotransduction in cardiac hypertrophy and failure. Circ Res 116:1462–1476. https://doi.org/10.1161/CIRCRESAHA.116.304937
6. Boycott HE, Nguyen MN, Vrellaku B, Gehmlich K, Robinson P (2020) Nitric oxide and mechano-electrical transduction in cardiomyocytes. Front Physiol 11:606740. https://doi.org/10.3389/fphys.2020.606740
7. Chen-Izu Y, Izu LT (2017) Mechano-chemo-transduction in cardiac myocytes. J Physiol 595:3949–3958. https://doi.org/10.1113/JP273101
8. Gaetani R et al (2020) When stiffness matters: mechanosensing in heart development and disease. Front Cell Dev Biol 8:334. https://doi.org/10.3389/fcell.2020.00334
9. Haque ZK, Wang DZ (2017) How cardiomyocytes sense pathophysiological stresses for cardiac remodeling. Cell Mol Life Sci 74:983–1000. https://doi.org/10.1007/s00018-016-2373-0
10. Parra V, Rothermel BA (2017) Calcineurin signaling in the heart: the importance of time and place. J Mol Cell Cardiol 103:121–136. https://doi.org/10.1016/j.yjmcc.2016.12.006

11. Pasqualini FS, Nesmith AP, Horton RE, Sheehy SP, Parker KK (2016) Mechanotransduction and metabolism in cardiomyocyte microdomains. Biomed Res Int 2016:4081638. https://doi.org/10.1155/2016/4081638
12. Peyronnet R, Nerbonne JM, Kohl P (2016) Cardiac mechano-gated ion channels and arrhythmias. Circ Res 118:311–329. https://doi.org/10.1161/CIRCRESAHA.115.305043
13. Citi S (2019) The mechanobiology of tight junctions. Biophys Rev 11:783–793. https://doi.org/10.1007/s12551-019-00582-7
14. Cabassi A, Miragoli M (2017) Altered mitochondrial metabolism and mechanosensation in the failing heart: focus on intracellular calcium signaling. Int J Mol Sci 18. https://doi.org/10.3390/ijms18071487
15. Jones PP, MacQuaide N, Louch WE (2018) Dyadic plasticity in cardiomyocytes. Front Physiol 9:1773. https://doi.org/10.3389/fphys.2018.01773
16. Koser F, Loescher C, Linke WA (2019) Posttranslational modifications of titin from cardiac muscle: how, where, and what for? FEBS J 286:2240–2260. https://doi.org/10.1111/febs.14854
17. Gautel M (2011) The sarcomeric cytoskeleton: who picks up the strain? Curr Opin Cell Biol 23:39–46. https://doi.org/10.1016/j.ceb.2010.12.001
18. Nakamura M, Sadoshima J (2018) Mechanisms of physiological and pathological cardiac hypertrophy. Nat Rev Cardiol 15:387–407. https://doi.org/10.1038/s41569-018-0007-y
19. Takahashi K, Kakimoto Y, Toda K, Naruse K (2013) Mechanobiology in cardiac physiology and diseases. J Cell Mol Med 17:225–232. https://doi.org/10.1111/jcmm.12027
20. Ward M, Iskratsch T (2020) Mix and (mis-)match – the mechanosensing machinery in the changing environment of the developing, healthy adult and diseased heart. Biochim Biophys Acta Mol Cell Res 1867:118436. https://doi.org/10.1016/j.bbamcr.2019.01.017
21. Buyandelger B et al (2011) MLP (muscle LIM protein) as a stress sensor in the heart. Pflugers Arch 462:135–142. https://doi.org/10.1007/s00424-011-0961-2
22. Mao Z, Nakamura F (2020) Structure and function of filamin C in the muscle Z-disc. Int J Mol Sci 21. https://doi.org/10.3390/ijms21082696
23. Frank D, Frey N (2011) Cardiac Z-disc signaling network. J Biol Chem 286:9897–9904. https://doi.org/10.1074/jbc.R110.174268
24. Le S et al (2018) Dystrophin as a molecular shock absorber. ACS Nano 12:12140–12148. https://doi.org/10.1021/acsnano.8b05721
25. Graham DM, Burridge K (2016) Mechanotransduction and nuclear function. Curr Opin Cell Biol 40:98–105. https://doi.org/10.1016/j.ceb.2016.03.006
26. Heffler J et al (2020) A balance between intermediate filaments and microtubules maintains nuclear architecture in the cardiomyocyte. Circ Res 126:e10–e26. https://doi.org/10.1161/CIRCRESAHA.119.315582
27. Knoll R et al (2002) The cardiac mechanical stretch sensor machinery involves a Z disc complex that is defective in a subset of human dilated cardiomyopathy. Cell 111:943–955. https://doi.org/10.1016/s0092-8674(02)01226-6
28. Vafiadaki E, Arvanitis DA, Sanoudou D (2015) Muscle LIM protein: master regulator of cardiac and skeletal muscle functions. Gene 566:1–7. https://doi.org/10.1016/j.gene.2015.04.077
29. Herzog W (2018) The multiple roles of titin in muscle contraction and force production. Biophys Rev 10:1187–1199. https://doi.org/10.1007/s12551-017-0395-y
30. Linke WA (2008) Sense and stretchability: the role of titin and titin-associated proteins in myocardial stress-sensing and mechanical dysfunction. Cardiovasc Res 77:637–648. https://doi.org/10.1016/j.cardiores.2007.03.029
31. Hamdani N, Herwig M, Linke WA (2017) Tampering with springs: phosphorylation of titin affecting the mechanical function of cardiomyocytes. Biophys Rev 9:225–237. https://doi.org/10.1007/s12551-017-0263-9

32. Swist S et al (2020) Maintenance of sarcomeric integrity in adult muscle cells crucially depends on Z-disc anchored titin. Nat Commun 11:4479. https://doi.org/10.1038/s41467-020-18131-2
33. Kruger M, Kotter S (2016) Titin, a central mediator for hypertrophic signaling, exercise-induced mechanosignaling and skeletal muscle remodeling. Front Physiol 7:76. https://doi.org/10.3389/fphys.2016.00076
34. Kruger M, Linke WA (2009) Titin-based mechanical signalling in normal and failing myocardium. J Mol Cell Cardiol 46:490–498. https://doi.org/10.1016/j.yjmcc.2009.01.004
35. Radke MH et al (2019) Deleting full length titin versus the titin M-band region leads to differential mechanosignaling and cardiac phenotypes. Circulation 139:1813–1827. https://doi.org/10.1161/CIRCULATIONAHA.118.037588
36. Tonino P et al (2017) The giant protein titin regulates the length of the striated muscle thick filament. Nat Commun 8:1041. https://doi.org/10.1038/s41467-017-01144-9
37. Granzier HL et al (2014) Deleting titin's I-band/A-band junction reveals critical roles for titin in biomechanical sensing and cardiac function. Proc Natl Acad Sci U S A 111:14589–14594. https://doi.org/10.1073/pnas.1411493111
38. Knoll R et al (2010) A common MLP (muscle LIM protein) variant is associated with cardiomyopathy. Circ Res 106:695–704. https://doi.org/10.1161/CIRCRESAHA.109.206243
39. Kuhn C et al (2012) Cardiac remodeling is not modulated by overexpression of muscle LIM protein (MLP). Basic Res Cardiol 107:262. https://doi.org/10.1007/s00395-012-0262-8
40. Weins A et al (2001) Differentiation- and stress-dependent nuclear cytoplasmic redistribution of myopodin, a novel actin-bundling protein. J Cell Biol 155:393–404. https://doi.org/10.1083/jcb.200012039
41. Faul C, Dhume A, Schecter AD, Mundel P (2007) Protein kinase A, Ca2+/calmodulin-dependent kinase II, and calcineurin regulate the intracellular trafficking of myopodin between the Z-disc and the nucleus of cardiac myocytes. Mol Cell Biol 27:8215–8227. https://doi.org/10.1128/MCB.00950-07
42. Ulbricht A et al (2013) Cellular mechanotransduction relies on tension-induced and chaperone-assisted autophagy. Curr Biol 23:430–435. https://doi.org/10.1016/j.cub.2013.01.064
43. Creamer TP (2020) Calcineurin. Cell Commun Signal 18:137. https://doi.org/10.1186/s12964-020-00636-4
44. Molkentin JD et al (1998) A calcineurin-dependent transcriptional pathway for cardiac hypertrophy. Cell 93:215–228. https://doi.org/10.1016/s0092-8674(00)81573-1
45. Bueno OF et al (2002) Impaired cardiac hypertrophic response in Calcineurin Abeta-deficient mice. Proc Natl Acad Sci U S A 99:4586–4591. https://doi.org/10.1073/pnas.072647999
46. Dierck F et al (2017) The novel cardiac z-disc protein CEFIP regulates cardiomyocyte hypertrophy by modulating calcineurin signaling. J Biol Chem 292:15180–15191. https://doi.org/10.1074/jbc.M117.786764
47. Frey N et al (2004) Mice lacking calsarcin-1 are sensitized to calcineurin signaling and show accelerated cardiomyopathy in response to pathological biomechanical stress. Nat Med 10:1336–1343. https://doi.org/10.1038/nm1132
48. Jeong D et al (2008) PICOT attenuates cardiac hypertrophy by disrupting calcineurin-NFAT signaling. Circ Res 102:711–719. https://doi.org/10.1161/CIRCRESAHA.107.165985
49. Kielbasa OM et al (2011) Myospryn is a calcineurin-interacting protein that negatively modulates slow-fiber-type transformation and skeletal muscle regeneration. FASEB J 25:2276–2286. https://doi.org/10.1096/fj.10-169219
50. Kuwahara K et al (2006) TRPC6 fulfills a calcineurin signaling circuit during pathologic cardiac remodeling. J Clin Invest 116:3114–3126. https://doi.org/10.1172/JCI27702
51. Nakamura A, Yoshida K, Takeda S, Dohi N, Ikeda S (2002) Progression of dystrophic features and activation of mitogen-activated protein kinases and calcineurin by physical exercise, in hearts of mdx mice. FEBS Lett 520:18–24. https://doi.org/10.1016/s0014-5793(02)02739-4

52. Nakayama H, Wilkin BJ, Bodi I, Molkentin JD (2006) Calcineurin-dependent cardiomyopathy is activated by TRPC in the adult mouse heart. FASEB J 20:1660–1670. https://doi.org/10.1096/fj.05-5560com
53. Cha H et al (2008) PICOT is a critical regulator of cardiac hypertrophy and cardiomyocyte contractility. J Mol Cell Cardiol 45:796–803. https://doi.org/10.1016/j.yjmcc.2008.09.124
54. McLendon PM, Robbins J (2011) Desmin-related cardiomyopathy: an unfolding story. Am J Physiol Heart Circ Physiol 301:H1220–H1228. https://doi.org/10.1152/ajpheart.00601.2011
55. Klauke B et al (2010) De novo desmin-mutation N116S is associated with arrhythmogenic right ventricular cardiomyopathy. Hum Mol Genet 19:4595–4607. https://doi.org/10.1093/hmg/ddq387
56. Goldfarb LG et al (1998) Missense mutations in desmin associated with familial cardiac and skeletal myopathy. Nat Genet 19:402–403. https://doi.org/10.1038/1300
57. Kreplak L, Herrmann H, Aebi U (2008) Tensile properties of single desmin intermediate filaments. Biophys J 94:2790–2799. https://doi.org/10.1529/biophysj.107.119826
58. Lange S, Pinotsis N, Agarkova I, Ehler E (2020) The M-band: the underestimated part of the sarcomere. Biochim Biophys Acta Mol Cell Res 1867:118440. https://doi.org/10.1016/j.bbamcr.2019.02.003
59. Lange S et al (2005) The kinase domain of titin controls muscle gene expression and protein turnover. Science 308:1599–1603. https://doi.org/10.1126/science.1110463
60. Gotthardt M et al (2003) Conditional expression of mutant M-line titins results in cardiomyopathy with altered sarcomere structure. J Biol Chem 278:6059–6065. https://doi.org/10.1074/jbc.M211723200
61. Gramlich M et al (2009) Stress-induced dilated cardiomyopathy in a knock-in mouse model mimicking human titin-based disease. J Mol Cell Cardiol 47:352–358. https://doi.org/10.1016/j.yjmcc.2009.04.014
62. Will RD et al (2010) Myomasp/LRRC39, a heart- and muscle-specific protein, is a novel component of the sarcomeric M-band and is involved in stretch sensing. Circ Res 107:1253–1264. https://doi.org/10.1161/CIRCRESAHA.110.222372
63. Pruna M, Ehler E (2020) The intercalated disc: a mechanosensing signalling node in cardiomyopathy. Biophys Rev 12:931–946. https://doi.org/10.1007/s12551-020-00737-x
64. Borrmann CM et al (2006) The area composita of adhering junctions connecting heart muscle cells of vertebrates. II. Colocalizations of desmosomal and fascia adherens molecules in the intercalated disk. Eur J Cell Biol 85:469–485. https://doi.org/10.1016/j.ejcb.2006.02.009
65. Franke WW, Borrmann CM, Grund C, Pieperhoff S (2006) The area composita of adhering junctions connecting heart muscle cells of vertebrates. I. Molecular definition in intercalated disks of cardiomyocytes by immunoelectron microscopy of desmosomal proteins. Eur J Cell Biol 85:69–82. https://doi.org/10.1016/j.ejcb.2005.11.003
66. McCain ML, Lee H, Aratyn-Schaus Y, Kleber AG, Parker KK (2012) Cooperative coupling of cell-matrix and cell-cell adhesions in cardiac muscle. Proc Natl Acad Sci U S A 109:9881–9886. https://doi.org/10.1073/pnas.1203007109
67. Sit B, Gutmann D, Iskratsch T (2019) Costameres, dense plaques and podosomes: the cell matrix adhesions in cardiovascular mechanosensing. J Muscle Res Cell Motil 40:197–209. https://doi.org/10.1007/s10974-019-09529-7
68. McDaniel DP et al (2007) The stiffness of collagen fibrils influences vascular smooth muscle cell phenotype. Biophys J 92:1759–1769. https://doi.org/10.1529/biophysj.106.089003
69. Anthis NJ et al (2009) The structure of an integrin/talin complex reveals the basis of inside-out signal transduction. EMBO J 28:3623–3632. https://doi.org/10.1038/emboj.2009.287
70. Gingras AR et al (2009) Structural determinants of integrin binding to the talin rod. J Biol Chem 284:8866–8876. https://doi.org/10.1074/jbc.M805937200
71. Israeli-Rosenberg S, Manso AM, Okada H, Ross RS (2014) Integrins and integrin-associated proteins in the cardiac myocyte. Circ Res 114:572–586. https://doi.org/10.1161/CIRCRESAHA.114.301275

72. Pham CG et al (2000) Striated muscle-specific beta(1D)-integrin and FAK are involved in cardiac myocyte hypertrophic response pathway. Am J Physiol Heart Circ Physiol 279: H2916–H2926. https://doi.org/10.1152/ajpheart.2000.279.6.H2916
73. Bauer MS et al (2019) Structural and mechanistic insights into mechanoactivation of focal adhesion kinase. Proc Natl Acad Sci U S A 116:6766–6774. https://doi.org/10.1073/pnas.1820567116
74. Brancaccio M et al (2003) Melusin, a muscle-specific integrin beta1-interacting protein, is required to prevent cardiac failure in response to chronic pressure overload. Nat Med 9:68–75. https://doi.org/10.1038/nm805
75. Frangogiannis NG (2017) The extracellular matrix in myocardial injury, repair, and remodeling. J Clin Invest 127:1600–1612. https://doi.org/10.1172/JCI87491
76. Frangogiannis NG (2019) The extracellular matrix in ischemic and nonischemic heart failure. Circ Res 125:117–146. https://doi.org/10.1161/CIRCRESAHA.119.311148
77. Chin IL, Hool L, Choi YS (2019) A review of in vitro platforms for understanding cardiomyocyte mechanobiology. Front Bioeng Biotechnol 7:133. https://doi.org/10.3389/fbioe.2019.00133
78. Takeda N, Manabe I (2011) Cellular interplay between cardiomyocytes and nonmyocytes in cardiac remodeling. Int J Inflam 2011:535241. https://doi.org/10.4061/2011/535241
79. Imanaka-Yoshida K, Aoki H (2014) Tenascin-C and mechanotransduction in the development and diseases of cardiovascular system. Front Physiol 5:283. https://doi.org/10.3389/fphys.2014.00283
80. del Rio A et al (2009) Stretching single talin rod molecules activates vinculin binding. Science 323:638–641. https://doi.org/10.1126/science.1162912
81. Goult BT, Yan J, Schwartz MA (2018) Talin as a mechanosensitive signaling hub. J Cell Biol 217:3776–3784. https://doi.org/10.1083/jcb.201808061
82. Manso AM et al (2017) Loss of mouse cardiomyocyte talin-1 and talin-2 leads to beta-1 integrin reduction, costameric instability, and dilated cardiomyopathy. Proc Natl Acad Sci U S A 114:E6250–E6259. https://doi.org/10.1073/pnas.1701416114
83. Lapidos KA, Kakkar R, McNally EM (2004) The dystrophin glycoprotein complex: signaling strength and integrity for the sarcolemma. Circ Res 94:1023–1031. https://doi.org/10.1161/01.RES.0000126574.61061.25
84. Peter AK, Cheng H, Ross RS, Knowlton KU, Chen J (2011) The costamere bridges sarcomeres to the sarcolemma in striated muscle. Prog Pediatr Cardiol 31:83–88. https://doi.org/10.1016/j.ppedcard.2011.02.003
85. Fratter C et al (2020) EMQN best practice guidelines for genetic testing in dystrophinopathies. Eur J Hum Genet 28:1141–1159. https://doi.org/10.1038/s41431-020-0643-7
86. Finsterer J, Stollberger C (2003) The heart in human dystrophinopathies. Cardiology 99:1–19. https://doi.org/10.1159/000068446
87. Mavrogeni SI, Markousis-Mavrogenis G, Papavasiliou A, Papadopoulos G, Kolovou G (2018) Cardiac involvement in Duchenne muscular dystrophy and related dystrophinopathies. Methods Mol Biol 1687:31–42. https://doi.org/10.1007/978-1-4939-7374-3_3
88. Catanzaro JN et al (2020) Proarrhythmic manifestations of neuromuscular dystrophinopathies. Cardiol Rev. https://doi.org/10.1097/CRD.0000000000000305
89. Amoasii L et al (2018) Gene editing restores dystrophin expression in a canine model of Duchenne muscular dystrophy. Science 362:86–91. https://doi.org/10.1126/science.aau1549
90. Chopra A, Tabdanov E, Patel H, Janmey PA, Kresh JY (2011) Cardiac myocyte remodeling mediated by N-cadherin-dependent mechanosensing. Am J Physiol Heart Circ Physiol 300: H1252–H1266. https://doi.org/10.1152/ajpheart.00515.2010
91. Kanoldt V et al (2020) Metavinculin modulates force transduction in cell adhesion sites. Nat Commun 11:6403. https://doi.org/10.1038/s41467-020-20125-z
92. Zhang JQ et al (2001) Ultrastructural and biochemical localization of N-RAP at the interface between myofibrils and intercalated disks in the mouse heart. Biochemistry 40:14898–14906. https://doi.org/10.1021/bi0107445

93. Baddam SR et al (2018) The desmosomal cadherin desmoglein-2 experiences mechanical tension as demonstrated by a FRET-based tension biosensor expressed in living cells. Cells 7. https://doi.org/10.3390/cells7070066
94. Teng J, Loukin S, Kung C (2014) Mechanosensitive ion channels in cardiovascular physiology. Exp Clin Cardiol 20:6550–6560
95. Krajnik A et al (2020) Phosphoinositide signaling and mechanotransduction in cardiovascular biology and disease. Front Cell Dev Biol 8:595849. https://doi.org/10.3389/fcell.2020.595849
96. Cordova-Casanova A et al (1865) Mechanical stretch increases L-type calcium channel stability in cardiomyocytes through a polycystin-1/AKT-dependent mechanism. Biochim Biophys Acta Mol Cell Res 289-296:2018. https://doi.org/10.1016/j.bbamcr.2017.11.001
97. Kuo IY et al (2014) The number and location of EF hand motifs dictates the calcium dependence of polycystin-2 function. FASEB J 28:2332–2346. https://doi.org/10.1096/fj.13-247106
98. Paavola J et al (2013) Polycystin-2 mutations lead to impaired calcium cycling in the heart and predispose to dilated cardiomyopathy. J Mol Cell Cardiol 58:199–208. https://doi.org/10.1016/j.yjmcc.2013.01.015
99. Balbo BE et al (2016) Cardiac dysfunction in Pkd1-deficient mice with phenotype rescue by galectin-3 knockout. Kidney Int 90:580–597. https://doi.org/10.1016/j.kint.2016.04.028
100. Boulter C et al (2001) Cardiovascular, skeletal, and renal defects in mice with a targeted disruption of the Pkd1 gene. Proc Natl Acad Sci U S A 98:12174–12179. https://doi.org/10.1073/pnas.211191098
101. Eden M et al (2016) Myoscape controls cardiac calcium cycling and contractility via regulation of L-type calcium channel surface expression. Nat Commun 7:11317. https://doi.org/10.1315/11317
102. Eder P, Molkentin JD (2011) TRPC channels as effectors of cardiac hypertrophy. Circ Res 108:265–272. https://doi.org/10.1161/CIRCRESAHA.110.225888
103. Wu X, Eder P, Chang B, Molkentin JD (2010) TRPC channels are necessary mediators of pathologic cardiac hypertrophy. Proc Natl Acad Sci U S A 107:7000–7005. https://doi.org/10.1073/pnas.1001825107
104. Seth M et al (2009) TRPC1 channels are critical for hypertrophic signaling in the heart. Circ Res 105:1023–1030. https://doi.org/10.1161/CIRCRESAHA.109.206581
105. Gottlieb P et al (2008) Revisiting TRPC1 and TRPC6 mechanosensitivity. Pflugers Arch 455: 1097–1103. https://doi.org/10.1007/s00424-007-0359-3
106. Lin BL et al (2019) In vivo selective inhibition of TRPC6 by antagonist BI 749327 ameliorates fibrosis and dysfunction in cardiac and renal disease. Proc Natl Acad Sci U S A 116:10156–10161. https://doi.org/10.1073/pnas.1815354116
107. Xie J et al (2012) Cardioprotection by Klotho through downregulation of TRPC6 channels in the mouse heart. Nat Commun 3:1238. https://doi.org/10.1038/ncomms2240
108. Numaga-Tomita T, Nishida M (2020) TRPC channels in cardiac plasticity. Cells 9. https://doi.org/10.3390/cells9020454
109. Xian Tao L et al (2006) The stretch-activated potassium channel TREK-1 in rat cardiac ventricular muscle. Cardiovasc Res 69:86–97. https://doi.org/10.1016/j.cardiores.2005.08.018
110. Brohawn SG (2015) How ion channels sense mechanical force: insights from mechanosensitive K2P channels TRAAK, TREK1, and TREK2. Ann N Y Acad Sci 1352: 20–32. https://doi.org/10.1111/nyas.12874
111. Brohawn SG, Campbell EB, MacKinnon R (2014) Physical mechanism for gating and mechanosensitivity of the human TRAAK K+ channel. Nature 516:126–130. https://doi.org/10.1038/nature14013
112. Brohawn SG, Su Z, MacKinnon R (2014) Mechanosensitivity is mediated directly by the lipid membrane in TRAAK and TREK1 K+ channels. Proc Natl Acad Sci U S A 111:3614–3619. https://doi.org/10.1073/pnas.1320768111
113. Wong TY et al (2018) Mechanical stretching simulates cardiac physiology and pathology through mechanosensor Piezo1. J Clin Med 7. https://doi.org/10.3390/jcm7110410

114. Caporizzo MA, Chen CY, Salomon AK, Margulies KB, Prosser BL (2018) Microtubules provide a viscoelastic resistance to myocyte motion. Biophys J 115:1796–1807. https://doi.org/10.1016/j.bpj.2018.09.019
115. Robison P, Prosser BL (2017) Microtubule mechanics in the working myocyte. J Physiol 595: 3931–3937. https://doi.org/10.1113/JP273046
116. Chen CY et al (2018) Suppression of detyrosinated microtubules improves cardiomyocyte function in human heart failure. Nat Med 24:1225–1233. https://doi.org/10.1038/s41591-018-0046-2
117. Robison P et al (2016) Detyrosinated microtubules buckle and bear load in contracting cardiomyocytes. Science 352:aaf0659. https://doi.org/10.1126/science.aaf0659
118. Dahl KN, Ribeiro AJ, Lammerding J (2008) Nuclear shape, mechanics, and mechanotransduction. Circ Res 102:1307–1318. https://doi.org/10.1161/CIRCRESAHA.108.173989
119. Cho S et al (2019) Mechanosensing by the lamina protects against nuclear rupture, DNA damage, and cell-cycle arrest. Dev Cell 49:920–935 e925. https://doi.org/10.1016/j.devcel.2019.04.020
120. Dahl KN, Kahn SM, Wilson KL, Discher DE (2004) The nuclear envelope lamina network has elasticity and a compressibility limit suggestive of a molecular shock absorber. J Cell Sci 117: 4779–4786. https://doi.org/10.1242/jcs.01357
121. Ahn J et al (2019) Structural basis for lamin assembly at the molecular level. Nat Commun 10: 3757. https://doi.org/10.1038/s41467-019-11684-x
122. Makarov AA et al (2019) Lamin A molecular compression and sliding as mechanisms behind nucleoskeleton elasticity. Nat Commun 10:3056. https://doi.org/10.1038/s41467-019-11063-6
123. Davidson PM, Lammerding J (2014) Broken nuclei–lamins, nuclear mechanics, and disease. Trends Cell Biol 24:247–256. https://doi.org/10.1016/j.tcb.2013.11.004
124. Lammerding J et al (2004) Lamin A/C deficiency causes defective nuclear mechanics and mechanotransduction. J Clin Invest 113:370–378. https://doi.org/10.1172/JCI19670
125. Gonzalez JM, Navarro-Puche A, Casar B, Crespo P, Andres V (2008) Fast regulation of AP-1 activity through interaction of lamin A/C, ERK1/2, and c-Fos at the nuclear envelope. J Cell Biol 183:653–666. https://doi.org/10.1083/jcb.200805049
126. Tesson F et al (2014) Lamin A/C mutations in dilated cardiomyopathy. Cardiol J 21:331–342. https://doi.org/10.5603/CJ.a2014.0037
127. Lu JT, Muchir A, Nagy PL, Worman HJ (2011) LMNA cardiomyopathy: cell biology and genetics meet clinical medicine. Dis Model Mech 4:562–568. https://doi.org/10.1242/dmm.006346
128. Donnaloja F, Carnevali F, Jacchetti E, Raimondi MT (2020) Lamin A/C mechanotransduction in laminopathies. Cells 9. https://doi.org/10.3390/cells9051306
129. Dupont S et al (2011) Role of YAP/TAZ in mechanotransduction. Nature 474:179–183. https://doi.org/10.1038/nature10137
130. Buxboim A et al (2014) Matrix elasticity regulates lamin-A,C phosphorylation and turnover with feedback to actomyosin. Curr Biol 24:1909–1917. https://doi.org/10.1016/j.cub.2014.07.001
131. Machowska M, Piekarowicz K, Rzepecki R (2015) Regulation of lamin properties and functions: does phosphorylation do it all? Open Biol 5. https://doi.org/10.1098/rsob.150094
132. Zhang X et al (2009) SUN1/2 and Syne/Nesprin-1/2 complexes connect centrosome to the nucleus during neurogenesis and neuronal migration in mice. Neuron 64:173–187. https://doi.org/10.1016/j.neuron.2009.08.018
133. Banerjee I et al (2014) Targeted ablation of nesprin 1 and nesprin 2 from murine myocardium results in cardiomyopathy, altered nuclear morphology and inhibition of the biomechanical gene response. PLoS Genet 10:e1004114. https://doi.org/10.1371/journal.pgen.1004114

134. Chapman MA et al (2014) Disruption of both nesprin 1 and desmin results in nuclear anchorage defects and fibrosis in skeletal muscle. Hum Mol Genet 23:5879–5892. https://doi.org/10.1093/hmg/ddu310
135. Lu W et al (2012) Nesprin interchain associations control nuclear size. Cell Mol Life Sci 69: 3493–3509. https://doi.org/10.1007/s00018-012-1034-1
136. Rajgor D, Mellad JA, Autore F, Zhang Q, Shanahan CM (2012) Multiple novel nesprin-1 and nesprin-2 variants act as versatile tissue-specific intracellular scaffolds. PLoS One 7:e40098. https://doi.org/10.1371/journal.pone.0040098
137. Lombardi ML et al (2011) The interaction between nesprins and sun proteins at the nuclear envelope is critical for force transmission between the nucleus and cytoskeleton. J Biol Chem 286:26743–26753. https://doi.org/10.1074/jbc.M111.233700
138. Driscoll TP, Cosgrove BD, Heo SJ, Shurden ZE, Mauck RL (2015) Cytoskeletal to nuclear strain transfer regulates YAP signaling in mesenchymal stem cells. Biophys J 108:2783–2793. https://doi.org/10.1016/j.bpj.2015.05.010
139. Saucerman JJ, Tan PM, Buchholz KS, McCulloch AD, Omens JH (2019) Mechanical regulation of gene expression in cardiac myocytes and fibroblasts. Nat Rev Cardiol 16:361–378. https://doi.org/10.1038/s41569-019-0155-8
140. Frank D et al (2008) Gene expression pattern in biomechanically stretched cardiomyocytes: evidence for a stretch-specific gene program. Hypertension 51:309–318. https://doi.org/10.1161/HYPERTENSIONAHA.107.098046
141. Kempf T et al (2006) The transforming growth factor-beta superfamily member growth-differentiation factor-15 protects the heart from ischemia/reperfusion injury. Circ Res 98: 351–360. https://doi.org/10.1161/01.RES.0000202805.73038.48
142. Eggers KM et al (2008) Growth-differentiation factor-15 for early risk stratification in patients with acute chest pain. Eur Heart J 29:2327–2335. https://doi.org/10.1093/eurheartj/ehn339
143. Kempf T et al (2007) Growth-differentiation factor-15 improves risk stratification in ST-segment elevation myocardial infarction. Eur Heart J 28:2858–2865. https://doi.org/10.1093/eurheartj/ehm465
144. Kempf T et al (2007) Prognostic utility of growth differentiation factor-15 in patients with chronic heart failure. J Am Coll Cardiol 50:1054–1060. https://doi.org/10.1016/j.jacc.2007.04.091
145. Kempf T, Wollert KC (2009) Growth-differentiation factor-15 in heart failure. Heart Fail Clin 5:537–547. https://doi.org/10.1016/j.hfc.2009.04.006
146. Frank D et al (2012) MicroRNA-20a inhibits stress-induced cardiomyocyte apoptosis involving its novel target Egln3/PHD3. J Mol Cell Cardiol 52:711–717. https://doi.org/10.1016/j.yjmcc.2011.12.001
147. Boriani G et al (2018) Cardiolaminopathies from bench to bedside: challenges in clinical decision-making with focus on arrhythmia-related outcomes. Nucleus 9:442–459. https://doi.org/10.1080/19491034.2018.1506680
148. Kubanek M et al (2020) Desminopathy: novel desmin variants, a new cardiac phenotype, and further evidence for secondary mitochondrial dysfunction. J Clin Med 9. https://doi.org/10.3390/jcm9040937
149. Peretto G et al (2018) Updated clinical overview on cardiac laminopathies: an electrical and mechanical disease. Nucleus 9:380–391. https://doi.org/10.1080/19491034.2018.1489195
150. Ahlberg G et al (2018) Rare truncating variants in the sarcomeric protein titin associate with familial and early-onset atrial fibrillation. Nat Commun 9:4316. https://doi.org/10.1038/s41467-018-06618-y
151. Brayson D, Shanahan CM (2017) Current insights into LMNA cardiomyopathies: existing models and missing LINCs. Nucleus 8:17–33. https://doi.org/10.1080/19491034.2016.1260798
152. Cingolani HE, Ennis IL (2007) Sodium-hydrogen exchanger, cardiac overload, and myocardial hypertrophy. Circulation 115:1090–1100. https://doi.org/10.1161/CIRCULATIONAHA.106.626929

153. Previtali SC et al (2020) Rimeporide as a fi rst-in-class NHE-1 inhibitor: results of a phase Ib trial in young patients with Duchenne Muscular Dystrophy. Pharmacol Res 159:104999. https://doi.org/10.1016/j.phrs.2020.104999
154. Wiedmann F et al (2020) The experimental TASK-1 potassium channel inhibitor A293 can be employed for rhythm control of persistent atrial fibrillation in a translational large animal model. Front Physiol 11:629421. https://doi.org/10.3389/fphys.2020.629421
155. Wiedmann F et al (2021) Mechanosensitive TREK-1 two-pore-domain potassium (K2P) channels in the cardiovascular system. Prog Biophys Mol Biol 159:126–135. https://doi.org/10.1016/j.pbiomolbio.2020.05.007
156. Zhang B (2021) CRISPR/Cas gene therapy. J Cell Physiol 236:2459–2481. https://doi.org/10.1002/jcp.30064
157. Stadtmauer EA et al (2020) CRISPR-engineered T cells in patients with refractory cancer. Science 367. https://doi.org/10.1126/science.aba7365
158. Carroll KJ et al (2016) A mouse model for adult cardiac-specific gene deletion with CRISPR/Cas9. Proc Natl Acad Sci U S A 113:338–343. https://doi.org/10.1073/pnas.1523918113
159. Landmesser U et al (2020) From traditional pharmacological towards nucleic acid-based therapies for cardiovascular diseases. Eur Heart J 41:3884–3899. https://doi.org/10.1093/eurheartj/ehaa229
160. Olivaes J, Bonamino MH, Markoski MM (2019) CRISPR/Cas 9 system for the treatment of dilated cardiomyopathy: a hypothesis related to function of a MAP kinase. Med Hypotheses 128:91–93. https://doi.org/10.1016/j.mehy.2019.05.013

Mechanotransduction in Heart Development

Alexandra E. Giovou and Vincent M. Christoffels

1 Introduction

Architectural integrity of the heart is necessary to ensure efficient blood circulation throughout the body. Cardiac structural malformations, as seen in congenital heart disease, can lead to diverse pathologies [1, 2]. Although different parts of the heart can be affected, including the cardiac chambers and the great vessels, a common outcome from such defects is the loss of cardiac functionality, which requires corrective surgery to prevent death after birth and which may ultimately lead to heart failure and sudden cardiac death [3]. Despite the advances in surgery, the high prevalence of congenital heart disease, and the reoccurring complications in young adults, highlight the importance of understanding the origins of these developmental defects. Heart development follows a series of strictly regulated processes until it reaches its full maturation [4]. These well-orchestrated processes are established through transcriptional regulation and molecular signaling (i.e., patterning and morphogenesis).

In the early embryo, a simple tube is formed that contracts slowly in a unidirectional peristaltic pattern, pumping fluid through the early embryo to keep it alive and maintain growth and development. Soon after the tube has been established, it rapidly increases in length, loops, expands dramatically in diameter at regions where the atria and ventricles develop, and forms unidirectional valves in between the expansions (atrioventricular valves) and at the outlet. Chamber walls form trabeculae and increase in thickness, contraction forces and displaced volumes increase, the vascular bed grows and increases in complexity and in its resistance,

A. E. Giovou · V. M. Christoffels (✉)
Department of Medical Biology, Amsterdam Cardiovascular Sciences, Amsterdam University Medical Centers, University of Amsterdam, Amsterdam, Netherlands
e-mail: v.m.christoffels@amsterdamumc.nl

M. Hecker, D. J. Duncker (eds.), *Cardiac Mechanobiology in Physiology and Disease*, Cardiac and Vascular Biology 9,
https://doi.org/10.1007/978-3-031-23965-6_2

and the activation and contraction patterns of the chambers begin to resemble the synchronized serial pattern of the mature heart. In mammals and birds, the ventricular and atrial septa will form and close to establish parallel flows (left systemic, right pulmonary), and the outflow tract (OFT) will transform into an aorta connected to the left ventricle, and a pulmonary artery connected to the right ventricle [4, 5].

Throughout development, the heart continuously contracts in a regular pattern in order to pump blood to maintain embryonic development and growth. To facilitate its pumping efficiency and adjust it to the demands of the growing organism, the myocardium has to constantly adapt by modulating cardiomyocyte contractility and its coordination with the environmental forces, including alterations in pressure and in hemodynamics. Parallel with heart development, blood circulation also undergoes dynamic alterations in order to match the increasing demands of the growing embryo and fetus. Blood flow has been established as an essential modulator of heart development, and hemodynamic alterations have an impact on cardiogenesis [6]. Blood flow is able to instruct cardiac development not only by directly affecting cells upon contact, but it can also trigger a response to distal cell populations within the developing myocardium. These spatiotemporal interactions take place through various mechanosensitive pathways, in which mechanical stimuli are converted into a biological output [7]. Therefore, proper stage-to-stage transition, together with precise coordination to the adjusting circulatory system, is required to ensure the progression and stability of cardiac development. In this chapter we provide an overview of the mechanosensitive pathways that dynamically modulate the process of development of components of the heart.

2 Forces and Mechanotransduction in Heart Development

In the early embryo, bilateral pools of mesodermal progenitors form the cardiac crescent, which subsequently forms the linear heart tube. The first calcium oscillations have been observed in the crescent, while the first contractions of the cardiomyocytes have been observed prior to and during the formation of the linear heart tube. These activities have been found to be required for cardiac differentiation and morphogenesis [8]. Soon thereafter, blood starts to flow through the contracting heart tube and begins to influence its morphogenesis. Specifically, hemodynamic forces participate in the shaping of the chambers and septation, valve formation, and maturation of the heart [9, 10]. As the endocardial cells come into direct contact with blood flow, they are considered to be the critical mechanosensors detecting hemodynamic alterations under normal or pathological conditions [11]. They sense shear stress from blood flow through different sensors including cilia and mechanosensitive ion channels and relay the information through mechanotransduction pathways that in turn act through signaling pathways like Notch, transforming growth factor beta/bone morphogenetic protein (TGFβ/BMP), Neuregulin, and vascular endothelial growth factor (VEGF). These pathways, in turn, direct proliferation rates, differentiation, size, and shape of the endocardium itself and the underlying myocardium [9, 10]. Moreover, during the cardiac activation-contraction cycle, all cardiac cells (endocardial, myocardial, mesenchymal) also undergo multiple mechanical stresses caused by contractions and expansion (e.g., stretch, pressure), which are also sensed and transduced, thus adding to the complexity of how mechanical forces regulate cardiac development and functionality [12].

3 Hemodynamic Influence on Cardiac Looping and Chamber Development

The linear tube elongates and loops by the addition of myocardial and endocardial cells deriving from the second heart field. Epithelial second heart field cells in the splanchnic mesoderm dorsal of the heart tube undergo YAP/TAZ-mediated proliferation (yes-associated protein/WW domain-containing transcription regulator 1) and oriented elongation and display polarized actomyosin distribution, indicating that epithelial tension contributes to the extension of the heart tube [13]. The linear heart tube loops rightward during its elongation, thus breaking left-right symmetry. This process is evolutionary conserved and underlies the proper alignment of the ventricles and great vessels and morphogenesis of the atria and sinus venosus. During gastrulation, symmetry is broken by rotating cilia in the left-right organizer (Kupffer's vesicle in fish, node in mammals), which generate a leftward fluid flow and Pkd2-dependent asymmetric Ca^{2+} transients [14, 15]. This leads to asymmetric expression of nodal signaling components [16]. A number of downstream targets including Pitx2c and other factors independent of Pitx2c have been identified that regulate rightward rotation of the arterial pole and left-shift of the venous pole of the tube, which results in the asymmetric morphogenesis of the looping heart [17]. Consistent with the importance of cilia in this process, a genetic screen in mice revealed the large contribution of defective cilia/cilia-related genes and congenital heart defects [18].

The heart tube consists of two distinct layers, the outer myocardial and the inner endocardial layer. The endocardial layer has multiple roles in the developing heart as it regulates trabeculation, chamber septation, valve formation, and coronary vascularization and makes a material contribution to the valves, coronary vessels, and septa. These two cardiac layers are separated by an acellular network of extracellular matrix (ECM) proteins also known as cardiac jelly [4]. Robust cell proliferation of the myocardial cells at the outer curvatures of the looping tube results in the expansion of the cardiac chambers [4]. Importantly, the initiation of the looping process coincides with the onset of blood circulation [19]. From that stage onward, there is a dramatic increase in heart rate, blood pressure, and blood volume [9, 10]. Therefore, cardiac looping and chamber formation all occur under blood flow conditions that change during the progression of heart development.

Studies in chicken and transgenic mice revealed that blood flow is required for chamber morphogenesis [20, 21]. The underlying mechanisms have been further explored in zebrafish (Fig. 1) [22]. Both flow and contractility appear to

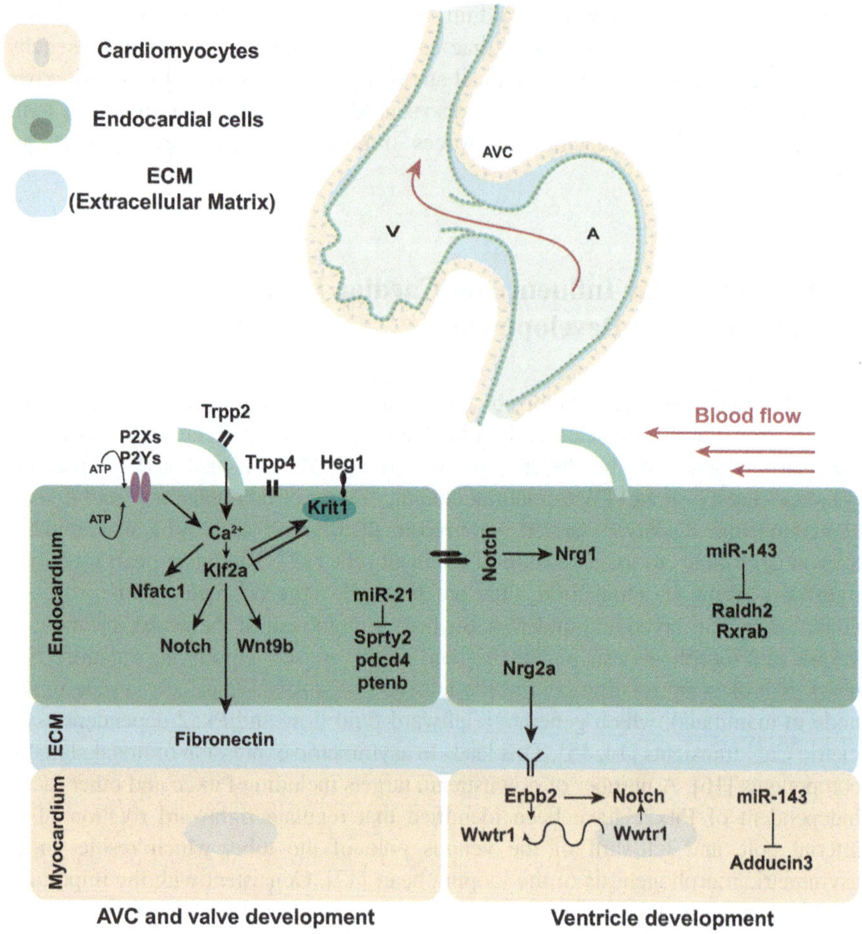

Fig. 1 Flow- and shear stress-responsive pathways regulating the atrioventricular canal (AVC), valve, and ventricular development in zebrafish. Adapted from [22]. Abbreviations used: AV, Atrioventricular; AVC, Atrioventricular canal; CHD, Congenital heart defect/disease; ECM, Extracellular matrix; EndoMT, Endothelial-Mesenchymal Transition; KLF2, Krüppel-like Factor 2; OFT, Outflow tract; WNT, Fusion of "Wingless" and "Int-1"; Wnt9b, Wnt Family Member 9b

independently regulate cell behavior in chamber morphogenesis in an endocardium-dependent manner [23]. Zebrafish lacking endocardium show compromised chamber morphogenesis [24]. Both endocardium and myocardium increase their proliferation rates to drive chamber expansion at the outer curvature [23]. Shear stress modulates endocardial cell characteristics. Reduction of shear stress has been shown to increase endocardial cell size and reduce their proliferation rate [25]. Through sensed shear stress, blood flow provides the endocardial cells with input directing their proliferation rates, differentiation, and shape [10, 11, 19, 25]. The flow-responsive transcription factor KLF2 (Krüpper-like factor 2) in the endocardium contributes to chamber formation. For example, restricted blood flow results in decreased klf2a (zebrafish homologue of KLF2) expression, and genetic ablation of the gene promotes an increase in cell size. In contrast, *Klf2a* overexpression reduces the endocardial cell surface area [25]. Klf2 has been shown to modulate myocardial heart wall integrity via FGF (fibroblast growth factor) signaling. Loss of function of *klf2a* in zebrafish impairs myocardial wall integrity, while endocardial overexpression efficiently reverses the observed phenotype [26].

Micro-RNAs can be regulated rapidly in response to flow. Heartbeat- and/or blood flow-dependent miR-143 expression impacts the process of ventricular development by targeting different processes in the endocardium and myocardium, including inhibiting retinoic acid signaling (*raldh2* and *rxrab*) and *adductin3 (add3)*, respectively, in order to drive chamber formation (Fig. 1) [27, 28]. In addition, miR-143 may regulate endocardial cell number. It is to be expected that future studies will identify additional mechanosensitive genes and downstream effectors that execute the patterns of proliferation and cell behavior of the endocardium and the myocardium underlying chamber morphogenesis.

Chemically blocking cardiac contraction in zebrafish embryos was shown to promote chamber growth and cardiomyocyte enlargement [29]. On the other hand, after reducing the blood flow by tail amputation, the hypertrophic phenotype was partially reversed. However, in zebrafish mutants lacking endocardium, the hypertrophic response was not rescued, suggesting that the hypertrophy is not the result of shear stress, but is caused by transmural pressure [29].

Hypoplastic left heart syndrome is a congenital heart disease involving an underdeveloped (hypoplastic) left ventricle that does not support efficient circulation. This puts the right ventricle under increased work load causing it to stretch [30]. The ventricular hypoplasia has been attributed to restricted blood flow, causing growth reduction, although cardiac cell intrinsic disturbances of particular gene programs were found to contribute to the pathogenesis, arguing against a solely hemodynamic origin [31]. In a recent study, miR-486 was identified as a stretch-induced miRNA that is upregulated in the right ventricle and is able to promote ventricular growth. Interestingly, neonatal mice treated with miR-486 showed enhanced cardiomyocyte proliferation, suggesting that miR-486 can be beneficial for promoting a hyperplastic response in underdeveloped ventricles caused by hemodynamic alterations [30].

4 Chamber Development, Trabeculation, and Hemodynamic Conditions

During chamber development, the ventricular myocardial walls form trabeculations and a compact outer layer. Trabeculae formation is the process in which the myocardium, covered by an endocardial layer, forms ridges that appear to protrude into the luminal space of the ventricular chambers [4]. Functionally, the trabeculae increase oxygen and nutrient uptake in the developing heart, prior to the establishment of the coronary vasculature [32, 33]. Ectothermic animals (fish, amphibians, reptiles) maintain the trabecular wall architecture into adulthood; endotherms (birds, mammals) obtain a compact wall and a Purkinje fiber network derived from the trabecules [34]. In fish, trabeculation is driven by directional migration of cardiomyocytes toward the lumen (toward the endocardium). The cells delaminate from the epithelial-like myocardial layer ("compact" layer) of the early developing heart [35, 36]. In hearts that do not contract, myocardial protrusions still form, but trabeculae do not, indicative of cardiac function being required for trabeculation [36, 37]. Flow (shear stress) is likely to play a key role in trabeculation, as experimental reduction of blood flow inhibits this process [36, 38–40]. Blood flow patterns through the ventricle may instruct both the initiation of luminal protrusions and the subsequent formation of stable trabeculae [36, 40].

The initiation of cardiomyocyte protrusion during trabeculation appears to be orchestrated by the Notch and Neuregulin pathways, and its progression depends on optimal hemodynamic conditions [38, 40, 41]. For example, blood flow/contractility is required for Nrg2a (Neuregulin2a) expression, which induces trabeculation [41]. In addition, Neuregulin signaling activates Notch signaling in the myocardium [42] and negatively regulates nuclear localization (activity) of the Hippo effector Wwtr1 (otherwise known as TAZ) in order to enable trabeculation [43]. Ablation of *Notch1* in mice resulted in underdeveloped trabeculae [44]. The ventricular endocardium has cilia that act as flow (shear stress) sensors (Fig. 1). Mice lacking cilia exhibit thinner compact myocardium and impaired trabeculation [45]. Blood flow sensed by primary cilia causes activation of *notch1b* in the zebrafish ventricular endocardium, and Notch1 activation induces the expression of *efnb2a* (ephrin b2a) and *nrg1* (Neuregulin 1), which are both required for trabeculation [46]. Zebrafish mutants with reduced shear stress (by *gata1* knockdown to reduce blood viscosity) also show reduced Notch signaling and attenuated trabeculation [47]. Injection of *nrg1* mRNA in these embryos rescued Notch-related expression and trabeculation.

5 Flow Forces Drive Outflow Tract Development

Abrogation of the primary cilia in mice results in abnormal OFT development, indicating a role for mechanosensing and Hedgehog signaling in OFT morphogenesis [48]. Developmental expansion of the OFT in zebrafish embryos involves accrual (proliferation and addition) of both endocardial and myocardial cells [49]. However, in conditions of disrupted heart function, endocardial growth ceases. The flow-sensitive receptor activin A receptor-like type 1 (Acvrl1) (TGFβ receptor) is required for the addition, but not for the proliferation of endocardial cells [49].

The establishment of the asymmetric aortic arch from the bilaterally symmetric branchial arch arteries requires unilateral expression of *Pitx2* (paired-like homeodomain 2). Pitx2 was found to drive the asymmetric morphogenesis of the upstream OFT, causing preferential blood supply to the left sixth branchial arch artery. The uneven distribution of the flow of blood causes left-sided PDGFR (platelet-derived growth factor receptor) and VEGFR2 signaling and stabilization of the left sixth branchial arch artery, while the right counterpart regresses, thus enabling the asymmetric remodeling of the great arteries [50].

In a recent study, it was demonstrated that the ECM protein Thrombospondin-1 (*Thbs1*) is induced by mechanical stretch and involved in the remodeling of the aorta [51]. *Thbs1* was found to mediate dynamic interactions between mechanical stress and the YAP-mediated transcriptional response in the blood vessel wall. Deletion of *Thbs1* in mice resulted in altered vascular remodeling in response to flow cessation and pressure overload [51]. Piezo1 also impacts the development of the OFT. Morpholino-mediated knockdown of the mechanosensitive ion channel Piezo1 in zebrafish embryos led to loss of *elnb* (elastin B) expression in the OFT and structural abnormalities with the OFT appearing constricted [52].

6 Role of Mechanical Forces in Valve Development

Within the atrioventricular canal (AVC) and outflow tract (OFT), the endocardial cells will undergo EndoMT to form the cardiac cushions and later mature into unidirectional cardiac valves. Cushion formation is initiated by signals deriving from the underlying AVC and OFT myocardium. Zebrafish mutants with impaired myocardial contractility do not develop cardiac AV cushions (endocardial ring) [53]. This study indicated blood flow was not required for the initial steps of cushion formation. However, obstruction of blood flow in chicken or zebrafish embryos impairs cushion and atrioventricular valve formation [6, 37, 54, 55]. Moreover, deletion of cardiac cilia also causes anomalies in the cardiac cushion development [45]. These findings directly link hemodynamic alterations and shear stress to valve morphogenesis.

Before valves develop, reversing (or oscillatory) blood flow occurs between the atrium and ventricle due to regurgitation. The retrograde flow is particularly strong at the AVC. Experimental reduction of the retrograde flow fraction in zebrafish resulted in failure to develop AV valves [55, 56]. The accumulation of endocardial cells at the AVC, which precedes AV valve formation in zebrafish, was also disrupted, indicating the retrograde, or oscillatory flow at the AVC is also required for the onset of AV valve development [55]. The transcription factor KLF2 is expressed in the endocardium overlying the developing valves and is dependent on hemodynamic forces [57–60]. Loss of *klf2a* in zebrafish results in valve defects similar to those observed with loss of flow or shear stress [55, 61, 62]. A KLF2-WNT9B signaling mechanism conserved in zebrafish and mouse was identified by which fluid forces sensed by endocardial cells direct heart valve development [60] (Fig. 2). Klf2 induces *wnt9b* in the endocardium and canonical WNT signaling in the underlying cushion mesenchyme. Endocardial *wnt9a* expression depends on Klf2 and on shear forces and, like *klf2*, is required for valve development [60]. Other targets of Klf2 include *notch1b*,

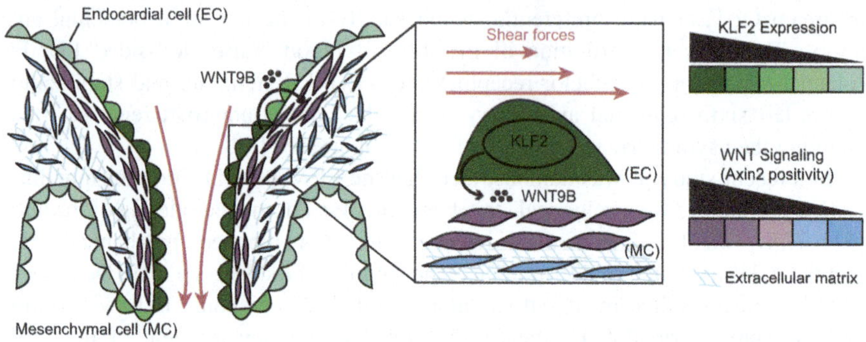

Fig. 2 Remodeling of cardiac cushions into valves. Hemodynamic forces activate expression of the transcription factor Klf2 in the endocardium, which indirectly activates Wnt9b expression. Endocardial Wnt9b activates canonical WNT signaling in the adjacent mesenchymal cells to orchestrate remodeling and valve formation. During valve formation, the hemodynamic forces will change, providing feedback information to modulate Klf2 expression, WNT signaling, and valve formation itself. Modified from [60]

nrg1, and *edn1* in the AV endocardium [55] (Fig. 1). In addition, Notch signaling activity is also increased upon high shear stress and together with KLF2 signaling controls valvulogenesis. More specifically, Notch-mediated lateral inhibition causes Delta-like-4 positive endothelial cells to ingress into the cardiac jelly and form a subendocardial cell population. These Dll4-positive cells ingress in response to Wnt9a, produced in response to shear stress-activated Klf2 signaling [63]. In luminal endocardial cells that are subjected to shear stress, two independent mechanosensitive pathways activate Klf2a- and Notch1b-mediated signaling, respectively, which in turn repress angiogenesis receptor Vegfr3/Flt4. As a result, Flt4 becomes restricted to the subendocardial/abluminal EC population, which is not subjected to fluid shear stress, where it inhibits Notch activity. Together, these studies highlight the antagonistic activities between shear stress-activated Notch and subendocardial Flt4 underlying localized differences in EC fate and valvulogenesis [64]. Manipulation of *klf2a* expression resulted in reduced levels of *bmp4* expression in the myocardium, supporting the involvement of a shear stress-mediated interaction of endocardium and myocardium in AVC specification and cushion development [55]. Additionally, *klf2b* and *egr1* (early growth response 1) were identified as flow-responsive transcription factor-encoding genes, capable of regulating valvugenesis, with zebrafish mutants for either gene showing valve malformation [65].

The valves originating from endocardial cells in the AVC form the valve leaflets, a process involving coordinated endocardial tissue movements [62]. These cell movements and tissue reorganization depend on *fibronectin1b* and other ECM proteins. Flow and Klf2a control these movements and the expression of *fibronectin1b* [62]. This illustrates how endocardial mechanotransduction and valve morphogenesis are linked via tissue reorganization mediated by ECM protein synthesis.

While alterations in shear stress have been established to have an impact on the transcriptional regulation of *klf2a*, the mechanisms of how the cells sense and translate this action into a nuclear output have not been fully resolved. One proposed mechanism involves the shear stress-mediated activation of the mechanosensitive calcium channels rpp2 (polycystin 2, transient receptor potential cation channel) and Trpv4 (transient receptor potential cation channel subfamily V member 4) regulating *klf2a* expression (Fig. 1). Zebrafish mutants of both *trpp2* or *trpv4* exhibit reduced *klf2a* expression and valve malformation [61]. It is hypothesized that these channels, present on the endocardial cell surface, regulate *klf2a* expression by promoting the calcium influx into the endocardial cells of the AVC, triggered by the retrograde blood flow. A recent study in zebrafish showed that in addition to the established TRPP-KLF2a axis, an independent shear stress-responsive mechanism is involved in valvulogenesis. Specifically, it was shown that shear stress induces the extracellular ATP-mediated activation of the purinergic receptor groups P2X and P2Y, in the endocardial cells of the AVC. This triggers pulsatile calcium ion influx, which in turn activates nuclear translocation and activation of transcription factor Nfatc1 (nuclear factor of activated T cells 1) and targets the gene program driving valve formation (Fig. 1) [66].

In addition, blood flow causes upregulation of *hegl (heart development protein with EGF-like domains 1)*, which in turn stabilizes the protein levels of Krit1 (KRIT1 Ankyrin repeat containing) [67]. Together, Hegl and Krit1 lower the expression levels of *klf2a*. In zebrafish *kritl* mutants, *klf2a* and *notch1b* are upregulated in the endocardium causing inhibition of valve leaflet formation. Thus, a balance between flow-mediated *klf2a* induction and suppression mechanisms drives valve leaflet morphology [67].

While the formation of the AVC and the OFT valves is generally conserved, cells deriving from the neural crest have a unique participation in the formation of the OFT valves [68]. These valves will form at the base of each of the great arteries, to prevent the blood from leaking back into the ventricles. While in static culture conditions, the vascular endothelial cells and endothelial valvular cells are morphologically similar, upon subjection to unidirectional shear stress, they show differences in spatial organization [69]. Instead of the parallel alignment of the vascular endothelial cells to the direction of shear stress, the valvular endothelial cells orient vertically toward the blood flow [69]. With the application of shear stress, this different spatial organization between the aortic and the aortic valvular endothelial cells appears to be the result of focal adhesion rearrangement, which is mediated by PI3K-dependent and PI3K-independent signaling, respectively. This finding suggests that hemodynamic forces can impact the differently localized endothelial populations by driving different molecular responses.

Endocardial loss of the mechanosensitive ion channel Piezo1 in mice impairs cytoskeletal alignment upon shear stress [70], and haploinsufficient *Piezo1* mice present disorganized alignment of the endothelial cells, linking hemodynamic forces to cellular morphology [71]. Moreover, lowering the mRNA levels of *piezo1* in zebrafish leads to OFT and aortic valve defects [52]. Of note, while the AVC valvular progenitors are surrounded by myocardial cells, their OFT counterparts contact smooth muscle cells, possibly creating differences in the impact of mechanosensitive forces [72]. In zebrafish, upon shear stress alterations, the Piezo channels are able to coordinate the development of the OFT valves, through distinct mechanisms. Piezo channels together with the TRP channels can induce *Klf2a* and Notch1 activity in the endothelium, while in the underlying smooth muscle cells, Piezo modulates the localization of Yap1 [72].

The expression of *Cad11* (cadherin 11) is hemodynamic force-dependent [73], and during cardiac maturation its expression is progressively restricted within the endocardium of the OFT valves [74]. Genetic ablation of *Cad11* results in underdeveloped OFT valves by disrupting cellular migration [75].

Myocardial contractility also participates in the proper development of the OFT valves. Specifically, increasing myocardial contractility in zebrafish increases shear stress and *notch1b* expression and leads to OFT valve hyperplasia. By impairing contractility, OFT valve formation was inhibited. Furthermore, solely increasing shear stress leads to increased Notch1 signaling and OFT leaflet growth. However, decreasing shear stress, without manipulating contractility, had no effect on leaflet formation. These findings suggest that contractility is essential for the OFT valve development, and by further coordinating with shear stress, it promotes OFT leaflet growth in a *notchb1*-mediated manner [76].

Mechanical forces also regulate epigenetic factors and microRNAs that contribute to valve development. The histone-deacetylase HDAC5 is a negative regulator of *klf2a*. Within the endocardial cells of the AVC, PKD2 (protein kinase D2) inhibits the activity of HDCA5 and ensures the upregulation of *klf2a*. In endocardial *prkd2* zebrafish mutants, the expression of *klf2a* and *notch1b* is decreased, and valve formation is disrupted [77]. In addition, expression of *klf2a* has been described to regulate the abundance of the ECM protein Fibronectin1b within the AVC cardiac jelly. Expression of *fibronectin1b* together with endocardial cell migration into the cushions is required for the formation of the valves. Zebrafish mutants lacking *klf2a* exhibited impairment of both these two processes [62].

In zebrafish, *miR-21* is expressed in the AVC and OFT endocardium, areas that are subject to high blood flow and shear stress. The expression of *pri-miR-21-1* disappears in conditions of disrupted flow (heart beat) and quickly restores when flow (heart beat) is resumed. Flow-dependent expression of *miR-21* governs valve development by regulating the expression of targets (e.g., *sprouty*, *pdcd4*, *ptenb*) and by inducing cell proliferation in the endocardium of the AVC and OFT where shear stress is highest [78].

7 Conclusions

It is evident that heart development requires the interplay between mechanical forces and transcriptional programs. The KLF2-WNT9B-dependent formation of the AV valves provides a beautiful example illustrating this process. Here, AV valve modeling is driven by a transcriptional program the activity of which is constantly tuned by transduced forces that change with the progression of valve development and function itself. More in general, heart and vascular development in its entirety depends on the dynamic interplay between transcriptional programs and spatially and temporally dynamic forces produced by heart function. It is likely that even minor errors in the series of precisely graded signals and transduction mechanisms may result in coronary heart disease (CHD). Given the prevalence of CHD, unravelling how hemodynamic and mechanical forces control molecular pathways in a spatiotemporal manner could ultimately prove beneficial for understanding the pathologies of CHD and open new avenues for possible therapeutic applications.

Funding

We acknowledge the support from the Netherlands Cardiovascular Research Initiative: an initiative with support of the Dutch Heart Foundation and Hartekind, CVON2019-002 OUTREACH.

Authors' Contributions Alexandra Giovou and Vincent Christoffels wrote the paper.

Conflict of Interest The authors declare that they have no conflict of interest.

Ethical Approval This article does not contain any studies with animals or human participants performed by any of the authors.

Informed Consent Not applicable.

Consent for Publication All authors declare that they permit publication of this manuscript in its present form.

References

1. Hoffman JI, Kaplan S (2002) The incidence of congenital heart disease. J Am Coll Cardiol 39(12):1890–1900
2. Woudstra OI et al (2017) Origins and consequences of congenital heart defects affecting the right ventricle. Cardiovasc Res 113(12):1509–1520
3. Norozi K et al (2006) Incidence and risk distribution of heart failure in adolescents and adults with congenital heart disease after cardiac surgery. Am J Cardiol 97(8):1238–1243
4. Moorman AFM, Christoffels VM (2003) Cardiac chamber formation: development, genes and evolution. Physiol Rev 83(4):1223–1267
5. Christoffels V, Jensen B (2020) Cardiac morphogenesis: specification of the four-chambered heart. Cold Spring Harb Perspect Biol 12(10):1–18
6. Hove JR et al (2003) Intracardiac fluid forces are an essential epigenetic factor for embryonic cardiogenesis. Nature 421(6919):172–177
7. Culver JC, Dickinson ME (2010) The effects of hemodynamic force on embryonic development. Microcirculation 17(3):164–178
8. Tyser RC et al (2016) Calcium handling precedes cardiac differentiation to initiate the first heartbeat. Elife 5:e17113
9. Andres-Delgado L, Mercader N (2016) Interplay between cardiac function and heart development. Biochim Biophys Acta 1863(7 Pt B):1707–1716
10. Haack T, Abdelilah-Seyfried S (2016) The force within: endocardial development, mechanotransduction and signalling during cardiac morphogenesis. Development 143(3): 373–386

11. Hahn C, Schwartz MA (2009) Mechanotransduction in vascular physiology and atherogenesis. Nat Rev Mol Cell Biol 10(1):53–62
12. Gaetani R et al (2020) When stiffness matters: Mechanosensing in heart development and disease. Front Cell Dev Biol 8:334
13. Francou A, De Bono C, Kelly RG (2017) Epithelial tension in the second heart field promotes mouse heart tube elongation. Nat Commun 8:14770
14. Shiratori H, Hamada H (2006) The left-right axis in the mouse: from origin to morphology. Development 133(11):2095–2104
15. Mizuno K et al (2020) Role of Ca(2+) transients at the node of the mouse embryo in breaking of left-right symmetry. Sci Adv 6(30):eaba1195
16. Ferreira RR, Vermot J (2017) The balancing roles of mechanical forces during left-right patterning and asymmetric morphogenesis. Mech Dev 144(Pt A):71–80
17. Desgrange A et al (2020) Transient nodal Signaling in left precursors coordinates opposed asymmetries shaping the heart loop. Dev Cell 55(4):413–431 e6
18. Li Y et al (2015) Global genetic analysis in mice unveils central role for cilia in congenital heart disease. Nature 521(7553):520–524
19. Courchaine K, Rykiel G, Rugonyi S (2018) Influence of blood flow on cardiac development. Prog Biophys Mol Biol 137:95–110
20. Huang C et al (2003) Embryonic atrial function is essential for mouse embryogenesis, cardiac morphogenesis and angiogenesis. Development 130(24):6111–6119
21. Broekhuizen ML et al (1999) Altered hemodynamics in chick embryos after extraembryonic venous obstruction. Ultrasound Obstet Gynecol 13(6):437–445
22. Sidhwani P, Yelon D (2019) Fluid forces shape the embryonic heart: insights from zebrafish. Curr Top Dev Biol 132:395–416
23. Auman HJ et al (2007) Functional modulation of cardiac form through regionally confined cell shape changes. PLoS.Biol 5(3):e53
24. Stainier DY et al (1995) Cloche, an early acting zebrafish gene, is required by both the endothelial and hematopoietic lineages. Development 121(10):3141–3150
25. Dietrich AC et al (2014) Blood flow and bmp signaling control endocardial chamber morphogenesis. Dev Cell 30(4):367–377
26. Rasouli SJ et al (2018) The flow responsive transcription factor Klf2 is required for myocardial wall integrity by modulating Fgf signaling. Elife 7:e38889
27. Miyasaka KY et al (2011) Heartbeat regulates cardiogenesis by suppressing retinoic acid signaling via expression of miR-143. Mech Dev 128(1–2):18–28
28. Deacon DC et al (2010) The miR-143-adducin3 pathway is essential for cardiac chamber morphogenesis. Development 137(11):1887–1896
29. Yang J et al (2014) Cessation of contraction induces cardiomyocyte remodeling during zebrafish cardiogenesis. Am J Physiol Heart Circ Physiol 306(3):H382–H395
30. Lange S et al (2019) miR-486 is modulated by stretch and increases ventricular growth. JCI Insight 4(19):e125507
31. Krane M et al (2021) Sequential defects in cardiac lineage commitment and maturation cause Hypoplastic left heart syndrome. Circulation 144(17):1409–1428
32. Samsa LA, Yang B, Liu J (2013) Embryonic cardiac chamber maturation: Trabeculation, conduction, and cardiomyocyte proliferation. Am J Med Genet C Semin Med Genet 163C(3): 157–168
33. Wu M (2018) Mechanisms of trabecular formation and specification during Cardiogenesis. Pediatr Cardiol 39(6):1082–1089
34. Jensen B et al (2013) Evolution and development of the building plan of the vertebrate heart. Biochim Biophys Acta 1833(4):783–794
35. Liu J et al (2010) A dual role for ErbB2 signaling in cardiac trabeculation. Development 137(22):3867–3875
36. Staudt DW et al (2014) High-resolution imaging of cardiomyocyte behavior reveals two distinct steps in ventricular trabeculation. Development 141(3):585–593

37. Sedmera D et al (1999) Remodeling of chick embryonic ventricular myoarchitecture under experimentally changed loading conditions. Anat Rec 254:238–252
38. Lee J et al (2016) 4-dimensional light-sheet microscopy to elucidate shear stress modulation of cardiac trabeculation. J Clin Invest 126(5):1679–1690
39. Vedula V et al (2017) A method to quantify mechanobiologic forces during zebrafish cardiac development using 4-D light sheet imaging and computational modeling. PLoS Comput Biol 13(10):e1005828
40. Peshkovsky C, Totong R, Yelon D (2011) Dependence of cardiac trabeculation on neuregulin signaling and blood flow in zebrafish. Dev Dyn 240(2):446–456
41. Rasouli SJ, Stainier DYR (2017) Regulation of cardiomyocyte behavior in zebrafish trabeculation by Neuregulin 2a signaling. Nat Commun 8:15281
42. Jimenez-Amilburu V et al (2016) In vivo visualization of cardiomyocyte Apicobasal polarity reveals epithelial to mesenchymal-like transition during cardiac Trabeculation. Cell Rep 17(10): 2687–2699
43. Lai JKH et al (2018) The hippo pathway effector Wwtr1 regulates cardiac wall maturation in zebrafish. Development 145(10):dev159210
44. Grego-Bessa J et al (2007) Notch signaling is essential for ventricular chamber development. Dev Cell 12(3):415–429
45. Slough J, Cooney L, Brueckner M (2008) Monocilia in the embryonic mouse heart suggest a direct role for cilia in cardiac morphogenesis. Dev Dyn 237(9):2304–2314
46. Samsa LA et al (2015) Cardiac contraction activates endocardial notch signaling to modulate chamber maturation in zebrafish. Development 142(23):4080–4091
47. Messerschmidt V et al (2018) Light-sheet fluorescence microscopy to capture 4-dimensional images of the effects of modulating shear stress on the developing zebrafish heart. J Vis Exp 15(139):57769
48. Clement CA et al (2009) The primary cilium coordinates early cardiogenesis and hedgehog signaling in cardiomyocyte differentiation. J Cell Sci 122(Pt 17):3070–3082
49. Sidhwani P et al (2020) Cardiac function modulates endocardial cell dynamics to shape the cardiac outflow tract. Development 147(12):dev185900
50. Yashiro K, Shiratori H, Hamada H (2007) Haemodynamics determined by a genetic programme govern asymmetric development of the aortic arch. Nature 450(7167):285–288
51. Yamashiro Y et al (2020) Matrix mechanotransduction mediated by thrombospondin-1/ integrin/YAP in the vascular remodeling. Proc Natl Acad Sci USA 117(18):9896–9905
52. Faucherre A et al (2020) Piezo1 is required for outflow tract and aortic valve development. J Mol Cell Cardiol 143:51–62
53. Bartman T et al (2004) Early myocardial function affects endocardial cushion development in zebrafish. PLoS Biol 2(5):E129
54. Hogers B et al (1999) Extraembryonic venous obstructions lead to cardiovascular malformations and can be embryolethal. Cardio vasc Res 41(1):87–99
55. Vermot J et al (2009) Reversing blood flows act through klf2a to ensure normal valvulogenesis in the developing heart. PLoS Biol 7(11):e1000246
56. Kalogirou S et al (2014) Intracardiac flow dynamics regulate atrioventricular valve morphogenesis. Cardiovasc Res 104(1):49–60
57. Dekker RJ et al (2002) Prolonged fluid shear stress induces a distinct set of endothelial cell genes, most specifically lung Kruppel-like factor (KLF2). Blood 100(5):1689–1698
58. Dekker RJ et al (2005) Endothelial KLF2 links local arterial shear stress levels to the expression of vascular tone-regulating genes. Am J Pathol 167(2):609–618
59. Lee JS et al (2006) Klf2 is an essential regulator of vascular hemodynamic forces in vivo. Dev Cell 11(6):845–857
60. Goddard LM et al (2017) Hemodynamic forces sculpt developing heart valves through a KLF2-WNT9B paracrine Signaling Axis. Dev Cell 43(3):274–289 e5
61. Heckel E et al (2015) Oscillatory flow modulates mechanosensitive klf2a expression through trpv4 and trpp2 during heart valve development. Curr Biol 25(10):1354–1361

62. Steed E et al (2016) klf2a couples mechanotransduction and zebrafish valve morphogenesis through fibronectin synthesis. Nat Commun 7:11646
63. Paolini A et al (2021) Mechanosensitive notch-Dll4 and Klf2-Wnt9 signaling pathways intersect in guiding valvulogenesis in zebrafish. Cell Rep 37(1):109782
64. Fontana F et al (2020) Antagonistic activities of Vegfr3/Flt4 and Notch1b fine-tune mechanosensitive Signaling during zebrafish cardiac Valvulogenesis. Cell Rep 32(2):107883
65. Faggianelli-Conrozier N (2019) In: Aikaterini Polyzou RC, Roth S, Trompouki E, Vermot J (eds) Complementary functions of the mechanosensitive factors egr1, klf2b and klf2a instruct the endocardial program. bioRxiv: Cold Spring Harbor Laboratory
66. Fukui H et al (2021) Bioelectric signaling and the control of cardiac cell identity in response to mechanical forces. Science 374(6565):351–354
67. Donat S et al (2018) Heg1 and Ccm1/2 proteins control endocardial mechanosensitivity during zebrafish valvulogenesis. Elife 7:e28939
68. Menon V et al (2015) Altered Hemodynamics in the embryonic heart affects outflow valve development. J Cardiovasc Dev Dis 2(2):108–124
69. Butcher JT et al (2004) Unique morphology and focal adhesion development of valvular endothelial cells in static and fluid flow environments. Arterioscler Thromb Vasc Biol 24(8): 1429–1434
70. Ranade SS et al (2014) Piezo1, a mechanically activated ion channel, is required for vascular development in mice. Proc Natl Acad Sci USA 111(28):10347–10352
71. Li J et al (2014) Piezo1 integration of vascular architecture with physiological force. Nature 515(7526):279–282
72. Duchemin AL, Vignes H, Vermot J (2019) Mechanically activated piezo channels modulate outflow tract valve development through the Yap1 and Klf2-notch signaling axis. Elife 8: e44706
73. Butcher JT et al (2006) Transcriptional profiles of valvular and vascular endothelial cells reveal phenotypic differences: influence of shear stress. Arterioscler Thromb Vasc Biol 26(1):69–77
74. Zhou J et al (2013) Cadherin-11 expression patterns in heart valves associate with key functions during embryonic cushion formation, valve maturation and calcification. Cells Tissues Organs 198(4):300–310
75. Bowen CJ et al (2015) Cadherin-11 coordinates cellular migration and extracellular matrix remodeling during aortic valve maturation. Dev Biol 407(1):145–157
76. Hsu JJ et al (2019) Contractile and hemodynamic forces coordinate Notch1b-mediated outflow tract valve formation. JCI Insight 5(10):e124460
77. Just S et al (2011) Protein kinase D2 controls cardiac valve formation in zebrafish by regulating histone deacetylase 5 activity. Circulation 124(3):324–334
78. Banjo T et al (2013) Haemodynamically dependent valvulogenesis of zebrafish heart is mediated by flow-dependent expression of miR-21. Nat Commun 4:1978

Mechanobiology of Cardiac Growth in Health and Disease

Gloria Garoffolo and Maurizio Pesce

1 Molecular Control of Cardiac Mechanotransduction

Cells are able to transduce mechanical stimuli deriving from their environment to control gene expression and cell fate. The concept of "mechanotransduction" has been extensively studied in various cell types and in several pathological contexts [1]. Mechanical stimuli, such as extracellular matrix (ECM) stiffness, shear stress, and pressure overload, play important roles in both tissue development and disease. In the heart, in particular, mechanical regulation is intimately connected to cell differentiation and organ development from embryonic to adult life [2]. Indeed, cardiac tissue is under a constant, self-generated mechanical stress, which promotes cardiac cell maturation and tissue homeostasis. However, altered mechanoperception can activate intracellular signalling cascades driving cells toward pro-inflammatory/pro-fibrotic phenotypes [3]. External cues are also important to determine the final size of the heart. In mammals, size regulation of the heart is biphasic. During embryo development and until shortly after birth, cardiac growth is affected by cardiomyocyte proliferation; by contrast, at postnatal stages, the growth is governed by myocyte hypertrophy (Fig. 1).

Even if several studies have been reported the role of growth factors and cell cycle regulators in cardiac proliferation during cardiogenesis, recent investigators have discovered that heart size is strictly controlled by mechanically regulated signalling. One example is the Hippo signalling pathway, whose function has been correlated to organ growth during fetal life [4, 5]. Hippo pathway is an organ size and growth regulator, which is evolutionarily and functionally conserved among species. In

G. Garoffolo · M. Pesce (✉)
Unità di Ingegneria Tissutale Cardiovascolare, Centro cardiologico Monzino, IRCCS, Milan, Italy
e-mail: maurizio.pesce@ccfm.it

M. Hecker, D. J. Duncker (eds.), *Cardiac Mechanobiology in Physiology and Disease*, Cardiac and Vascular Biology 9,
https://doi.org/10.1007/978-3-031-23965-6_3

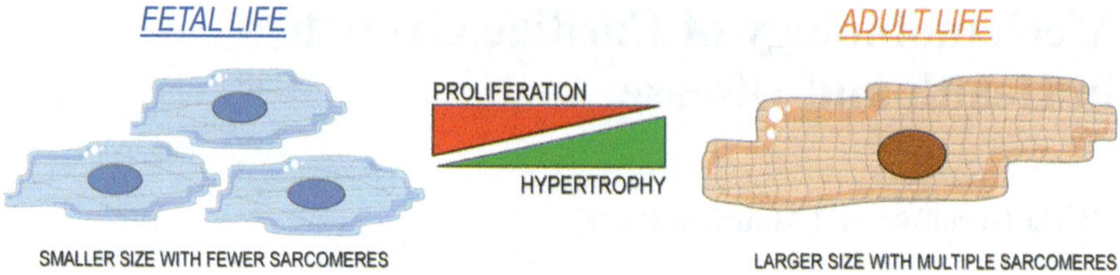

Fig. 1 Biphasic regulation of heart size. Neonatal cardiomyocyte is characterized by high replicative activity, while adult cardiomyocyte undergoes maturation by increasing in size and in number of organized sarcomeres

keeping with these evidences, Hippo pathway component-deficient embryos have cardiomegaly during mouse development, as a result of elevated cardiomyocyte proliferation [6], thus suggesting an involvement of the pathway in cardiac size regulation. In particular, it has been observed that Hippo mutations did not affect hypertrophy of cardiomyocytes but hyperplasia [6]. Among the best-characterized downstream effectors of Hippo signalling, there are YAP (Yes-associated protein) and TAZ (WW-domain-containing transcription regulator 1), two transcriptional co-activators that can partner in the nucleus with DNA-binding proteins, such as members of the TEAD (transcriptional enhancer factor (TEA) domain) transcription factor family, to regulate the expression of genes involved in cell proliferation and survival [7]. YAP-TEAD transcriptional complex promotes cardiac proliferation and growth in the embryo, without affecting cardiomyocyte size. YAP1 protein expression, found in both myocyte and nonmyocyte cells, was detected in neonatal and juvenile mouse heart and declined with age, becoming undetectable after 12 weeks of age [8]. Indeed, the fetal heart grows through cardiomyocyte proliferation, but by postnatal day 4, cardiomyocytes stop increasing in number, and the growth occurs by increasing their size due to maturation of their contractile apparatus, associated with an increment of ECM deposition by cardiac fibroblasts [9]. Fetal YAP1 inactivation in heart development induced lethal myocardial hypoplasia and decreased cardiomyocyte proliferation [8]. By contrast, YAP1 activation increased fetal cardiomyocyte cell cycle activity, regulating the transcription of genes related to cell cycle [8]. In addition, the relevance of YAP/TAZ signalling is evolutionarily conserved in cardiac development. For example, in zebrafish, the expression of these two ortholog genes has been described in cardiac progenitor cells, where they are involved, in combination with the ZF TEAD ortholog, in the migration of cardiac precursors to the midline [10].

The shift from the "hyperplasia" to "hypertrophy" stages in cardiac development occurring shortly after birth is regulated not only by multiple paracrine signals [11] but also by changes in ECM composition and biophysical characteristics, and this strongly affects YAP transcriptional activity and cardiomyocyte proliferation (Fig. 2) [12]. Indeed, the reduction in tissue stiffness induced cardiomyocyte cell cycle activity through YAP-dependent pathway and actin stability. In particular, matrix rigidity regulated Agrin expression, a large extracellular heparin sulfate

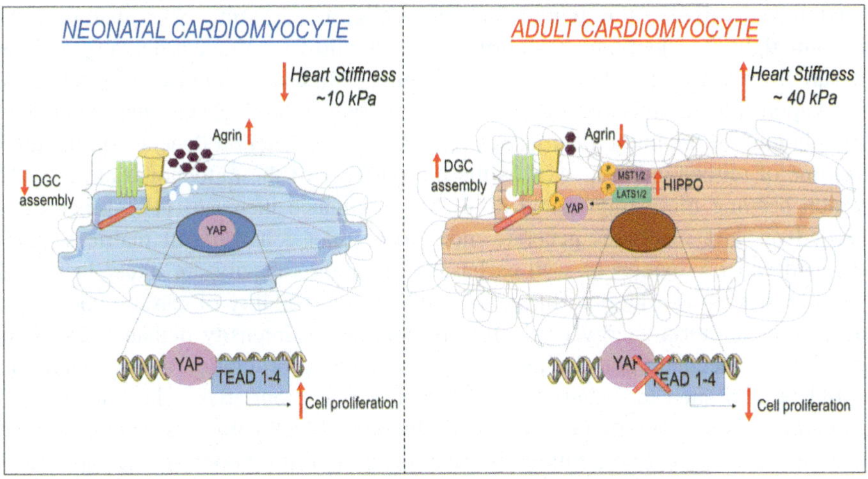

Fig. 2 ECM stiffness regulates cardiomyocyte proliferation through the activation of dystrophin glycoprotein complex (DGC)-mediated YAP signalling. High levels of Agrin induce the disassembly of DGC, thus promoting YAP nuclear translocation and activity

proteoglycan, which also plays as a mechanosensitive transducer of YAP activity [13]. In particular, Agrin suppressed cardiomyocyte maturation at neonatal stages and promoted proliferation by inducing the disassembly of dystrophin glycoprotein complex (DGC) (Fig. 2) [12, 14]. This complex, linking actin cytoskeleton to ECM, is essential for cardiomyocyte homeostasis and inhibits their proliferation by sequestering YAP (Fig. 2) [14]. In particular, the Hippo pathway and DGC act cooperatively preventing YAP nuclear localization and, thus, its transcriptional activity. In fact, YAP phosphorylation by components of the Hippo kinase pathway favors the interaction with DGC and its sequestration (Fig. 2) [14]. The levels of Agrin decline sharply after the first week of life, coincident with the loss of cardiac regenerative potential (Fig. 2).

Mechanical properties of the cardiac matrix affect cardiomyocyte maturation. For example, mouse neonatal cardiomyocytes cultured on stiff substrate showed increased myofibrillar organization and reduced cell cycle activity, reflecting the change in the heart tissue elastic modulus from ~10 kPa to ~40 kPa observed in the transition from neonatal to the adult stages [15] and justifying the increase in the contraction force. In contrast, soft matrices induced cardiomyocyte dedifferentiation, characterized by partial disassembly of the sarcomeres and an increased proliferation rate [16]. ECM rigidity is able to affect also the mitochondrial architecture in cardiomyocytes [17]. These cells are one of the most metabolically active cell types, and most of the ATP is generated by the mitochondria [18]. In early fetal cardiogenesis, mitochondria are rare with few cristae, while, during development, they increase in number and density, reflecting the switch between glycolysis and oxidative phosphorylation [19]. Indeed, in neonatal rat myocytes, mitochondria are randomly distributed throughout the cell, while, as cardiomyocytes mature, they become tightly packed into myofibril bundles to allow an efficient delivery of ATP to sarcomeres, favoring the metabolic switching [20]. Changes in ECM rigidity may be responsible for this mitochondrial remodelling, affecting cell shape and thus morphology of the mitochondrial network. Energy production depends on the total volume of mitochondria as well as the morphology of the mitochondrial network, which become more fragmented as ECM compliance increases [17]. Since cells seeded on stiff substrates generate more force than cells on soft matrices, ATP demand and production as well as basal respiration capacity are likely increased onto substrates with high stiffness, thus reflecting the ability of cardiomyocytes to adapt mitochondria energy production to contraction intensity demand [21]. It is possible to speculate that decline in cardiomyocyte proliferation and regeneration could be attributed to changes in ECM mechanical proprieties, which induce the assembly of long and aligned sarcomeric bundles, block cardiomyocyte cell cycle activity, and favor binucleation. In keeping with this hypothesis, disruption of cardiomyocyte cytoskeleton using myosin inhibitor such as Blebbistatin determined dedifferentiation of the cardiomyocyte contractile apparatus and induced cell cycle re-entry [16].

2 Mechanical Regulation of Primary Cardiomyocyte Differentiation

During cardiogenesis, mechanical regulation is involved in both first and second heart field cell migration and differentiation. In particular, for the formation of the cardiac crescent, mechanical compliance of ECM of the endoderm is necessary to direct the first heart field cardiac progenitors toward their appropriate anterior lateral position caudal to the headfolds [22]. In this position, the cells of the first heart field are exposed to various stimuli (BMP-mediated cytokines, fibroblast growth factor, WNT inhibitors), which induce cardiac differentiation [23, 24]. As development proceeds, the linear heart tube expands through the recruitment of additional cells, the second heart field progenitors, which remain in a proliferative, undifferentiated state until they enter the heart tube and give rise to myocytes and vascular cells [25, 26]. In a fascinating hypothesis supported by experimental findings, it has been speculated that primary differentiation of cardiomyocytes could be supported by mechanical signalling before the onset of the coordinated beating of the primordial heart related to electromechanical coupling [27]. This hypothesis stems from the original observation that shortly after the beginning of cardiac looping, the myocardium undergoes significant and constant stiffening, and this is predicted to affect the contractility and the coordinated beating of cells by regulating the opening of mechanoperceptive Ca^{2+} channels. Since the opening of the Ca^{2+} channels may occur in a quantitatively relevant fashion beyond a low stiffness threshold, the mechanical maturation of the extracellular matrix could represent a powerful signal for the transition from a "shivering" phase to an organized contraction of the myocardial primitive cells. This event might, in turn, allow "syncytial" transmission of the contraction forces from the contracting cells to their neighbors via the viscoelastic properties of the matrix, thus resulting in propagation of the coordinated beating before the onset of electromechanical coupling [28]. The importance of mechanical communication for the regulation of beating in primitive myocytes has been also demonstrated using experimental settings able to stimulate mechanically neonatal myocytes onto matrices with specific mechanical features and frequencies, showing the susceptibility of the cells to adapt their contraction with respect to that of the adjacent cells [29].

The mechanosensitivity of cardiomyocytes is finally demonstrated by their ability to adjust their morphology to align in perpendicular direction to strain [30]. In particular, angiotensin II, released by cyclic stretch, induces the expression of connexin 43 (Cx43), a gap junction protein, thus suggesting that mechanical stimulation is also implicated in the formation of gap junctions between cardiomyocytes and in the propagation of electrical signalling, once it is established in the myocardial tissue [31]. Two signalling cues present in the myocardium, ECM and mechanical strain, are able to affect also cardiac progenitor cell proliferation and differentiation. In particular, cyclic strain promoted cardiomyogenesis of embryonic stem cells by increasing cardiac-specific gene expression, both in vitro and in vivo [32], occurring in particular through an increase in mRNA expression of MEF2C and GATA-4, two transcription factors involved in cardiac cell differentiation [33]. ECM composition can finally direct cardiac progenitor differentiation. For example, the commitment of stem cells to differentiate into cardiomyocytes depends not only on ECM structure and elasticity but also on the relative abundance of its components, such as collagens, fibronectin, and laminin [34], and their post-translational modifications, which are strictly related to its mechanical characteristics. For example, native heart ECM drives cardiomyocyte differentiation of pluripotent cells [35]. Indeed, embryonic stem cells seeded on native heart ECM expressed high levels of cardiac genes (Troponin I, cardiac myosin heavy chain) compared to cells seeded on liver ECM [35]. This result suggests that, although heart and liver share the same ECM components and no tissue-specific component in these tissues was found, different quantities of each ECM molecule as well as their mechanical characteristics and geometric arrangement are crucial for instructing proper cardiac differentiation signalling on embryonic cells [35]. The biophysical properties of the ECM can be affected by ECM cross-linking mechanisms, which are known to promote disease progression by altering cellular responses. Abnormalities in ECM remodeling due to altered collagen deposition were associated with risk conditions and myocardial injury promoting a fibrotic response. Cardiac fibrosis, the earliest hallmarks of heart failure, is characterized by excessive deposition of ECM proteins, in particular collagen. This accumulation is the result of increased collagen synthesis associated with a decreased proteolytic activity due to an imbalance between the levels of matrix metalloproteinases (MMPs) and tissue inhibitors of metalloproteinases (TIMPs) [36]. These events are mediated by cardiac fibroblasts, a normally quiescent cell population that differentiates into activated myofibroblasts following an injury

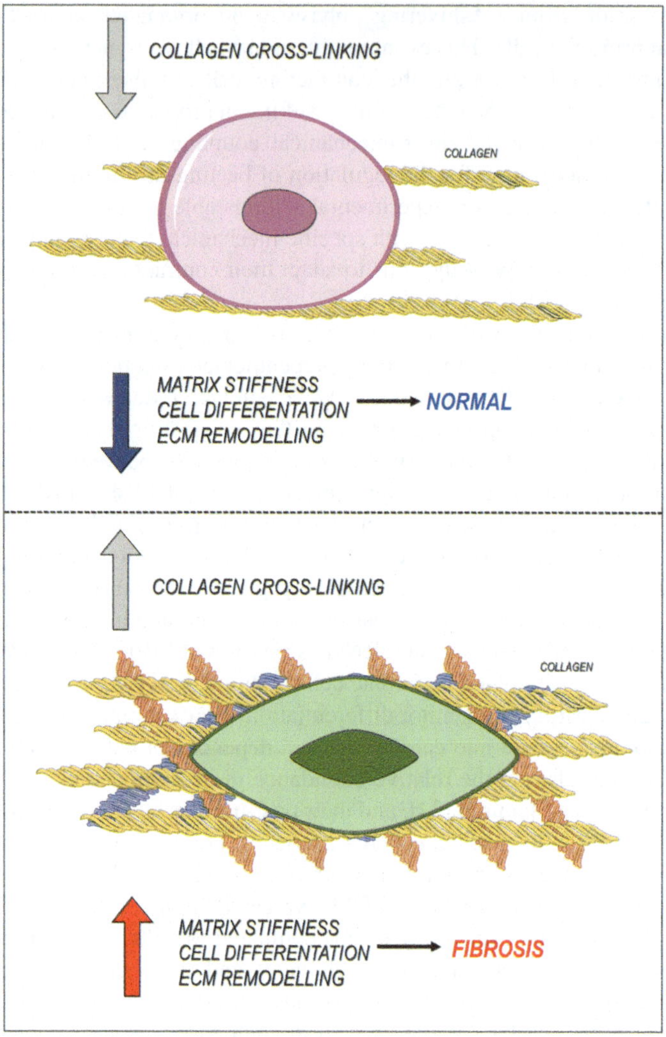

Fig. 3 Example of mechanical regulation of cardiac fibroblasts. Increased collagen cross-linking induces myofibroblast differentiation and ECM remodelling, affecting biophysical properties of the tissue

[1]. Myofibroblasts are involved in pro-inflammatory response and in excessive ECM deposition. These reparative mechanisms are necessary to maintain the structural integrity of the heart and to prevent myocardial rupture [37]. Among the factors contributing to this pathological condition, altered collagen cross-linking may represent a crucial contributing player (Fig. 3). In fact, a strong correlation exists between increased collagen cross-linking and ventricular stiffness in patients with heart failure, as well as levels of circulating lysyl oxidase (LOX), a primary collagen

cross-linking enzyme [38]. The degree of collagen cross-linking determines the solubility, the stiffness, and the resistance to degradation of the resulting ECM [39], and this has an impact on the passive mechanical properties of the myocardium. In particular cardiac tissue with high degree of cross-linking is stiffer than tissue with lower level of cross-linked collagen [40] (Fig. 3).

Not only pro-inflammatory/pro-fibrotic stimuli can alter cardiac matrix stiffness, but also aging is able to affect ECM composition by increasing collagen synthesis and cross-linking [41]. Indeed, hearts from autopsies of 80-year-old patients expressed higher levels of collagen I compared to younger individuals [42]. This limits cardiac function by decreasing the tensile strength of the heart. In addition, aged myofibroblasts showed an impaired capacity for dedifferentiation and a resistance to apoptosis, preventing the normal physiological responses to tissue injury and promoting the progression of fibrotic disorders [43].

While the composition and mechanical characteristics of the cardiac matrix are principally regulated by the function of cardiac fibroblasts and myofibroblasts, also under control of mechanically activated pathways [44], cardiomyocytes also respond to changes in ECM cross-linking by modifying their contractility. Specifically, while cells are able to contract to a larger extent on soft surfaces [45], the maximal contractile forces are developed on substrates with higher stiffness. Importantly, stiffness of the ECM has an effect on maturation of the myocytes' contractile apparatus and even the geometry of the cells that, from irregularly shaped, acquire the typical rod-shaped morphology [46]. The relationship between the sarcomeric structure or cellular geometry and the mechanical properties of the ECM has finally an important emergence in the two different forms of heart failure with preserved or reduced ejection fraction (HFpEF and HFrEF, respectively) where the difference in sensing the mechanical load by the cardiac matrix can alternatively affect the sarcomeric structure or the geometry of the myocytes inducing changes in the structure of the left ventricle [47]. At the basis of this variation in the two pathological settings is, probably, the rearrangement of the contractile units of the myocyte cytoskeleton in response to the increased preload, such as after an ischemic insult vs. the increase in afterload in consequence of elevated pressure. Although these responses are far from being totally understood, the changes in the amount and the mechanical behavior of the components of the costameres, the lateral cytoskeletal structures connecting the contractile apparatus to the junctional complexes, may play a major role with consequences for variations in the structural remodelling of the myocytes [48].

3 Conclusions

There is an intricate relationship between the mechanical properties of the heart and its cellular and functional maturation. The evolution toward pathologic states seems also associated to variations in mechanical sensing and translation of mechanical cues in maladaptive remodelling processes. The understanding of this complex relationship is very important, not only to elaborate novel strategies to protect the heart from acute and chronic injuries or the effect of modifiable/unmodifiable risk conditions but also to be able some day to prevent remodelling and instruct regeneration.

Funding

This study was funded by the Ministero della Salute (Ricerca Corrente) (to MP).

Conflict of Interest The authors declare that they have no conflicts of interest.

Ethical Approval This article does not contain any studies with human participants or animals performed by any of the authors.

References

1. Garoffolo G, Ferrari S, Rizzi S, Barbuto M, Bernava G, Pesce M (2020) Harnessing Mechanosensation in next generation cardiovascular tissue engineering. Biomol Ther 10:10
2. Garoffolo G, Pesce M (2019) Mechanotransduction in the cardiovascular system: from developmental origins to homeostasis and pathology. Cell 8:8
3. Garoffolo G, Madonna R, de Caterina R, Pesce M (2018) Cell based mechanosensing in vascular patho-biology: more than a simple go-with the flow. Vasc Pharmacol 111:7–14
4. Harvey K, Tapon N (2007) The Salvador-Warts-Hippo pathway—an emerging tumour-suppressor network. Nat Rev Cancer 7:182–191
5. Halder G, Johnson RL (2011) Hippo signaling: growth control and beyond. Development 138: 9–22
6. Heallen T, Zhang M, Wang J, Bonilla-Claudio M, Klysik E, Johnson RL, Martin JF (2011) Hippo pathway inhibits Wnt signaling to restrain cardiomyocyte proliferation and heart size. Science 332:458–461
7. Piccolo S, Dupont S, Cordenonsi M (2014) The biology of YAP/TAZ: hippo signaling and beyond. Physiol Rev 94:1287–1312
8. von Gise A, Lin Z, Schlegelmilch K, Honor LB, Pan GM, Buck JN, Ma Q, Ishiwata T, Zhou B, Camargo FD et al (2012) YAP1, the nuclear target of Hippo signaling, stimulates heart growth through cardiomyocyte proliferation but not hypertrophy. Proc Natl Acad Sci USA 109:2394–2399
9. Li F, Wang X, Capasso JM, Gerdes AM (1996) Rapid transition of cardiac myocytes from hyperplasia to hypertrophy during postnatal development. J Mol Cell Cardiol 28:1737–1746
10. Miesfeld JB, Link BA (2014) Establishment of transgenic lines to monitor and manipulate yap/Taz-Tead activity in zebrafish reveals both evolutionarily conserved and divergent functions of the hippo pathway. Mech Dev 133:177–188
11. Leu M, Ehler E, Perriard JC (2001) Characterisation of postnatal growth of the murine heart. Anat Embryol 204:217–224
12. Bassat E, Mutlak YE, Genzelinakh A, Shadrin IY, Baruch Umansky K, Yifa O, Kain D, Rajchman D, Leach J, Riabov Bassat D et al (2017) The extracellular matrix protein agrin promotes heart regeneration in mice. Nature 547:179–184
13. Wang X, Senapati S, Akinbote A, Gnanasambandam B, Park PS, Senyo SE (2020) Microenvironment stiffness requires decellularized cardiac extracellular matrix to promote heart regeneration in the neonatal mouse heart. Acta Biomater 113:380–392
14. Morikawa Y, Heallen T, Leach J, Xiao Y, Martin JF (2017) Dystrophin-glycoprotein complex sequesters Yap to inhibit cardiomyocyte proliferation. Nature 547:227–231

15. Bhana B, Iyer RK, Chen WL, Zhao R, Sider KL, Likhitpanichkul M, Simmons CA, Radisic M (2010) Influence of substrate stiffness on the phenotype of heart cells. Biotechnol Bioeng 105: 1148–1160
16. Yahalom-Ronen Y, Rajchman D, Sarig R, Geiger B, Tzahor E (2015) Reduced matrix rigidity promotes neonatal cardiomyocyte dedifferentiation, proliferation and clonal expansion. Elife 4: 4
17. Lyra-Leite DM, Petersen AP, Ariyasinghe NR, Cho N, McCain ML (2021) Mitochondrial architecture in cardiac myocytes depends on cell shape and matrix rigidity. J Mol Cell Cardiol 150:32–43
18. Barth E, Stammler G, Speiser B, Schaper J (1992) Ultrastructural quantitation of mitochondria and myofilaments in cardiac muscle from 10 different animal species including man. J Mol Cell Cardiol 24:669–681
19. Dorn GW 2nd, Vega RB, Kelly DP (2015) Mitochondrial biogenesis and dynamics in the developing and diseased heart. Genes Dev 29:1981–1991
20. Chung S, Dzeja PP, Faustino RS, Perez-Terzic C, Behfar A, Terzic A (2007) Mitochondrial oxidative metabolism is required for the cardiac differentiation of stem cells. Nat Clin Pract Cardiovasc Med 4(Suppl 1):S60–S67
21. Lyra-Leite DM, Andres AM, Petersen AP, Ariyasinghe NR, Cho N, Lee JA, Gottlieb RA, McCain ML (2017) Mitochondrial function in engineered cardiac tissues is regulated by extracellular matrix elasticity and tissue alignment. Am J Physiol Heart Circ Physiol 313: H757–H767
22. Wiens DJ (1996) An alternative model for cell sheet migration on fibronectin during heart formation. J Theor Biol 179:33–39
23. Schultheiss TM, Burch JB, Lassar AB (1997) A role for bone morphogenetic proteins in the induction of cardiac myogenesis. Genes Dev 11:451–462
24. Marvin MJ, Di Rocco G, Gardiner A, Bush SM, Lassar AB (2001) Inhibition of Wnt activity induces heart formation from posterior mesoderm. Genes Dev 15:316–327
25. Kelly RG (2012) The second heart field. Curr Top Dev Biol 100:33–65
26. van den Berg G, Abu-Issa R, de Boer BA, Hutson MR, de Boer PA, Soufan AT, Ruijter JM, Kirby ML, van den Hoff MJ, Moorman AF (2009) A caudal proliferating growth center contributes to both poles of the forming heart tube. Circ Res 104:179–188
27. Majkut S, Idema T, Swift J, Krieger C, Liu A, Discher DE (2013) Heart-specific stiffening in early embryos parallels matrix and myosin expression to optimize beating. Curr Biol 23:2434–2439
28. Chiou KK, Rocks JW, Chen CY, Cho S, Merkus KE, Rajaratnam A, Robison P, Tewari M, Vogel K, Majkut SF et al (2016) Mechanical signaling coordinates the embryonic heartbeat. Proc Natl Acad Sci USA 113:8939–8944
29. Viner H, Nitsan I, Sapir L, Drori S, Tzlil S (2019) Mechanical communication acts as a noise filter. iScience 14:58–68
30. Salameh A, Wustmann A, Karl S, Blanke K, Apel D, Rojas-Gomez D, Franke H, Mohr FW, Janousek J, Dhein S (2010) Cyclic mechanical stretch induces cardiomyocyte orientation and polarization of the gap junction protein connexin 43. Circ Res 106:1592–1602
31. Wang TL, Tseng YZ, Chang H (2000) Regulation of connexin 43 gene expression by cyclical mechanical stretch in neonatal rat cardiomyocytes. Biochem Biophys Res Commun 267:551–557
32. Gwak SJ, Bhang SH, Kim IK, Kim SS, Cho SW, Jeon O, Yoo KJ, Putnam AJ, Kim BS (2008) The effect of cyclic strain on embryonic stem cell-derived cardiomyocytes. Biomaterials 29: 844–856
33. Schmelter M, Ateghang B, Helmig S, Wartenberg M, Sauer H (2006) Embryonic stem cells utilize reactive oxygen species as transducers of mechanical strain-induced cardiovascular differentiation. FASEB J 20:1182–1184

34. Johnson TD, Hill RC, Dzieciatkowska M, Nigam V, Behfar A, Christman KL, Hansen KC (2016) Quantification of decellularized human myocardial matrix: a comparison of six patients. Proteomics Clin Appl 10:75–83
35. Higuchi S, Lin Q, Wang J, Lim TK, Joshi SB, Anand GS, Chung MC, Sheetz MP, Fujita H (2013) Heart extracellular matrix supports cardiomyocyte differentiation of mouse embryonic stem cells. J Biosci Bioeng 115:320–325
36. Trojanowska M, LeRoy EC, Eckes B, Krieg T (1998) Pathogenesis of fibrosis: type 1 collagen and the skin. J Mol Med (Berl) 76:266–274
37. van den Borne SW, Diez J, Blankesteijn WM, Verjans J, Hofstra L, Narula J (2010) Myocardial remodeling after infarction: the role of myofibroblasts. Nat Rev Cardiol 7:30–37
38. Kasner M, Westermann D, Lopez B, Gaub R, Escher F, Kuhl U, Schultheiss HP, Tschope C (2011) Diastolic tissue Doppler indexes correlate with the degree of collagen expression and cross-linking in heart failure and normal ejection fraction. J Am Coll Cardiol 57:977–985
39. Brower GL, Gardner JD, Forman MF, Murray DB, Voloshenyuk T, Levick SP, Janicki JS (2006) The relationship between myocardial extracellular matrix remodeling and ventricular function. Eur J Cardiothorac Surg 30:604–610
40. Pinnell SR, Martin GR (1968) The cross-linking of collagen and elastin: enzymatic conversion of lysine in peptide linkage to alpha-aminoadipic-delta-semialdehyde (allysine) by an extract from bone. Proc Natl Acad Sci USA 61:708–716
41. Ferrari S, Pesce M (2021) Stiffness and aging in cardiovascular diseases: the dangerous relationship between force and senescence. Int J Mol Sci 22:22
42. Gazoti Debessa CR, Mesiano Maifrino LB, Rodrigues de Souza R (2001) Age related changes of the collagen network of the human heart. Mech Ageing Dev 122:1049–1058
43. Kato K, Logsdon NJ, Shin YJ, Palumbo S, Knox A, Irish JD, Rounseville SP, Rummel SR, Mohamed M, Ahmad K et al (2020) Impaired Myofibroblast dedifferentiation contributes to nonresolving fibrosis in aging. Am J Respir Cell Mol Biol 62:633–644
44. Pesce M, Messina E, Chimenti I, Beltrami AP (2017) Cardiac Mechanoperception: a life-long story from early beats to aging and failure. Stem Cells Dev 26:77–90
45. Pandey P, Hawkes W, Hu J, Megone WV, Gautrot J, Anilkumar N, Zhang M, Hirvonen L, Cox S, Ehler E et al (2018) Cardiomyocytes sense matrix rigidity through a combination of muscle and non-muscle myosin contractions. Dev Cell 45:661
46. Jacot JG, McCulloch AD, Omens JH (2008) Substrate stiffness affects the functional maturation of neonatal rat ventricular myocytes. Biophys J 95:3479–3487
47. Katz AM, Rolett EL (2016) Heart failure: when form fails to follow function. Eur Heart J 37: 449–454
48. Sessions AO, Engler AJ (2016) Mechanical regulation of cardiac aging in model systems. Circ Res 118:1553–1562

The Role of Mechanosensitive Signaling Cascades in Repair and Fibrotic Remodeling of the Infarcted Heart

Claudio Humeres, Harikrishnan Venugopal, and Nikolaos G. Frangogiannis

1 Introduction

Myocardial infarction typically results from rupture of a vulnerable coronary plaque and can lead to sudden loss of hundreds of millions of cardiomyocytes [1]. The adult mammalian heart lacks regenerative capacity and heals through formation of a collagen-based scar [2], which is necessary to prevent catastrophic complications, such as ventricular rupture [3]. The massive loss of cardiomyocytes associated with a large infarction results in significant reduction in ventricular contractile capacity, leading to marked elevations in filling pressures. Thus, in comparison to other organs, repair of the infarcted heart has some unique characteristics, as the injured myocardium continues to contract and attempts to compensate to maintain the stroke volume through remodeling. In the infarcted ventricle, repair and remodeling are closely intertwined: faulty repair is associated with worse remodeling and progressive dysfunction, leading to post-infarction heart failure [4]. Repair and remodeling of the infarcted heart involve the cooperation of several different cell types, including surviving cardiomyocytes, immune cells, vascular cells, and fibroblasts. All these cell types sense biochemical and mechanical changes in their microenvironment and respond by modulating their phenotype and function. In the healing infarct, there is a bidirectional relation between mechanical forces and biochemical pathways: mechanical stress transduces signaling cascades, whereas the changes in the extracellular matrix network and in the intracellular cytoskeletal and myofilament networks regulate transmission of mechanical forces.

C. Humeres · H. Venugopal · N. G. Frangogiannis (✉)
Department of Medicine (Cardiology), Albert Einstein College of Medicine, The Wilf Family Cardiovascular Research Institute, Bronx, NY, USA
e-mail: nikolaos.frangogiannis@einsteinmed.edu

M. Hecker, D. J. Duncker (eds.), *Cardiac Mechanobiology in Physiology and Disease*, Cardiac and Vascular Biology 9,
https://doi.org/10.1007/978-3-031-23965-6_4

This chapter deals with the role of mechanosensitive signaling cascades in repair and remodeling of the infarcted heart. We discuss the interplay between mechanical stress and bioactive mediators in regulating phenotypic changes of immune and reparative cells in the healing infarct. Moreover, we examine how the dramatic changes in the cellular and matrix environment in the healing infarct affect ventricular function. Finally, we propose promising therapeutic interventions targeting mechanosensitive pathways to attenuate adverse post-infarction remodeling.

2 The Functional Consequences of Myocardial Ischemia

Myocardial infarction results from prolonged ischemia due to sudden occlusion of a coronary artery. Ischemia of the territory subserved by the vessel is associated with rapid onset of systolic dysfunction. Contractile force ceases after only 60 s of ischemia [5, 6], despite the availability of energy reserves. Early dysfunction is caused by the rapid generation of inorganic phosphate that inhibits contractile protein function [7, 8] and by the effects of intracellular acidosis on calcium binding to contractile proteins [9, 10]. In addition to systolic dysfunction, myocardial ischemia also increases ventricular stiffness, causing diastolic dysfunction. The effects of ischemia on ventricular compliance have been attributed to edema, triggered by generation of metabolites that increase osmolarity (such as lactate) [11] and to the energy-requiring nature of cardiomyocyte relaxation [12].

Although the functional consequences of a brief ischemic episode are reversible, prolonged ischemia triggers cardiomyocyte death, leading to persistent contractile dysfunction. In large animal models, death of vulnerable cardiomyocytes in the subendocardial area occurs after 20 min of ischemia. Longer ischemic intervals trigger a wavefront of cardiomyocyte death [13] that extends from the subendocardium to the less vulnerable subepicardial cardiomyocytes, eventually leading to a transmural infarct after 6 h of coronary occlusion. It has been suggested that during this period, the ischemic myocardium may behave as a passive elastic material [14]. However, active relaxation requires energy, in order to export Ca^{2+} ions back into the sarcoplasmic reticulum against a concentration gradient and to hydrolyze myosin-bound ATP to decrease myosin-actin cross-bridging [12, 15]. Thus, the dramatic perturbations in myocardial energetics during ischemia lead to the early development of diastolic dysfunction in infarcted hearts.

3 The Phases of Infarct Healing

Death of cardiomyocytes in the ischemic myocardial segments triggers an inflammation-driven reparative response that ultimately results in formation of a collagen-based scar. Thus, repair of the infarcted myocardium can be divided into three distinct, but overlapping phases: the inflammatory phase, the proliferative phase, and the maturation phase. During the inflammatory phase, cardiomyocyte necrosis triggers systemic and myocardial inflammation, leading to recruitment of leukocytes in the infarct. As professional macrophages phagocytose dead cells and matrix debris, inflammation is suppressed. Release of fibrogenic mediators activates reparative interstitial cells and marks the transition to the proliferative phase, leading to infiltration of the infarct with activated myofibroblasts and vascular cells. Finally, during the maturation phase, structural collagen becomes cross-linked, leading to formation of a stable scar, populated by de-activated fibroblasts and mature vessels. In the healing infarct, the well-orchestrated transitions between the phases ensure timely replacement of the dead cells with scar. Although the endogenous reparative process does not restore the contractile properties of the lost myocardium, formation of a scar prevents catastrophic complications, such as ventricular rupture. Moreover, the cellular and extracellular matrix compositions of the scar are key determinants of the severity of adverse post-infarction remodeling.

4 The Inflammatory Phase of Infarct Healing

Release of damage-associated molecular patterns (DAMPs) by necrotic cardiomyocytes triggers toll-like receptor (TLR)-mediated innate immune responses [16], inducing release of pro-inflammatory cytokines and chemokines and leading to the recruitment of neutrophils and pro-inflammatory monocyte subsets in the healing infarct. Inflammatory mediators also stimulate synthesis and activation of proteases (such as matrix metalloproteinases (MMPs)) [17], thus degrading the collagenous extracellular matrix network and releasing pro-inflammatory matrix fragments [18]. A large body of evidence suggests that myocardial compliance is diminished during the inflammatory phase of infarct healing. Several distinct mechanisms may account for increased stiffness. First, induction of pro-inflammatory cytokines, such as interleukin (IL)-6, has been suggested to induce a rapid increase in cardiomyocyte passive tension [19] through modulation of the phosphorylation status of titin, a myofilament protein that acts as a molecular spring in the cardiac sarcomere [19]. Second, release and activation of proteases (such as MMP-2) may cleave titin, further increasing myocardial stiffness [20]. Third, changes in the composition of the extracellular matrix may also contribute to reduced compliance. Although the inflammatory phase is associated with prominent degradation of the collagen matrix network [21], disrupted organization of the matrix may have a profound impact on the mechanical properties of the myocardium and may also perturb relaxation of surviving cardiomyocytes.

5 The Proliferative Phase of Infarct Healing

In the infarcted myocardium, inflammatory cells serve to clear the infarct from dead cells and matrix debris and also set the stage for activation of reparative fibroblasts and vascular cells. The post-infarction inflammatory reaction is tightly regulated; after an early peak, recruitment of pro-inflammatory leukocytes is reduced, anti-inflammatory subsets of monocytes [22] and T cells [23–25] infiltrate the myocardium, and macrophages acquire a reparative phenotype. Clearance of apoptotic cells and matrix fragments by phagocytic macrophages triggers release of anti-inflammatory mediators (such as transforming growth factor (TGF)-βs, growth differentiation factor (GDF) 15, and interleukin (IL)-10) [26–28] that suppress chemokine synthesis and attenuate expression of leukocyte integrins, thus inhibiting leukocyte recruitment. TGF-βs also exert prominent fibrogenic actions, activating fibroblast to myofibroblast conversion [29, 30] and stimulating collagen synthesis and deposition [31] in the infarcted heart. Tight regulation of the post-infarction inflammatory response, and rapid activation of reparative signals, is critical in order to prevent ventricular rupture [32], a fatal acute complication that is typically driven by overactive leukocyte-mediated inflammation [33], or perturbed fibroblast activation [34, 35]. The cellular events involved in scar formation following infarction require activation of mechanosensitive signaling pathways.

6 Myofibroblasts: Central Effectors of Repair in the Infarcted Heart

Expansion and activation of matrix-producing interstitial fibroblasts that express contractile proteins, such as α-smooth muscle actin (α-SMA), are the hallmark of the proliferative phase of infarct healing (Fig. 1) [36, 37]. These cells are called myofibroblasts and are the main cellular source of collagens in the healing infarct [38]. Most infarct myofibroblasts are derived from conversion and proliferation of resident cardiac fibroblasts. Although some investigations suggested that a wide range of cell types, including endothelial cells [39], bone marrow-derived progenitors [40], and macrophages [41], may contribute to the expanding infarct myofibroblast population through direct transdifferentiation, robust lineage tracing-based evidence showed that the contributions from non-fibroblasts is limited [42, 43]. In the infarcted myocardium, secreted mediators and mechanosensitive signaling cascades cooperate to trigger formation of stress fibers and induction of α-SMA in activated fibroblasts. The exact contribution of α-SMA to the mechanical properties and matrix-synthetic capacity of myofibroblasts remains poorly understood. Incorporation of α-SMA into the stress fibers may contribute to generation of contractile forces by fibroblasts, thus participating in scar remodeling [44]. However, in vitro experiments have suggested that α-SMA overexpression in cardiac fibroblasts populating collagen lattices is not sufficient to promote contraction of the matrix [45]. On the other hand, α-SMA expression may have broad effects on the fibroblast behavior, regulating proliferation [45] and motility [46].

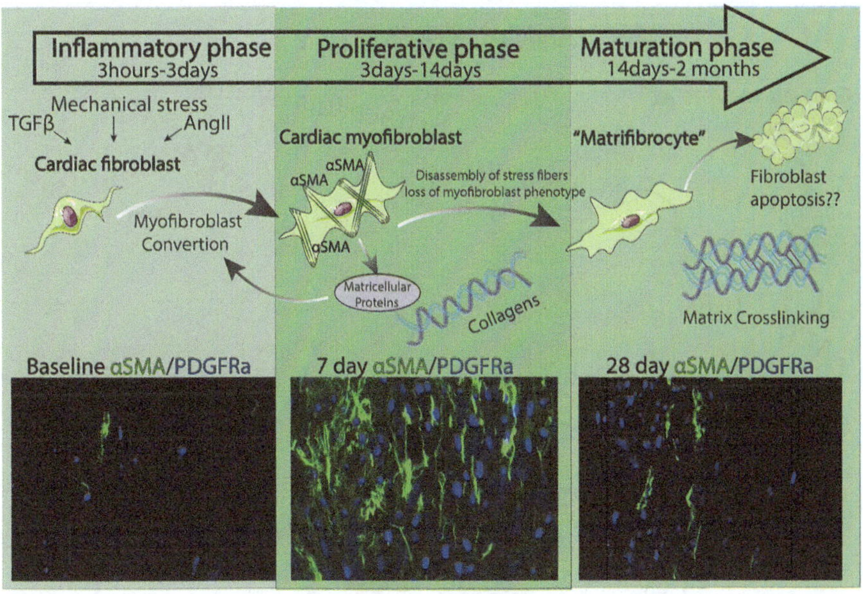

Fig. 1 State transitions of fibroblasts during the phases of cardiac repair. The heart contains a significant population of resident cardiac fibroblasts. During the inflammatory phase of infarct healing, these cells may secrete inflammatory mediators and proteases. During the proliferative phase, secreted mediators such as TGFβ1 and angiotensin II and mechanosensitive signaling cascades cooperate to trigger cardiac fibroblast expansion and conversion to myofibroblasts, which incorporate contractile proteins, such as α-smooth muscle actin (α-SMA) into cytoskeletal stress fibers. In the healing infarct, α-SMA-expressing myofibroblasts are the main cellular source of matricellular proteins and structural extracellular matrix proteins, such as collagens. Secretion of structural matrix proteins by activated myofibroblasts is followed by induction of matrix cross-linking enzymes that contribute to scar maturation. As the scar matures, fibroblasts exhibit disassembly of α-SMA stress fibers and markedly reduce their numbers. Current evidence suggests that depletion of myofibroblasts from the mature scar may reflect apoptosis of fibroblasts or acquisition of a distinct fibroblast phenotype, characterized by high expression of bone-cartilage genes, known as matrifibrocyte

7 The Mechanisms of Fibroblast and Myofibroblast Activation in the Infarcted Myocardium: From the Cell Surface to the Nucleus (Fig. 2)

7.1 Neurohumoral Pathways and the Renin-Angiotensin-Aldosterone System (RAAS)

Neurohumoral pathways (including adrenergic signaling and the RAAS) are prominently involved in activation of fibroblasts in the infarcted and remodeling heart. The bulk of in vitro and in vivo evidence supports a central role for the RAAS in activation of infarct myofibroblasts [47]. Moreover, the clinical benefit of

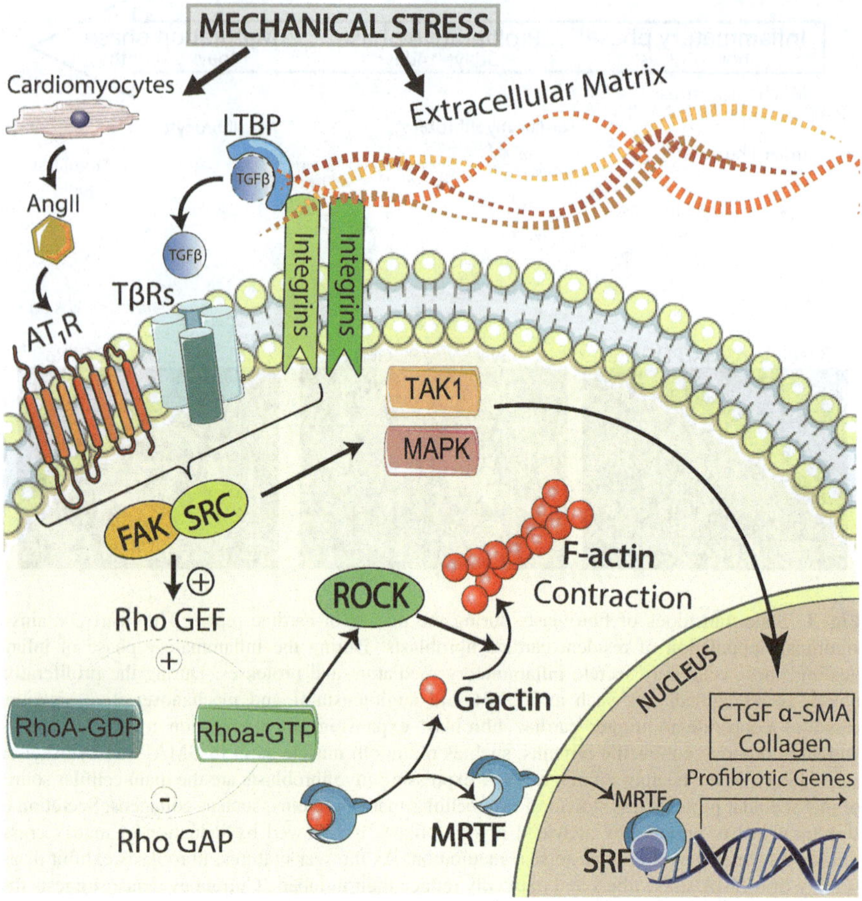

Fig. 2 Mechanosensitive signaling pathways involved in cardiac fibroblast activation in the infarcted and remodeling myocardium. Mechanical stretch plays a major role in fibroblast activation by promoting the release of bio-active TGF-β from the matrix-bound latent complex with the LAP and LTBP (large latent complex), through mechanosensitive integrin-mediated actions. In addition, mechanical stress can also trigger angiotensin II release by neighboring cardiomyocytes, activating fibroblasts through the type 1 angiotensin receptor (AT1). Neurohumoral and mechanical stress-mediated activation of tyrosine kinases, G-protein-coupled receptors, and integrins in cardiac fibroblasts leads to activation of focal adhesion kinases (FAK) and Src family tyrosine kinases that subsequently recruit Rho guanine nucleotide-exchange factors (RhoGEF). RhoGEFs catalyze the exchange of GDP for GTP, leading to the activation of the small GTP-binding protein RhoA. When RhoA activation by RhoGEFs is favored over its inactivation by RhoGAPs, downstream signaling occurs through the Rho-associated coiled coil containing kinases (ROCKs). ROCK activation results in the polymerization of globular G-actin into filamentous F-actin required for the formation of contractile stress fibers in myofibroblasts. F-actin polymerization results in disruption of the association of Myocardin-related transcription factor (MRTF) with G-actin and subsequent MRTF translocation to the nucleus, where it can promote profibrotic gene expression through its interaction with serum response factor (SRF). In addition to their effects on Rho/ROCK pathways, mechanoreceptor activation can also promote cardiac fibroblast activation through p38 MAPK, ERK MAPK, and TAK1 signaling

pharmacologic strategies targeting the RAAS may involve inhibition of fibrosis, thus highlighting the central role of neurohumoral cascades in the pathogenesis of post-infarction remodeling. Mechanical stress is known to trigger angiotensin II (Ang II) release by cardiomyocytes in vitro and in vivo [48, 49]. In the infarct border zone, surviving cardiomyocytes are subjected to mechanical stretch that may generate angiotensin II, thus contributing to fibrosis and cardiac hypertrophy. Angiotensin exerts a wide range of activating effects on cardiac fibroblasts, stimulating proliferation [50], triggering a migratory response [51, 52], inducing expression of integrins [53], promoting fibroblast conversion to α-SMA+ myofibroblasts [54], and potently upregulating synthesis and secretion of collagens and other structural matrix proteins [49, 55]. The activating actions of angiotensin II on fibroblasts involve the type 1 angiotensin receptor (AT1), whereas AT2 (the type 2 receptor) has been suggested to inhibit angiotensin II-induced fibroblast proliferation and matrix synthesis [56, 57]. Some fibrogenic angiotensin II actions are direct, whereas other effects may require the activation of downstream mediators, such as aldosterone [58, 59] or TGF-β [60–62]. In vitro evidence in isolated cardiomyocytes suggests that in addition to its effects on angiotensin II release, mechanical stress may also directly activate AT1 receptor signaling in an angiotensin II-independent manner [63]; however, the significance of this mechanism in activation of infarct myofibroblasts remains unknown.

7.2 The Role of Cytokines and Chemokines in Fibroblast Activation

Cytokines and chemokines play an important role in activation of fibroblasts in the infarcted heart, both through direct effects and by stimulating recruitment of macrophages and lymphocytes with fibroblast-activating properties. The pro-inflammatory chemokine C-C motif chemokine ligand 2 (CCL2)/Monocyte Chemoattractant Protein (MCP)-1 plays an important role in clearance of the infarct from dead cells and matrix debris but is also involved in the pathogenesis of post-infarction remodeling through effects on recruitment of macrophages expressing the C-C motif chemokine receptor 2 (CCR2). These CCR2+ macrophages secrete fibrogenic mediators, such as TGF-β and osteopontin [64–66], thus promoting myofibroblast activation. Although some studies have suggested direct activating effects of CCL2 on hepatic, cutaneous, and pulmonary fibroblasts [67–69], experiments in isolated cardiac fibroblasts did not identify any significant CCL2 actions on matrix gene expression and on proliferative activity [64].

The pro-inflammatory cytokines Tumor Necrosis Factor (TNF)-α, IL-1β, and IL-6 have also been implicated in phenotypic modulation of infarct fibroblasts and in the pathogenesis of post-infarction remodeling. TNF-α and IL-1β do not directly induce a matrix-synthetic program in cardiac fibroblasts but may rather decrease collagen synthesis and inhibit myofibroblast conversion [17, 70–72]. However, stimulation with TNF-α or IL-1β ultimately triggers myofibroblast activation and fibrotic remodeling through indirect actions, stimulating TGF-β synthesis [73], and accentuating the fibrogenic effects of angiotensin II, through upregulation of AT1 receptors [74]. Moreover, TNF-α and IL-1β potently stimulate MMP synthesis [17, 75, 76], thus promoting collagen degradation. Thus, cytokine-mediated generation of matrix fragments in the infarct may further accentuate fibroblast activation [18]. Some members of the IL-6 family of cytokines (also known as the gp130 family, in reference to their common signal transducer molecule) have also been implicated in the activation of fibroblasts in infarcted and remodeling hearts. In vivo studies have suggested that genetic absence of IL-6 may attenuate cardiac fibrosis and dysfunction in a model of myocardial infarction [77]. IL-6-mediated fibroblast activation has been attributed to signal transducer and activator of transcription (STAT)3-dependent stimulation of collagen synthesis [78], or to induction of TGF-β [79]. IL-11, another member of the gp130 family of cytokines, has been suggested to play a critical role in the activation of cardiac fibroblasts [80], serving as a downstream fibrogenic signal, stimulated by TGF-β.

It should be emphasized that induction of cytokines and chemokines in the infarcted heart is predominantly driven by innate immune pathways, activated by necrotic cardiomyocytes. However, mechanosensitive pathways have also been suggested to promote pro-inflammatory cytokine induction [81, 82]. Mechanical strain induces CCL2 synthesis in endothelial cells [81]. In a co-culture system, cyclical endothelial stretch activated monocytes, promoting monocyte STAT3 activation and synthesis of pro-inflammatory cytokines [83]. Moreover, stretch of cardiac fibroblasts stimulates NF-κB activation [84], triggering IL-1β release [82]. In human cardiac fibroblasts harvested from heart failure patients, mechanical stress induced inflammatory pathways, and the cell culture supernatant from stretched fibroblasts stimulated monocyte transendothelial migration [85]. Thus, in the infarcted heart, mechanical stress may amplify the pro-inflammatory effects of innate immune signaling, accentuating fibroblast activation.

7.3 The Role of TGF-β in the Activation of Infarct Myofibroblasts

TGF-βs are central effectors of myofibroblast activation and have been implicated in the pathogenesis of a wide range of fibrotic conditions, involving several different organs [86]. The well-documented relation between mechanical stress and activation of TGF-β signaling cascades suggests that TGF-β may be a crucial molecular link between mechanotransduction and fibroblast activation. All three TGF-β isoforms (TGF-β1, TGF-β2, and TGF-β3) are induced and activated in the infarcted heart, exhibiting distinct temporal patterns of expression [27, 87]. TGF-β1 and TGF-β2 are upregulated early, whereas TGF-β3 exhibits a late pattern of upregulation, peaking 7 days after coronary occlusion [27, 87]. Although several different cell types (including platelets, macrophages, fibroblasts, and cardiomyocytes) can produce TGF-βs, their relative contribution to TGF-β upregulation in the infarcted myocardium is unclear. Considering the presence of significant amounts of latent, but activatable, TGF-β in normal hearts, bound to the extracellular matrix, the significance of de novo synthesis of TGF-β in the healing and remodeling infarct is not known.

TGF-βs are synthesized as proproteins that contain both the mature growth factor and its propeptide (also known as the latency-associated peptide/LAP). In the endoplasmic reticulum, the proprotein is linked to a member of the latent TGF-β binding protein (LTBP) family. The complex is subsequently transported to the Golgi, where furin cleaves the mature TGF-β from the LAP. However, TGF-β and LAP remain tightly bound through non-covalent interactions. TGF-β secretion by various cell types occurs in the form of the large latent complex (LLC, TGF-β/LAP/LTBP) or the small latent complex (SLC, TGF-β/LAP) [88]. The LAP is sufficient to confer latency, preventing TGF-β binding to its receptors [89].

In the infarcted heart, proteases, matricellular proteins, and integrins cooperate to release the mature TGF-β dimer from the latent complexes in close proximity to the cell surface, thus promoting TGF-β/TGF-β receptor interactions, and activation of downstream signaling cascades. Several different families of proteases, including calpains, cathepsins, serine proteases, MMPs, and members of the ADAM with thrombospondin type I motif (ADAMTS) family, are capable of activating TGF-β in vitro and, when released in the infarcted heart, may trigger TGF-β activation [90–95]. Specialized matrix proteins, such as ED-A fibronectin (ED-A Fn) [96] and thrombospondin (TSP)-1 [97], have also been implicated in TGF-β activation in the infarcted heart. The TGF-β-activating effects of ED-A Fn have been attributed to localization of activatable TGF-β in the area of injury through LTBP immobilization into the matrix [98, 99]. On the other hand, TSP-1 may act through interactions with the LAP, leading to release of bioactive TGF-β [100].

The biochemical pathways responsible for generation of bioactive TGF-β in tissues can be stimulated by mechanical stress. Experimental evidence suggests that mechanical stretch plays a major role in TGF-β activation, promoting sustained myofibroblast activation [101, 102], predominantly through integrin-mediated actions. αv integrins have been implicated in mediating spatially restricted TGF-β activation in fibrotic tissues [103, 104], through protease-dependent or protease-independent mechanisms. Protease-mediated TGF-β activation requires recruitment of the membrane-bound metalloproteinase MMP14, which releases TGF-β from latent complexes through proteolytic actions [105]. Non-proteolytic activation may involve αV-integrin-mediated mechanical forces, exerted by the actin cytoskeleton, that induce conformational changes of the latent complex, resulting in presentation of active TGF-β to its receptors [106–108]. In vivo studies in a wide range of fibrotic conditions support the central role of αV integrin as a mechanosensor that mediates mechanical activation of TGF-β [109–111]. Based on this evidence, mechanical activation of αV integrin in the infarct border zone represents a highly plausible mechanism for TGF-β activation in healing infarcts. However, the role of the αV integrin/TGF-β axis in myocardial infarction has not been documented in vivo.

In addition to its effects on TGF-β activation, mechanical stress also regulates TGF-β transcription in several different cell types, including cardiomyocytes [112, 113] and fibroblasts [112]. It should be emphasized that in vitro studies suggest that the composition of the extracellular matrix plays a central role in regulation of TGF-β responses in fibroblasts. The compliance of collagen gels profoundly affects transcription of genes encoding contractile proteins (such as α-SMA) in TGF-β-stimulated fibroblasts. Fibroblasts populating floating collagen lattices have low intracellular tension and exhibit low level α-SMA upregulation in response to TGF-β stimulation. In contrast, in anchored collagen lattices (typically associated with moderate intracellular tension), TGF-β significantly induces α-SMA expression. In fibroblasts cultured on thin films of collagen-coated plastic, intracellular tension is high and is associated with accentuated α-SMA induction upon TGF-β stimulation [114].

Following activation, TGF-β signals through heterotetrameric TGF-β receptor complexes transducing cascades involving a series of intracellular effectors, the Smads, or activating Smad-independent cascades [115, 116]. In vitro and in vivo evidence suggests that both canonical Smad-dependent cascades and non-canonical Smad-independent signaling contribute to fibroblast activation in fibrotic hearts. In vivo studies have demonstrated that Smad3 signaling is critically involved in activation of reparative fibroblasts following myocardial infarction, inducing ECM protein synthesis, integrin transcription, and α-SMA expression [31, 117] and contributing to the formation of an organized scar [35]. Smad3 signaling in infarct myofibroblasts restrains cell proliferation [31, 35, 117], generating well-aligned arrays of activated myofibroblasts that preserve the structural integrity of the infarcted heart through activation of a reparative integrin-Reactive Oxygen Species (ROS) axis [35]. These findings highlight the critical reparative function of activated fibroblasts in myocardial infarction. In contrast, Smad2 does not seem to play a major role in mediating fibroblast activation in infarcted hearts [118]. The contrasting effects of Smad2 and Smad3 may reflect differences in their patterns of activation and nuclear translocation following TGF-β stimulation or distinct interactions with transcriptional regulators in the nucleus. On the other hand, some actions of TGF-β on cardiac fibroblasts may involve activation of Smad-independent p38 mitogen-activated protein kinase (MAPK), extracellular signal-regulated kinase (ERK) MAPK, Ras homologue gene family/Rho-associated coiled-coil containing kinases (RhoA/ROCK), or transforming growth factor-β-activated kinase 1 (TAK1) signaling [119–121]. However, considering the wide range of mediators that can activate these common kinase pathways, the relative contribution of TGF-β in their activation is unclear.

TGF-β exerts a broad range of direct effects on fibroblasts that contribute to scar-mediated repair of the infarcted heart but can also be involved (if excessive, or prolonged) in the pathogenesis of post-infarction fibrosis. TGF-β stimulation potently induces myofibroblast conversion [29], and increases ECM protein synthesis by activated fibroblasts [31]. Moreover, TGF-β increases expression of integrins [35] and shifts the protease/anti-protease balance toward a matrix-preserving phenotype by inducing protease inhibitors, such as Plasminogen Activator Inhibitor

(PAI)-1 and tissue inhibitor of metalloproteinase 1 (TIMP1) [117, 122], and by suppressing MMP synthesis [123]. It should be noted that the effects of TGF-β on cardiac fibroblast proliferation are dependent on the context. For example, some studies have reported that TGF-β stimulates proliferation of cardiac fibroblasts [124], whereas other investigations demonstrated anti-proliferative effects [31, 125]. The contrasting findings likely reflect the context-dependent actions of TGF-β, which exerts distinct effects on various cell types, depending on their state of differentiation, the presence or absence of other bioactive mediators, and the biochemical composition of the matrix environment.

In addition to the direct actions of TGF-β on transcription of fibrosis-associated genes, some TGF-β-mediated effects on infarct myofibroblasts may be mediated through induction of other secreted mediators, such as the matricellular protein CCN2/connective tissue growth factor (CTGF) [126] and the gp130 cytokine IL-11 [80]. CCN2 induction has been demonstrated in border zone cardiomyocytes and in fibroblasts or myofibroblasts infiltrating the infarct [127]. Although CCN2 has been suggested to mediate activating effects of TGF-β on cardiac fibroblasts [128], experiments using both loss- and gain-of-function approaches have challenged the in vivo significance of this pathway [129, 130]. On the other hand, studies on the effects of IL-11 have produced conflicting results. Although infusion of recombinant mouse IL-11 was found to aggravate fibrosis in a mouse model of myocardial infarction [131], presumably through actions on fibroblasts, treatment with human IL-11 was found to be cardioprotective and anti-fibrotic, through activation of STAT3-dependent anti-apoptotic pathways [132].

The activating effects of TGF-β on cardiac fibroblasts are restrained through induction of the inhibitory Smad, Smad7. In the infarcted heart, Smad7 is upregulated in myofibroblasts and inhibits both Smad2/3-mediated signaling and Smad-independent cascades, through actions downstream of the TGF-β receptors. The anti-fibrotic effects of Smad7 play a crucial role in protecting the infarcted ventricle from excessive fibrosis, adverse remodeling, and heart failure. Moreover, Smad7 has TGF-β-independent actions, suppressing fibrogenic effects of receptor tyrosine kinases, such as Erbb2 [133]. Although studies in mesangial cells showed that mechanical stretch induces Smad7 expression [134], whether mechanosensitive signaling is implicated in the induction of inhibitory Smads in cardiac fibroblasts is not known.

7.4 Components of the Provisional Matrix as Regulators of Fibroblast Activation

Changes in the extracellular matrix environment during the proliferative phase of infarct healing play a central role in fibroblast activation. During the proliferative phase of healing, formation of a provisional matrix, comprised of fibrin, fibronectin, and proteoglycans, may be important for fibroblast migration, proliferation, and conversion to myofibroblasts [135]. These matrix-bound proteins cooperate with growth factors, such as TGF-β, to transduce signals in fibroblasts. In vitro studies suggested that the ED-A splice variant of fibronectin is critical for TGF-β1-induced α-SMA upregulation in fibroblasts, an indicator of myofibroblast conversion [99]. Consistently with these findings, ED-A fibronectin loss attenuated myofibroblast transdifferentiation in healing myocardial infarcts [96]. The specific interactions between the ED-A segment and the TGF-β signaling cascade remain unknown. Although formation of the fibronectin-based provisional matrix in the healing infarct reflects extravasation of plasma fibronectin through hyperpermeable cytokine-stimulated microvessels [136, 137], induction of cellular fibronectin and assembly of fibronectin fibrils may be, at least in part, mechanically stimulated. Cyclic strain in fibroblasts has been reported to upregulate fibronectin fibrillogenesis [138].

Hyaluronan and versican have also been implicated in myofibroblast conversion. In vitro studies showed that pericellular hyaluronan was involved in maintenance of a myofibroblast phenotype by TGF-β-stimulated cells [139], and in vivo experiments suggested that CD44, the main receptor for hyaluronan, mediated collagen synthesis in infarct fibroblasts [140]. The role of versican in activation of infarct myofibroblasts is more speculative. In dermal fibroblasts, versican was implicated in myofibroblast conversion [141]. Although versican is upregulated in the infarcted myocardium [142], whether it mediates activation of infarct myofibroblasts is unknown.

7.5 *Specialized Matrix Proteins Transduce Reparative and Fibrogenic Signals in the Infarcted Heart*

In addition to the effects of provisional matrix components that originate from extravasated plasma proteins (such as fibrinogen-derived fibrin and fibronectin), the extracellular matrix in the infarct is enriched through de novo synthesis of matricellular proteins [143]. These specialized matrix proteins do not play a primary structural role but bind to the structural matrix and modulate signaling cascades, serving to localize growth factor and protease activity in the area of injury [144, 145]. Many matricellular macromolecules are induced in response to mechanical stress, thus representing an important link between mechanical stimulation and tissue remodeling [146]. In addition to the effects of mechanosensitive pathways (including neurohumoral cascades), cytokines, growth factors, and oxidative stress can also stimulate synthesis of matricellular proteins in the infarcted and remodeling heart [147, 148].

Several members of the matricellular family (including thrombospondins, tenascin-C, osteopontin, secreted protein acidic and rich in cysteine (SPARC), and periostin) have been implicated in repair and remodeling of the infarcted heart through effects on fibroblast activation (Table 1). Thrombospondin (TSP)-1 is markedly upregulated following myocardial infarction and is selectively deposited in the infarct border zone [97], where it may activate TGF-β, promoting fibroblast activation while inhibiting angiogenesis through direct CD36-dependent [189] effects on endothelial cells. Tenascin-C is also markedly upregulated in the healing infarct [190] and may promote fibroblast migration [156] while stimulating integrin-dependent macrophage activation [157]. Osteopontin upregulation in the healing infarct may promote fibroblast-mediated collagen deposition [171]. Induction of SPARC plays a role in formation of an organized scar, stimulating granulation tissue formation and collagen maturation, and protecting from left ventricular rupture [165]. Secretion of periostin by activated infarct myofibroblasts also plays an important role in activation of a reparative matrix-secreting program by fibroblasts [184], preventing cardiac rupture [184, 185], but also contributes to chronic progression of fibrosis and adverse post-infarction remodeling [185]. Although several robust in vivo studies have documented the critical role of members of the matricellular family in both repair and adverse remodeling of the infarcted heart, the specific molecular mechanisms responsible for these actions remain unclear. Matricellular proteins have several functional domains that can modulate receptor-mediated signaling cascades or regulate availability of active growth factors and proteases. Thus, dissection of the relative significance of specific molecular mechanisms and interactions is challenging.

Moreover, it is unclear to what extent induction of matricellular proteins in the infarct involves activation of mechanosensitive cascades. Some members of the matricellular family, such as tenascin-C, are highly inducible upon mechanical stimulation [147, 148]. Thus, the selective deposition of tenascin-C in the border zone and in the remote remodeling myocardium may reflect the accentuated mechanical stretch of fibroblasts and cardiomyocytes in these areas. Other matricellular proteins may be upregulated predominantly through cytokine and growth factor-mediated pathways, without significant involvement of mechanically activated pathways. For example, osteopontin is predominantly expressed in macrophages infiltrating the infarcted heart [137, 191], and its upregulation is dependent on stimulation by inflammatory signals [65].

7.6 Integrins

Integrins are a superfamily of cell adhesion receptors that bind primarily to extracellular matrix ligands, serving as molecular links between the matrix environment and intracellular pathways, thus modulating a wide range of cellular responses [192, 193]. On the cell surface, integrins exist as heterodimers, comprised of non-covalently interacting combinations of α and β subunits. Some of the integrin subunits appear in only a single heterodimer, whereas others are more promiscuous, forming combinations with several other subunits. For example, the α5 subunit is present only in combination with β1; in contrast, the αV subunit has 5 different

Table 1 The role of the prototypical matricellular proteins in myocardial infarction

Matricellular protein	Cellular localization	Mechanism of mechanical activation	Biochemical properties	In vivo role in myocardial infarction	Proposed cellular functions and molecular effects	Refs.
Thrombospondin (TSP)-1	Deposited in the extracellular matrix of the infarct border zone. Cellular sources include vascular cells, macrophages, fibroblasts, and border zone cardiomyocytes	Mechanical stretch upregulates TSP-1 expression in vascular cells, fibroblasts	Multidomain homotrimeric glycoprotein	Protects from adverse post-infarction remodeling by inhibiting inflammation in the border zone	Suppresses inflammation through activation of TGF-β Stimulates myofibroblast conversion through TGF-β activation Inhibits MMPs and stabilizes the matrix network Inhibits angiogenesis Has direct effects on collagen fibril formation Regulates cardiomyocyte size under conditions of stress through activation of the endoplasmic reticulum stress effector PERK	[97, 149–155]
Tenascin C	Deposited in the matrix network in the infarct and remodeling myocardium. Cellular sources include macrophages, fibroblasts,	Mechanical stress markedly upregulates tenascin-C expression in fibroblasts and in cardiomyocytes	Hexameric multimodular glycoprotein (hexabrachion).	Induces ventricular dilation and accelerates adverse remodeling after myocardial infarction	Suggested to act as an endogenous danger signal that stimulates toll-like receptor (TLR) signaling and promotes inflammation	[148, 156–164]

	and vascular cells and possibly also cardiomyocytes				Promotes a migratory phenotype in fibroblasts Activates the fibrogenic properties of macrophages through integrin signaling Exerts fibrogenic actions though accentuation of growth factor signaling Has been reported to have regenerative effects in fish and amphibian hearts	
SPARC (secreted protein acidic and rich in cysteine)	Expressed by fibroblasts, vascular cells, and macrophages	Although SPARC induction in response to mechanical stress has been reported in podocytes, mechanosensitive regulation of SPARC in myocardial cells has not been studied	Multimeric calcium-binding glycoprotein with high affinity to collagen	SPARC induction in the infarcted myocardium protects from cardiac rupture and reduces systolic dysfunction, presumably by preserving the extracellular matrix	Enhances collagen fibril assembly and matrix organization and regulates collagen interaction with the cell surface of fibroblasts Exerts fibrogenic actions through ADAMTS1 upregulation Accentuates growth factor signaling, promoting fibroblast activation and angiogenesis	[165–170]

(continued)

Table 1 (continued)

Matricellular protein	Cellular localization	Mechanism of mechanical activation	Biochemical properties	In vivo role in myocardial infarction	Proposed cellular functions and molecular effects	Refs.
Osteopontin	Expressed predominantly in activated infarct macrophages but is also upregulated in fibroblasts and vascular cells upon stress	Mechanical stress induces expression of osteopontin in bone cells; however, evidence for similar effects in myocardial cells is lacking	Acidic glycoprotein, expressed predominantly in macrophages, acting through a central integrin-binding RGD sequence and via CD44	Reduces adverse remodeling following infarction, presumably by preserving the matrix network and by stimulating angiogenesis	Inhibits fibroblast apoptosis and stimulates fibroblast activation through miR-21. Promotes collagen synthesis and accumulation Regulates extracellular matrix assembly Promotes myofibroblast conversion Stimulate integrin-dependent fibrogenic actions May modulate cytokine-induced MMP activation May induce expression and activity of the cross-linking enzyme lysyl-oxidase Upregulates galectin-3 expression in macrophages	[171–183]

Periostin	Expressed by activated myofibroblasts	Mechanical stretch induces periostin synthesis in cardiac fibroblasts	Fasciclin domain-containing glycoprotein, upregulated in activated fibroblasts and (as a typical matricellular protein) secreted into the matrix following injury	Protects from cardiac rupture and mediates repair of the infarcted heart but is also involved in chronic fibrotic remodeling and progression of dysfunction	Regulates matrix assembly Stimulates acquisition of a pro-fibrotic phenotype by fibroblasts, by increasing growth factor activity Promotes fibroblast migration through interactions with integrins Upregulates expression of matrix cross-linking enzymes, such as lysyl-oxidase May have regenerative functions in neonatal mouse hearts	[184–188]

partners, and β1 can form combinations with 12 different α subunits. The complexity of the integrin system reflects the wide range of roles of the superfamily in development, homeostasis, tissue injury, and repair.

In the healing infarct, integrins serve as key mechanosensors and function as molecular bridges between the extracellular matrix and the cells. TGF-β and angiotensin II induce and activate integrins in cardiac fibroblasts [35, 53]. Subsequently, activated integrins interact with matrix-bound binding partners (such as fibronectin, collagens, and a wide range of matricellular macromolecules) and transduce downstream signaling cascades. Regardless of the mechanism responsible for their activation, integrin cascades typically involve focal adhesion kinase (FAK), Src family tyrosine kinases, Rho/ROCK, and MAPK pathways [194]. Integrin-mediated activation plays a central role in a wide range of fibroblast responses, including migration, proliferation, extracellular matrix protein synthesis, and myofibroblast conversion. The effects of integrins in the infarcted heart are not limited to fibroblast activation but also involve other cell types, such as cardiomyocytes, immune cells, and vascular cells. β1, β2, and β3 integrins are the best-studied members of the family in experimental models of myocardial infarction. Expression of integrin β1A is highly expressed in fibroblasts and in leukocytes infiltrating the infarct, whereas integrin β1D is predominantly localized in cardiomyocytes [195]. β2 integrins are critically involved in recruitment of leukocytes in the healing infarct [196]. On the other hand, expression of β3 integrins is localized predominantly in infarct vascular cells [195, 197]. Cell-specific loss-of-function experiments in vivo suggested that cardiomyocyte β1 integrin signaling exerts protective actions on the ischemic heart, by preserving mitochondrial function [198]. In vitro studies and associative in vivo data suggest that β1 integrin may also be implicated in fibroblast proliferation [199] and in induction of matrix synthesis [200, 201] following infarction. Unfortunately, robust studies examining the role of fibroblast-specific β1 integrin responses in infarction models are lacking. αV integrin (which can associate with both β1 and β3 chains) is also rapidly activated in response to mechanical stress and may promote fibroblast activation through direct actions or via activation of latent TGF-β [202, 203].

7.7 *Mechanosensitive Ion Channels*

Emerging evidence suggests an important role for mechanosensitive ion channels in activation of fibroblasts in infarcted and remodeling hearts [204]. Fibroblasts express a wide range of cation channels that enable cation flux across their membrane, contributing to activation of downstream signaling cascades. Some of the ion channels expressed by fibroblasts are mechanically gated, thus regulating ion conductance and cellular activation in response to mechanical stress. The best studied of these channels are the members of the Transient Receptor Potential (TRP) family (such as TRPC6, TRPV4, and TRPM7), the potassium-selective channels TREK-1 and KATP, and Piezo1 (Fig. 3). These mechanically activated channels may serve as

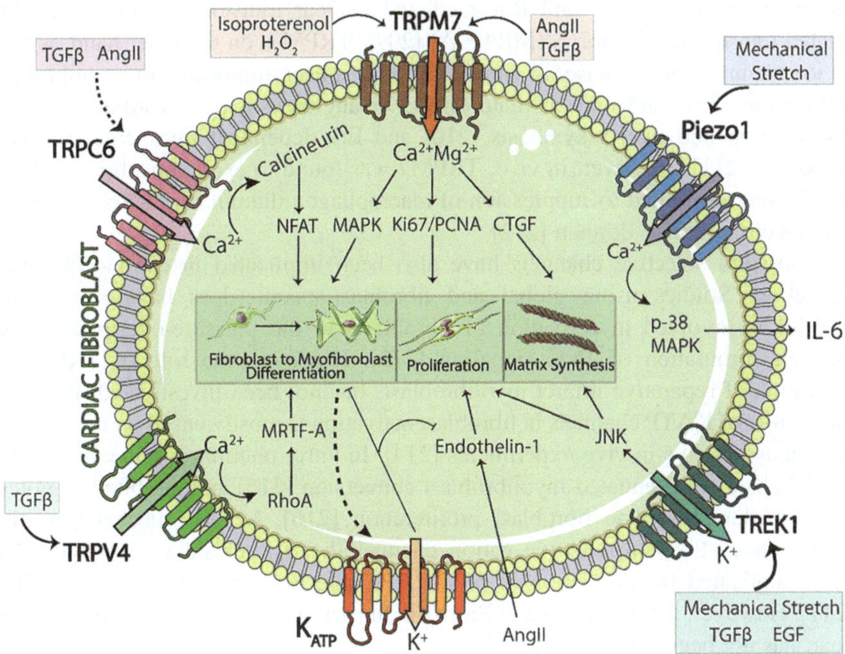

Fig. 3 Mechanosensitive ion channels' role in cardiac fibroblast activation. Fibroblasts express a wide range of cation channels that enable cation flux across their membrane, contributing to activation of downstream signaling cascades that stimulate cardiac fibroblast proliferation, matrix synthesis, and myofibroblast conversion. Some of these ion channels are mechanically gated, thus regulating ion conductance and cellular activation in response to mechanical stress. Members of the Transient Receptor Potential (TRP) family (TRPC6, TRPV4, and TRPM7) are activated by mechanical stress and by fibrogenic mediators, such as angiotensin II and TGF-β. TRPC6 and TRPV4 stimulate cardiac fibroblast to myofibroblast differentiation via intracellular Ca+2 increase-dependent activation of Calcineurin/NFAT and via RhoA/MRTF pathways. The potassium selective channel TREK1 and the non-selective channel Piezo1 have also been suggested as molecular links between mechanical stress and fibrogenic activation. Piezo1 activation in cardiac fibroblasts has been suggested to increase expression and secretion of IL-6 via p38 MAPK pathways. TREK1 is involved in fibroblast activation through JNK activation, whereas KATP channels can act as negative feedback mechanisms, attenuating angiotensin II-induced fibroblast proliferation and endothelin-1 expression while also inhibiting further myofibroblast conversion

direct links between mechanical stress and fibroblast activation or may participate in transduction of signals generated by other mechanosensitive pathways. The TRP channels have been implicated in myofibroblast conversion [205–208]. TRPC6-mediated increases in calcium permeability were implicated in myofibroblast transdifferentiation through activation of the calcineurin-nuclear factor of activated T cells (NFAT) signaling pathway [205]. Experiments using a global loss-of-function model suggested that TRPV4 is implicated in adverse remodeling and in progression of fibrosis following myocardial infarction [209]. The fibrogenic effects of TRPV4 may also involve accentuated myofibroblast conversion through increased calcium influx and Rho-mediated nuclear translocation of myocardin-related transcription factor A (MRTF-A) [209]. TRPM7, on the other hand, is both an ion channel and a kinase. In vitro, TRPM7 has been implicated in myofibroblast conversion [207] and may mediate effects of angiotensin II on cardiac fibroblast proliferation and matrix synthesis [210] and Erk-dependent fibrogenic oxidative responses [211]. However, in vivo, TRPM7 was found to exert anti-fibrotic effects, which were attributed to suppression of macrophage inflammatory activity through actions of the kinase domain [212].

Potassium selective channels have also been implicated in cardiac fibroblast activation. Studies using global and fibroblast-specific loss have shown that TREK1 is involved in activation of fibroblasts in the pressure-overloaded heart through stimulation of JNK signaling [213]. However, the role of TREK1 in activation of reparative infarct myofibroblasts has not been investigated. Evidence on the role of KATP channels in fibroblast activation is mostly based on in vitro data and on associative in vivo experiments [214]. In vitro, pharmacologic inhibition of KATP channels attenuated myofibroblast conversion [215] and inhibited angiotensin II-mediated cardiac fibroblast proliferation [216]. Finally, mechanosensitive activation of the non-selective cation channel Piezo1 in cardiac fibroblasts has been implicated in stimulation of p38 MAPK and downstream secretion of IL-6 [217]. However, the in vivo role of Piezo1 in fibrosis and remodeling of the infarcted heart has not been investigated.

7.8 Focal Adhesion Kinase (FAK)

The tyrosine kinase FAK plays an important role in activation of matrix-synthetic myofibroblasts in response to mechanical stress, serving as a link between integrin activation and stimulation of a fibrogenic program [218–221]. Although the involvement of FAK in activation of infarct myofibroblasts is plausible, robust documentation of fibroblast-specific actions of FAK in myocardial infarction is lacking. Administration of FAK inhibitors has been suggested to attenuate progression of fibrosis in myocardial infarction models [222, 223]. However, FAK has broad effects on several different cell types, including cardiomyocytes and interstitial and vascular cells [224, 225]. Thus, any protective effects of pharmacologic FAK inhibition on the remodeling infarcted heart may not involve exclusively attenuation of fibroblast activity.

7.9 MAPKs

In vitro studies have documented that MAPKs are activated in cardiac fibroblasts in response to mechanical stress [226]. Considering the broad effects of MAPKs on many different cell types, dissecting their fibroblast-specific actions is challenging.

Moreover, fibroblast MAPKs are not activated only by mechanical stress but also in response to a variety of biochemical stimuli, including neurohumoral mediators, cytokines, growth factors, and matricellular proteins. p38α MAPK is the predominant MAPK isoform expressed in cardiac fibroblasts [227] and has been implicated in myofibroblast conversion upon ischemic injury or neurohumoral stimulation.

7.10 *The RhoA/ROCK Pathway*

Mechanical stress activates the small GTP-binding protein RhoA, stimulating downstream signaling through ROCKs. The RhoA/ROCK system plays an important role in generation of actin-myosin contractility and in regulation of cytoskeletal dynamics. Subsequently, stress-induced RhoA-dependent actin filament assembly triggers nuclear translocation of MRTF, thus inducing α-SMA transcription and mediating myofibroblast conversion [228].

RhoA/ROCK has profound effects on all cell types involved in cardiac repair, remodeling, and fibrosis [229]. Studies in pressure overload models suggest that pharmacologic inhibition of RhoA-ROCK, ROCK1 disruption, or ROCK2 loss attenuates fibrotic remodeling [230–232]. Considering the paucity of data from cell-specific interventions, whether these effects are mediated through attenuation of fibroblast activity, or involve actions on other cell types, is unclear. In fact, studies in pressure overload models suggested cardiomyocyte-mediated fibrogenic actions of RhoA signaling, likely involving release of fibroblast-activating paracrine mediators [233]. The role of fibroblast-specific activation of RhoA signaling in the infarcted heart has not been investigated. The RhoA/ROCK system may be involved in both reparative and maladaptive fibrogenic responses following infarction.

7.11 *The YAP/TAZ Pathway in Mechanosensitive Activation of Fibroblasts*

Activation of the homologous transcriptional coactivators Yes-associated protein (YAP) and transcriptional coactivator with PDZ-binding motif (TAZ) is a central molecular event mediating cellular responses to mechanical stress [234]. Extensive evidence has implicated YAP/TAZ in activation of fibrotic responses in many different tissues [235, 236]. In the myocardium, a growing body of evidence supports the role of YAP/TAZ in regulation of fibroblast phenotype, not only in pathologic conditions but also in homeostasis. Fibroblast-specific deletion of the Hippo large tumor suppressor kinases (LATS1 and LATS2), which are essential negative regulators of YAP/TAZ, caused fibrosis in the absence of injury in mice and accentuated matrix deposition following myocardial infarction [237]. Moreover, YAP/TAZ overexpression in fibroblasts was found to accentuate synthesis of fibrosis-associated genes and of inflammatory mediators [238]. On the other hand, fibroblast-specific deletion of YAP/TAZ attenuated fibrotic remodeling of the infarcted heart and improved ventricular function [239, 240]. Thus, on the basis of these findings, the YAP/TAZ system appears to play an important role in maladaptive fibrotic remodeling after myocardial infarction, without significantly affecting the reparative response.

7.12 Mechanosensitive Activation of MRTF in Infarct Fibroblast Activation

Mechanical stress-induced activation of the RhoA/ROCK axis triggers F-actin polymerization, stimulating nuclear translocation of the transcription factor MRTF. In the nucleus, MRTF interacts with the ubiquitously expressed transcription factor serum response factor (SRF), initiating transcription of fibrosis-associated genes, such as α-SMA, thus leading to myofibroblast conversion [241–243]. The role of MRTF in activation of infarct myofibroblasts has been suggested by experiments demonstrating attenuated post-infarction fibrosis in mice with global loss of MRTF-A [244]. Although consistent with the in vitro effects of MRTF in cardiac fibroblasts, attenuation of fibrosis in global knockout mice may also reflect actions on other cell types, such as cardiomyocytes and vascular cells [245, 246].

8 The Extracellular Matrix in the Maturation Phase of Infarct Healing

Scar maturation is associated with cross-linking of the extracellular matrix in the infarct zone and with profound changes in the cellular composition of the scar. As the scar matures, the density of infarct myofibroblasts is markedly reduced [247]. Although some infarct myofibroblasts may undergo apoptosis [248, 249], the majority transition to a specialized scar-preserving type of fibroblast, which has been termed "matrifibrocyte" [250]. These cells do not exhibit myofibroblast characteristics, lacking expression of α-SMA and periostin, but synthesize genes associated with bone, cartilage, and tendon development [250]. Although the mechanisms responsible for myofibroblast to matrifibrocyte transition remain unknown, it is tempting to hypothesize that changes in the mechanical properties of the scar, triggered by the deposition and cross-linking of collagen, may play a role in modulation of cellular phenotype. Moreover, recent studies have suggested that the collagen in the mature scar exhibits extensive denaturation, in the absence of protease activity, presumably related to the continuous stretch of the matrix in the remodeling scar [251].

The maturation phase of infarct healing is also associated with changes in microvascular phenotype. As the scar matures, many infarct microvessels acquire a coat of mural cells, whereas uncoated vessels regress. Microvessel coating in the healing infarct requires activation of platelet-derived growth factor receptor (PDGFR)-β signaling [247, 252]. Changes in the mechanical properties of the healing infarct may play an important role in the regulation of endothelial-mural cell interactions.

9 Mechanical Stress May Play a Major Role in Mediating Adverse Remodeling in the Non-infarcted Myocardium

In the infarcted ventricle, the dramatic structural and morphological changes are not limited to the infarct zone. In the presence of a large infarction, the viable myocardium is subjected to pressure and volume loads, as intraventricular pressures are significantly increased, and the chamber dilates. Moreover, surviving cardiomyocytes become hypertrophic, attempting to compensate for the loss of contractility. Increased wall stress generates mechanical stimuli that have been suggested to activate macrophages [253] and fibroblasts in the non-infarcted areas, resulting in development of interstitial fibrosis and contributing to heart failure progression. Interstitial changes are accentuated in the infarct border zone, presumably due to the increased stretch of cardiomyocytes in contact with non-contractile scar and possibly also in response to diffusion of inflammatory and fibrogenic mediators from the infarcted area [254, 255].

10 Translational Perspectives

10.1 Ventricular Unloading in Myocardial Infarction

Pharmacological interventions using angiotensin converting enzyme (ACE) inhibitors or AT1 receptor blockers to target the RAAS may exert beneficial actions, at least in part, through effects on hemodynamic loads. However, load-independent mechanisms are likely to play a prominent role in mediating the protective effects of RAAS inhibition, considering that comparable afterload reduction with other agents (such as calcium channel blockers) provides limited or no benefit [256]. Unloading through mechanical support is the simplest strategy to attenuate mechanical stress-induced changes in myocardial infarction. In patients with cardiogenic shock after myocardial infarction, early initiation of mechanical circulatory support prior to percutaneous intervention (PCI) improves survival [257]. Although the potential beneficial effects of primary unloading in patients without overt cardiogenic shock have not been established, some limited clinical evidence suggests that early circulatory support may afford lasting benefits in patients with a large myocardial infarction. Mechanical support prior to PCI through intra-aortic balloon counterpulsation did not affect infarct size in patients with acute anterior ST elevation MI (STEMI) but was associated with a significant reduction in the exploratory composite endpoint of time to death, shock, or new or worsening heart failure, assessed 6 months after the acute event [258]. Beneficial actions of ventricular unloading may involve protection of susceptible cardiomyocytes from death through attenuation of their metabolic activity [259], inhibition of mechanosensitive pro-apoptotic cascades [260], or preservation of calcium handling [261]. Moreover, reduced end-diastolic pressures may improve coronary arteriolar flow, thus enhancing myocardial perfusion and limiting infarct size [262, 263]. In addition to these effects, unloading may exert important actions on inflammatory and fibrogenic pathways, attenuating infiltration with injurious pro-inflammatory macrophages, and reducing MMP activity [264]. These actions may preserve the matrix that surrounds injured cardiomyocytes, transducing important pro-survival signals [123]. Regardless of the underlying protective mechanism of early unloading, prolonged mechanical support is neither practical nor justified in most patients with myocardial infarction [265]. Thus, attempts to target mechanically activated signals in the remodeling infarcted heart should focus on identification of specific maladaptive mechanosensitive cascades.

10.2 Targeting Mechanosensitive Cascades to Improve Repair and to Attenuate Fibrotic Post-infarction Remodeling

Targeting fibrotic remodeling of the infarcted heart poses major challenges. Because infarct fibroblasts and myofibroblasts play both reparative and maladaptive roles, suppression of fibrogenic pathways carries significant risks. Inhibition of the reparative functions of fibroblasts would be expected to increase the risk of cardiac rupture [35]. Moreover, excessive reduction of the collagen content of the scar may reduce tensile strength, thus increasing ventricular dilation and worsening adverse remodeling [35, 266]. Experimental evidence suggests that mechanosensitive pathways are involved in both reparative and maladaptive fibrogenic pathways following myocardial infarction. Thus, design of strategies targeting mechanosensitive cascades to attenuate fibrotic remodeling of the infarcted heart should take into account the temporal patterns of reparative and maladaptive responses and the patient-specific characteristics of the fibrotic response. When implemented after formation of a mature scar, strategies inhibiting mechanical activation of infarct fibroblasts are unlikely to have deleterious actions on the reparative process. However, such approaches could be beneficial, only in subjects exhibiting prolonged or excessive fibrogenic responses. Specialized imaging approaches [267, 268] and suitable biomarkers [269] reflecting continuous fibrogenic activation are needed to identify these patients.

11 Conclusions

During repair and remodeling of the infarcted heart, cardiomyocytes, immune cells, fibroblasts, and vascular cells sense mechanical stress. Although mechanosensitive cascades play an important role in activation of inflammatory, migratory, proliferative, and matrix remodeling responses, dissection of the relative role of mechanical vs soluble signals is challenging. Prolonged mechanical unloading is impractical for most patients surviving a myocardial infarction. Identification of specific molecular cascades stimulated by mechanical stress in fibroblasts, vascular cells, and macrophages is of critical significance in order to develop therapeutic targets. Dissection of reparative and maladaptive signals triggered by mechanical stress may allow development of therapeutic strategies that selectively recapitulate the beneficial effects of ventricular unloading on reparative myocardial cells.

Sources of Funding: Dr. Frangogiannis' laboratory is supported by NIH grants R01 HL76246, R01 HL85440, and R01 HL149407 and by US Department of Defense grant PR181464. Dr. Humeres was supported by an American Heart Association post-doctoral award.

Disclosures

None.

Funding *This study was funded* by NIH grants R01 HL76246, R01 HL85440, and R01 HL149407 and by US Department of Defense grant PR181464 (to Dr. Frangogiannis) and by an American Heart Association post-doctoral award (to Dr. Humeres).

Conflict of Interest The authors declare that they have no conflicts of interest.

Ethical Approval This article does not contain any studies with human participants or animals performed by any of the authors.

References

1. Frangogiannis NG (2015) Pathophysiology of myocardial infarction. Compr Physiol 5:1841–1875
2. Prabhu SD, Frangogiannis NG (2016) The biological basis for cardiac repair after myocardial infarction: from inflammation to fibrosis. Circ Res 119:91–112
3. Gao XM, White DA, Dart AM, Du XJ (2012) Post-infarct cardiac rupture: recent insights on pathogenesis and therapeutic interventions. Pharmacol Ther 134:156–179
4. Frangogiannis NG (2014) The inflammatory response in myocardial injury, repair, and remodelling. Nat Rev Cardiol 11:255–265

5. Jennings RB (2013) Historical perspective on the pathology of myocardial ischemia/reperfusion injury. Circ Res 113:428–438
6. Sayen JJ, Sheldon WF, Peirce G, Kuo PT (1958) Polarographic oxygen, the epicardial electrocardiogram and muscle contraction in experimental acute regional ischemia of the left ventricle. Circ Res 6:779–798
7. Elliott AC, Smith GL, Eisner DA, Allen DG (1992) Metabolic changes during ischaemia and their role in contractile failure in isolated ferret hearts. J Physiol 454:467–490
8. Kentish JC (1986) The effects of inorganic phosphate and creatine phosphate on force production in skinned muscles from rat ventricle. J Physiol 370:585–604
9. Steenbergen C, Deleeuw G, Rich T, Williamson JR (1977) Effects of acidosis and ischemia on contractility and intracellular pH of rat heart. Circ Res 41:849–858
10. Solaro RJ, Lee JA, Kentish JC, Allen DG (1988) Effects of acidosis on ventricular muscle from adult and neonatal rats. Circ Res 63:779–787
11. Steenbergen C, Hill ML, Jennings RB (1985) Volume regulation and plasma membrane injury in aerobic, anaerobic, and ischemic myocardium in vitro. Effects of osmotic cell swelling on plasma membrane integrity. Circ Res 57:864–875
12. Pouleur H (1990) Diastolic dysfunction and myocardial energetics. Eur Heart J 11(Suppl C):30–34
13. Reimer KA, Lowe JE, Rasmussen MM, Jennings RB (1977) The wavefront phenomenon of ischemic cell death. 1. Myocardial infarct size vs duration of coronary occlusion in dogs. Circulation 56:786–794
14. Villarreal FJ, Lew WY, Waldman LK, Covell JW (1991) Transmural myocardial deformation in the ischemic canine left ventricle. Circ Res 68:368–381
15. Thune JJ, Solomon SD (2006) Left ventricular diastolic function following myocardial infarction. Curr Heart Fail Rep 3:170–174
16. Arslan F, Smeets MB, O'Neill LA, Keogh B, McGuirk P, Timmers L, Tersteeg C, Hoefer IE, Doevendans PA, Pasterkamp G et al (2010) Myocardial ischemia/reperfusion injury is mediated by leukocytic toll-like receptor-2 and reduced by systemic administration of a novel anti-toll-like receptor-2 antibody. Circulation 121:80–90
17. Saxena A, Chen W, Su Y, Rai V, Uche OU, Li N, Frangogiannis NG (2013) IL-1 Induces proinflammatory leukocyte infiltration and regulates fibroblast phenotype in the infarcted myocardium. J Immunol 191:4838–4848
18. de Castro Bras LE, Frangogiannis NG (2020) Extracellular matrix-derived peptides in tissue remodeling and fibrosis. Matrix Biol 91-92:176–187
19. Kotter S, Kazmierowska M, Andresen C, Bottermann K, Grandoch M, Gorressen S, Heinen A, Moll JM, Scheller J, Godecke A et al (2016) Titin-based cardiac myocyte stiffening contributes to early adaptive ventricular remodeling after myocardial infarction. Circ Res 119:1017–1029
20. Ali MA, Cho WJ, Hudson B, Kassiri Z, Granzier H, Schulz R (2010) Titin is a target of matrix metalloproteinase-2: implications in myocardial ischemia/reperfusion injury. Circulation 122: 2039–2047
21. Takahashi S, Barry AC, Factor SM (1990) Collagen degradation in ischaemic rat hearts. Biochem J 265:233–241
22. Nahrendorf M, Swirski FK, Aikawa E, Stangenberg L, Wurdinger T, Figueiredo JL, Libby P, Weissleder R, Pittet MJ (2007) The healing myocardium sequentially mobilizes two monocyte subsets with divergent and complementary functions. J Exp Med 204:3037–3047
23. Hofmann U, Frantz S (2015) Role of lymphocytes in myocardial injury, healing, and remodeling after myocardial infarction. Circ Res 116:354–367
24. Weirather J, Hofmann UD, Beyersdorf N, Ramos GC, Vogel B, Frey A, Ertl G, Kerkau T, Frantz S (2014) Foxp3+ CD4+ T cells improve healing after myocardial infarction by modulating monocyte/macrophage differentiation. Circ Res 115:55–67

25. Dobaczewski M, Xia Y, Bujak M, Gonzalez-Quesada C, Frangogiannis NG (2010) CCR5 signaling suppresses inflammation and reduces adverse remodeling of the infarcted heart, mediating recruitment of regulatory T cells. Am J Pathol 176:2177–2187
26. Kempf T, Zarbock A, Widera C, Butz S, Stadtmann A, Rossaint J, Bolomini-Vittori M, Korf-Klingebiel M, Napp LC, Hansen B et al (2011) GDF-15 is an inhibitor of leukocyte integrin activation required for survival after myocardial infarction in mice. Nat Med 17:581–588
27. Chen B, Huang S, Su Y, Wu YJ, Hanna A, Brickshawana A, Graff J, Frangogiannis NG (2019) Macrophage Smad3 protects the infarcted heart, stimulating phagocytosis and regulating inflammation. Circ Res 125:55–70
28. Frangogiannis NG, Mendoza LH, Lindsey ML, Ballantyne CM, Michael LH, Smith CW, Entman ML (2000) IL-10 is induced in the reperfused myocardium and may modulate the reaction to injury. J Immunol 165:2798–2808
29. Desmouliere A, Geinoz A, Gabbiani F, Gabbiani G (1993) Transforming growth factor-beta 1 induces alpha-smooth muscle actin expression in granulation tissue myofibroblasts and in quiescent and growing cultured fibroblasts. J Cell Biol 122:103–111
30. Pakshir P, Noskovicova N, Lodyga M, Son DO, Schuster R, Goodwin A, Karvonen H, Hinz B (2020) The myofibroblast at a glance. J Cell Sci 133:jcs227900
31. Dobaczewski M, Bujak M, Li N, Gonzalez-Quesada C, Mendoza LH, Wang XF, Frangogiannis NG (2010) Smad3 signaling critically regulates fibroblast phenotype and function in healing myocardial infarction. Circ Res 107:418–428
32. Hanna A, Shinde AV, Frangogiannis NG (2020) Validation of diagnostic criteria and histopathological characterization of cardiac rupture in the mouse model of nonreperfused myocardial infarction. Am J Physiol Heart Circ Physiol 319:H948–H964
33. Tsujita K, Kaikita K, Hayasaki T, Honda T, Kobayashi H, Sakashita N, Suzuki H, Kodama T, Ogawa H, Takeya M (2007) Targeted deletion of class A macrophage scavenger receptor increases the risk of cardiac rupture after experimental myocardial infarction. Circulation 115: 1904–1911
34. Maruyama S, Nakamura K, Papanicolaou KN, Sano S, Shimizu I, Asaumi Y, van den Hoff MJ, Ouchi N, Recchia FA, Walsh K (2016) Follistatin-like 1 promotes cardiac fibroblast activation and protects the heart from rupture. EMBO Mol Med 8:949–966
35. Kong P, Shinde AV, Su Y, Russo I, Chen B, Saxena A, Conway SJ, Graff JM, Frangogiannis NG (2018) Opposing actions of fibroblast and cardiomyocyte Smad3 signaling in the infarcted myocardium. Circulation 137:707–724
36. Willems IE, Havenith MG, De Mey JG, Daemen MJ (1994) The alpha-smooth muscle actin-positive cells in healing human myocardial scars. Am J Pathol 145:868–875
37. Frangogiannis NG, Michael LH, Entman ML (2000) Myofibroblasts in reperfused myocardial infarcts express the embryonic form of smooth muscle myosin heavy chain (SMemb). Cardiovasc Res 48:89–100
38. Cleutjens JP, Verluyten MJ, Smiths JF, Daemen MJ (1995) Collagen remodeling after myocardial infarction in the rat heart. Am J Pathol 147:325–338
39. Aisagbonhi O, Rai M, Ryzhov S, Atria N, Feoktistov I, Hatzopoulos AK (2011) Experimental myocardial infarction triggers canonical Wnt signaling and endothelial-to-mesenchymal transition. Dis Model Mech 4:469–483
40. Mollmann H, Nef HM, Kostin S, von Kalle C, Pilz I, Weber M, Schaper J, Hamm CW, Elsasser A (2006) Bone marrow-derived cells contribute to infarct remodelling. Cardiovasc Res 71:661–671
41. Haider N, Bosca L, Zandbergen HR, Kovacic JC, Narula N, Gonzalez-Ramos S, Fernandez-Velasco M, Agrawal S, Paz-Garcia M, Gupta S et al (2019) Transition of macrophages to fibroblast-like cells in healing myocardial infarction. J Am Coll Cardiol 74:3124–3135
42. Kanisicak O, Khalil H, Ivey MJ, Karch J, Maliken BD, Correll RN, Brody MJ, Suh-Chin JL, Aronow BJ, Tallquist MD et al (2016) Genetic lineage tracing defines myofibroblast origin and function in the injured heart. Nat Commun 7:12260

43. Moore-Morris T, Cattaneo P, Guimaraes-Camboa N, Bogomolovas J, Cedenilla M, Banerjee I, Ricote M, Kisseleva T, Zhang L, Gu Y et al (2018) Infarct fibroblasts do not derive from bone marrow lineages. Circ Res 122:583–590
44. van Putten S, Shafieyan Y, Hinz B (2016) Mechanical control of cardiac myofibroblasts. J Mol Cell Cardiol 93:133–142
45. Shinde AV, Humeres C, Frangogiannis NG (2017) The role of alpha-smooth muscle actin in fibroblast-mediated matrix contraction and remodeling. Biochim Biophys Acta 1863:298–309
46. Ronnov-Jessen L, Petersen OW (1996) A function for filamentous alpha-smooth muscle actin: retardation of motility in fibroblasts. J Cell Biol 134:67–80
47. Weber KT, Sun Y (2000) Recruitable ACE and tissue repair in the infarcted heart. J Renin-Angiotensin-Aldosterone Syst 1:295–303
48. Sadoshima J, Xu Y, Slayter HS, Izumo S (1993) Autocrine release of angiotensin II mediates stretch-induced hypertrophy of cardiac myocytes in vitro. Cell 75:977–984
49. Sadoshima J, Izumo S (1993) Molecular characterization of angiotensin II--induced hypertrophy of cardiac myocytes and hyperplasia of cardiac fibroblasts. Critical role of the AT1 receptor subtype. Circ Res 73:413–423
50. Schorb W, Booz GW, Dostal DE, Conrad KM, Chang KC, Baker KM (1993) Angiotensin II is mitogenic in neonatal rat cardiac fibroblasts. Circ Res 72:1245–1254
51. Ehanire T, Ren L, Bond J, Medina M, Li G, Bashirov L, Chen L, Kokosis G, Ibrahim M, Selim A et al (2015) Angiotensin II stimulates canonical TGF-beta signaling pathway through angiotensin type 1 receptor to induce granulation tissue contraction. J Mol Med (Berl) 93: 289–302
52. Siddesha JM, Valente AJ, Sakamuri SS, Yoshida T, Gardner JD, Somanna N, Takahashi C, Noda M, Chandrasekar B (2013) Angiotensin II stimulates cardiac fibroblast migration via the differential regulation of matrixins and RECK. J Mol Cell Cardiol 65:9–18
53. Stawowy P, Margeta C, Blaschke F, Lindschau C, Spencer-Hansch C, Leitges M, Biagini G, Fleck E, Graf K (2005) Protein kinase C epsilon mediates angiotensin II-induced activation of beta1-integrins in cardiac fibroblasts. Cardiovasc Res 67:50–59
54. Swaney JS, Roth DM, Olson ER, Naugle JE, Meszaros JG, Insel PA (2005) Inhibition of cardiac myofibroblast formation and collagen synthesis by activation and overexpression of adenylyl cyclase. Proc Natl Acad Sci U S A 102:437–442
55. Crabos M, Roth M, Hahn AW, Erne P (1994) Characterization of angiotensin II receptors in cultured adult rat cardiac fibroblasts. Coupling to signaling systems and gene expression. J Clin Invest 93:2372–2378
56. Ohkubo N, Matsubara H, Nozawa Y, Mori Y, Murasawa S, Kijima K, Maruyama K, Masaki H, Tsutumi Y, Shibazaki Y et al (1997) Angiotensin type 2 receptors are reexpressed by cardiac fibroblasts from failing myopathic hamster hearts and inhibit cell growth and fibrillar collagen metabolism. Circulation 96:3954–3962
57. Kurisu S, Ozono R, Oshima T, Kambe M, Ishida T, Sugino H, Matsuura H, Chayama K, Teranishi Y, Iba O et al (2003) Cardiac angiotensin II type 2 receptor activates the kinin/NO system and inhibits fibrosis. Hypertension 41:99–107
58. Lijnen P, Petrov V (2000) Induction of cardiac fibrosis by aldosterone. J Mol Cell Cardiol 32: 865–879
59. Campbell SE, Janicki JS, Weber KT (1995) Temporal differences in fibroblast proliferation and phenotype expression in response to chronic administration of angiotensin II or aldosterone. J Mol Cell Cardiol 27:1545–1560
60. Zhou Y, Poczatek MH, Berecek KH, Murphy-Ullrich JE (2006) Thrombospondin 1 mediates angiotensin II induction of TGF-beta activation by cardiac and renal cells under both high and low glucose conditions. Biochem Biophys Res Commun 339:633–641
61. Kagami S, Border WA, Miller DE, Noble NA (1994) Angiotensin II stimulates extracellular matrix protein synthesis through induction of transforming growth factor-beta expression in rat glomerular mesangial cells. J Clin Invest 93:2431–2437

62. Campbell SE, Katwa LC (1997) Angiotensin II stimulated expression of transforming growth factor-beta1 in cardiac fibroblasts and myofibroblasts. J Mol Cell Cardiol 29:1947–1958
63. Zou Y, Akazawa H, Qin Y, Sano M, Takano H, Minamino T, Makita N, Iwanaga K, Zhu W, Kudoh S et al (2004) Mechanical stress activates angiotensin II type 1 receptor without the involvement of angiotensin II. Nat Cell Biol 6:499–506
64. Frangogiannis NG, Dewald O, Xia Y, Ren G, Haudek S, Leucker T, Kraemer D, Taffet G, Rollins BJ, Entman ML (2007) Critical role of monocyte chemoattractant protein-1/CC chemokine ligand 2 in the pathogenesis of ischemic cardiomyopathy. Circulation 115:584–592
65. Dewald O, Zymek P, Winkelmann K, Koerting A, Ren G, Abou-Khamis T, Michael LH, Rollins BJ, Entman ML, Frangogiannis NG (2005) CCL2/Monocyte Chemoattractant Protein-1 regulates inflammatory responses critical to healing myocardial infarcts. Circ Res 96:881–889
66. Sakai N, Wada T, Furuichi K, Shimizu K, Kokubo S, Hara A, Yamahana J, Okumura T, Matsushima K, Yokoyama H et al (2006) MCP-1/CCR2-dependent loop for fibrogenesis in human peripheral CD14-positive monocytes. J Leukoc Biol 79:555–563
67. Kruglov EA, Nathanson RA, Nguyen T, Dranoff JA (2006) Secretion of MCP-1/CCL2 by bile duct epithelia induces myofibroblastic transdifferentiation of portal fibroblasts. Am J Physiol Gastrointest Liver Physiol 290:G765–G771
68. Yamamoto T, Eckes B, Mauch C, Hartmann K, Krieg T (2000) Monocyte chemoattractant protein-1 enhances gene expression and synthesis of matrix metalloproteinase-1 in human fibroblasts by an autocrine IL-1 alpha loop. J Immunol 164:6174–6179
69. Gharaee-Kermani M, Denholm EM, Phan SH (1996) Costimulation of fibroblast collagen and transforming growth factor beta1 gene expression by monocyte chemoattractant protein-1 via specific receptors. J Biol Chem 271:17779–17784
70. Siwik DA, Chang DL, Colucci WS (2000) Interleukin-1beta and tumor necrosis factor-alpha decrease collagen synthesis and increase matrix metalloproteinase activity in cardiac fibroblasts in vitro. Circ Res 86:1259–1265
71. Bronnum H, Eskildsen T, Andersen DC, Schneider M, Sheikh SP (2013) IL-1beta suppresses TGF-beta-mediated myofibroblast differentiation in cardiac fibroblasts. Growth Factors 31: 81–89
72. van Nieuwenhoven FA, Hemmings KE, Porter KE, Turner NA (2013) Combined effects of interleukin-1alpha and transforming growth factor-beta1 on modulation of human cardiac fibroblast function. Matrix Biol 32:399–406
73. Voloshenyuk TG, Hart AD, Khoutorova E, Gardner JD (2011) TNF-alpha increases cardiac fibroblast lysyl oxidase expression through TGF-beta and PI3Kinase signaling pathways. Biochem Biophys Res Commun 413:370–375
74. Gurantz D, Cowling RT, Varki N, Frikovsky E, Moore CD, Greenberg BH (2005) IL-1beta and TNF-alpha upregulate angiotensin II type 1 (AT1) receptors on cardiac fibroblasts and are associated with increased AT1 density in the post-MI heart. J Mol Cell Cardiol 38:505–515
75. Bujak M, Dobaczewski M, Chatila K, Mendoza LH, Li N, Reddy A, Frangogiannis NG (2008) Interleukin-1 receptor type I signaling critically regulates infarct healing and cardiac remodeling. Am J Pathol 173:57–67
76. Li J, Schwimmbeck PL, Tschope C, Leschka S, Husmann L, Rutschow S, Reichenbach F, Noutsias M, Kobalz U, Poller W et al (2002) Collagen degradation in a murine myocarditis model: relevance of matrix metalloproteinase in association with inflammatory induction. Cardiovasc Res 56:235–247
77. Jing R, Long TY, Pan W, Li F, Xie QY (2019) IL-6 knockout ameliorates myocardial remodeling after myocardial infarction by regulating activation of M2 macrophages and fibroblast cells. Eur Rev Med Pharmacol Sci 23:6283–6291
78. Datta R, Bansal T, Rana S, Datta K, Datta Chaudhuri R, Chawla-Sarkar M, Sarkar S (2017) Myocyte-derived Hsp90 modulates collagen upregulation via biphasic activation of STAT-3 in fibroblasts during cardiac hypertrophy. Mol Cell Biol 37:e00611–e00616

79. Zhang Y, Wang JH, Zhang YY, Wang YZ, Wang J, Zhao Y, Jin XX, Xue GL, Li PH, Sun YL et al (2016) Deletion of interleukin-6 alleviated interstitial fibrosis in streptozotocin-induced diabetic cardiomyopathy of mice through affecting TGFbeta1 and miR-29 pathways. Sci Rep 6:23010
80. Schafer S, Viswanathan S, Widjaja AA, Lim WW, Moreno-Moral A, DeLaughter DM, Ng B, Patone G, Chow K, Khin E et al (2017) IL-11 is a crucial determinant of cardiovascular fibrosis. Nature 552:110–115
81. Wang DL, Wung BS, Shyy YJ, Lin CF, Chao YJ, Usami S, Chien S (1995) Mechanical strain induces monocyte chemotactic protein-1 gene expression in endothelial cells. Effects of mechanical strain on monocyte adhesion to endothelial cells. Circ Res 77:294–302
82. Honsho S, Nishikawa S, Amano K, Zen K, Adachi Y, Kishita E, Matsui A, Katsume A, Yamaguchi S, Nishikawa K et al (2009) Pressure-mediated hypertrophy and mechanical stretch induces IL-1 release and subsequent IGF-1 generation to maintain compensative hypertrophy by affecting Akt and JNK pathways. Circ Res 105:1149–1158
83. Loperena R, Van Beusecum JP, Itani HA, Engel N, Laroumanie F, Xiao L, Elijovich F, Laffer CL, Gnecco JS, Noonan J et al (2018) Hypertension and increased endothelial mechanical stretch promote monocyte differentiation and activation: roles of STAT3, interleukin 6 and hydrogen peroxide. Cardiovasc Res 114:1547–1563
84. Gan W, Li T, Ren J, Li C, Liu Z, Yang M (2018) Serum-glucocorticoid-regulated kinase 1 contributes to mechanical stretch-induced inflammatory responses in cardiac fibroblasts. Mol Cell Biochem 445:67–78
85. Lindner D, Zietsch C, Tank J, Sossalla S, Fluschnik N, Hinrichs S, Maier L, Poller W, Blankenberg S, Schultheiss HP et al (2014) Cardiac fibroblasts support cardiac inflammation in heart failure. Basic Res Cardiol 109:428
86. Frangogiannis NG (2020) Transforming growth factor (TGF)-beta in tissue fibrosis. J Exp Med 217:e20190103. https://doi.org/20190110.20191084/jem.20190103
87. Dewald O, Ren G, Duerr GD, Zoerlein M, Klemm C, Gersch C, Tincey S, Michael LH, Entman ML, Frangogiannis NG (2004) Of mice and dogs: species-specific differences in the inflammatory response following myocardial infarction. Am J Pathol 164:665–677
88. Robertson IB, Rifkin DB (2016) Regulation of the bioavailability of TGF-beta and TGF-beta-related proteins, vol 8. Cold Spring Harb Perspect Biol, p a021907
89. Annes JP, Chen Y, Munger JS, Rifkin DB (2004) Integrin alphaVbeta6-mediated activation of latent TGF-beta requires the latent TGF-beta binding protein-1. J Cell Biol 165:723–734
90. Shea BS, Probst CK, Brazee PL, Rotile NJ, Blasi F, Weinreb PH, Black KE, Sosnovik DE, Van Cott EM, Violette SM et al (2017) Uncoupling of the profibrotic and hemostatic effects of thrombin in lung fibrosis. JCI Insight 2:e86608
91. Briassouli P, Rifkin D, Clancy RM, Buyon JP (2011) Binding of anti-SSA antibodies to apoptotic fetal cardiocytes stimulates urokinase plasminogen activator (uPA)/uPA receptor-dependent activation of TGF-beta and potentiates fibrosis. J Immunol 187:5392–5401
92. Bourd-Boittin K, Bonnier D, Leyme A, Mari B, Tuffery P, Samson M, Ezan F, Baffet G, Theret N (2011) Protease profiling of liver fibrosis reveals the ADAM metallopeptidase with thrombospondin type 1 motif, 1 as a central activator of transforming growth factor beta. Hepatology 54:2173–2184
93. Edgtton KL, Gow RM, Kelly DJ, Carmeliet P, Kitching AR (2004) Plasmin is not protective in experimental renal interstitial fibrosis. Kidney Int 66:68–76
94. Okuno M, Akita K, Moriwaki H, Kawada N, Ikeda K, Kaneda K, Suzuki Y, Kojima S (2001) Prevention of rat hepatic fibrosis by the protease inhibitor, camostat mesilate, via reduced generation of active TGF-beta. Gastroenterology 120:1784–1800
95. Khalil N, Corne S, Whitman C, Yacyshyn H (1996) Plasmin regulates the activation of cell-associated latent TGF-beta 1 secreted by rat alveolar macrophages after in vivo bleomycin injury. Am J Respir Cell Mol Biol 15:252–259
96. Arslan F, Smeets MB, Riem Vis PW, Karper JC, Quax PH, Bongartz LG, Peters JH, Hoefer IE, Doevendans PA, Pasterkamp G et al (2011) Lack of fibronectin-EDA promotes survival

and prevents adverse remodeling and heart function deterioration after myocardial infarction. Circ Res 108:582–592
97. Frangogiannis NG, Ren G, Dewald O, Zymek P, Haudek S, Koerting A, Winkelmann K, Michael LH, Lawler J, Entman ML (2005) The critical role of endogenous Thrombospondin (TSP)-1 in preventing expansion of healing myocardial infarcts. Circulation 111:2935–2942
98. Klingberg F, Chau G, Walraven M, Boo S, Koehler A, Chow ML, Olsen AL, Im M, Lodyga M, Wells RG et al (2018) The fibronectin ED-A domain enhances recruitment of latent TGF-beta-binding protein-1 to the fibroblast matrix. J Cell Sci 131:jcs201293
99. Serini G, Bochaton-Piallat ML, Ropraz P, Geinoz A, Borsi L, Zardi L, Gabbiani G (1998) The fibronectin domain ED-A is crucial for myofibroblastic phenotype induction by transforming growth factor-beta1. J Cell Biol 142:873–881
100. Ribeiro SM, Poczatek M, Schultz-Cherry S, Villain M, Murphy-Ullrich JE (1999) The activation sequence of thrombospondin-1 interacts with the latency-associated peptide to regulate activation of latent transforming growth factor-beta. J Biol Chem 274:13586–13593
101. Walker M, Godin M, Pelling AE (2020) Mechanical stretch sustains myofibroblast phenotype and function in microtissues through latent TGF-beta1 activation. Integr Biol (Camb) 12:199–210
102. Lee AA, Delhaas T, McCulloch AD, Villarreal FJ (1999) Differential responses of adult cardiac fibroblasts to in vitro biaxial strain patterns. J Mol Cell Cardiol 31:1833–1843
103. Munger JS, Huang X, Kawakatsu H, Griffiths MJ, Dalton SL, Wu J, Pittet JF, Kaminski N, Garat C, Matthay MA et al (1999) The integrin alpha v beta 6 binds and activates latent TGF beta 1: a mechanism for regulating pulmonary inflammation and fibrosis. Cell 96:319–328
104. Hakkinen L, Koivisto L, Gardner H, Saarialho-Kere U, Carroll JM, Lakso M, Rauvala H, Laato M, Heino J, Larjava H (2004) Increased expression of beta6-integrin in skin leads to spontaneous development of chronic wounds. Am J Pathol 164:229–242
105. Mu D, Cambier S, Fjellbirkeland L, Baron JL, Munger JS, Kawakatsu H, Sheppard D, Broaddus VC, Nishimura SL (2002) The integrin alpha(v)beta8 mediates epithelial homeostasis through MT1-MMP-dependent activation of TGF-beta1. J Cell Biol 157:493–507
106. Wipff PJ, Hinz B (2008) Integrins and the activation of latent transforming growth factor beta1 - an intimate relationship. Eur J Cell Biol 87:601–615
107. Margadant C, Sonnenberg A (2010) Integrin-TGF-beta crosstalk in fibrosis, cancer and wound healing. EMBO Rep 11:97–105
108. Wipff PJ, Rifkin DB, Meister JJ, Hinz B (2007) Myofibroblast contraction activates latent TGF-beta1 from the extracellular matrix. J Cell Biol 179:1311–1323
109. Froese AR, Shimbori C, Bellaye PS, Inman M, Obex S, Fatima S, Jenkins G, Gauldie J, Ask K, Kolb M (2016) Stretch-induced activation of transforming growth factor-beta1 in pulmonary fibrosis. Am J Respir Crit Care Med 194:84–96
110. Noskovicova N, Schuster R, van Putten S, Ezzo M, Koehler A, Boo S, Coelho NM, Griggs D, Ruminski P, McCulloch CA et al (2021) Suppression of the fibrotic encapsulation of silicone implants by inhibiting the mechanical activation of pro-fibrotic TGF-beta. Nat Biomed Eng 5(12):1437–1456
111. Sarrazy V, Koehler A, Chow ML, Zimina E, Li CX, Kato H, Caldarone CA, Hinz B (2014) Integrins alphavbeta5 and alphavbeta3 promote latent TGF-beta1 activation by human cardiac fibroblast contraction. Cardiovasc Res 102:407–417
112. van Wamel AJ, Ruwhof C, van der Valk-Kokshoorn LJ, Schrier PI, van der Laarse A (2002) Stretch-induced paracrine hypertrophic stimuli increase TGF-beta1 expression in cardiomyocytes. Mol Cell Biochem 236:147–153
113. Zheng W, Seftor EA, Meininger CJ, Hendrix MJ, Tomanek RJ (2001) Mechanisms of coronary angiogenesis in response to stretch: role of VEGF and TGF-beta. Am J Physiol Heart Circ Physiol 280:H909–H917
114. Arora PD, Narani N, McCulloch CA (1999) The compliance of collagen gels regulates transforming growth factor-beta induction of alpha-smooth muscle actin in fibroblasts. Am J Pathol 154:871–882

115. Hanna A, Humeres C, Frangogiannis NG (2021) The role of Smad signaling cascades in cardiac fibrosis. Cell Signal 77:109826
116. Frangogiannis NG (2022) Transforming growth factor-beta in myocardial disease. Nat Rev Cardiol 19(7):435–455. https://doi.org/10.1038/s41569-41021-00646-w
117. Bujak M, Ren G, Kweon HJ, Dobaczewski M, Reddy A, Taffet G, Wang XF, Frangogiannis NG (2007) Essential role of Smad3 in infarct healing and in the pathogenesis of cardiac remodeling. Circulation 116:2127–2138
118. Huang S, Chen B, Su Y, Alex L, Humeres C, Shinde AV, Conway SJ, Frangogiannis NG (2019) Distinct roles of myofibroblast-specific Smad2 and Smad3 signaling in repair and remodeling of the infarcted heart. J Mol Cell Cardiol 132:84–97
119. Molkentin JD, Bugg D, Ghearing N, Dorn LE, Kim P, Sargent MA, Gunaje J, Otsu K, Davis J (2017) Fibroblast-specific genetic manipulation of p38 mitogen-activated protein kinase in vivo reveals its central regulatory role in fibrosis. Circulation 136:549–561
120. Bageghni SA, Hemmings KE, Zava N, Denton CP, Porter KE, Ainscough JFX, Drinkhill MJ, Turner NA (2018) Cardiac fibroblast-specific p38alpha MAP kinase promotes cardiac hypertrophy via a putative paracrine interleukin-6 signaling mechanism. FASEB J 32(9):4941–4954
121. Wang H, Leinwand LA, Anseth KS (2014) Roles of transforming growth factor-beta1 and OB-cadherin in porcine cardiac valve myofibroblast differentiation. FASEB J 28:4551–4562
122. Schiller M, Javelaud D, Mauviel A (2004) TGF-beta-induced SMAD signaling and gene regulation: consequences for extracellular matrix remodeling and wound healing. J Dermatol Sci 35:83–92
123. Russo I, Cavalera M, Huang S, Su Y, Hanna A, Chen B, Shinde AV, Conway SJ, Graff J, Frangogiannis NG (2019) Protective effects of activated myofibroblasts in the pressure-overloaded myocardium are mediated through Smad-dependent activation of a matrix-preserving program. Circ Res 124:1214–1227
124. Kapoun AM, Liang F, O'Young G, Damm DL, Quon D, White RT, Munson K, Lam A, Schreiner GF, Protter AA (2004) B-type natriuretic peptide exerts broad functional opposition to transforming growth factor-beta in primary human cardiac fibroblasts: fibrosis, myofibroblast conversion, proliferation, and inflammation. Circ Res 94:453–461
125. Petrov VV, Fagard RH, Lijnen PJ (2000) Transforming growth factor-beta(1) induces angiotensin-converting enzyme synthesis in rat cardiac fibroblasts during their differentiation to myofibroblasts. J Renin-Angiotensin-Aldosterone Syst 1:342–352
126. Mori T, Kawara S, Shinozaki M, Hayashi N, Kakinuma T, Igarashi A, Takigawa M, Nakanishi T, Takehara K (1999) Role and interaction of connective tissue growth factor with transforming growth factor-beta in persistent fibrosis: a mouse fibrosis model. J Cell Physiol 181:153–159
127. Ohnishi H, Oka T, Kusachi S, Nakanishi T, Takeda K, Nakahama M, Doi M, Murakami T, Ninomiya Y, Takigawa M et al (1998) Increased expression of connective tissue growth factor in the infarct zone of experimentally induced myocardial infarction in rats. J Mol Cell Cardiol 30:2411–2422
128. Chen MM, Lam A, Abraham JA, Schreiner GF, Joly AH (2000) CTGF expression is induced by TGF- beta in cardiac fibroblasts and cardiac myocytes: a potential role in heart fibrosis. J Mol Cell Cardiol 32:1805–1819
129. Accornero F, van Berlo JH, Correll RN, Elrod JW, Sargent MA, York A, Rabinowitz JE, Leask A, Molkentin JD (2015) Genetic analysis of connective tissue growth factor as an effector of transforming growth factor beta signaling and cardiac remodeling. Mol Cell Biol 35:2154–2164
130. Zhang Y, Yan H, Guang GC, Deng ZR (2017) Overexpressed connective tissue growth factor in cardiomyocytes attenuates left ventricular remodeling induced by angiotensin II perfusion. Clin Exp Hypertens 39:168–174
131. Ng B, Dong J, D'Agostino G, Viswanathan S, Widjaja AA, Lim WW, Ko NSJ, Tan J, Chothani SP, Huang B et al (2019) Interleukin-11 is a therapeutic target in idiopathic pulmonary fibrosis. Sci Transl Med 11:eaaw1237

132. Obana M, Maeda M, Takeda K, Hayama A, Mohri T, Yamashita T, Nakaoka Y, Komuro I, Takeda K, Matsumiya G et al (2010) Therapeutic activation of signal transducer and activator of transcription 3 by interleukin-11 ameliorates cardiac fibrosis after myocardial infarction. Circulation 121:684–691
133. Humeres C, Shinde AV, Hanna A, Alex L, Hernandez SC, Li R, Chen B, Conway SJ, Frangogiannis NG (2022) Smad7 effects on TGF-beta and Erbb2 restrain myofibroblast activation, and protect from post-infarction heart failure. J Clin Invest 132:e146926. https://doi.org/10.1172/JCI146926
134. Hayashi Y, Katoh T, Asano K, Onozaki A, Sakurai K, Asahi K, Nakayama M, Watanabe T (2012) Mechanical stretch down-regulates expression of the Smad6 gene in cultured rat mesangial cells. Clin Exp Nephrol 16:690–696
135. Frangogiannis NG, Kovacic JC (2020) Extracellular matrix in ischemic heart disease, part 4/4: JACC focus seminar. J Am Coll Cardiol 75:2219–2235
136. Casscells W, Kimura H, Sanchez JA, Yu ZX, Ferrans VJ (1990) Immunohistochemical study of fibronectin in experimental myocardial infarction. Am J Pathol 137:801–810
137. Dobaczewski M, Bujak M, Zymek P, Ren G, Entman ML, Frangogiannis NG (2006) Extracellular matrix remodeling in canine and mouse myocardial infarcts. Cell Tissue Res 324:475–488
138. Zhang Y, Lin Z, Foolen J, Schoen I, Santoro A, Zenobi-Wong M, Vogel V (2014) Disentangling the multifactorial contributions of fibronectin, collagen and cyclic strain on MMP expression and extracellular matrix remodeling by fibroblasts. Matrix Biol 40:62–72
139. Webber J, Meran S, Steadman R, Phillips A (2009) Hyaluronan orchestrates transforming growth factor-beta1-dependent maintenance of myofibroblast phenotype. J Biol Chem 284:9083–9092
140. Huebener P, Abou-Khamis T, Zymek P, Bujak M, Ying X, Chatila K, Haudek S, Thakker G, Frangogiannis NG (2008) CD44 is critically involved in infarct healing by regulating the inflammatory and fibrotic response. J Immunol 180:2625–2633
141. Hattori N, Carrino DA, Lauer ME, Vasanji A, Wylie JD, Nelson CM, Apte SS (2011) Pericellular versican regulates the fibroblast-myofibroblast transition: a role for ADAMTS5 protease-mediated proteolysis. J Biol Chem 286:34298–34310
142. Toeda K, Nakamura K, Hirohata S, Hatipoglu OF, Demircan K, Yamawaki H, Ogawa H, Kusachi S, Shiratori Y, Ninomiya Y (2005) Versican is induced in infiltrating monocytes in myocardial infarction. Mol Cell Biochem 280:47–56
143. Frangogiannis NG (2019) The extracellular matrix in ischemic and nonischemic heart failure. Circ Res 125:117–146
144. Bornstein P (2009) Matricellular proteins: an overview. J Cell Commun Signal 3:163–165
145. Murphy-Ullrich JE (2001) The de-adhesive activity of matricellular proteins: is intermediate cell adhesion an adaptive state? J Clin Invest 107:785–790
146. Chiquet M, Matthisson M, Koch M, Tannheimer M, Chiquet-Ehrismann R (1996) Regulation of extracellular matrix synthesis by mechanical stress. Biochem Cell Biol 74:737–744
147. Imanaka-Yoshida K, Aoki H (2014) Tenascin-C and mechanotransduction in the development and diseases of cardiovascular system. Front Physiol 5:283
148. Yamamoto K, Dang QN, Kennedy SP, Osathanondh R, Kelly RA, Lee RT (1999) Induction of tenascin-C in cardiac myocytes by mechanical deformation. Role of reactive oxygen species. J Biol Chem 274:21840–21846
149. Xia Y, Dobaczewski M, Gonzalez-Quesada C, Chen W, Biernacka A, Li N, Lee DW, Frangogiannis NG (2011) Endogenous thrombospondin 1 protects the pressure-overloaded myocardium by modulating fibroblast phenotype and matrix metabolism. Hypertension 58:902–911
150. Gonzalez-Quesada C, Cavalera M, Biernacka A, Kong P, Lee DW, Saxena A, Frunza O, Dobaczewski M, Shinde A, Frangogiannis NG (2013) Thrombospondin-1 induction in the diabetic myocardium stabilizes the cardiac matrix in addition to promoting vascular rarefaction through angiopoietin-2 upregulation. Circ Res 113:1331–1344

151. Rosini S, Pugh N, Bonna AM, Hulmes DJS, Farndale RW, Adams JC (2018) Thrombospondin-1 promotes matrix homeostasis by interacting with collagen and lysyl oxidase precursors and collagen cross-linking sites. Sci Signal 11:eaar2566
152. Belmadani S, Bernal J, Wei CC, Pallero MA, Dell'italia L, Murphy-Ullrich JE, Berecek KH (2007) A thrombospondin-1 antagonist of transforming growth factor-beta activation blocks cardiomyopathy in rats with diabetes and elevated angiotensin II. Am J Pathol 171:777–789
153. Vanhoutte D, Schips TG, Vo A, Grimes KM, Baldwin TA, Brody MJ, Accornero F, Sargent MA, Molkentin JD (2021) Thbs1 induces lethal cardiac atrophy through PERK-ATF4 regulated autophagy. Nat Commun 12:3928
154. Yamashiro Y, Thang BQ, Shin SJ, Lino CA, Nakamura T, Kim J, Sugiyama K, Tokunaga C, Sakamoto H, Osaka M et al (2018) Role of thrombospondin-1 in mechanotransduction and development of thoracic aortic aneurysm in mouse and humans. Circ Res 123:660–672
155. Graf R, Freyberg M, Kaiser D, Friedl P (2002) Mechanosensitive induction of apoptosis in fibroblasts is regulated by thrombospondin-1 and integrin associated protein (CD47). Apoptosis 7:493–498
156. Tamaoki M, Imanaka-Yoshida K, Yokoyama K, Nishioka T, Inada H, Hiroe M, Sakakura T, Yoshida T (2005) Tenascin-C regulates recruitment of myofibroblasts during tissue repair after myocardial injury. Am J Pathol 167:71–80
157. Shimojo N, Hashizume R, Kanayama K, Hara M, Suzuki Y, Nishioka T, Hiroe M, Yoshida T, Imanaka-Yoshida K (2015) Tenascin-C may accelerate cardiac fibrosis by activating macrophages via the integrin alphaVbeta3/nuclear factor-kappaB/interleukin-6 Axis. Hypertension 66:757–766
158. Chablais F, Jazwinska A (2012) The regenerative capacity of the zebrafish heart is dependent on TGFbeta signaling. Development 139:1921–1930
159. Mercer SE, Odelberg SJ, Simon HG (2013) A dynamic spatiotemporal extracellular matrix facilitates epicardial-mediated vertebrate heart regeneration. Dev Biol 382:457–469
160. Kimura T, Tajiri K, Sato A, Sakai S, Wang Z, Yoshida T, Uede T, Hiroe M, Aonuma K, Ieda M et al (2019) Tenascin-C accelerates adverse ventricular remodelling after myocardial infarction by modulating macrophage polarization. Cardiovasc Res 115:614–624
161. Podesser BK, Kreibich M, Dzilic E, Santer D, Forster L, Trojanek S, Abraham D, Krssak M, Klein KU, Tretter EV et al (2018) Tenascin-C promotes chronic pressure overload-induced cardiac dysfunction, hypertrophy and myocardial fibrosis. J Hypertens 36:847–856
162. Abbadi D, Laroumanie F, Bizou M, Pozzo J, Daviaud D, Delage C, Calise D, Gaits-Iacovoni F, Dutaur M, Tortosa F et al (2018) Local production of tenascin-C acts as a trigger for monocyte/macrophage recruitment that provokes cardiac dysfunction. Cardiovasc Res 114:123–137
163. Santer D, Nagel F, Goncalves IF, Kaun C, Wojta J, Fagyas M, Krssak M, Balogh A, Papp Z, Toth A et al (2020) Tenascin-C aggravates ventricular dilatation and angiotensin-converting enzyme activity after myocardial infarction in mice. ESC Heart Fail 7:2113–2122
164. Asparuhova MB, Ferralli J, Chiquet M, Chiquet-Ehrismann R (2011) The transcriptional regulator megakaryoblastic leukemia-1 mediates serum response factor-independent activation of tenascin-C transcription by mechanical stress. FASEB J 25:3477–3488
165. Schellings MW, Vanhoutte D, Swinnen M, Cleutjens JP, Debets J, van Leeuwen RE, d'Hooge J, Van de Werf F, Carmeliet P, Pinto YM et al (2009) Absence of SPARC results in increased cardiac rupture and dysfunction after acute myocardial infarction. J Exp Med 206: 113–123
166. McCurdy SM, Dai Q, Zhang J, Zamilpa R, Ramirez TA, Dayah T, Nguyen N, Jin YF, Bradshaw AD, Lindsey ML (2011) SPARC mediates early extracellular matrix remodeling following myocardial infarction. Am J Physiol Heart Circ Physiol 301:H497–H505
167. Toba H, de Castro Bras LE, Baicu CF, Zile MR, Lindsey ML, Bradshaw AD (2016) Increased ADAMTS1 mediates SPARC-dependent collagen deposition in the aging myocardium. Am J Physiol Endocrinol Metab 310:E1027–E1035
168. Bradshaw AD, Baicu CF, Rentz TJ, Van Laer AO, Boggs J, Lacy JM, Zile MR (2009) Pressure overload-induced alterations in fibrillar collagen content and myocardial diastolic

function: role of secreted protein acidic and rich in cysteine (SPARC) in post-synthetic procollagen processing. Circulation 119:269–280
169. Harris BS, Zhang Y, Card L, Rivera LB, Brekken RA, Bradshaw AD (2011) SPARC regulates collagen interaction with cardiac fibroblast cell surfaces. Am J Physiol Heart Circ Physiol 301: H841–H847
170. Durvasula RV, Shankland SJ (2005) Mechanical strain increases SPARC levels in podocytes: implications for glomerulosclerosis. Am J Physiol Renal Physiol 289:F577–F584
171. Trueblood NA, Xie Z, Communal C, Sam F, Ngoy S, Liaw L, Jenkins AW, Wang J, Sawyer DB, Bing OH et al (2001) Exaggerated left ventricular dilation and reduced collagen deposition after myocardial infarction in mice lacking osteopontin. Circ Res 88:1080–1087
172. Yousefi K, Irion CI, Takeuchi LM, Ding W, Lambert G, Eisenberg T, Sukkar S, Granzier HL, Methawasin M, Lee DI et al (2019) Osteopontin promotes left ventricular diastolic dysfunction through a mitochondrial pathway. J Am Coll Cardiol 73:2705–2718
173. Caballero EP, Santamaria MH, Corral RS (2018) Endogenous osteopontin induces myocardial CCL5 and MMP-2 activation that contributes to inflammation and cardiac remodeling in a mouse model of chronic Chagas heart disease. Biochim Biophys Acta Mol basis Dis 1864:11–23
174. Lopez B, Gonzalez A, Lindner D, Westermann D, Ravassa S, Beaumont J, Gallego I, Zudaire A, Brugnolaro C, Querejeta R et al (2013) Osteopontin-mediated myocardial fibrosis in heart failure: a role for lysyl oxidase? Cardiovasc Res 99:111–120
175. Dahiya S, Givvimani S, Bhatnagar S, Qipshidze N, Tyagi SC, Kumar A (2011) Osteopontin-stimulated expression of matrix metalloproteinase-9 causes cardiomyopathy in the mdx model of Duchenne muscular dystrophy. J Immunol 187:2723–2731
176. Psarras S, Mavroidis M, Sanoudou D, Davos CH, Xanthou G, Varela AE, Panoutsakopoulou V, Capetanaki Y (2012) Regulation of adverse remodelling by osteopontin in a genetic heart failure model. Eur Heart J 33:1954–1963
177. Sam F, Xie Z, Ooi H, Kerstetter DL, Colucci WS, Singh M, Singh K (2004) Mice lacking osteopontin exhibit increased left ventricular dilation and reduced fibrosis after aldosterone infusion. Am J Hypertens 17:188–193
178. Collins AR, Schnee J, Wang W, Kim S, Fishbein MC, Bruemmer D, Law RE, Nicholas S, Ross RS, Hsueh WA (2004) Osteopontin modulates angiotensin II-induced fibrosis in the intact murine heart. J Am Coll Cardiol 43:1698–1705
179. Matsui Y, Jia N, Okamoto H, Kon S, Onozuka H, Akino M, Liu L, Morimoto J, Rittling SR, Denhardt D et al (2004) Role of osteopontin in cardiac fibrosis and remodeling in angiotensin II-induced cardiac hypertrophy. Hypertension 43:1195–1201
180. Lenga Y, Koh A, Perera AS, McCulloch CA, Sodek J, Zohar R (2008) Osteopontin expression is required for myofibroblast differentiation. Circ Res 102:319–327
181. Zhao X, Johnson JN, Singh K, Singh M (2007) Impairment of myocardial angiogenic response in the absence of osteopontin. Microcirculation 14:233–240
182. Lorenzen JM, Schauerte C, Hubner A, Kolling M, Martino F, Scherf K, Batkai S, Zimmer K, Foinquinos A, Kaucsar T et al (2015) Osteopontin is indispensible for AP1-mediated angiotensin II-related miR-21 transcription during cardiac fibrosis. Eur Heart J 36:2184–2196
183. Xie Z, Singh M, Siwik DA, Joyner WL, Singh K (2003) Osteopontin inhibits interleukin-1beta-stimulated increases in matrix metalloproteinase activity in adult rat cardiac fibroblasts: role of protein kinase C-zeta. J Biol Chem 278:48546–48552
184. Shimazaki M, Nakamura K, Kii I, Kashima T, Amizuka N, Li M, Saito M, Fukuda K, Nishiyama T, Kitajima S et al (2008) Periostin is essential for cardiac healing after acute myocardial infarction. J Exp Med 205:295–303
185. Oka T, Xu J, Kaiser RA, Melendez J, Hambleton M, Sargent MA, Lorts A, Brunskill EW, Dorn GW 2nd, Conway SJ et al (2007) Genetic manipulation of periostin expression reveals a role in cardiac hypertrophy and ventricular remodeling. Circ Res 101:313–321
186. Chen Z, Xie J, Hao H, Lin H, Wang L, Zhang Y, Chen L, Cao S, Huang X, Liao W et al (2017) Ablation of periostin inhibits post-infarction myocardial regeneration in neonatal mice

mediated by the phosphatidylinositol 3 kinase/glycogen synthase kinase 3beta/cyclin D1 signalling pathway. Cardiovasc Res 113:620–632
187. Kumar P, Smith T, Raeman R, Chopyk DM, Brink H, Liu Y, Sulchek T, Anania FA (2018) Periostin promotes liver fibrogenesis by activating lysyl oxidase in hepatic stellate cells. J Biol Chem 293:12781–12792
188. Iekushi K, Taniyama Y, Azuma J, Katsuragi N, Dosaka N, Sanada F, Koibuchi N, Nagao K, Ogihara T, Morishita R (2007) Novel mechanisms of valsartan on the treatment of acute myocardial infarction through inhibition of the antiadhesion molecule periostin. Hypertension 49:1409–1414
189. Dawson DW, Pearce SF, Zhong R, Silverstein RL, Frazier WA, Bouck NP (1997) CD36 mediates the in vitro inhibitory effects of thrombospondin-1 on endothelial cells. J Cell Biol 138:707–717
190. Willems IE, Arends JW, Daemen MJ (1996) Tenascin and fibronectin expression in healing human myocardial scars. J Pathol 179:321–325
191. Murry CE, Giachelli CM, Schwartz SM, Vracko R (1994) Macrophages express osteopontin during repair of myocardial necrosis. Am J Pathol 145:1450–1462
192. Kechagia JZ, Ivaska J, Roca-Cusachs P (2019) Integrins as biomechanical sensors of the microenvironment. Nat Rev Mol Cell Biol 20:457–473
193. Takada Y, Ye X, Simon S (2007) The integrins. Genome Biol 8:215
194. Mitra SK, Schlaepfer DD (2006) Integrin-regulated FAK-Src signaling in normal and cancer cells. Curr Opin Cell Biol 18:516–523
195. Sun M, Opavsky MA, Stewart DJ, Rabinovitch M, Dawood F, Wen WH, Liu PP (2003) Temporal response and localization of integrins beta1 and beta3 in the heart after myocardial infarction: regulation by cytokines. Circulation 107:1046–1052
196. Dreyer WJ, Smith CW, Michael LH, Rossen RD, Hughes BJ, Entman ML, Anderson DC (1989) Canine neutrophil activation by cardiac lymph obtained during reperfusion of ischemic myocardium. Circ Res 65:1751–1762
197. Frangogiannis NG, Mendoza LH, Lewallen M, Michael LH, Smith CW, Entman ML (2001) Induction and suppression of interferon-inducible protein 10 in reperfused myocardial infarcts may regulate angiogenesis. FASEB J 15:1428–1430
198. Okada H, Lai NC, Kawaraguchi Y, Liao P, Copps J, Sugano Y, Okada-Maeda S, Banerjee I, Schilling JM, Gingras AR et al (2013) Integrins protect cardiomyocytes from ischemia/reperfusion injury. J Clin Invest 123:4294–4308
199. Saraswati S, Lietman CD, Li B, Mathew S, Zent R, Young PP (2020) Small proline-rich repeat 3 is a novel coordinator of PDGFRbeta and integrin beta1 crosstalk to augment proliferation and matrix synthesis by cardiac fibroblasts. FASEB J 34:7885–7904
200. Valiente-Alandi I, Potter SJ, Salvador AM, Schafer AE, Schips T, Carrillo-Salinas F, Gibson AM, Nieman ML, Perkins C, Sargent MA et al (2018) Inhibiting fibronectin attenuates fibrosis and improves cardiac function in a model of heart failure. Circulation 138(12):1236–1252
201. Takawale A, Zhang P, Patel VB, Wang X, Oudit G, Kassiri Z (2017) Tissue inhibitor of matrix metalloproteinase-1 promotes myocardial fibrosis by mediating CD63-integrin beta1 interaction. Hypertension 69:1092–1103
202. Shi M, Zhu J, Wang R, Chen X, Mi L, Walz T, Springer TA (2011) Latent TGF-beta structure and activation. Nature 474:343–349
203. Murray IR, Gonzalez ZN, Baily J, Dobie R, Wallace RJ, Mackinnon AC, Smith JR, Greenhalgh SN, Thompson AI, Conroy KP et al (2017) alphav integrins on mesenchymal cells regulate skeletal and cardiac muscle fibrosis. Nat Commun 8:1118
204. Stewart L, Turner NA (2021) Channelling the force to reprogram the matrix: mechanosensitive ion channels in cardiac fibroblasts. Cells 10:990
205. Nishida M, Onohara N, Sato Y, Suda R, Ogushi M, Tanabe S, Inoue R, Mori Y, Kurose H (2007) Galpha12/13-mediated up-regulation of TRPC6 negatively regulates endothelin-1-induced cardiac myofibroblast formation and collagen synthesis through nuclear factor of activated T cells activation. J Biol Chem 282:23117–23128

206. Davis J, Burr AR, Davis GF, Birnbaumer L, Molkentin JD (2012) A TRPC6-dependent pathway for myofibroblast transdifferentiation and wound healing in vivo. Dev Cell 23:705–715
207. Du J, Xie J, Zhang Z, Tsujikawa H, Fusco D, Silverman D, Liang B, Yue L (2010) TRPM7-mediated Ca2+ signals confer fibrogenesis in human atrial fibrillation. Circ Res 106:992–1003
208. Adapala RK, Thoppil RJ, Luther DJ, Paruchuri S, Meszaros JG, Chilian WM, Thodeti CK (2013) TRPV4 channels mediate cardiac fibroblast differentiation by integrating mechanical and soluble signals. J Mol Cell Cardiol 54:45–52
209. Adapala RK, Kanugula AK, Paruchuri S, Chilian WM, Thodeti CK (2020) TRPV4 deletion protects heart from myocardial infarction-induced adverse remodeling via modulation of cardiac fibroblast differentiation. Basic Res Cardiol 115:14
210. Li S, Li M, Yi X, Guo F, Zhou Y, Chen S, Wu X (2017) TRPM7 channels mediate the functional changes in cardiac fibroblasts induced by angiotensin II. Int J Mol Med 39:1291–1298
211. Guo JL, Yu Y, Jia YY, Ma YZ, Zhang BY, Liu PQ, Chen SR, Jiang JM (2014) Transient receptor potential melastatin 7 (TRPM7) contributes to H2O2-induced cardiac fibrosis via mediating Ca(2+) influx and extracellular signal-regulated kinase 1/2 (ERK1/2) activation in cardiac fibroblasts. J Pharmacol Sci 125:184–192
212. Rios FJ, Zou ZG, Harvey AP, Harvey KY, Nosalski R, Anyfanti P, Camargo LL, Lacchini S, Ryazanov AG, Ryazanova L et al (2020) Chanzyme TRPM7 protects against cardiovascular inflammation and fibrosis. Cardiovasc Res 116:721–735
213. Abraham DM, Lee TE, Watson LJ, Mao L, Chandok G, Wang HG, Frangakis S, Pitt GS, Shah SH, Wolf MJ et al (2018) The two-pore domain potassium channel TREK-1 mediates cardiac fibrosis and diastolic dysfunction. J Clin Invest 128:4843–4855
214. Benamer N, Vasquez C, Mahoney VM, Steinhardt MJ, Coetzee WA, Morley GE (2013) Fibroblast KATP currents modulate myocyte electrophysiology in infarcted hearts. Am J Physiol Heart Circ Physiol 304:H1231–H1239
215. Pertiwi KR, Hillman RM, Scott CA, Chilton EL (2019) Ischemia reperfusion injury produces, and ischemic preconditioning prevents, rat cardiac fibroblast differentiation: role of KATP channels. J Cardiovasc Dev Dis 6:22
216. Liou JY, Hong HJ, Sung LC, Chao HH, Chen PY, Cheng TH, Chan P, Liu JC (2011) Nicorandil inhibits angiotensin-II-induced proliferation of cultured rat cardiac fibroblasts. Pharmacology 87:144–151
217. Blythe NM, Muraki K, Ludlow MJ, Stylianidis V, Gilbert HTJ, Evans EL, Cuthbertson K, Foster R, Swift J, Li J et al (2019) Mechanically activated Piezo1 channels of cardiac fibroblasts stimulate p38 mitogen-activated protein kinase activity and interleukin-6 secretion. J Biol Chem 294:17395–17408
218. Leask A (2013) Focal adhesion kinase: a key mediator of transforming growth factor beta signaling in fibroblasts. Adv Wound Care (New Rochelle) 2:247–249
219. Zhang P, Wang W, Wang X, Wang X, Song Y, Zhang J, Zhao H (2013) Focal adhesion kinase mediates atrial fibrosis via the AKT/S6K signaling pathway in chronic atrial fibrillation patients with rheumatic mitral valve disease. Int J Cardiol 168:3200–3207
220. Chan MW, Arora PD, Bozavikov P, McCulloch CA (2009) FAK, PIP5KIgamma and gelsolin cooperatively mediate force-induced expression of alpha-smooth muscle actin. J Cell Sci 122: 2769–2781
221. Dalla Costa AP, Clemente CF, Carvalho HF, Carvalheira JB, Nadruz W Jr, Franchini KG (2010) FAK mediates the activation of cardiac fibroblasts induced by mechanical stress through regulation of the mTOR complex. Cardiovasc Res 86:421–431
222. Fan GP, Wang W, Zhao H, Cai L, Zhang PD, Yang ZH, Zhang J, Wang X (2015) Pharmacological inhibition of focal adhesion kinase attenuates cardiac fibrosis in mice cardiac fibroblast and post-myocardial-infarction models. Cell Physiol Biochem 37:515–526

223. Zhang J, Fan G, Zhao H, Wang Z, Li F, Zhang P, Zhang J, Wang X, Wang W (2017) Targeted inhibition of focal adhesion kinase attenuates cardiac fibrosis and preserves heart function in adverse cardiac remodeling. Sci Rep 7:43146
224. Cheng Z, Sundberg-Smith LJ, Mangiante LE, Sayers RL, Hakim ZS, Musunuri S, Maguire CT, Majesky MW, Zhou Z, Mack CP et al (2011) Focal adhesion kinase regulates smooth muscle cell recruitment to the developing vasculature. Arterioscler Thromb Vasc Biol 31: 2193–2202
225. Hakim ZS, DiMichele LA, Rojas M, Meredith D, Mack CP, Taylor JM (2009) FAK regulates cardiomyocyte survival following ischemia/reperfusion. J Mol Cell Cardiol 46:241–248
226. Wang J, Chen H, Seth A, McCulloch CA (2003) Mechanical force regulation of myofibroblast differentiation in cardiac fibroblasts. Am J Physiol Heart Circ Physiol 285:H1871–H1881
227. Sinfield JK, Das A, O'Regan DJ, Ball SG, Porter KE, Turner NA (2013) p38 MAPK alpha mediates cytokine-induced IL-6 and MMP-3 expression in human cardiac fibroblasts. Biochem Biophys Res Commun 430:419–424
228. Zhao XH, Laschinger C, Arora P, Szaszi K, Kapus A, McCulloch CA (2007) Force activates smooth muscle alpha-actin promoter activity through the Rho signaling pathway. J Cell Sci 120:1801–1809
229. Shimizu T, Liao JK (2016) Rho kinases and cardiac remodeling. Circ J 80:1491–1498
230. Phrommintikul A, Tran L, Kompa A, Wang B, Adrahtas A, Cantwell D, Kelly DJ, Krum H (2008) Effects of a Rho kinase inhibitor on pressure overload induced cardiac hypertrophy and associated diastolic dysfunction. Am J Physiol Heart Circ Physiol 294:H1804–H1814
231. Rikitake Y, Oyama N, Wang CY, Noma K, Satoh M, Kim HH, Liao JK (2005) Decreased perivascular fibrosis but not cardiac hypertrophy in ROCK1+/− haploinsufficient mice. Circulation 112:2959–2965
232. Shimizu T, Narang N, Chen P, Yu B, Knapp M, Janardanan J, Blair J, Liao JK (2017) Fibroblast deletion of ROCK2 attenuates cardiac hypertrophy, fibrosis, and diastolic dysfunction. JCI Insight 2:e93187
233. Lauriol J, Keith K, Jaffre F, Couvillon A, Saci A, Goonasekera SA, McCarthy JR, Kessinger CW, Wang J, Ke Q et al (2014) RhoA signaling in cardiomyocytes protects against stress-induced heart failure but facilitates cardiac fibrosis. Sci Signal 7:ra100
234. Dasgupta I, McCollum D (2019) Control of cellular responses to mechanical cues through YAP/TAZ regulation. J Biol Chem 294:17693–17706
235. Szeto SG, Narimatsu M, Lu M, He X, Sidiqi AM, Tolosa MF, Chan L, De Freitas K, Bialik JF, Majumder S et al (2016) YAP/TAZ are mechanoregulators of TGF-beta-Smad signaling and renal fibrogenesis. J Am Soc Nephrol 27:3117–3128
236. Jorgenson AJ, Choi KM, Sicard D, Smith KM, Hiemer SE, Varelas X, Tschumperlin DJ (2017) TAZ activation drives fibroblast spheroid growth, expression of profibrotic paracrine signals, and context-dependent ECM gene expression. Am J Physiol Cell Physiol 312:C277–C285
237. Xiao Y, Hill MC, Li L, Deshmukh V, Martin TJ, Wang J, Martin JF (2019) Hippo pathway deletion in adult resting cardiac fibroblasts initiates a cell state transition with spontaneous and self-sustaining fibrosis. Genes Dev 33:1491–1505
238. Francisco J, Zhang Y, Nakada Y, Jeong JI, Huang CY, Ivessa A, Oka S, Babu GJ, Del Re DP (2021) AAV-mediated YAP expression in cardiac fibroblasts promotes inflammation and increases fibrosis. Sci Rep 11:10553
239. Francisco J, Zhang Y, Jeong JI, Mizushima W, Ikeda S, Ivessa A, Oka S, Zhai P, Tallquist MD, Del Re DP (2020) Blockade of fibroblast YAP attenuates cardiac fibrosis and dysfunction through MRTF-A inhibition. JACC Basic Transl Sci 5:931–945
240. Mia MM, Cibi DM, Binte Abdul Ghani SA, Singh A, Tee N, Sivakumar V, Bogireddi H, Cook SA, Mao J, Singh MK (2022) Loss of Yap/taz in cardiac fibroblasts attenuates adverse remodeling and improves cardiac function. Cardiovasc Res 118(7):1785–1804. https://doi.org/10.1093/cvr/cvab205

241. Tomasek JJ, McRae J, Owens GK, Haaksma CJ (2005) Regulation of alpha-smooth muscle actin expression in granulation tissue myofibroblasts is dependent on the intronic CArG element and the transforming growth factor-beta1 control element. Am J Pathol 166:1343–1351
242. Lighthouse JK, Small EM (2016) Transcriptional control of cardiac fibroblast plasticity. J Mol Cell Cardiol 91:52–60
243. Small EM (2012) The actin-MRTF-SRF gene regulatory axis and myofibroblast differentiation. J Cardiovasc Transl Res 5:794–804
244. Small EM, Thatcher JE, Sutherland LB, Kinoshita H, Gerard RD, Richardson JA, Dimaio JM, Sadek H, Kuwahara K, Olson EN (2010) Myocardin-related transcription factor-a controls myofibroblast activation and fibrosis in response to myocardial infarction. Circ Res 107:294–304
245. Weng X, Yu L, Liang P, Chen D, Cheng X, Yang Y, Li L, Zhang T, Zhou B, Wu X et al (2015) Endothelial MRTF-A mediates angiotensin II induced cardiac hypertrophy. J Mol Cell Cardiol 80:23–33
246. Trembley MA, Quijada P, Agullo-Pascual E, Tylock KM, Colpan M, Dirkx RA Jr, Myers JR, Mickelsen DM, de Mesy Bentley K, Rothenberg E et al (2018) Mechanosensitive gene regulation by myocardin-related transcription factors is required for cardiomyocyte integrity in load-induced ventricular hypertrophy. Circulation 138(17):1864–1878
247. Ren G, Michael LH, Entman ML, Frangogiannis NG (2002) Morphological characteristics of the microvasculature in healing myocardial infarcts. J Histochem Cytochem 50:71–79
248. Zhao W, Lu L, Chen SS, Sun Y (2004) Temporal and spatial characteristics of apoptosis in the infarcted rat heart. Biochem Biophys Res Commun 325:605–611
249. Takemura G, Ohno M, Hayakawa Y, Misao J, Kanoh M, Ohno A, Uno Y, Minatoguchi S, Fujiwara T, Fujiwara H (1998) Role of apoptosis in the disappearance of infiltrated and proliferated interstitial cells after myocardial infarction. Circ Res 82:1130–1138
250. Fu X, Khalil H, Kanisicak O, Boyer JG, Vagnozzi RJ, Maliken BD, Sargent MA, Prasad V, Valiente-Alandi I, Blaxall BC et al (2018) Specialized fibroblast differentiated states underlie scar formation in the infarcted mouse heart. J Clin Invest 128:2127–2143
251. Hanna A, Shinde AV, Li R, Alex L, Humeres C, Balasubramanian P, Frangogiannis NG (2021) Collagen denaturation in the infarcted myocardium involves temporally distinct effects of MT1-MMP-dependent proteolysis and mechanical tension. Matrix Biol 99:18–42
252. Zymek P, Bujak M, Chatila K, Cieslak A, Thakker G, Entman ML, Frangogiannis NG (2006) The role of platelet-derived growth factor signaling in healing myocardial infarcts. J Am Coll Cardiol 48:2315–2323
253. Sager HB, Hulsmans M, Lavine KJ, Moreira MB, Heidt T, Courties G, Sun Y, Iwamoto Y, Tricot B, Khan OF et al (2016) Proliferation and recruitment contribute to myocardial macrophage expansion in chronic heart failure. Circ Res 119:853–864
254. Sager HB, Hulsmans M, Lavine KJ, Beltrami Moreira MB, Heidt T, Courties G, Sun Y, Iwamoto Y, Tricot B, Khan OF et al (2016) Proliferation and recruitment contribute to myocardial macrophage expansion in chronic heart failure. Circ Res 119(7):853–864
255. Chen B, Frangogiannis NG (2016) Macrophages in the remodeling failing heart. Circ Res 119: 776–778
256. Jugdutt BI, Musat-Marcu S (2000) Opposite effects of amlodipine and enalapril on infarct collagen and remodelling during healing after reperfused myocardial infarction. Can J Cardiol 16:617–625
257. Miyashita S, Banlengchit R, Marbach JA, Chweich H, Kawabori M, Kimmelstiel CD, Kapur NK (2022) Left ventricular unloading before percutaneous coronary intervention is associated with improved survival in patients with acute myocardial infarction complicated by cardiogenic shock: A systematic review and meta-analysis. Cardiovasc Revasc Med 39:28–35
258. Patel MR, Smalling RW, Thiele H, Barnhart HX, Zhou Y, Chandra P, Chew D, Cohen M, French J, Perera D et al (2011) Intra-aortic balloon counterpulsation and infarct size in patients

with acute anterior myocardial infarction without shock: the CRISP AMI randomized trial. JAMA 306:1329–1337
259. Weil BR, Konecny F, Suzuki G, Iyer V, Canty JM Jr (2016) Comparative hemodynamic effects of contemporary percutaneous mechanical circulatory support devices in a porcine model of acute myocardial infarction. JACC Cardiovasc Interv 9:2292–2303
260. Liao XD, Tang AH, Chen Q, Jin HJ, Wu CH, Chen LY, Wang SQ (2003) Role of Ca2+ signaling in initiation of stretch-induced apoptosis in neonatal heart cells. Biochem Biophys Res Commun 310:405–411
261. Wei X, Li T, Hagen B, Zhang P, Sanchez PG, Williams K, Li S, Bianchi G, Son HS, Wu C et al (2013) Short-term mechanical unloading with left ventricular assist devices after acute myocardial infarction conserves calcium cycling and improves heart function. JACC Cardiovasc Interv 6:406–415
262. Saku K, Kakino T, Arimura T, Sunagawa G, Nishikawa T, Sakamoto T, Kishi T, Tsutsui H, Sunagawa K (2018) Left ventricular mechanical unloading by total support of impella in myocardial infarction reduces infarct size, preserves left ventricular function, and prevents subsequent heart failure in dogs. Circ Heart Fail 11:e004397
263. Kapur NK, Paruchuri V, Urbano-Morales JA, Mackey EE, Daly GH, Qiao X, Pandian N, Perides G, Karas RH (2013) Mechanically unloading the left ventricle before coronary reperfusion reduces left ventricular wall stress and myocardial infarct size. Circulation 128: 328–336
264. Esposito ML, Zhang Y, Qiao X, Reyelt L, Paruchuri V, Schnitzler GR, Morine KJ, Annamalai SK, Bogins C, Natov PS et al (2018) Left ventricular unloading before reperfusion promotes functional recovery after acute myocardial infarction. J Am Coll Cardiol 72:501–514
265. Hanna A, Frangogiannis NG (2019) The cell biological basis for primary unloading in acute myocardial infarction. Int J Cardiol 293:45–47
266. Bujak M, Kweon HJ, Chatila K, Li N, Taffet G, Frangogiannis NG (2008) Aging-related defects are associated with adverse cardiac remodeling in a mouse model of reperfused myocardial infarction. J Am Coll Cardiol 51:1384–1392
267. Ibanez B, Aletras AH, Arai AE, Arheden H, Bax J, Berry C, Bucciarelli-Ducci C, Croisille P, Dall'Armellina E, Dharmakumar R et al (2019) Cardiac MRI endpoints in myocardial infarction experimental and clinical trials: JACC scientific expert panel. J Am Coll Cardiol 74:238–256
268. Baues M, Dasgupta A, Ehling J, Prakash J, Boor P, Tacke F, Kiessling F, Lammers T (2017) Fibrosis imaging: current concepts and future directions. Adv Drug Deliv Rev 121:9–26
269. Lopez B, Ravassa S, Moreno MU, Jose GS, Beaumont J, Gonzalez A, Diez J (2021) Diffuse myocardial fibrosis: mechanisms, diagnosis and therapeutic approaches. Nat Rev Cardiol 18: 479–498

Mechanobiology of Cardiac Fibroblasts in Cardiac Remodeling

Luke R. Perreault and Lauren D. Black III

1 Introduction

Cardiac fibroblasts (CFs), long considered to be passive contributors to cardiac mechanics and function, have undergone a critical reevaluation in the past two decades. CFs are now considered to be a critical cell type within the mammalian heart and a key source of extracellular matrix proteins and are the predominant cell type responsible for the maintenance, remodeling, and regulation of ECM. Because of this critical function, they contribute a great deal to the structural and biochemical properties of heart tissue and consequently its tissue mechanics. Beyond this, CFs are responsible for a variety of roles in cardiac tissue development [1], cell signaling [2–4], immune support [5], and, particularly, response to inflammation [2, 6–9].

CFs are functionally and morphologically like fibroblasts from other tissues within the body and similarly express different phenotypes depending on tissue developmental age, mechanics, and associated biochemical and physiological cues. Fibroblasts can be broadly defined as cells of mesenchymal origin that express and secrete ECM proteins and are morphologically flat, spindle-like cells with multiple protrusions, lacking a basement membrane and exhibiting a prominent Golgi apparatus and rough endoplasmic reticulum [10]. Estimations of their abundance in

L. R. Perreault
Department of Biomedical Engineering, Tufts University, Medford, MA, USA

Department of Engineering, Boston College, Chestnut Hill, MA, USA

L. D. Black III (✉)
Department of Biomedical Engineering, Tufts University, Medford, MA, USA

Cellular, Molecular and Developmental Biology Program, Graduate School for Biomedical Sciences, Tufts University School of Medicine, Boston, MA, USA
e-mail: Lauren.black@tufts.edu

M. Hecker, D. J. Duncker (eds.), *Cardiac Mechanobiology in Physiology and Disease*, Cardiac and Vascular Biology 9,
https://doi.org/10.1007/978-3-031-23965-6_5

cardiac tissue, depending on species and developmental age, range from 20 to 60% of total cell count [11, 12].

CFs play a critical, oft-underappreciated role in cardiac mechanobiology, both through their actions mediating the tissue composition of the heart in normal maintenance and injury repair and their own physiological sensitivity to their mechanical environment. CFs arise early in cardiac development, proliferating extensively and depositing ECM proteins and aiding in the formation of the heart by forming an extensive, mechanically supportive protein network to anchor cardiomyocytes (CMs) [4].

CF contribution to cardiac mechanobiology is perhaps most evident in the case of cardiac injury, which induces inflammation, CM necrosis, and a cascade of cytokines that signal changes in cellular behavior [3, 7, 8, 12–15]. Under these conditions, CFs phenotypically "activate" into myofibroblasts, highly active, proliferative, and invasive cells that release matrix metalloproteinases (MMPs), enzymes that break down damaged cECM, increasing collagen turnover and leading to deposition of fibrotic scar tissue. This, in turn, can lead to progressive stiffening of cardiac tissue and reduction of tissue compliancy that impedes heart contractility. This progressive remodeling can lead over time to heart failure [7, 16–19]. Conversely, CFs exhibit phenotypic plasticity arising from sensitivity to their mechanical environment. Increased substrate stiffness can drive CFs to transition to myofibroblasts and perpetuate the fibrotic process, leading to excessive matrix deposition and functional tissue loss [20, 21].

In the following chapter, the significance of fibroblasts in cardiac mechanobiology and cardiac remodeling will be elucidated further. Specifically, CF roles in cardiac development and, more broadly, the role of mechanical signaling in heart formation will be discussed, along with CF responses and contributions to the cardiac injury and disease. Last, therapeutic approaches directed at modulating mechanotransduction of (and by) CFs will be explored in detail, both in the context of ECM effects and impacts on CFs, CMs, and other cell types.

2 Overview of Cardiac Fibroblasts

The role of cardiac fibroblasts in the heart is predominantly that of regulators of ECM, maintaining its structure and function. They are responsible for the production of growth factors, cytokines, MMPs, and ECM proteins that are all necessary to maintain a balance between the degradation of ECM and synthesis of new components and work in concert with immune cells and processes to regulate cardiac tissue degradation and repair in injury and disease. As a result, CFs play a critical role in regulating tissue mechanics across developmental ages and disease states and are in turn impacted by mechanobiological cues that motivate the expression of these various biological components [12, 22–25]. This section will go into detail on the developmental origin and contribution of cardiac fibroblasts to heart formation and their general function and physiology as it pertains to tissue mechanics.

2.1 Cardiac Fibroblasts in the Context of Cell and Tissue Development

The roles of CFs in heart development are like their homeostatic role in the adult heart: synthesis of ECM and crosstalk with other resident cardiac cells [26, 27]. CFs are derived from a mesenchymal lineage originating in the proepicardial organ, assisting in developing a compact myocardial tissue via interaction with early cardiomyocytes [28]. These mesenchymal cells contribute to the continued formation of the proepicardium and later differentiate into interstitial fibroblasts through epithelial-mesenchymal transition (EMT) [4]. This is not the only origin for CFs, however: they appear to possess several different lineages, and evidence suggests some may originate from progenitor stem cells in circulation and within the heart [12, 29]. Endothelial cells (ECs) can also differentiate into a CF phenotype in response to inflammatory signals [30, 31], further complicating the developmental origin of this cell type. Fibroblast lineage has also been shown to be location-dependent, with fractions derived from endothelial and epicardial lineages and a subset mostly localized to the right atrium derived from the neural crest [32]. In all, CFs are extensively heterogeneous in their developmental origin, complicating their identification in culture and their experimental study due to a lack of a definitive marker [32].

Several soluble factors contribute to the establishment of a fibroblast phenotype, including platelet-derived growth factor (PDGF), fibroblast growth factor (FGF), and TGF-β1 [4, 33]. Embryonic CFs then degrade components of ECM to migrate through the epicardium to the mesenchymal layer and help to form the endocardial cushions, transient structures critical to the formation of the functional heart [34]. CFs then proliferate extensively, depositing ECM proteins and aiding in the formation of the cardiac organ and following birth aid in the remodeling and transition to a more mature cECM, depositing an array of collagens, fibronectin, and vitronectin to form an extensive, mechanically supportive protein network [4].

Embryonic cardiac fibroblasts appear to express a phenotype geared toward cardiac tissue growth and development: in addition to the fact that they express a different array of ECM components compared to adult cardiac fibroblasts (e.g., increased Col4, Col6, fibrillin-2, and periostin, among others [35]), they appear to motivate the actions of other cells throughout development. A study by Ieda et al. (2009) showed evidence that embryonic CFs promote CM proliferation and ventricular compaction, via a β1 integrin signaling pathway, versus adult CFs, which appeared to promote CM maturation and hypertrophy in the same study [26]. Additionally, a study comparing neonatal and adult CF transcriptomes appeared to identify transcriptomic changes that were concomitant with CM maturation [36]. In the context of cardiac mechanobiology, the maturation of cardiac tissue and shift in ECM from a developmental to a homeostatic protein composition is correlated with increased mechanical strength and elasticity [4, 16, 37, 38].

These effects also appear to extend to work involving engineered cardiac tissues. In studies of CF and CM interplay in engineered tissues, it was shown that adult CFs negatively affect engineered tissue growth and tissue properties, deteriorating electromechanical function and downregulating genes associated with calcium handling and contraction, whereas fetal CFs improved or did not appreciably change the same parameters [39]. Taken together, the data above lends critical evidence that not only do CFs guide the deposition and remodeling of cECM throughout development, but age-dependent signaling from cardiac fibroblasts is a crucial driver cardiac development and maturation.

2.2 Cardiac Fibroblasts in ECM Maintenance and Remodeling

In their role as ECM regulators in the heart, CFs produce growth factors, cytokines, MMPs, and ECM proteins. These functions work in concert to maintain a balance between degradation of ECM and synthesis of new components [12]. CFs are sensitive to both mechanical and biochemical changes within the myocardium and readily adapt their phenotype in response to these cues. These phenotypic responses can have significant impacts on the mechanobiology of heart tissue.

This is most prevalent in cardiac injury: injury to heart tissue induces inflammation, CM necrosis, and a cascade of cytokines that signal changes in cellular behavior [3, 7, 8, 12–15], inducing CFs to activate to a myofibroblast phenotype, characterized by high activity, proliferation, and increased migration. Myofibroblasts release an array of MMPs, break down damaged ECM, and increase collagen turnover to ultimately facilitate the deposition of stiff, noncompliant fibrotic scar tissue, representing a major reduction in myocardial elasticity and macroscopic tissue mechanics in the heart [7, 16–19].

A variety of cues signal CFs to transition to myofibroblasts, including the β1-integrin-mediated response to increased substrate stiffness [2, 40, 41], the cytokine PDGF [17, 42], DAMPs [43], and TGF-β1 [14, 17, 40, 42, 44]. Of critical importance to the CF-to-myofibroblast transition is TGF-β1, which appears to be necessary for myofibroblast activation both in vivo and in vitro [14, 17, 21, 45–47].

In addition to the marked upregulation of collagen and MMPs, the myofibroblast phenotype is characterized by upregulated expression of α-smooth muscle actin (αSMA), a protein critical in cellular traction force on ECM: this upregulation facilitates myofibroblast roles in collagen remodeling and wound contraction after an injury. Inclusion of αSMA within stress fibers represents a key step in myofibroblast activation and the cell response to adverse ECM alterations [22, 48]. However, CFs can begin to express actin stress fibers apart from αSMA when exposed to mechanical tension or stiff substrates. This represents a "proto-myofibroblast" phenotype and can be a precursor to full differentiation to a myofibroblast [20, 49] (Fig. 1). Myofibroblasts release an array of MMPs degrading fibrillar collagen, enabling migration and active proliferation into the site of a wound area, enabling invading immune cells to remove necrotic cell debris, and later

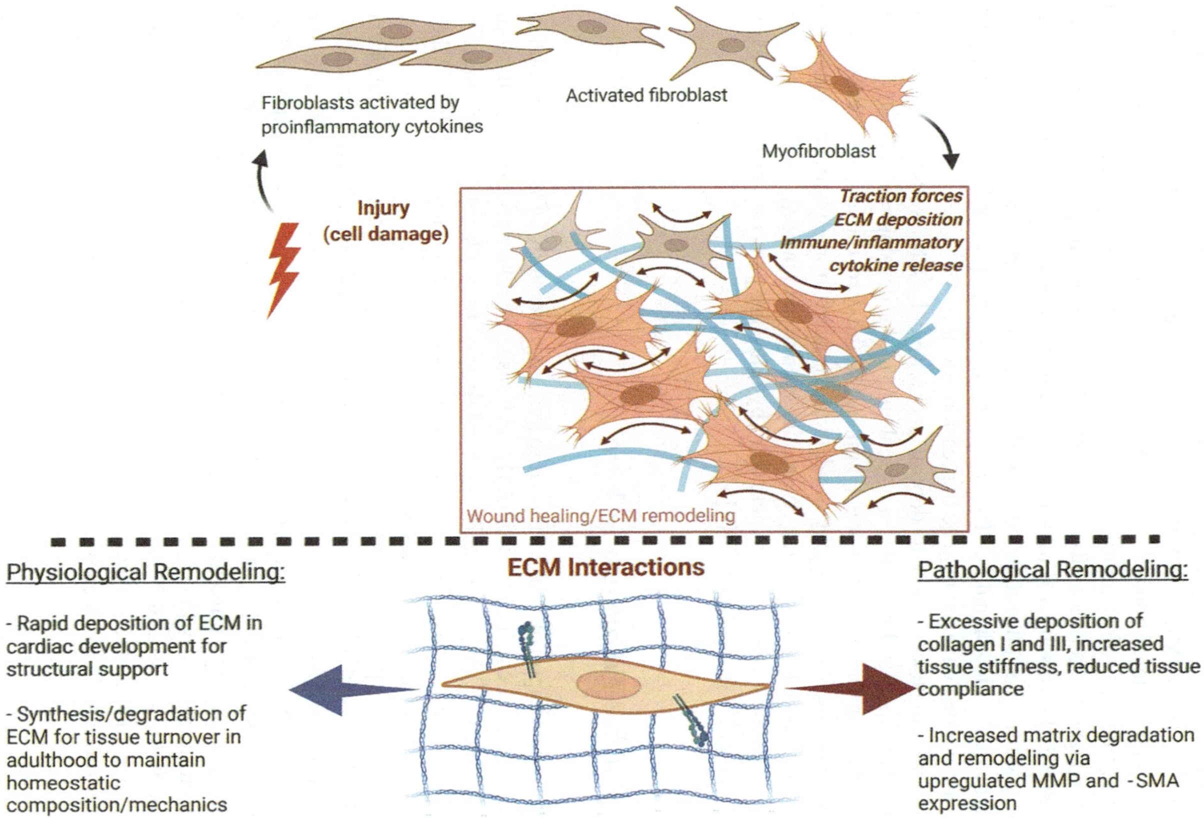

Fig. 1 Cardiac fibroblast deposition and remodeling of extracellular matrix play a significant role in mediating cardiac mechanobiology. In development, deposition of ECM proteins and turnover of ECM components strengthen and stabilize the developing heart. In adulthood, fibroblast activation and ECM

facilitating fibrotic stabilization of the wound bed, via collagen deposition and wound contraction that increases the stiffness of cardiac tissue [12].

It is critical to note that CFs within the injured heart can activate and assume a range of myofibroblast phenotypes that exhibit extensive functional variation. An extensive review by Michelle Tallquist (2020) [50] describes the range of functional roles CFs can assume upon activation. Myofibroblasts can assume proinflammatory, proangiogenic, or profibrotic roles, among others, and these population subsets can be distinguished experimentally via single-cell RNA-seq. Hinz and Lagares (2020) [48] extend this discussion to senescent myofibroblasts within damaged tissue, which acquire this phenotype and evade apoptosis in response to pro-survival biomechanical and signaling factors (activation of pro-fibrotic TGF-β1, β1 integrin-mediated changes in gene expression, etc.).

This phenotype continues to deposit and remodel ECM well beyond what is required for repair and may play a critical role in the development of pathological fibrosis. To compensate for the noncompliance of the scar, viable CMs are forced to undergo pathological hypertrophy to adapt to the needs of the restructured myocardium, but ultimately reducing the contractile strength and integrity of cardiac tissue. CFs additionally contribute to hypertrophy by releasing paracrine signals that assist in triggering this response in CMs [51].

2.3 *Cardiac Fibroblasts in Vitro*

Despite broad importance in cardiac research, in vitro study of cardiac fibroblasts remains a distinct challenge, particularly with respect to their positive identification in assays due to a lack of a definitive fibroblast-specific marker and maintenance of their phenotype in conventional tissue culture, the latter in large part due to their aggressive sensitivity to changes in environmental mechanics.

With respect to potential markers, it is necessary to evaluate the expression of two or more markers to identify a fibroblast population. Candidates include fibroblast-specific protein-1 (FSP-1), which appears to be expressed on hematopoietic and endothelial cells in the heart, vimentin, and periostin (specifically expressed in myofibroblasts), in addition to genes *Pdgfra* and *Tcf21* [10, 52, 53]. One promising marker is discoidin domain receptor 2 (DDR2), a surface collagen receptor expressed in fibroblasts [54] but also fibrocytes, a mesenchymal-lineage cell found in blood with evidence of being a fibroblast precursor. The absence of a definitive marker due to the heterogeneity of CFs represents a critical limitation in cardiac research and remains an area of active study and exploration [15].

Fig. 1 (continued) remodeling post-injury can lead to pathological remodeling, characterized by inflammation, aggressive matrix degradation, excessive deposition of collagen, and increased tissue stiffness. Figure made with Biorender

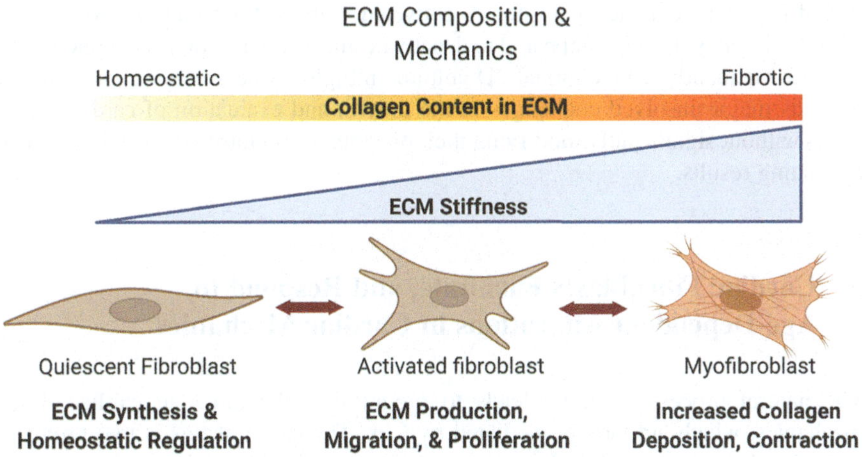

Fig. 2 ECM composition and stiffness contribute to activated fibroblast phenotypic heterogeneity. Low, homeostatic stiffness is associated with a quiescent fibroblast phenotype expressing genes associated with ECM deposition, regulation, and non-pathological remodeling. High collagen content and subsequent increase in substrate stiffness can lead to persistent activation and a myofibroblast phenotype associated with high contractility and increased fibrotic collagen expression. Figure made with Biorender

The phenotypic plasticity of CFs and their sensitivity to their mechanical environment presents major in vitro culture challenges as well. The effects of high substrate stiffness and extensive passaging, conditions found on conventional tissue culture plastic in a 2D culture environment, can lead fibroblasts to begin to undergo the activation process toward a myofibroblast phenotype and increase their expression of αSMA [4, 18, 55], rendering long-term 2D culture of these cells challenging without significantly altering their phenotype.

There are two critical reasons for this: first, culture plastic represents a high-stiffness substrate, which is on the order of GPa [22], and is exponentially higher than conventional cardiac tissue, which is about 8–11 kPa [56], and therefore culture plastic represents a substrate not physiologically relevant. High stiffness is well-established to induce phenotypic changes in CFs, inducing a myofibroblast phenotype and increased αSMA expression (Fig. 2). A study of fibroblasts in 2D and 3D in vitro cancer models of tumor-stromal cell interactions established that the 3D model increases production of paracrine factors that can drive cancer transition to an invasive phenotype [57]. These factors were largely absent in 2D culture, underscoring not only the importance of 3D culture in producing physiologically relevant models but the risks of drawing broad scientific conclusions strictly from observation of cells in vitro within 2D models.

A recent study by Walker et al. took this view a step further, establishing that aortic valve fibroblasts cultured on hydrogels capable of softening varied in their phenotype post-softening depending on the extent of time they were cultured in a stiff microenvironment. Myofibroblasts cultured on stiff substrates for an extended period (7 days) remained persistently activated, versus those cultured for shorter periods (1–3 days), which appeared to deactivate and return to a quiescent phenotype [55]. Consequently, conventional 2D culture on high-stiffness tissue culture plastic presents major unsolved challenges in the growth and evaluation of cardiac fibroblasts without significantly modifying their phenotype and must be considered when evaluating results.

3 Cardiac Fibroblasts Facilitate, and Respond to, Age-Dependent Alterations in Cardiac Mechanics

The natural process of aging leads to progressive alterations in cardiac tissue mechanics, which are largely mediated by CFs. The cause of this is two-pronged: first, CF phenotype and ECM expression alter considerably with respect to developmental age and facilitate alterations in ECM and tissue mechanics [5, 43, 58]. Second, aging and senescing cardiomyocytes and other cardiac cells, in addition to progressive, lifelong ECM alteration and age-related ECM degradation, impact CF function and can trigger a fibrotic phenotype [59–63].

Studies of cardiac development and ECM alterations have indicated that, versus young developmental age (i.e., fetal, neonatal) CFs, CFs from aging cardiac tissue can drive cardiac hypertrophy-like phenotypes in engineered cardiac tissues, representing an adverse mechanical remodeling response [58, 64]. Similarly, a single-cell RNA-seq study established that transcriptome differences in mouse CFs from a neonatal to adult state are tied to CM maturation, implying CF protein expression may help drive cardiac tissue maturation and consequently contractility and tissue function [36]. A study by our lab (2021) also provided evidence of age-related changes in CF inflammatory response. In an RNA-seq analysis of rat CFs, in adult CFs versus fetal and neonatal CFs, we observed an upregulation in inflammatory mediators (*Il-1β*, *Il-6*) and chemokines such as *Ccl3* and *Ccl4* [65].

With advancing age, some CF structural and compositional changes of the cECM are associated with decreased wound healing effectiveness and functional loss. A study of induced HF in young (~18 months) and old (>8 years) sheep by Horn et al. (2012) indicated that significant differences in matrix remodeling in response to HF could be observed in the two developmental ages. Whereas an increase in collagen accumulation was observed in young sheep, older sheep had an increase in collagen degradation, an increase in ECM-degrading enzyme expression, and increased cardiac hypertrophy leading to functional loss [61]. A similar study by Lindsay et al. (2005) in young and old mice indicated cECM of hearts of older mice had an increased degradative capacity, greater wall thickness, and blunted cardiac fibroblast proliferation and functional capacity [62].

Studies have tied this increase in adverse mechanical remodeling and functional loss to both the effects of cardiomyocyte senescence and CF senescence on the ECM-remodeling functions of CFs. Senescent and injured CMs undergo apoptosis and cell death and in aging tissue are left predominantly not replaced by new, functional CMs. This can lead to an increase in inflammation and production of reactive oxygen species, which can drive CFs to a persistent myofibroblast phenotype and upregulate ECM deposition, negatively impacting cardiac mechanics [5, 63, 66, 67]. Further, a study by Zhu et al. (2013) showed evidence that P53-mediated senescent CFs contributed to an increase in adverse cardiac matrix degradation and remodeling, which can lead to cardiac rupture [59].

4 Mechanobiology of Fibroblasts in Cardiac Injury and Disease

Coronary heart disease (CHD) is an endemic problem in the United States with approximately 720,000 Americans experiencing a new coronary event (defined as a first hospitalized MI or CHD death) and 325,000 experiencing a recurrent event this year alone [68]. The primary etiology for the development of MI is an occlusion of the coronary artery due to the embolization of arterial plaque. This induces rapid hypoxia in the myocardium perfused by the artery, and consequently the most effective method of treatment for MI is rapid intervention to restore the blood supply and mitigate ischemia-induced necrosis of CMs [69–71].

Cardiac fibroblasts play a critical role in these processes. The adult mammalian heart has little regenerative capacity, and consequently healing is primarily dependent on CF ECM deposition to synthesize a collagen-based scar to stabilize the damaged tissue. This inevitably leads to a loss of cardiac function due to critical alterations in tissue mechanics: specifically, the noncompliance of the fibrotic scar tissue. This necessitates functional remodeling of the heart to adapt, which over time can lead to alteration in cardiac function, arrhythmia [72], and ultimately heart failure [14, 60]. Similarly, conditions such as hypertension and increased cardiac pressure, diabetic cardiomyopathy, and age-related ECM alterations such as accumulation of collagen can all promote and contribute to fibrosis [73, 74]. The section below will highlight the major contributions of CFs to disease and injury progression in cardiac tissue and the mechanical changes produced therein.

5 CFs Role in Pathophysiology

The response of the CFs to pathophysiological environments is often specific to the type of trauma or injury inflicted on the organ. Even as the specific properties of the pathological environment may vary, the majority of cardiac pathologies result in some form of myocardial fibrosis, which can lead to both systolic and diastolic dysfunction of the heart. Excessive fibrosis hinders contraction by inhibiting the electrical and mechanical coupling of neighboring CMs with one another, and this is often evidenced by a decrease in ejection fraction even in the absence of other obvious injuries [75]. In addition, the increased stiffness of the myocardium decreases the force generated by individual CMs during contraction as CMs are tuned to optimize contraction at specific substrate stiffnesses [76]. The overall decrease in tissue compliance also leads to diastolic dysfunction by impeding relaxation of the CMs which in turn significantly reduces diastolic ventricular filling [77]. Fibrosis, broadly, can be organized into reparative and reactive fibrosis. Reparative fibrosis is a characteristic response to ischemia and CM death and is required to stabilize large regions of myocardium and protect against rupture during systole. Reactive fibrosis, which can occur as a response to physical stressors, is often dispersed in regions surrounding viable CMs. A more detailed accounting of CF-specific responses to broad categories of cardiac injury is detailed in the sections that follow.

5.1 Congenital Heart Disease

The more severe congenital heart defects present clinically with defects of the compact layer of ventricular myocardium, including both thinning and a lack of the region altogether [78–80]. Considering that epicardial-myocardial signaling has been indicated as a critical factor in CF development, alterations in myocardial thickness in CHD likely impact development and differentiation of epicardial derived precursor cells into CFs [81]. Whether the origination of CFs is affected by CHDs is an active area of research. Downregulation of PDGF and FGF has been observed in rodent models of left heart hypoplasia on day 22 of development [82]. These factors are implicated in EMT, which is required for CF dispersal through developing myocardial tissue. However, the impacts CHD-affected CF populations may have on CMs in vivo, and their influence on disease progression, remain ambiguous and require further study.

Many CHDs are associated with fibrosis initiated as a response to cardiac hypertrophy. A PCR microarray analysis of patients with Tetralogy of Fallot (TF), a combination of four different structural defects in the heart, showed upregulation of the ECM proteins fibronectin and collagen I 12 months after birth. In contrast, genes associated with ECM proteolysis were unaffected, implying synthesis of ECM proteins by CFs is increased due to exposure to the altered mechanical loading caused by CHD-related structural abnormalities [83].

Patients with CHDs often experience pressure overload (PO) or volume overload (VO) (because of altered blood flow), depending on the type of defect present in their pathology. PO is often a characteristic of TF, septated right ventricle, and aortic or pulmonary arterial stenosis—all examples where the ventricles are exposed to abnormally high pressures. Conversely, VO may be symptomatic of aortic or pulmonary regurgitation, as well as atrial septal defects. In a study by Chaturvedi et al., biopsies were taken from the right and left ventricular tissue from patients presenting PO or VO, which manifested due to congenital defects. These tissues were analyzed with respect to passive ECM stiffness, which indicated that tissue strips isolated from CHD patients with PO were stiffer than those isolated from patients with VO. Increased tissue stiffness was not related to alterations in the composition of the myocardium, as VO patients had a higher collagen content per gram of tissue than did PO patients.

While more work is needed to understand this discrepancy between ECM composition and tissue stiffness, potential explanations could be related to alterations in matrix cross-linking or within the gross structure and alignment of the ECM, both of which could impact uniaxial tension measurements. Regardless, these studies highlight how CHDs induce CFs to alter the extracellular environment [84].

5.2 *Pressure Overload*

PO occurs in the heart, when blood flow through the respective outflow tract is impeded via a blockage or due to some other increase in vascular resistance, which induces an increase in ventricular wall stress during systole. Chronic PO usually initiates CM hypertrophy or fibrosis (or both). CFs have critical compensatory roles within PO, but their phenotype is also significantly altered during the early stages of PO prior to the initiation of secondary signaling cascades. In particular, CF-ECM interactions are altered as CFs increase $\alpha 1$ integrin expression, promoting more extensive adhesion to collagen and laminin. However, they do not initiate significant remodeling of matrix proteins. Further, migration—but not proliferation—is increased in response to increased pressure [85]. Over time, prolonged stimulation can induce activation of hypertrophic signaling pathways.

In a pulmonary artery banding model of PO, CFs demonstrated an increase in total collagen expression 2 weeks post-banding, likely as an effort to respond to the initial mechanical stress. When the pressure overload becomes chronic at 4 weeks, the cell response is altered, and CFs begin to increase insoluble collagen production, which ultimately leads to increases in end-diastolic pressure, representing how dynamic CF responses to injury and alterations in environmental mechanics can influence whole organ function [86].

5.3 *Hypertrophy*

Hypertrophy of cardiomyocytes in the heart can be a response to both physiological and pathological cardiac remodeling in response to increased mechanical stressors. Physiological hypertrophy is beneficial and reversible and occurs when CMs increase size to increase their functional capacity and enhance cardiac function. CFs appear to play a largely homeostatic role in this process, as indicated in an exercise-induced model of physiological hypertrophy: CFs cultured in vitro displayed increased migration when seeded on fibronectin but decreased migration with collagen. In a similar vein, α5 integrin subunit expression, which binds to the RGD sequence in the III_{10} domain on type III fibronectin on, increased, while expression of the α1 and α2 integrin subunits was reduced. Further, cells did not alter collagen gel compaction, suggesting alterations to ECM remodeling were minimal [87].

Pathological hypertrophy, the compensatory remodeling to adapt to functional decreases due to stress or injury, is functionally adverse and effectively irreversible. A study by Burgess et al. used an animal model of early hypertension caused by aortic constriction, which indicated CF-ECM interactions were altered via adhesion, but not through remodeling. This was demonstrated through an increase in β1 integrin expression (associated with collagen binding) and a decrease in α1, α2, and α5 integrin subunit expression. Isolated CFs cultured in vitro displayed reduced migration on both collagen and fibronectin coatings when compared to CFs isolated from untreated animals but showed no significant variation in collagen gel compaction. This conclusion was expanded upon by Stewart et al., in a study that indicated that aortic constriction resulted in pathological hypertrophy and modified CF behavior. Cellular proliferation and migration were increased 7 days following constriction, and this alteration was sustained over time. The variations in CF response to different hypertrophy mechanisms reinforce knowledge of CF-CM communication, as they indicate that CF responses to acute and chronic pathological stimuli are associated with CM reaction to the altered biophysical environment [85].

5.4 *Myocardial Infarction and Heart Failure*

CFs are integral in the pathophysiological events that proceed following an acute myocardial infarction. Immediately following coronary artery occlusion, CFs respond to hypoxia by deploying chemokines that recruit neutrophils to the damaged tissue. Inflammatory mediators signal CFs to initiate matrix remodeling as tissue necrosis progresses, replacing infarcted tissue with a fibrotic scar that is predominantly composed of dense collagen fibers. Remodeling begins thereafter, as CFs upregulate their expression of matrix metalloproteinases and accelerate degradation of the existing collagen matrix, accommodating the deposition of newly synthesized proteins [88]. This response is reflected in in vitro work evaluating CF responses to oxidative stress by stimulating them with delivery of H_2O_2. The CFs react by increasing MMP expression and decreasing synthesis of collagen fibers [89], which is followed by a shift to excessive, progressive collagen I, collagen III, and fibronectin deposition [90].

Immediately post-MI, CFs and a slew of other cell types including endothelial cells, neutrophils, and macrophages are recruited to the injury site by cytokines, liberated intracellular contents, ECM fragments, and reactive oxygen species (ROS) [14, 74, 91–93]. This rapid inflammatory response marks the beginning of the cardiac repair process, which can be broken down into three overlapping phases: the inflammatory phase, proliferative phase, and maturation phase. Proinflammatory cytokines, such as IL-1β, tumor necrosis factor (TNF), and IL-6, are vital to the progression of the early inflammatory phase and are consistently upregulated in experimental MI models. These cytokines contribute to the progression of inflammation by triggering cellular upregulation of inflammatory genes, extravasation of immune cells and platelets into the infarct region, activation of CFs, and production of MMPs and chemokines [66, 74, 93, 94].

The signaling of neutrophils, macrophages, and lymphocytes into the ischemic zone begins within hours of injury, peaks several days later, and marks the initiation of significant structural remodeling. Inflammatory cells, CFs, and necrotic CMs secrete a variety of MMPs to degrade cell and matrix tissue and assist in macrophage phagocytosis. In the absence of a mechanically robust ECM matrix, granulation tissue consisting of fibrin, fibronectin, laminin, and glycosaminoglycans is deposited, providing provisional mechanical support until myofibroblasts begin upregulating deposition of collagen [95].

The proliferative, or fibrotic, phase of wound healing ensues and lasts one to several weeks. This phase is dominated by the action of myofibroblasts, which increase their concentration in the infarct zone via aggressive migration from surrounding myocardium, proliferation, and differentiation from a variety of other cells types including endothelial cells and fibrocytes [43, 95–98]. If infarct size is large, or ongoing myocardial stress factors such as hypertension or cardiomyopathy are present, production of inflammatory cytokines and influx of immune cells can fail to resolve, resulting in chronic inflammation and persistent cardiac remodeling, which can lead to heart failure [93, 99].

Inflammatory cell expression of MMPs decreases, coinciding with myofibroblast upregulation of collagen (predominantly type I and III), markedly increasing the collagen content of the cECM in the infarct region, necessary to support and stabilize the damaged myocardium [4, 95, 100]. Following this is the maturation or remodeling phase, during which myofibroblast numbers reduce within the fibrotic scar and collagen content begins to stabilize, and collagen cross-linking further strengthens the resultant scar tissue [4, 95, 100].

To stabilize the damaged tissue with newly synthesized ECM proteins, CFs are required to proliferate rapidly, and the doubling time subsequently decreases by 50% as compared to CFs isolated from control animals [101]. While some remodeling is required to stabilize the thin and unstable ventricle wall, the CFs are often unable to regulate the extent of fibrosis, and subsequent excessive remodeling disrupts normal heart function. Atomic force microscopy of the left ventricle 30 days following injury in a mouse model of MI reveals a nearly three-fold increase in the elastic modulus of the scar region as compared to the neighboring non-infarcted region [102]. Alterations in scar mechanics are dependent upon collagen content and organization within the infarct as demonstrated by Fomovsky et al. Biaxial mechanical testing revealed an increase in the elastic modulus of the infarcted myocardium as a function of time from 1 to 6 weeks post-MI [103]. This increased stiffness was primarily caused by an increase in overall collagen content and not alterations in cross-linking as previously described. Interestingly, the collagen orientation in the rat infarct model used by Fomovsky et al. was isotropic, which was in contrast to larger animal models that show alignment of the remodeled collagen scar in the circumferential direction of the heart. In a follow-up study, they demonstrated that location of the infarct is the primary determinant of the scar fiber orientation as this effects the directionality of the stress placed on the healing infarct [104]. In particular, infarcts located near the equator of the heart were stretched more circumferentially and thus were circumferentially aligned, while infarcts located near the apex are stretched isotropically, and thus their fiber orientation is isotropic. More recently, our own group has assessed the structure-function relation of the ECM as a function of remodeling time following myocardial infarction and demonstrated that while there are increases in overall collagen content, the maturity of the collagen, the degree of cross-linking of the collagen, and the overall stiffness of the ECM are significantly reduced as compared to healthy ECM [105]. Indeed, the decrease in cross-linking and mechanics is significantly correlated, indicating that while there is an overabundance of collagen, at the cellular level, the structure is actually weaker which may have important implications for continued remodeling of the infarct region after stabilization.

While CFs are responsible for these changes to the matrix, it is important to consider how this altered environment influences CF function through secondary feedback mechanisms. For example, CFs respond to increasing matrix stiffness by enhancing their rates of migration [106]. This correlates to studies which have demonstrated an increase in CF migration rate by over 170% following MI [107]. Therefore, with enhanced migration, remodeling is rarely localized to the region of the scar, but occurs globally throughout the heart [108].

The disruption of the electrical and mechanical continuum of the myocardium following MI prevents efficient pumping of blood systemically throughout the body [109]. To compensate for the impedance generated by the scar, viable CMs undergo pathological hypertrophy to meet the altered needs of the organ [110]. CFs help initiate this transition via the release of paracrine signals. While it is clear that the initiation of pathological hypertrophy may be detrimental, it is actually required for survival and potentially indicates a role for CFs as a stress sensor to neighboring CMs [51].

Although the development of pathological hypertrophy allows the heart to adapt to the altered environment, the heart will eventually be unable to compensate and enter a state of congestive heart failure. CFs isolated from a failing left ventricle have an altered integrin profile primarily marked by integrin shedding, which occurs in cardiac disease states [111, 112]. While β1 adhesion is decreased by over 50%, adhesion to collagen is still increased, and α1, α2, α5, and β3 integrin expression is unchanged. Cells remodel the existing matrix by releasing MMPs and increasing the deposition of fibrillar collagen by over 130%. Moreover, the CFs further mature to adopt a fully differentiated myofibroblast phenotype illustrated by an increase in intracellular expression of vimentin [101].

6 Conclusions and Critical Outstanding Questions

CF contributions to cardiac mechanobiology represent critical areas of research in cardiac tissue mechanics, physiology, and potential cardiac therapeutics. Fibroblasts are major drivers of cECM composition, structure, and function in various contexts, including cardiac development, cardiac homeostasis, injury repair, and inflammation. This makes them an attractive target for therapeutics aimed at modulating adverse changes in cardiac tissue that negatively impact mechanics, such as excessive fibrosis after an infarction event. However, key questions in both in CF physiology and in vitro CF research remain crucial to elucidate. We suggest some critical areas for consideration here: first, a deeper understanding of signaling pathways and biochemical and mechanical stimuli that drive fibroblast phenotype and activation to myofibroblasts is necessary and would likely provide a pathway to determining targets to modulate CF behavior therapeutically. Additionally, more research must be done in determining viable markers for fibroblasts and myofibroblasts, as the lack of a definitive marker severely limits the efficacy of CF research and therapeutic potential. Lastly, development of cell culture systems beyond conventional 2D growth of tissue culture plastic represents a vitally necessary improvement to CF research. CF sensitivity to environmental mechanics and their phenotypic plasticity hinder research into CF phenotype, mechanical and biochemical signaling, and generation of CF-targeting therapeutics.

Acknowledgments The authors would like to acknowledge funding sources for this work including the Congressionally Directed Medical Research Program at the Department of Defense (W81XWH-16-1-0304 to LDB) and the American Heart Association (20TPA35500082 to LDB and 18PRE33960362 to LRP).

Funding

This work was funded by the Congressionally Directed Medical Research Program at the Department of Defense (W81XWH-16-1-0304 to LDB) and the American Heart Association (20TPA35500082 to LDB and 18PRE33960362 to LRP).

Conflict of Interest The authors declare that they have no conflict of interest.

Ethical approval This article does not contain any studies with human participants or animals performed by any of the authors.

References

1. Furtado MB et al (2014) Cardiogenic genes expressed in cardiac fibroblasts contribute to heart development and repair. Circ Res 114(9):1422–1434
2. Balasubramanian S et al (2012) β3 integrin in cardiac fibroblast is critical for extracellular matrix accumulation during pressure overload hypertrophy in mouse. PLoS One 7(9):e45076
3. Dobaczewski M, Frangogiannis NG (2009) Chemokines and cardiac fibrosis. Front Biosci (Schol Ed) 1:391–405

4. Sullivan KE, Black LD (2013) The role of cardiac fibroblasts in extracellular matrix-mediated signaling during normal and pathological cardiac development. J Biomech Eng 135(7):71001
5. Díaz-Araya G et al (2015) Cardiac fibroblasts as sentinel cells in cardiac tissue: receptors, signaling pathways and cellular functions. Pharmacol Res 101:30–40
6. Choong OK et al (2019) Hypoxia-induced H19/YB-1 cascade modulates cardiac remodeling after infarction. Theranostics 9(22):6550–6567
7. Fan D et al (2012) Cardiac fibroblasts, fibrosis and extracellular matrix remodeling in heart disease. Fibrogenesis Tissue Repair 5(1):15
8. Gao AE, Sullivan KE, Black Iii LD (2016) Lysyl oxidase expression in cardiac fibroblasts is regulated by α2β1 integrin interactions with the cellular microenvironment. Biochem Biophys Res Commun 475(1):70–75
9. Kawaguchi M et al (2011) Inflammasome activation of cardiac fibroblasts is essential for myocardial ischemia/reperfusion injury. Circulation 123(6):594–604
10. Dostal D, Glaser S, Baudino TA (2015) Cardiac fibroblast physiology and pathology. Compr Physiol 5(2):887–909
11. Silva AC et al (2020) Bearing my heart: the role of extracellular matrix on cardiac development, homeostasis and injury response. Front Cell Dev Biol 8:1705
12. Porter KE, Turner NA (2009) Cardiac fibroblasts: at the heart of myocardial remodeling. Pharmacol Ther 123(2):255–278
13. Aránguiz-Urroz P et al (2011) Beta2-adrenergic receptor regulates cardiac fibroblast autophagy and collagen degradation. Biochim Biophys Acta (BBA)—Mol Basis Dis 1812(1):23–31
14. Biernacka A, Dobaczewski M, Frangogiannis NG (2011) TGF-β signaling in fibrosis. Growth Factors 29(5):196–202
15. Brown RD et al (2005) THE CARDIAC FIBROBLAST: therapeutic target in myocardial remodeling and failure. Annu Rev Pharmacol Toxicol 45(1):657–687
16. Cho N, Razipour SE, McCain ML (2018) Featured article: TGF-beta1 dominates extracellular matrix rigidity for inducing differentiation of human cardiac fibroblasts to myofibroblasts. Exp Biol Med (Maywood) 243(7):601–612
17. Jester JV et al (2002) TGFβ induced myofibroblast differentiation of rabbit keratocytes requires synergistic TGFβ, PDGF and integrin signaling. Exp Eye Res 75(6):645–657
18. Masur SK et al (1996) Myofibroblasts differentiate from fibroblasts when plated at low density. Proc Natl Acad Sci USA 93(9):4219–4223
19. Serini G et al (1998) The fibronectin domain ED-A is crucial for myofibroblastic phenotype induction by transforming growth factor-β1. J Cell Biol 142(3):873–881
20. Tomasek JJ et al (2002) Myofibroblasts and mechano-regulation of connective tissue remodelling. Nat Rev Mol Cell Biol 3(5):349–363
21. Hinz B (2009) Tissue stiffness, latent TGF-β1 activation, and mechanical signal transduction: implications for the pathogenesis and treatment of fibrosis. Curr Rheumatol Rep 11(2):120
22. Achterberg VF et al (2014) The Nano-scale mechanical properties of the extracellular matrix regulate dermal fibroblast function. J Investig Dermatol 134(7):1862–1872
23. Avery D et al (2018) Extracellular matrix directs phenotypic heterogeneity of activated fibroblasts. Matrix Biol 67:90–106
24. Chen WY et al (1989) Differences between adult and foetal fibroblasts in the regulation of hyaluronate synthesis: correlation with migratory activity. J Cell Sci 94(3):577–584
25. Costa-Almeida R, Soares R, Granja PL (2018) Fibroblasts as maestros orchestrating tissue regeneration. J Tissue Eng Regen Med 12(1):240–251
26. Ieda M et al (2009) Cardiac fibroblasts regulate myocardial proliferation through β1 integrin signaling. Dev Cell 16(2):233–244
27. Moore-Morris T et al (2015) Cardiac fibroblasts: from development to heart failure. J Mol Med 93(8):823–830
28. Moorman AF, Christoffels VM (2003) Cardiac chamber formation: development, genes, and evolution. Physiol Rev 83:1223

29. Moore-Morris T et al (2016) Origins of cardiac fibroblasts. J Mol Cell Cardiol 91:1–5
30. Gray GA et al (2018) Resident cells of the myocardium: more than spectators in cardiac injury, repair and regeneration. Curr Opin Physio 1:46–51
31. Tirziu D, Giordano FJ, Simons MJC (2010) Cell communications in the heart. Circulation 122(9):928–937
32. Ali SR et al (2014) Developmental heterogeneity of cardiac fibroblasts does not predict pathological proliferation and activation. Circ Res 115(7):625–635
33. Krenning G, Zeisberg EM, Kalluri R (2010) The origin of fibroblasts and mechanism of cardiac fibrosis. J Cell Physiol 225(3):631–637
34. Bowers SL, Banerjee I, Baudino TA (2010) The extracellular matrix: at the center of it all. J Mol Cell Cardiol 48(3):474–482
35. Williams C et al (2014) Young developmental age cardiac extracellular matrix promotes the expansion of neonatal cardiomyocytes in vitro. Acta Biomater 10(1):194–204
36. Wang Y et al (2020) Single-cell analysis of murine fibroblasts identifies neonatal to adult switching that regulates cardiomyocyte maturation. Nat Commun 11(1):2585–2585
37. Eitan Y et al (2010) Acellular Cardiac Extracellular Matrix as a Scaffold for Tissue Engineering: In Vitro Cell Support, Remodeling and Biocompatibility. Tissue Eng Part C Methods 16(4):671–683
38. Herum KM et al (2017) The soft- and hard-heartedness of cardiac fibroblasts: Mechanotransduction signaling pathways in fibrosis of the heart. J Clin Med 6(5):53
39. Li YA (2017) Huda; Bursac, Nenad, *age-dependent functional crosstalk between cardiac fibroblasts and cardiomyocytes in a 3D engineered tissue*. Acta Biomater 55:120–130
40. Cai T et al (2000) TGF-β1 modulated the expression of α5β1 integrin and integrin-mediated signaling in human Hepatocarcinoma cells. Biochem Biophys Res Commun 274(2):519–525
41. Chen D et al (2015) Fibronectin signals through integrin α5β1 to regulate cardiovascular development in a cell type-specific manner. Dev Biol 407(2):195–210
42. Leask A (2010) Potential therapeutic targets for cardiac fibrosis: TGFbeta, angiotensin, endothelin, CCN2, and PDGF, partners in fibroblast activation. Circ Res 106:1675
43. Turner NA (2016) Inflammatory and fibrotic responses of cardiac fibroblasts to myocardial damage associated molecular patterns (DAMPs). J Mol Cell Cardiol 94:189–200
44. Dobaczewski M, Chen W, Frangogiannis NG (2011) Transforming growth factor (TGF)-β signaling in cardiac remodeling. J Mol Cell Cardiol 51:600
45. Brown RD et al (2007) Cytokines regulate matrix metalloproteinases and migration in cardiac fibroblasts. Biochem Biophys Res Commun 362(1):200–205
46. Munger JS, Sheppard D (2011) Cross talk among TGF-β signaling pathways, integrins, and the extracellular matrix. Cold Spring Harb perspect biol 3(11):a005017–a005017
47. Silverio-Ruiz KG et al (2007) Opposite effects of bFGF and TGF-beta on collagen metabolism by human periodontal ligament fibroblasts. Cytokine 39:130
48. Hinz B, Lagares D (2020) Evasion of apoptosis by myofibroblasts: a hallmark of fibrotic diseases. Nat Rev Rheumatol 16(1):11–31
49. Hinz B et al (2001) Alpha-smooth muscle actin expression upregulates fibroblast contractile activity. Mol Biol Cell 12(9):2730–2741
50. Tallquist MD (2020) Cardiac fibroblast diversity. Annu Rev Physiol 82(1):63–78
51. Takeda N et al (2010) Cardiac fibroblasts are essential for the adaptive response of the murine heart to pressure overload. J Clin Invest 120(1):254–265
52. Tarbit E et al (2019) Biomarkers for the identification of cardiac fibroblast and myofibroblast cells. Heart Fail Rev 24(1):1–15
53. Farbehi N et al (2019) Single-cell expression profiling reveals dynamic flux of cardiac stromal, vascular and immune cells in health and injury. Elife 8:e43882
54. Camelliti P, Borg TK, Kohl P (2005) Structural and functional characterisation of cardiac fibroblasts. Cardiovasc Res 65(1):40–51
55. Walker CJ et al (2021) Nuclear mechanosensing drives chromatin remodelling in persistently activated fibroblasts. Nat Biomed Eng 5:1485

56. Lee S et al (2017) Contractile force generation by 3D hiPSC-derived cardiac tissues is enhanced by rapid establishment of cellular interconnection in matrix with muscle-mimicking stiffness. Biomaterials 131:111–120
57. Sung KE et al (2013) Understanding the impact of 2D and 3D fibroblast cultures on in vitro breast cancer models. PLoS One 8(10):e76373
58. Jonsson MKB et al (2016) A transcriptomic and Epigenomic comparison of fetal and adult human cardiac fibroblasts reveals novel key transcription factors in adult cardiac fibroblasts. JACC: Basic Transl Sci 1(7):590–602
59. Zhu F et al (2013) Senescent cardiac fibroblast is critical for cardiac fibrosis after myocardial infarction. PLoS One 8(9):e74535–e74535
60. Biernacka A, Frangogiannis NG (2011) Aging and cardiac fibrosis. Aging Dis 2(2):158
61. Horn MA et al (2012) Age-related divergent remodeling of the cardiac extracellular matrix in heart failure: collagen accumulation in the young and loss in the aged. J Mol Cell Cardiol 53(1):82–90
62. Lindsey ML et al (2005) Age-dependent changes in myocardial matrix metalloproteinase/tissue inhibitor of metalloproteinase profiles and fibroblast function. Cardiovasc Res 66(2): 410–419
63. Sheydina A, Riordon DR, Boheler KR (2011) Molecular mechanisms of cardiomyocyte aging. Clin Sci 121(8):315–329
64. Liau B et al (2017) Developmental stage-dependent effects of cardiac fibroblasts on function of stem cell-derived engineered cardiac tissues. Sci Rep 7(1):42290
65. Perreault LR et al (2021) RNA sequencing indicates age-dependent shifts in the cardiac fibroblast transcriptome between fetal, neonatal, and adult developmental ages. Physiol Genomics 53(10):414–429
66. Nian M et al (2004) Inflammatory cytokines and Postmyocardial infarction remodeling. Circ Res 94(12):1543–1553
67. Mays PK et al (1991) Age-related changes in collagen synthesis and degradation in rat tissues. Importance of degradation of newly synthesized collagen in regulating collagen production. Biochem J 276(2):307–313
68. Benjamin EJ et al (2018) Heart Disease and Stroke Statistics–2014;2018 Update: A Report From the American Heart Association. Circulation 137(12):e67–e492
69. Hinderer S, Schenke-Layland K (2019) Cardiac fibrosis—a short review of causes and therapeutic strategies. Adv Drug Deliv Rev 146:77–82
70. Lindsey ML, Mann DL, Entman ML, Spinale FG (2003) Extracellular matrix remodeling following myocardial injury. Ann Med 35(5):316–326
71. O'Neal W et al (2012) Cellular pathways of death and survival in acute myocardial infarction. J Clin Exp Cardiol 6:003
72. Kazbanov IV, ten Tusscher KHWJ, Panfilov AV (2016) Effects of heterogeneous diffuse fibrosis on arrhythmia dynamics and mechanism. Sci Rep 6(1):20835
73. Frangogiannis NG (2019) The extracellular matrix in ischemic and nonischemic heart failure. Circ Res 125(1):117–146
74. Frangogiannis NG, Rosenzweig A (2012) Regulation of the inflammatory response in cardiac repair. Circ Res 110(1):159–173
75. Li W et al (2009) Early manifestation of alteration in cardiac function in dystrophin deficient mdx mouse using 3D CMR tagging. J Cardiovasc Magn Reson 11:40
76. Tschöpe C et al (2004) Prevention of cardiac fibrosis and left ventricular dysfunction in diabetic cardiomyopathy in rats by transgenic expression of the human tissue kallikrein gene. FASEB J 18:828–835
77. Burlew BS, Weber KT (2002) Cardiac fibrosis as a cause of diastolic dysfunction. Herz 27(2): 92–98
78. Espinola-Zavaleta N et al (2006) Non-compacted cardiomyopathy: clinical-echocardiographic study. Cardiovasc Ultrasound 4:35

79. Ganame J, Ayres NA, Pignatelli RH (2012) Left ventricular noncompaction, a recently recognized form of cardiomyopathy. Circ J 1:119–124
80. Yang J et al (2012) Inhibition of Notch2 by numb/Numblike controls myocardial compaction in the heart. Cardiovasc Res 96:276–285
81. Snider P et al (2009) Origin of cardiac fibroblasts and the role of periostin. Circ Res 105:934–947
82. Guarino N, Shima H, Puri P (2000) The hypoplastic heart in congenital diaphragmatic hernia: reduced expression of basic fibroblast growth factor and platelet-derived growth factor. Pediatr Surg Int 16(4):243–246
83. Sharma HS et al (2006) DNA microarray analysis for human congenital heart disease. Cell Biochem Biophys 44:1–9
84. Chaturvedi RR et al (2010) Passive stiffness of myocardium from congenital heart disease and implications for diastole. Circulation 121(8):979–988
85. Stewart JA et al (2010) Temporal alterations in cardiac fibroblast function following induction of pressure overload. Cell Tissue Res 340:117–126
86. Baicu CF et al (2012) Time course of right ventricular pressure-overload induced myocardial fibrosis: relationship to changes in fibroblast postsynthetic procollagen processing. Am J Physiol Heart Circ Physiol 303(9):H1128–H1134
87. Burgess ML et al (2002) Differential integrin expression by cardiac fibroblasts from hypertensive and exercise-trained rat hearts. Cardiovasc Pathol 11(2):78–87
88. Manabe I (2002) Gene expression in fibroblasts and fibrosis: involvement in cardiac hypertrophy. Circ Res 91:1103–1113
89. Siwik DA, Pagano PJ, Colucci WS (2001) Oxidative stress regulates collagen synthesis and matrix metalloproteinase activity in cardiac fibroblasts. Am J Physiol Cell Physiol 280:C53–C60
90. Xie L et al (2010) Cystatin C increases in cardiac injury: a role in extracellular matrix protein modulation. Cardiovasc Res 87(4):628–635
91. Frangogiannis NG (2019) Cardiac fibrosis: cell biological mechanisms, molecular pathways and therapeutic opportunities. Mol Asp Med 65:70–99
92. Rienks M et al (2014) Myocardial extracellular matrix. Circ Res 114(5):872–888
93. Frangogiannis NG (2014) The inflammatory response in myocardial injury, repair, and remodelling. Nat Rev Cardiol 11(5):255–265
94. Frangogiannis NG (2008) The immune system and cardiac repair. Pharmacol Res 58(2):88–111
95. Richardson WJ et al (2015) Physiological implications of myocardial scar structure. Compr Physiol 5(4):1877–1909
96. Pakshir P, Hinz B (2018) The big five in fibrosis: macrophages, myofibroblasts, matrix, mechanics, and miscommunication. Matrix Biol 68-69:81–93
97. Shinde AV, Frangogiannis NG (2014) Fibroblasts in myocardial infarction: a role in inflammation and repair. J Mol Cell Cardiol 70:74–82
98. Travers JG et al (2016) Cardiac Fibrosis: The Fibroblast Awakens. Circ Res 118(6):1021–1040
99. Prabhu SD, Frangogiannis NG (2016) The biological basis for cardiac repair after myocardial infarction: from inflammation to fibrosis. Circ Res 119(1):91–112
100. Sullivan KE et al (2014) Extracellular matrix remodeling following myocardial infarction influences the therapeutic potential of mesenchymal stem cells. Stem Cell Res Ther 5(1):14
101. Flack EC et al (2006) Alterations in cultured myocardial fibroblast function following the development of left ventricular failure. J Mol Cell Cardiol 40:474–483
102. Jacot JG, Martin JC, Hunt DL (2010) Mechanobiology of cardiomyocyte development. J Biomech 43(1):93–98
103. Fomovsky GM, Holmes JW (2010) Evolution of scar structure, mechanics, and ventricular function after myocardial infarction in the rat. Am J Physiol Heart Circ Physiol 298(1):H221–H228

104. Fomovsky GM, Rouillard AD, Holmes JW (2012) Regional mechanics determine collagen fiber structure in healing myocardial infarcts. J Mol Cell Cardiol 52(5):1083–1090
105. Quinn KP et al (2016) Optical metrics of the extracellular matrix predict compositional and mechanical changes after myocardial infarction. Sci Rep 6:35823
106. Loftis MJ, Sexton D, Carver W (2003) Effects of collagen density on cardiac fibroblast behavior and gene expression. J Cell Physiol 196(3):504–511
107. Mitchell MD et al (2007) IL-1beta stimulates rat cardiac fibroblast migration via MAP kinase pathways. Am J Physiol Heart Circ Physiol 292:H1139–H1147
108. Sutton MGSJ, Sharpe N (2000) Left ventricular remodeling after myocardial infarction : pathophysiology and therapy. Circulation 101:2981–2988
109. Holmes JW, Borg TK, Covell JW (2005) Structure and mechanics of healing myocardial infarcts. Annu Rev Biomed Eng 7:223–253
110. Rubin SA, Fishbein MC, Swan HJ (1983) Compensatory hypertrophy in the heart after myocardial infarction in the rat. J Am Coll Cardiol 1:1435–1441
111. Goldsmith EC et al (2003) Integrin shedding as a mechanism of cellular adaptation during cardiac growth. Am J Physiol Heart Circ Physiol 284:H2227–H2234
112. Manso AM et al (2006) Integrins, membrane-type matrix metalloproteinases and ADAMs: potential implications for cardiac remodeling. Cardiovasc Res 69:574–584

Mechanobiology of Cardiac Remodelling in Cardiomyopathy

Thomas Randall and Elisabeth Ehler

Abbreviations

AAV9	Adeno-associated virus
BAG 3	Bcl2-associated athanogene 3
CARP1	Cardiac ankyrin repeat protein 1
CASA	Chaperone-assisted selective autophagy
DCM	Dilated cardiomyopathy
EC	Excitation-contraction
EH	Embryonic heart
FATZ	Filamin-, actinin- and telethonin-binding protein of the Z-disc
FHL1	Four and a half LIM domain protein 1
FHOD1	Formin-homology domain-containing protein 1
FRAP	Fluorescence recovery after photobleaching
HCM	Hypertrophic cardiomyopathy
iPSC-CM	Cardiomyocytes derived from induced pluripotent stem cells
LAD	Lamin-associated domains
LINC	Linker of nucleoskeleton and cytoskeleton
MAP 4	Microtubule-associated protein 4
MARK4	Microtubule-affinity regulating kinase
MARP	Muscle ankyrin repeat protein
	MLC Myosin light chain
MLCK	Myosin light chain kinase
MURF2	Muscle specific RING finger-2

T. Randall · E. Ehler (✉)
Randall Centre for Cell and Molecular Biophysics, School of Basic and Medical Biosciences, King's College London, London, UK

British Heart Foundation Centre of Research Excellence, School of Cardiovascular and Metabolic Medicine and Sciences, King's College London, London, UK
e-mail: elisabeth.ehler@kcl.ac.uk

M. Hecker, D. J. Duncker (eds.), *Cardiac Mechanobiology in Physiology and Disease*, Cardiac and Vascular Biology 9,
https://doi.org/10.1007/978-3-031-23965-6_6

PKC	Protein kinase C
PKP2	Plakophilin 2
ROCK	Rho-associated kinase
ROS	Reactive oxygen species
SRF	Serum response factor
TAZ	Transcriptional co-activator with PDZ-binding motif
VASH2	Tubulinyl-Tyr-carboxypeptidase 2
YAP	Yes-associated protein

The heart is the first fully developed organ during mammalian development, and its function is essential for supplying the entire body with oxygen and nutrients. To achieve this, electrical signals delivered via the conduction system are interpreted by co-ordinated contraction of the four different chambers of the heart. The heart is able to do this job with such efficiency thanks to its sophisticated design, which is only achieved fully after birth and which guarantees maximal force output with minimal electrical misfiring. However, this intricate organisation comes at a price, which is the loss of any significant regenerative potential in the mature heart. Thus, challenges that are imposed on the heart either by endogenous factors such as cardiomyopathy-causing mutations or by exogenous factors such as hypertension or loss of contractile tissue caused by a myocardial infarction cannot be simply met by a readily available extensive stem cell population, but are handled at the level of the cardiomyocyte. In addition to humoral signals, these challenges for the cardiomyocytes are often of a mechanical nature, where altered overall dynamics of the contractile tissue or changes in its consistency (e.g. increased fibrosis) are directly sensed by the cells. In this review we will discuss the cellular responses in the heart to mechanical stress at the level of the cardiomyocyte and highlight which multiprotein complexes and signalling pathways could be interesting drug targets to halt maladaptive responses.

The bulk of the ventricular tissue is made up by rod-shaped cardiomyocytes (Fig. 1), which are connected to each other at their bipolar ends, by specialised types of cell-cell contacts called the intercalated discs (for review see [1]). Laterally the strands of cells are ensheathed in extracellular matrix and make up a kind of contractile cable, where electricity only flows in a longitudinal direction, but is insulated from the cable next to it. This arrangement is necessary to prevent arrhythmias. One of the challenges of attempts to regenerate heart tissue following myocardial infarction is to re-establish this kind of organisation, and while several strategies were published that can trigger proliferation in cardiomyocytes or repopulate heart tissue with injected cardiomyocytes, arrhythmias are often the fatal consequence [2]. The cardiomyocytes interact with the insulating material directly, via costameres, which are vinculin-containing cell-matrix attachment sites at the level of the Z-disc of the myofibril [3]. The electrical signals at the plasma membrane are fed into the interior of the cardiomyocytes by regular invaginations also at Z-disc level, the T-tubules, which together with the sarcoplasmic reticulum (not shown in Fig. 1) make up the dyads that govern EC (excitation-contraction)

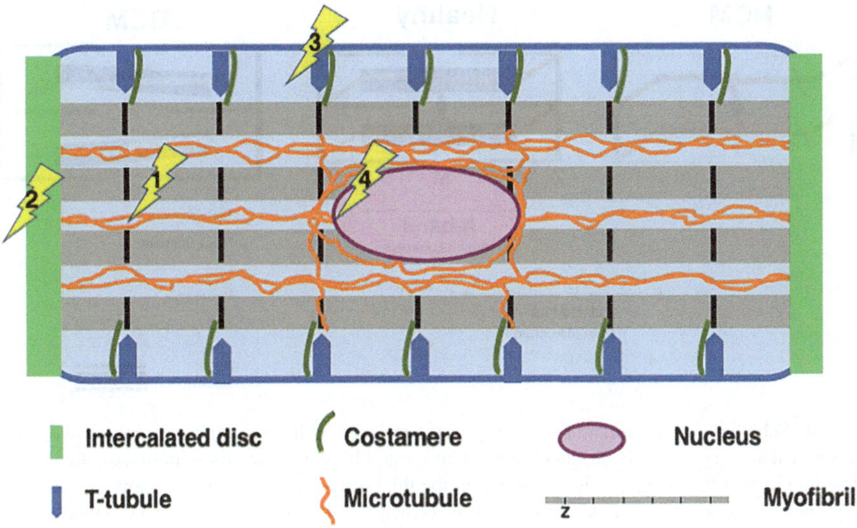

Fig. 1 Schematic representation of an isolated rod-shaped mature cardiomyocyte. Cell-cell contacts (intercalated discs) are restricted to the bipolar ends and shown in green. The myofibrils (shown in grey, with only the Z-discs of the sarcomeres indicated with black lines) are aligned in a paracrystalline fashion along the longitudinal axis of the cell. At the sides the cells are ensheathed in extracellular matrix (not shown), and lateral contacts between the myofibrils and the ECM are mediated by the costameres (dark green) at the level of the Z-disc. T-tubules are regularly spaced plasma membrane invaginations that together with the sarcomplasmic reticulum (not shown) make up the dyads required for EC coupling. Microtubules (orange) stretch mainly longitudinally between the myofibrils but are also concentrated around the nucleus and over the Z-discs. Intermediated filaments and mitochondria were omitted from the scheme for reasons of simplification. Yellow thunderbolts indicate potential focal points of mechanosignalling in the cells, i.e. at the intercalated disc, within the sarcomere (mainly mediated via titin), at the T-tubules (mediated via Piezo1) and at the nuclear lamina (mediated via the LINC and changes in nuclear pore size upon mechanical stress) that are expanded on later in the article

coupling [4]. This complex plasma membrane arrangement is only achieved after birth and characterises mature mammalian cardiomyocytes following physiological hypertrophy. Adult zebrafish cardiomyocytes that display a long and slim shape do not require this kind of organisation [5] and retain their proliferative and regenerative potential following myocardial injury [6]. The majority of the cytoplasm of the cardiomyocytes is made up by the myofibrils, which are integrated with each other by the intermediate filament protein desmin at the level of the Z-discs. The basic contractile unit of a myofibril is called a sarcomere and defined as the region between two Z-discs. Contraction is brought about by the tightly regulated interaction between actin (in the thin filaments) and myosin (in the thick filaments), and the paracrystalline assembly of these proteins to sarcomeres is due to longitudinal and transversal linking proteins. In the longitudinal direction, the link is provided by titin, with individual titin molecules stretching from the N-terminus in the Z-disc to their C-terminus in the middle of the sarcomere in a structure called the M-band

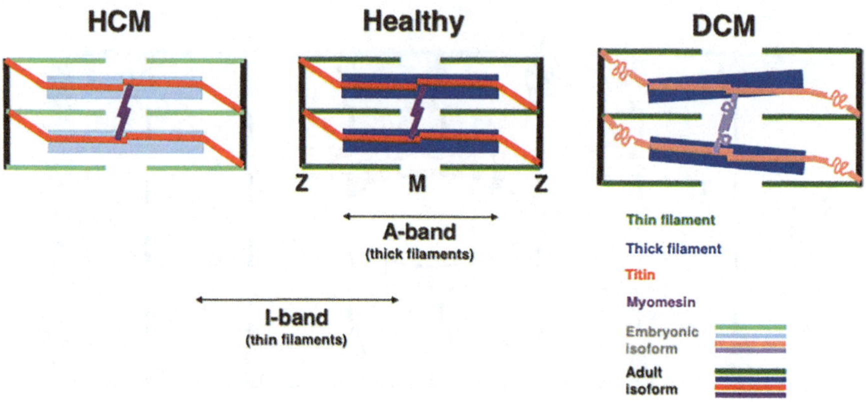

Fig. 2 Schematic representation of changes at the level of the sarcomere in cardiomyopathy. Simplified sketches of a single sarcomere, as delineated by the two Z-discs (in black), are shown. The thin (actin) filaments are shown in green, the thick (myosin) filaments are shown in blue, the titin (elastic) filaments are shown in red and the principal component of the M-band, myomesin, is shown in purple. In HCM (hypertrophic cardiomyopathy), a switch of the isoforms of contractile proteins to a more embryonic variant (shown in a lighter shade) is observed, while in DCM (dilated cardiomyopathy) this re-expression of embryonic isoforms is more prevalent for proteins that make up the sarcomeroskeleton such as titin and myomesin

(Fig. 2). There, titin is integrated with the myosin filaments via dimers made up by the protein myomesin, while in the Z-disc one of the main crosslinking proteins is the sarcomeric isoform of alpha-actinin [1]. Mitochondria are present in regular rows in-between the myofibrils (not shown in Fig. 1). In addition, microtubules stretch predominantly along the longitudinal axis of the cells (shown in orange in Fig. 1), but also show a concentration round the nucleus and horizontal extensions in the Z-disc regions [7]. Human cardiomyocytes often have one nucleus, while rodent cardiomyocytes are usually binucleated, but nuclei number varies dependent on species [8].

1 What Happens at the Cellular Level in Hypertrophic Cardiomyopathy (HCM) Versus Dilated Cardiomyopathy (DCM)?

The cytoarchitecture of the healthy cardiomyocyte is perfectly adapted to get the job done with maximum efficiency and minimal electrical noise. However, a variety of reasons can pose a challenge onto the heart and lead to heart disease. For reasons of simplicity, we will mainly consider hypertrophic (HCM) and dilated cardiomyopathy (DCM) in this review while being aware that these designations are by no means as strict as textbook wisdom would suggest. Usually, HCM is characterised by a thickening of the ventricular myocardium and a reduction in ventricle diameter. It can be caused by hypertension or by mutations in cardiac proteins, with the majority of cases being due to mutations in the MYH7 or the MYBPC3 gene, encoding for beta-myosin heavy chain or cardiac Myosin Binding protein-C, respectively [9]. The cardiomyocytes become bulkier and also often show myofibrillar disarray, which is a misalignment of myofibrils. This disarray can translate to the histological level as myocyte disarray and is often taken as bona fide proof of a HCM phenotype [10]. Currently the origin of the disarray is not very well understood, but a very elegant explanation proposes that varying levels of expression of mutant proteins between individual cardiomyocytes (usually adult cardiomyopathies are heterozygous) lead to force imbalances that eventually result in this organisation [11]. In addition to the disarray, HCM cardiomyocytes also lose the regular T-tubule invaginations [12] and the restriction of intercalated discs to the bipolar ends of the cell. Adherens junction proteins as well as gap junction proteins such as connexin-43, which make up intercellular ion channels, are found in aberrant localisations on the lateral surface of the cardiomyocytes, and this probably contributes together with T-tubule loss to the arrhythmic events, which are the main killers in HCM [13–15]. Both the microtubule network and the extent of the intermediate filaments made up of desmin tend to be increased in hypertrophic cardiomyopathy (reviewed in [16]), and in particular an increase in detyrosinated microtubules that bind more strongly to desmin was observed recently [7]. This stiffer cytoskeletal network impairs contractility, and any approach to break it up by depolymerising microtubules or interfering with detyrosination improves the contractile kinetics of failing cardiomyocytes [17].

DCM is described as a ballooning heart, with an increase in ventricle chamber diameter and a thinning of the ventricular walls. It is often accompanied by a pronounced elongation of the individual cardiomyocytes but can also be due to a general loss of size control, both in mouse models for DCM and in human patient samples [18, 19]. At the histological level, DCM can appear relatively unchanged compared to healthy heart [10], and thorough subcellular analysis was required to identify the intercalated disc as the subcellular area of major alterations, with an increased presence of actin-anchoring proteins and reduced expression of connexin-43 [20]. These initial observations from several mouse models for DCM were later confirmed also in human patient samples [19]. Investigations from other researchers suggest that a co-ordinated stoichiometry of intercalated disc proteins is crucial and that decreased or increased levels of one component often result in a DCM phenotype [21–25]. The other region of the cell that is dramatically affected in DCM is the costameres, which, instead of displaying a regular striated appearance on the surface of the cardiomyocytes, are arranged in a punctate fashion in isolated cardiomyocytes [20]. Within the last decade, it was shown that between 10 and 30% of mutations that lead to DCM are found in the titin gene, with a majority leading to a truncating variant [26]. As described above, titin molecules are central for sarcomere assembly and in addition to this structural role were also demonstrated to act as a signalling hub both in the I-band and the M-band, potentially also being responsive to mechanical signals. This was directly demonstrated for the titin kinase domain, where opening up via stretch leads to the exposure of a binding site for autophagosomal receptors such as Nbr1 and p62 (also called SQSTM1), part of a multiprotein signalling complex that also involves MURF2 (muscle-specific RING finger-2). Mechanical inactivity leads to the dissociation of this complex from the myofibril, its transfer to the nucleus, and the downregulation of SRF (serum response factor)-mediated gene expression [27]. Within the I-band, there are two multiprotein signalling complexes that were shown to be affected by mechanical stress, FHL1 and MARP mediated, although the direct mechanisms are less well understood at the moment (reviewed in [28]). Thus, reduced expression levels of titin due to truncating mutations are expected to impair these mechanosignalling pathways. For several years truncated titin molecules remained elusive, and their existence was only very recently demonstrated in the hearts of human patients with DCM mutations. Pioneering work demonstrated truncated titin in patient material as well as in cardiomyocytes derived from induced pluripotent stem cells (iPSC-CM), but the molecules fail to incorporate into the sarcomeres and instead associate with the cellular degradation machinery [29]. Previously it had been shown that iPSC-CM with titin-truncating mutations had less efficient sarcomere assembly and an impaired response to mechanical and beta-adrenergic stress [30].

2 Returning to Old Concepts: HCM Is a Disease of Force Production and DCM Is a Disease of Force Transmission?

More than 20 years ago, it was suggested in a review that HCM is solely due to hereditary mutations in proteins that make up the myofibrils, while DCM is caused by hereditary mutations in cytoskeletal proteins [31]. The advance of next-generation sequencing and the ensuing wealth of mutations that were attributed to either HCM or DCM have revealed that the case is more complex and not as clear-cut as initially assumed. However, looking at the bigger picture of cellular responses in cardiomyocytes to HCM and DCM and especially at alterations in sarcomere composition indicates that a general separation into HCM as a disease of force production and DCM as a disease of force transduction may still be a valid starting point. The sarcomere is the basic unit of a myofibril and defined as the region between two neighbouring Z-discs. It serves for the paracrystalline arrangement of the contractile proteins, actin and myosin, into thin and thick filaments (Fig. 2). To aid this integration, additional cytoskeletal proteins are required, which are titin, which stretches as elastic filaments from the Z-disc along the sarcomere to its middle, the M-band, where it links the thick filaments with another cytoskeletal protein called myomesin, which provides the main crosslink in a transversal direction. Other M-band proteins help to fine-tune this machinery [32]. As mentioned above, HCM is characterised by myofibril disarray, but ultrastructurally the individual sarcomeres are not much altered, at least in the electron microscope. This picture changes, when the isoform composition of the myofibrils is analysed, since the upregulation of the expression of embryonic isoforms of contractile proteins (e.g. beta- instead of alpha-myosin heavy chain in mouse myocardium and upregulation of alpha-smooth and alpha-skeletal actin instead of alpha-cardiac actin) is taken as a marker for an HCM phenotype [33] (see left-hand side of Fig. 2). In contrast, in DCM, the myofibrils are affected much more subtly, with little evidence for disarray in the light microscope but a loss of distinctive M-bands in the electron microscope [20, 34]. This is due to the upregulation of a more elastic version of myomesin, EH (embryonic heart)-myomesin, which is usually characteristic for the embryonic heart and for slow twitch skeletal muscles [35, 36], both types of muscle that display a slightly less paracrystalline arrangement and tolerate misalignments of the contractile filaments better [37]. The upregulation of the transverse connector, myomesin, to a more elastic isoform is also accompanied by an upregulation of expression of a more elastic variant of the longitudinal connector, titin, that is more similar to its foetal isoform [38] (right-hand side of Fig. 2). This suggests that at the level of the sarcomere, whichever the molecular mutation that causes the disease phenotype is, the responses to HCM and DCM are distinct with an attempt to improve contractile function by upregulating more "tolerant" versions of actin and myosin in HCM, while in DCM the focus is not so much on contractile output but on providing more elasticity both longitudinally and transversally by a more compliant sarcomeroskeleton. However, it has to be taken into account that these isoform changes do not necessarily translate into a more elastic ventricular wall, because especially in the case of titin, they are counteracted by posttranslational modifications [39]. In addition, interstitial fibrosis and the stiffer intercalated discs will contribute to an overall phenotype that is less flexible [19].

3 The Intercalated Disc: The Epicentre for Postnatal Cardiomyocyte Growth and Maladaptive Mechanosignalling?

The intercalated disc is the specialised type of cell-cell contact between cardiomyocytes that is responsible for anchoring the ends of the myofibrils in the adherens junctions, mechanical stabilisation via the desmosomes and intercellular communication via the gap junctions (reviewed in [40] and more recently in [41]). Apart from its direct involvement in cardiomyopathies such as arrhythmogenic cardiomyopathy [42], which is usually regarded as a disease of the desmosome and in DCM (see above), within the last decade, this subcellular region has attracted the attention of researchers as the most likely site in the cardiomyocyte for the insertion of new sarcomeres during postnatal cell growth. The myofibrils are anchored into adherens junctions at the plasma membrane of the intercalated disc via the barbed end of the actin filaments [43], like thin filaments which are anchored via their barbed ends into the Z-discs of the sarcomere. This means that the last sarcomere is actually a hybrid with a proper Z-disc at the end facing inwards but a kind of Z-disc "light" with a continued actin filament running through it at the end facing the plasma membrane. This Z-disc "light" has been termed the transitional junction by Bennett and colleagues [44] and contains classical Z-disc proteins such as sarcomeric alpha-actinin but lacks FATZ (filamin, actinin and telethonin binding protein of the Z-disc), CapZ and telethonin [45]. Telethonin is seen as the final glue that sticks titin N-termini to each other and makes the complex one of the most stable complexes that has been analysed [46, 47]. Therefore, insertion of additional sarcomeres into a Z-disc within a myofibril is probably an unfavourable event, and the transitional junction looks like a much better site to do this. Indeed, it was shown by Yoshida and coworkers that rabbit hearts respond to volume overload by repeated cycles of broadening and narrowing of the signal for the intercalated disc protein N-cadherin and one sarcomere can be inserted per day in this way [48]. Sarcomere lengths correlate exactly with membrane convolution in the healthy heart, and this correlation is lost in DCM [34]. Very recently another piece was added to the puzzle, when it was found that microtubules deliver mRNA encoding for sarcomeric proteins (alpha-cardiac actin, sarcomeric myosin) as well as ribosomes to peripheral regions of cardiomyocytes where sarcomere addition occurs [49]. It had been known for a while that depolymerisation of the microtubule network in neonatal rat cardiomyocytes resulted in smaller cells [50], and the recent data provide a mechanistic explanation for this observation. Taken together this suggests that the intercalated discs present the main site in the cardiomyocyte where new sarcomeres can be added following a mechanical challenge and that the microtubule transport system is necessary to provide the templates (i.e. the mRNA) as well as the production machinery (i.e. the ribosomes).

In cardiomyocytes in situ, microtubules also control the extremely regular spacing of ribosomes, flanking either side of the Z-disc [49]. This is probably needed for efficient protein turnover in the sarcomeres. The half-life of most proteins that associate with the sarcomere (troponin complex, tropomyosin, myosin light chain) was shown to be between 3 and 10 days [51], although more recent studies indicate a longer half-life for proteins such as titin and myosin heavy chain that are tightly integrated to make up the thick filaments [52]. When protein dynamics was investigated under experimental conditions in cultured cardiomyocytes, FRAP (fluorescence recovery after photobleaching) experiments have underlined the possibility for relatively rapid exchange of Z-disc components [53], and even the giant protein titin displays surprising rates of exchange within a couple of hours [54]. It is mind-boggling, how the mRNA that encodes for a 3000 kDa protein would be transported through a densely packed muscle cell, so ribosome location at the Z-disc, where the titin N-terminus is anchored and a reservoir of titin mRNAs there is a more efficient way to deal with replacement issues. These may be caused by potential damage due to the forces that are experienced during muscle contraction and necessitate a replacment of titin molecules. Maintaining proteostasis in cells that are subjected to high levels of mechanical stress such as striated muscle cells is a complex process that involves in particular the CASA (chaperone-assisted selective autophagy) machinery [55]. BAG3, which is at the centre of CASA, is downregulated in heart failure and is required for the turnover of proteins that are central to sarcomere assembly such as alpha-actinin, myomesin, desmin and MyBP-C [56]. Increasing

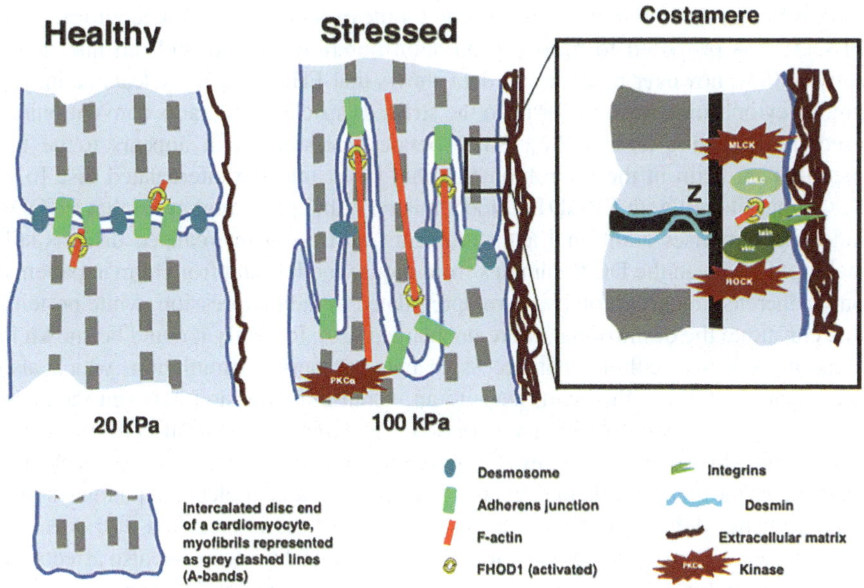

Fig. 3 Schematic representation of changes at the level of the cell-cell contacts (intercalated discs) and the cell-matrix contacts (costameres) in healthy versus stressed cells. Mechanical stress either caused by endogenous factors (i.e. cardiomyopathy mutations, activation of pathological signalling pathways) or by extracellular factors (e.g. stiffening of the environment from 20 kPa to 100 kPa) leads to alterations in cytoarchitecture with increased plasma membrane convolution at contact sites and more pronounced actin fibres and actin attachment sites. At the costameres, mechanical stress leads to the activation of integrin signalling, followed by downstream activation of talin, vinculin (vinc) and FHOD1. Activation of kinases such as PKCalpha, MLCK and ROCK mediates between stretch sensing and downstream cytoskeletal and transcriptional responses

BAG3 levels using the AAV9 (adeno-associated virus) system is sufficient to rescue CASA activity in failing hearts and has also beneficial effects on contractile function [56]. A failure to cope appropriately with the disposal of malfunctioning missense mutant proteins is often the biochemical basis of hereditary cardiomyopathy [57–59], so controlled protein synthesis and turnover are not only crucial for the maintenance of the healthy heart but even more so in the mechanically challenged heart.

The two opposing plasma membranes at the intercalated disc get more convoluted as animals age [40], most likely due to long-term coping with mechanical stress at this site (Fig. 3). Electron microscopy from DCM samples also showed that the increase in adherens junction proteins that is observed by confocal microscopy is not due to a thicker protein coat but due to increased plasma membrane convolution suggesting that the stresses experienced by the DCM intercalated disc accelerate its ageing process [20, 34]. The expanded coat of adherens junction proteins that lines these convoluted membranes is accompanied by an increased signal for F-actin at the intercalated disc [20]. The additional increased expression of a formin, FHOD1 (formin-homology domain-containing protein 1), at the intercalated discs in DCM

[60], indicates that it might be responsible for the excessive actin filament formation. FHOD1 was proposed to be solely an actin bundling protein with no nucleation activity [61]; however more recent data shows that FHOD1 can nucleate actin, but prefers cytoplasmic actin isoforms to the striated muscle actin that is conventionally used in nucleation assays [62]. Since beta-cytoplasmic actin appears to be the predominant actin in the microfilaments that insert into the intercalated disc [63], this would fit well with FHOD1's biochemistry, although formal proof that FHOD1 indeed polymerises additional actin filaments at stressed intercalated discs is still lacking (Fig. 3). In the DCM samples from mouse models and from human patients, only adherens junction proteins were upregulated in their expression, while proteins that constitute the desmosomes were unchanged [20]. Recently it could be shown in cardiomyocytes in culture that excessive beta-adrenergic stimulation, which also accompanies stress in the heart, leads to an increase in adhesion between the cells, which was mainly attributed to an increased extent and organisation of desmosomes [64]. Again, this supplies more evidence for the intercalated disc as a stress-sensitive element in the heart, with the precise time courses and molecular pathways still needing better definition. This stress-sensing role of the intercalated disc does not restrict itself to a changed composition of its cytoskeletal proteins but also affects the presence of notorious signalling molecules in cardiomyopathy. For example, it was shown that PKCalpha (protein kinase C), which is one of the many kinases to show increased activity in cardiomyopathy [65], is concentrated at the intercalated discs in failing hearts by a multiprotein complex that involves the scaffold protein CARP1 (cardiac ankyrin repeat protein 1) and that breaking up of this complex leads to decreased PKCalpha and the rescue of the functional phenotype in a mouse model for DCM [66]. PKCalpha needs lipids for its activation [67], and its proximity to the plasma membrane is likely to result in a constitutively active signalling molecule (Fig. 3). Currently, the exact nature of PKCalpha's downstream targets at the intercalated discs is unclear, although several candidates, such as PKP2 (Plakophilin 2), were suggested [68]. A recent review discusses the molecular basis for mechanosignalling of different intercalated disc components in more detail [69]. Increased mechanical stress at cell-cell contacts is not really a satisfactory explanation for the proposed mechanosensing of cardiomyocytes of their environment, which is expected to increase from a stiffness of 20 kPa in the adult heart tissue to beyond 100 kPa in fibrosis ([70]; Fig. 3). Because of the complex arrangement, the mechanoresponse of costameres, which link the cardiomyocytes laterally to the extracellular matrix, has not been studied in great molecular detail. A very nice substitute in this respect is to assume that the focal adhesions that cultured cardiomyocytes use to attach to the matrix they are grown on share similar characteristics to costameres that extend beyond their molecular composition, which includes proteins such as talin and vinculin. A recent study has linked the response of cardiomyocytes to increased matrix rigidity to talin activation and increased presence of phosphorylated MLC (myosin light chain), which is usually taken as a readout for non-muscle myosin assembly and activation (Fig. 3). Again, this is accompanied by increased filamentous actin and the activation of FHOD1, suggesting that similar mechanisms might be at play in the costameres as at the

intercalated disc, at least as far as actin filaments are concerned [71]. For a while it looked like the cardiac isoform of vinculin, metavinculin, might be at the centre of DCM signalling because missense mutations were detected in DCM patients [72, 73]; however recent data from metavinculin knockout mice failed to detect any dramatic phenotype in the heart [74].

4 Mechanical Stress and Channel Regulation

The sequential activity of ion channels defines the cardiac action potential, and impairments are known to accompany cardiomyopathies (for review see [75]). Several of these ion channels are known to be mechano-gated (volume-activated or stretch-activated) or mechano-modulated (voltage-gated potassium, calcium and sodium channels); however, their exact mechanism of activation in the heart is poorly understood (reviewed in [76]). The significant lag time of activation suggests the involvement of the cytoskeleton, e.g. actin filaments; however in some cases (e.g. TEK), actin is inhibitory, while in others (Piezo1) actin is thought to be required for activation [77, 78].

Most of these channels are not randomly distributed over the plasma membrane, but are preferentially associated with the T-tubules, leading to a striated appearance in immunostainings [76]. This kind of location was also shown for Piezo1, which is a mechanosensitive channel that was recently demonstrated to convert mechanical stretch into calcium and ROS (reactive oxygen species) signalling in cardiomyocytes with a lack of response in cardiac-specific Piezo1 knockout mice [79]. Upregulation of Piezo expression was detected in human hypertrophic cardiomyopathy and in rat hearts following myocardial infarction [79, 80]. Currently it is thought that the Piezo1 response may initially be beneficial and that it is a dysregulated activation that eventually leads to decompensation or accompanies cardiomyopathies. However, overexpression studies in cell lines have shown extremely rapid and complete inactivation of Piezo1 channels, so clearly more work is required to understand the exact mechanism of its regulation (reviewed in [81]).

5 How Are These Stress Signals Conveyed into Changes in Gene Expression?

While a lot of stress signals were characterised at cell-cell or cell-matrix contacts as well as emanating from within the sarcomere [82–84], how these signals would affect gene expression is not very well understood so far. It has been shown that the mechanosensor in titin's I-band region, CARP1, can translocate to the nucleus, where it may act as a negative regulator of cardiac gene expression [85]. Lack of mechanical inactivity affects the signalosome at titin's kinase domain and leads to

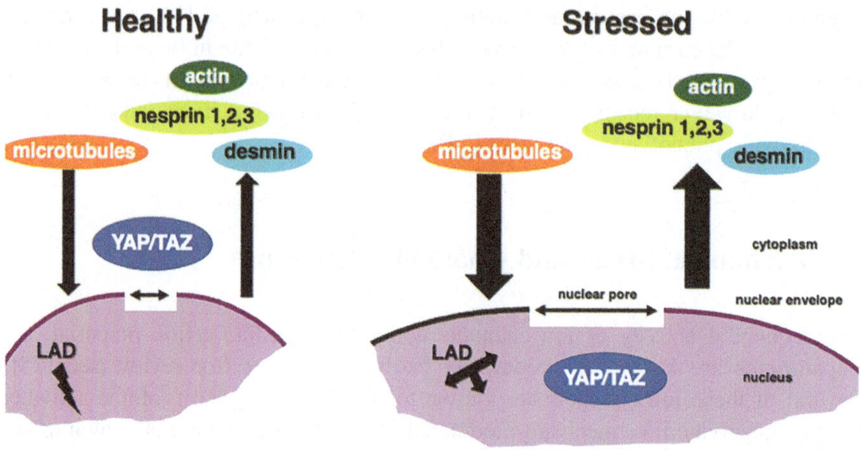

Fig. 4 Mechanosignalling between the cytoplasm and the nucleus may affect gene transcription in stressed cardiomyocytes. In a healthy cardiomyocyte, there is an intricate force balance between the cytoskeletal filament systems (pushing dynamic microtubules, filamentous actin pulling on nesprin and links from the myofibrils via desmin intermediate filaments). In stressed cardiomyocytes, this control gets lost, and dynamic microtubules can push towards the nuclear envelope, leading to invaginations of the nuclear lamina and an altered arrangement of LAD (lamin-associated domains). Increased cytoskeletal forces also pull on the nuclear envelope via nesprin and LINC-mediated interactions and lead to a stretching of nuclear pore diameter, which enables YAP/TAZ transcription factors to diffuse into the nucleoplasm

the nuclear export of SRF, which is likely to affect the transcription of genes encoding for thin filament proteins in a negative way [27]. Whenever mechanosignalling is explored in other tissues, the Hippo pathway is often found at its core, so it is of little surprise that recently its relevance was also shown for heart development and disease (reviewed in [86, 87]). Yap/Taz (Yes-associated protein; TAZ transcriptional co-activator with PDZ-binding motif) are the transcription factors that mediate Hippo signalling, and their entry into the nucleus was shown to be triggered by mechanical stimuli that lead to an increased nuclear pore size by activation of the interaction between filamentous actin filaments and nesprin proteins (Fig. 4), which are known to be able to bind to actin and microtubules as well as to provide a link to the nuclear envelope by binding to the LINC (linker of nucleoskeleton and cytoskeleton) complex [88]; for recent review of LINC, see [89]. There seems to be an intricate force balance between the different types of cytoskeletal filaments in the healthy cardiomyocytes. During stress this is lost, and, for example, microtubules can push against the nuclear envelope and lead to invaginations and stretching of the nuclear envelope. These imbalances in the dynamic behaviour of microtubules, their interaction with the nesprin and desmin cytoskeleton and the ensuing effect on the nuclear envelope will then translate to the LAD (lamin-associated domains) inside [90], which are usually assumed to play a repressive role in gene expression [91]. Interfering with this interaction by depolymerisation of the microtubules or disrupting the nesprin-LINC complex prevents disruption of the nuclear lamina, DNA damage and disease-associated transcriptional changes [90] Fig. 4). There may also be a link from intercalated disc sensed stress towards the nucleus via Yap. Cardiomyocytes that lack all alpha-catenin isoforms display increased nuclear localisation of Yap and increased proliferative behaviour in the neonate [92]. In conclusion, mechanolinks via cytoskeletal proteins from the plasma membrane to the nuclear envelope will affect gene transcription, which is in part due to the Hippo signalling pathway.

6 Will the Future Bring Drugs That Interfere with Mechanosignalling?

A better understanding how exactly mechanosignalling works from the cardiomyocyte surface to changes in gene expression is extremely likely to yield exciting drug targets that may help to attenuate maladaptive signalling in the future. The first pathway that was explored a while ago was interference with ROCK (Rho-associated kinase) signalling, which controls the assembly and activity of non-muscle myosin as well as leads to the activation of formins such as FHOD1. It was shown that a commonly used ROCK-inhibitor, Y-27632, could prevent the development of cardiomyopathy induced by hypertension in Dahl-sensitive rats [93] and also the interference with PKCalpha signalling has been employed for a short-term rescue in animal models of cardiomyopathy [94, 95]. Obviously ubiquitously expressed signalling molecules such as ROCK and PKCalpha are not very attractive drug targets, since their roles are not restricted to the heart and a plethora of side effects are likely. A more targeted attempt was to disrupt the signalling axis between the cytoskeleton and the LINC complex in the nuclear lamina by AAV9-mediated overexpression of a dominant negative LINC component [96]. AAV9 predominantly targets to heart tissue, and a prevention of a lamin mutation-induced cardiomyopathy was reported. An even more exciting avenue appears to be to tackle detyrosinated microtubes. As reported above, loosening up the microtubule network by drugs or interfering with detyrosination improved cardiac output [17], and very recently a kinase was identified, MARK4 (microtubule-affinity regulating kinase), that appears to be crucially involved in this process [97]. MARK4 expression is increased in the heart following myocardial infarction, and it exerts its effect on microtubules by phosphorylating MAP 4 (microtubule-associated protein 4), which in turn facilitates the access of the tubulin carboxypeptidase VASH2 (Tubulinyl-Tyr-carboxypeptidase 2) to polymerised microtubules and favours detyrosination. Interfering with this process by MARK4 downregulation improves contractile function following a mechanobiological insult [97]. Another interesting approach that addresses the turnover of mechanically damaged sarcomere proteins, which lead to impaired contractile function, is to stimulate CASA activity by overexpressing BAG3 using AAV9 [56].

The mechanosensor at the T-tubules, Piezo1, can be specifically activated by Yoda1 and inhibited by GsMTx4, a spider toxin [81]; however the latter seems to have distinct effects depending on the species and was demonstrated to be beneficial in a mouse model of cardiomyopathy but lacking any effect in a swine model [98].

XMU-MP1 was shown to inhibit the Hippo signalling pathway by blocking the activity of the upstream kinase Mst1/2, and this was shown to have cardioprotective effects in mice [99]. Treatment reduced apoptosis following pressure overload, which fits well with one of the better understood roles of Yap/Taz signalling, but had no effect on the proliferation of adult cardiomyocytes. Still, careful timing and limited administration must be considered for translation to humans, to ensure that no tumours develop in other tissues than the heart as a side effect.

Recent developments in gene therapy using CRISPR/Cas9-mediated repair look like an extremely exciting strategy for the future [100] and were shown to be extremely useful for the generation of missense mutated or corrected iPSC lines [29, 30]. However currently there are hindrances regarding delivery, for example, conventionally used Cas9 is too big for the most commonly used delivery vehicle, AAV9, and there is always the concern about potential off-target effects. In general, despite gene therapy and small molecule administration showing a lot of promise in small animal models, they often fail to translate effectively to humans, partly due to differences in physiology but mainly caused by the challenge of a much larger size of the heart that is to be repaired [101, 102]. Therefore, while scaling up from mouse models to larger animals will be the next challenge, these are exciting observations that are likely to result in translational approaches also for human patients. However, also a better understanding of mechanosignalling via the intercalated disc and the costameres is bound to increase our arsenal of heart failure drug targets beyond the current conventional strategies of mainly alleviating symptoms and not addressing causes as much.

7 Conclusion

In this review we have discussed the four epicentres of mechanosignalling in cardiomyocytes, the sarcomere, the intercalated disc, the plasma membrane in particular with regard to T-tubules and the nuclear lamina (see thunderbolt symbols in Fig. 1). It is clear that these do not act in isolation, but that crosstalk between them results from detection of mechanical stress to transmission of a stress signal to changes in gene expression. A more thorough understanding of these mechanosignalling pathways is likely to improve our arsenal of drugs that can slow down a cardiomyopathy phenotype.

Acknowledgements We are grateful for all the fruitful and exciting discussions we had with colleagues over the years. Work in the Ehler laboratory is funded by the UKRI-MRC and the BHF.

Compliance with Ethical Standards

The authors declare that there is no conflict of interest.

Funding This work was supported by UKRI-MRC and BHF grants to EE.

Conflict of Interest Authors declare that they have no conflicts of interest.

Ethical Approval This article does not contain any studies with human participants or animals performed by any of the authors.

References

1. Ehler E (2016) Cardiac cytoarchitecture – why the "hardware" is important for heart function! Biochim Biophys Acta 1863:1857–1863. https://doi.org/10.1016/j.bbamcr.2015.11.006
2. Gabisonia K et al (2019) MicroRNA therapy stimulates uncontrolled cardiac repair after myocardial infarction in pigs. Nature 569:418–422. https://doi.org/10.1038/s41586-019-1191-6
3. Samarel AM (2005) Costameres, focal adhesions, and cardiomyocyte mechanotransduction. Am J Physiol Heart Circ Physiol 289:H2291–H2301. https://doi.org/10.1152/ajpheart.00749.2005
4. Lu F, Pu WT (2020) The architecture and function of cardiac dyads. Biophys Rev 12:1007–1017. https://doi.org/10.1007/s12551-020-00729-x
5. Iorga B et al (2011) Micromechanical function of myofibrils isolated from skeletal and cardiac muscles of the zebrafish. J Gen Physiol 137:255–270. https://doi.org/10.1085/jgp.201010568
6. Poss KD, Wilson LG, Keating MT (2002) Heart regeneration in zebrafish. Science 298:2188–2190. https://doi.org/10.1126/science.1077857
7. Robison P et al (2016) Detyrosinated microtubules buckle and bear load in contracting cardiomyocytes. Science 352:aaf0659. https://doi.org/10.1126/science.aaf0659
8. Velayutham N, Agnew EJ, Yutzey KE (2019) Postnatal cardiac development and regenerative potential in large mammals. Pediatr Cardiol 40:1345–1358. https://doi.org/10.1007/s00246-019-02163-7
9. Frey N, Luedde M, Katus HA (2012) Mechanisms of disease: hypertrophic cardiomyopathy. Nat Rev Cardiol 9:91–100. https://doi.org/10.1038/nrcardio.2011.159
10. Seidman JG, Seidman C (2001) The genetic basis for cardiomyopathy: from mutation identification to mechanistic paradigms. Cell 104:557–567
11. Montag J et al (2017) Intrinsic MYH7 expression regulation contributes to tissue level allelic imbalance in hypertrophic cardiomyopathy. J Muscle Res Cell Motil 38:291–302. https://doi.org/10.1007/s10974-017-9486-4
12. Nikolaev VO et al (2010) Beta2-adrenergic receptor redistribution in heart failure changes cAMP compartmentation. Science 327:1653–1657. https://doi.org/10.1126/science.1185988
13. Masuelli L et al (2003) Beta-catenin accumulates in intercalated disks of hypertrophic cardiomyopathic hearts. Cardiovasc Res 60:376–387
14. Severs NJ, Bruce AF, Dupont E, Rothery S (2008) Remodelling of gap junctions and connexin expression in diseased myocardium. Cardiovasc Res 80:9–19
15. Hong T et al (2014) Cardiac BIN1 folds T-tubule membrane, controlling ion flux and limiting arrhythmia. Nat Med 20:624–632. https://doi.org/10.1038/nm.3543
16. Ehler E, Perriard JC (2000) Cardiomyocyte cytoskeleton and myofibrillogenesis in healthy and diseased heart. Heart Fail Rev 5:259–269. https://doi.org/10.1023/A:1009861504264

17. Chen CY et al (2018) Suppression of detyrosinated microtubules improves cardiomyocyte function in human heart failure. Nat Med 24:1225–1233. https://doi.org/10.1038/s41591-018-0046-2
18. Leu M, Ehler E, Perriard JC (2001) Characterisation of postnatal growth of the murine heart. Anat Embryol (Berl) 204:217–224
19. Pluess M, Daeubler G, dos Remedios CG, Ehler E (2015) Adaptations of cytoarchitecture in human dilated cardiomyopathy. Biophys Rev 7:25–32. https://doi.org/10.1007/s12551-014-0146-2
20. Ehler E et al (2001) Alterations at the intercalated disk associated with the absence of muscle LIM protein. J Cell Biol 153:763–772
21. Ferreira-Cornwell MC et al (2002) Remodeling the intercalated disc leads to cardiomyopathy in mice misexpressing cadherins in the heart. J Cell Sci 115:1623–1634
22. Kostetskii I et al (2005) Induced deletion of the N-cadherin gene in the heart leads to dissolution of the intercalated disc structure. Circ Res 96:346–354. https://doi.org/10.1161/01.RES.0000156274.72390.2c
23. Zemljic-Harpf AE et al (2007) Cardiac-myocyte-specific excision of the vinculin gene disrupts cellular junctions, causing sudden death or dilated cardiomyopathy. Mol Cell Biol 27:7522–7537
24. Hirschy A et al (2010) Stabilised beta-catenin in postnatal ventricular myocardium leads to dilated cardiomyopathy and premature death. Basic Res Cardiol 105:597–608. https://doi.org/10.1007/s00395-010-0101-8
25. Li J et al (2012) Loss of alphaT-catenin alters the hybrid adhering junctions in the heart and leads to dilated cardiomyopathy and ventricular arrhythmia following acute ischemia. J Cell Sci 125:1058–1067. https://doi.org/10.1242/jcs.098640
26. Herman DS et al (2012) Truncations of titin causing dilated cardiomyopathy. N Engl J Med 366:619–628. https://doi.org/10.1056/NEJMoa1110186
27. Lange S et al (2005) The kinase domain of titin controls muscle gene expression and protein turnover. Science 308:1599–1603
28. Lyon RC, Zanella F, Omens JH, Sheikh F (2015) Mechanotransduction in cardiac hypertrophy and failure. Circ Res 116:1462–1476. https://doi.org/10.1161/CIRCRESAHA.116.304937
29. Fomin A et al (2021) Truncated titin proteins and titin haploinsufficiency are targets for functional recovery in human cardiomyopathy due to TTN mutations. Sci Transl Med 13: eabd3079. https://doi.org/10.1126/scitranslmed.abd3079
30. Hinson JT et al (2015) Titin mutations in iPS cells define sarcomere insufficiency as a cause of dilated cardiomyopathy. Science 349:982–986. https://doi.org/10.1126/science.aaa5458
31. Chien K (1999) Stress pathways and heart failure. Cell 98:555–558
32. Lange S, Pinotsis N, Agarkova I, Ehler E (2020) The M-band: the underestimated part of the sarcomere. Biochim Biophys Acta, Mol Cell Res 1867:118440. https://doi.org/10.1016/j.bbamcr.2019.02.003
33. MacLellan WR, Schneider MD (2000) Genetic dissection of cardiac growth control pathways. Annu Rev Physiol 62:289–319
34. Wilson AJ, Schoenauer R, Ehler E, Agarkova I, Bennett PM (2014) Cardiomyocyte growth and sarcomerogenesis at the intercalated disc. Cell Mol Life Sci 71:165–181
35. Schoenauer R et al (2011) EH-myomesin splice isoform is a novel marker for dilated cardiomyopathy. Basic Res Cardiol 106:233–247. https://doi.org/10.1007/s00395-010-0131-2
36. Agarkova I et al (2004) The molecular composition of the sarcomeric M-band correlates with muscle fiber type. Eur J Cell Biol 83:1–12
37. Luther PK, Squire JM (2014) The intriguing dual lattices of the Myosin filaments in vertebrate striated muscles: evolution and advantage. Biology (Basel) 3:846–865. https://doi.org/10.3390/biology3040846
38. Makarenko I et al (2004) Passive stiffness changes caused by upregulation of compliant titin isoforms in human dilated cardiomyopathy hearts. Circ Res 95:708–716. https://doi.org/10.1161/01.RES.0000143901.37063.2f

39. Koser F, Loescher C, Linke WA (2019) Posttranslational modifications of titin from cardiac muscle: how, where, and what for? FEBS J 286:2240–2260. https://doi.org/10.1111/febs.14854
40. Forbes MS, Sperelakis., N. (1985) Intercalated discs of mammalian heart: a review of structure and function. Tissue Cell 17:605–648
41. Bennett PM (2015) In Ehler E (ed) Cardiac cytoarchitecture: how to maintain a working heart, vol 1. Springer, pp 41–70, Ch. 2
42. Vimalanathan AK, Ehler E, Gehmlich K (2018) Genetics of and pathogenic mechanisms in arrhythmogenic right ventricular cardiomyopathy. Biophys Rev 10:973–982. https://doi.org/10.1007/s12551-018-0437-0
43. Yamaguchi M, Yamano S, Muguruma M, Robson RM (1988) Polarity and length of actin filaments at the fascia adherens of the cardiac intercalated disk. J Ultrastruct Mol Struct Res 100:235–244
44. Bennett PM, Maggs AM, Baines AJ, Pinder JC (2006) The transitional junction: a new functional subcellular domain at the intercalated disc. Mol Biol Cell 17:2091–2100. https://doi.org/10.1091/mbc.E05-12-1109
45. Bennett PM (2018) Riding the waves of the intercalated disc of the heart. Biophys Rev 10: 955–959. https://doi.org/10.1007/s12551-018-0438-z
46. Zou P et al (2006) Palindromic assembly of the giant muscle protein titin in the sarcomeric Z-disk. Nature 439:229–233
47. Bertz M, Wilmanns M, Rief M (2009) The titin-telethonin complex is a directed, superstable molecular bond in the muscle Z-disk. Proc Natl Acad Sci U S A 106:13307–133310. https://doi.org/10.1073/pnas.0902312106
48. Yoshida M et al (2010) Weaving hypothesis of cardiomyocyte sarcomeres: discovery of periodic broadening and narrowing of intercalated disk during volume-load change. Am J Pathol 176:660–678. https://doi.org/10.2353/ajpath.2010.090348
49. Scarborough EA et al (2021) Microtubules orchestrate local translation to enable cardiac growth. Nat Commun 12:1547. https://doi.org/10.1038/s41467-021-21685-4
50. Rothen-Rutishauser BM, Ehler E, Perriard E, Messerli JM, Perriard J-C (1998) Different behaviour of the non-sarcomeric cytoskeleton in neonatal and adult rat cardiomyocytes. J Mol Cell Cardiol 30:19–31
51. Martin AF (1981) Turnover of cardiac troponin subunits. Kinetic evidence for a precursor pool of troponin-I. J Biol Chem 256:964–968
52. Fornasiero EF et al (2018) Precisely measured protein lifetimes in the mouse brain reveal differences across tissues and subcellular fractions. Nat Commun 9:4230. https://doi.org/10.1038/s41467-018-06519-0
53. Wang J et al (2005) Dynamics of Z-band based proteins in developing skeletal muscle cells. Cell Motil Cytoskeleton 61:34–48. https://doi.org/10.1002/cm.20063
54. da Silva Lopes K, Pietas A, Radke MH, Gotthardt M (2011) Titin visualization in real time reveals an unexpected level of mobility within and between sarcomeres. J Cell Biol 193:785–798. https://doi.org/10.1083/jcb.201010099
55. Höhfeld J et al (2021) Maintaining proteostasis under mechanical stress. EMBO Rep 22: e52507. https://doi.org/10.15252/embr.202152507
56. Martin TG et al (2021) Cardiomyocyte contractile impairment in heart failure results from reduced BAG3-mediated sarcomeric protein turnover. Nat Commun 12:2942. https://doi.org/10.1038/s41467-021-23272-z
57. Ehsan M et al (2018) Mutant muscle LIM Protein C58G causes cardiomyopathy through protein depletion. J Mol Cell Cardiol 121:287–296. https://doi.org/10.1016/j.yjmcc.2018.07.248
58. Jiang H et al (2021) Functional analysis of a gene-edited mouse model to gain insights into the disease mechanisms of a titin missense variant. Basic Res Cardiol 116:14. https://doi.org/10.1007/s00395-021-00853-z

59. Agarwal R et al (2021) Filamin C cardiomyopathy variants cause protein and lysosome accumulation. Circ Res 129(7):751–766. https://doi.org/10.1161/CIRCRESAHA.120.317076
60. Dwyer J, Pluess M, Iskratsch T, Dos Remedios CG, Ehler E (2014) The formin FHOD1 in cardiomyocytes. Anat Rec (Hoboken) 297:1560–1570. https://doi.org/10.1002/ar.22984
61. Schönichen A et al (2013) FHOD1 is a combined actin filament capping and bundling factor that selectively associates with actin arcs and stress fibers. J Cell Sci 126:1891–1901. https://doi.org/10.1242/jcs.126706
62. Patel AA, Oztug Durer ZA, van Loon AP, Bremer KV, Quinlan ME (2018) Drosophila and human FHOD family formin proteins nucleate actin filaments. J Biol Chem 293:532–540. https://doi.org/10.1074/jbc.M117.800888
63. Benz PM et al (2013) Mena/VASP and alphaII-Spectrin complexes regulate cytoplasmic actin networks in cardiomyocytes and protect from conduction abnormalities and dilated cardiomyopathy. Cell Commun Signal 11:56. https://doi.org/10.1186/1478-811X-11-56
64. Schinner C et al (2017) Adrenergic signaling strengthens cardiac myocyte cohesion. Circ Res 120:1305–1317. https://doi.org/10.1161/CIRCRESAHA.116.309631
65. van Berlo JH, Maillet M, Molkentin JD (2013) Signaling effectors underlying pathologic growth and remodeling of the heart. J Clin Invest 123:37–45. https://doi.org/10.1172/JCI62839
66. Lange S et al (2016) MLP and CARP are linked to chronic PKCalpha signalling in dilated cardiomyopathy. Nat Commun 7:12120. https://doi.org/10.1038/ncomms12120
67. Newton AC (2010) Protein kinase C: poised to signal. Am J Physiol Endocrinol Metab 298: E395–E402. https://doi.org/10.1152/ajpendo.00477.2009
68. Bass-Zubek AE et al (2008) Plakophilin 2: a critical scaffold for PKC alpha that regulates intercellular junction assembly. J Cell Biol 181:605–613. https://doi.org/10.1083/jcb.200712133
69. Pruna M, Ehler E (2020) The intercalated disc: a mechanosensing signalling node in cardiomyopathy. Biophys Rev 12:931–946. https://doi.org/10.1007/s12551-020-00737-x
70. Gaetani R et al (2020) When stiffness matters: mechanosensing in heart development and disease. Front Cell Dev Biol 8:334. https://doi.org/10.3389/fcell.2020.00334
71. Pandey P et al (2018) Cardiomyocytes sense matrix rigidity through a combination of muscle and non-muscle myosin contractions. Dev Cell 44:326–336 e323. https://doi.org/10.1016/j.devcel.2017.12.024
72. Maeda M, Holder E, Lowes B, Valent S, Bies RD (1997) Dilated cardiomyopathy associated with deficiency of the cytoskeletal protein metavinculin. Circulation 95:17–20
73. Olson TM et al (2002) Metavinculin mutations alter actin interaction in dilated cardiomyopathy. Circulation 105:431–437
74. Kanoldt V et al (2020) Metavinculin modulates force transduction in cell adhesion sites. Nat Commun 11:6403. https://doi.org/10.1038/s41467-020-20125-z
75. Rahm AK, Lugenbiel P, Schweizer PA, Katus HA, Thomas D (2018) Role of ion channels in heart failure and channelopathies. Biophys Rev 10:1097–1106. https://doi.org/10.1007/s12551-018-0442-3
76. Peyronnet R, Nerbonne JM, Kohl P (2016) Cardiac mechano-gated ion channels and arrhythmias. Circ Res 118:311–329. https://doi.org/10.1161/CIRCRESAHA.115.305043
77. Patel AJ, Honore E (2001) Anesthetic-sensitive 2P domain K+ channels. Anesthesiology 95: 1013–1021. https://doi.org/10.1097/00000542-200110000-00034
78. Gottlieb PA, Bae C, Sachs F (2012) Gating the mechanical channel Piezo1: a comparison between whole-cell and patch recording. Channels (Austin) 6:282–289. https://doi.org/10.4161/chan.21064
79. Jiang F et al (2021) The mechanosensitive Piezo1 channel mediates heart mechano-chemo transduction. Nat Commun 12:869. https://doi.org/10.1038/s41467-021-21178-4
80. Liang J et al (2017) Stretch-activated channel Piezo1 is up-regulated in failure heart and cardiomyocyte stimulated by AngII. Am J Transl Res 9:2945–2955
81. Beech DJ, Kalli AC (2019) Force sensing by Piezo channels in cardiovascular health and disease. Arterioscler Thromb Vasc Biol 39:2228–2239. https://doi.org/10.1161/ATVBAHA.119.313348

82. Sugden PH (2001) Signalling pathways in cardiac myocyte hypertrophy. Ann Med 33:611–622
83. Liang Y, Sheikh F (2016) Scaffold proteins regulating extracellular regulated kinase function in cardiac hypertrophy and disease. Front Pharmacol 7:37. https://doi.org/10.3389/fphar.2016.00037
84. Liang Y, Bradford WH, Zhang J, Sheikh F (2018) Four and a half LIM domain protein signaling and cardiomyopathy. Biophys Rev 10:1073–1085. https://doi.org/10.1007/s12551-018-0434-3
85. Miller MK et al (2003) The muscle ankyrin repeat proteins: CARP, ankrd2/Arpp and DARP as a family of titin filament-based stress response molecules. J Mol Biol 333:951–964
86. Mia MM, Singh MK (2019) The hippo signaling pathway in cardiac development and diseases. Front Cell Dev Biol 7:211. https://doi.org/10.3389/fcell.2019.00211
87. Xie J, Wang Y, Ai D, Yao L, Jiang H (2022) The role of the Hippo pathway in heart disease. FEBS J 289(19):5819–5833. https://doi.org/10.1111/febs.16092
88. Elosegui-Artola A et al (2017) Force triggers YAP nuclear entry by regulating transport across nuclear pores. Cell 171:1397–1410 e1314. https://doi.org/10.1016/j.cell.2017.10.008
89. Ross JA, Stroud MJ (2021) The nucleus: mechanosensing in cardiac disease. Int J Biochem Cell Biol 137:106035. https://doi.org/10.1016/j.biocel.2021.106035
90. Heffler J et al (2020) A balance between intermediate filaments and microtubules maintains nuclear architecture in the cardiomyocyte. Circ Res 126:e10–e26. https://doi.org/10.1161/CIRCRESAHA.119.315582
91. van Steensel B, Belmont AS (2017) Lamina-associated domains: links with chromosome architecture, heterochromatin, and gene repression. Cell 169:780–791. https://doi.org/10.1016/j.cell.2017.04.022
92. Vite A, Zhang C, Yi R, Emms S, Radice GL (2018) alpha-Catenin-dependent cytoskeletal tension controls Yap activity in the heart. Development 145:dev149823. https://doi.org/10.1242/dev.149823
93. Satoh S et al (2003) Chronic inhibition of Rho kinase blunts the process of left ventricular hypertrophy leading to cardiac contractile dysfunction in hypertension-induced heart failure. J Mol Cell Cardiol 35:59–70
94. Hambleton M et al (2006) Pharmacological- and gene therapy-based inhibition of protein kinase Calpha/beta enhances cardiac contractility and attenuates heart failure. Circulation 114:574–582. https://doi.org/10.1161/CIRCULATIONAHA.105.592550
95. Palaniyandi SS, Sun L, Ferreira JC, Mochly-Rosen D (2009) Protein kinase C in heart failure: a therapeutic target? Cardiovasc Res 82:229–239. https://doi.org/10.1093/cvr/cvp001
96. Chai RJ et al (2021) Disrupting the LINC complex by AAV mediated gene transduction prevents progression of Lamin induced cardiomyopathy. Nat Commun 12:4722. https://doi.org/10.1038/s41467-021-24849-4
97. Yu X et al (2021) MARK4 controls ischaemic heart failure through microtubule detyrosination. Nature 594:560–565. https://doi.org/10.1038/s41586-021-03573-5
98. Johnson DM, Antoons G (2018) Arrhythmogenic mechanisms in heart failure: linking beta-adrenergic stimulation, stretch, and calcium. Front Physiol 9:1453. https://doi.org/10.3389/fphys.2018.01453
99. Triastuti E et al (2019) Pharmacological inhibition of Hippo pathway, with the novel kinase inhibitor XMU-MP-1, protects the heart against adverse effects during pressure overload. Br J Pharmacol 176:3956–3971. https://doi.org/10.1111/bph.14795
100. Ma H et al (2017) Correction of a pathogenic gene mutation in human embryos. Nature 548:413–419. https://doi.org/10.1038/nature23305
101. Hulot JS, Ishikawa K, Hajjar RJ (2016) Gene therapy for the treatment of heart failure: promise postponed. Eur Heart J 37:1651–1658. https://doi.org/10.1093/eurheartj/ehw019
102. Verdonschot JAJ, Hazebroek MR, Ware JS, Prasad SK, Heymans SRB (2019) Role of targeted therapy in dilated cardiomyopathy: the challenging road toward a personalized approach. J Am Heart Assoc 8:e012514. https://doi.org/10.1161/JAHA.119.012514

Biophysical Stretch Induced Differentiation and Maturation of Induced Pluripotent Stem Cell-Derived Cardiomyocytes

Jiabin Qin, Alain van Mil, and Joost P. G. Sluijter

1 Introduction

The growing burden of cardiovascular disease, particularly due to ischemic heart disease as one of the leading causes of death and disability worldwide [1], gave rise to many efforts to improve our insights into post-myocardial infarction (MI) processes and to develop novel therapeutic interventions. Although progress has been made, still much effort is needed to translate research findings into clinical practice. Especially, the roles of the immune system [2], non-coding RNAs [3], extracellular vesicles [4], cardioprotection [5], and sex-specific effects [6] need to be further explored. Additionally, improved evaluation [7] and standardization of preclinical imaging [8] and specific influence of other non-cardiomyocyte cell types that are involved in the diseased heart need to be understood [9]. One major potential confounding factor for clear clinical translation [10] is the lack of proper human in vitro models to study cardiac disease, as intrinsic profound differences in electrophysiology and pharmacokinetics exist between cells of different species [11].

Induced pluripotent stem cell (iPSC) technology is attracting considerable interest in every major biomedical field since Takahashi et al. introduced the embryonic transcription factors Oct3/4, Sox2, Klf4, and c-Myc into adult human fibroblasts to convert them into pluripotent stem cells [12]. Due to the different physiology, disease progression, and possible immunogenicity, CMs derived from non-human

J. Qin · A. van Mil · J. P. G. Sluijter (✉)
Division of Heart and Lungs, Department of Cardiology, Laboratory of Experimental Cardiology, University Medical Center Utrecht, Utrecht, the Netherlands

Regenerative Medicine Center Utrecht, Circulatory Health Laboratory, University Utrecht and University Medical Center Utrecht, Utrecht, the Netherlands
e-mail: j.sluijter@umcutrecht.nl

M. Hecker, D. J. Duncker (eds.), *Cardiac Mechanobiology in Physiology and Disease*, Cardiac and Vascular Biology 9,
https://doi.org/10.1007/978-3-031-23965-6_7

PSCs [13, 14] may potentially limit human cardiac disease understanding and therapeutic discovery; moreover, the limited availability and difficulty of adult CMs to be isolated from other sources and species hampered cardiovascular research. Human iPSCs (hiPSCs) are comparable to human embryonic stem cells (hESCs), in which they can differentiate into a variety of organ-specific cell types and can thus be utilized to build up human patient- and disease-specific cell models at a scalable level. Due to the failure of postnatal CMs to reenter the cell cycle in vivo, the difficulty in obtaining large numbers of primary CMs, and the ethical concerns linked to hESCs, hiPSC-derived CMs (hiPSC-CMs) have provided an intriguing and alternative option to replace primary CMs and provide a compelling source for cardiac regeneration, particularly for remuscularization after myocardial infarction [15–17]. Several groups, led by pioneering work of Eschenhagen have created powerful hiPSC-derived engineered heart tissues (EHT) to improve cell therapy endurance and efficacy when implanted onto the surface of infarcted hearts [15, 18–20]. Moreover, the advent of CRISPR/Cas9 technology promotes establishing genetically defined hiPSC-CMs and further facilitates specific in vitro disease modeling, allowing potential therapeutic drugs to be tested on a personalized level [21–23]. Beyond cell therapy application or drug screening, in vitro progression of CM maturation from PSCs to fully differentiated CMs also provides insights into underlying mechanisms of heart development [24, 25]. In the last decade, significant progress has been achieved to differentiate bona fide CMs from iPSCs by generating embryoid bodies (EBs) in suspension [26] and modulating several critical signaling pathways in monolayered 2D-cultures, providing a CM population with high purity [27, 28]. Additionally, by conducting a simple modulation that includes synergistic Wnt signaling activation and reduced cell-cell contact, it is now possible to obtain hiPSC-CMs on a theoretically unlimited scale [29]. hiPSC-CMs differentiated by current protocols, on the other hand, display fetal instead of adult phenotypes, characterized by small-size, underdeveloped myofibrils, inadequate electro-conduction, and dependency on glycolysis-mediated metabolism [30, 31]. Consequently, applications of hiPSC-CMs have been hindered in the preclinical stage, and improving the maturity has become an acute interest. Many different directions and approaches, other than biophysical cues, including long-term culture [32], electrical stimulation [33, 34], metabolism modulation [35–39], 3D culture, co-culture with non-cardiomyocyte populations [40], and combined stimulation, have been investigated to enhance hiPSC-CM maturity [30, 41].

One of the earliest functional organs to develop in the human embryo is the heart, which starts beating and pumping blood to support circulation around 3 weeks after fertilization [42]. The whole cardiac embryogenesis indicates that myocardial growth and remodeling can occur consistently in dynamic mechanical environments. Proper cardiac cell coupling and interaction with the extracellular environment are required for myocardium formation. Vice versa, abnormal heart function during pathological conditions is associated with changes in the mechanical loading of the organ, resulting in altered mechanical stress on CMs. Various extracellular forces affecting CMs, including ECM properties, heart rhythmic stretch, and blood flow induced shear stress, have been shown to influence heart morphological and functional developments [43–47]. The potential maturation effects of mechanical stimulation were investigated on neonatal rat ventricular myocytes (NRVMs). Substrate stiffness comparable to rat myocardium was reported to promote NRVMs with improved morphology and elevated cardiac-associated protein expression [48–50]. Besides, mechanically loaded NRVM-based EHT, developed by Zimmermann et al. [51, 52], exhibited cardiac muscle bundles comparable to adult instead of immature native rat tissue. All this mounting evidence indicates that biophysical cues, which aim to mimic the natural myocardial mechanical environment, have the valuable potential to achieve CM differentiation and, more importantly, maturation in vitro, and these strategies have been utilized to enhance intercellular connection and improve functionality of hiPSC-CMs, both in two-dimensional (2D) and three-dimensional (3D) culture.

Research on mechanical cues on PSC-derived CMs may help to identify novel targets for improving their maturity and consequent functionality, enhancing their potential in cardiac cell therapy and for drug screening. This chapter will discuss the effects of state-of-the-art methods employing passive and dynamic biophysical cues to improve hiPSC-CM maturity. An overview of the hallmarks of the primary adult CM phenotype and how they differ from hiPSC-CMs is given, after which the methods and effects of external mechanical stimulation and mechanical properties are discussed, as well as the future perspectives of this field.

2 Changes in Cardiomyocyte Characteristics During Myocardial Development

In the native myocardium, CMs and other non-CM populations interact with the ECM to compose a sophisticated network. The ECM allows for mechanical integrity and acts as a structural conductor that dynamically transduces and facilitates biochemical and biophysical signals within the heart [53]. Due to its pivotal role, choices of suitable ECM for in vitro culture of CMs are primarily considered. Collagen is one of the predominant structural units in cardiac ECM, and laminin links CMs to their adjacent ECM acting as an adhesive protein [54, 55]. Therefore, Matrigel, which is a basement membrane preparation from mouse sarcoma primarily containing laminin and collagen IV [56], is the most generally used ECM coating for iPSC differentiation and iPSC-CM maintenance. On the other hand, the ECM acquired from decellularized hearts (dECM) can offer many distinct advantages. The dECM largely retain cardiac-specific ECM micro-architectures and mechanical properties which are difficult to reproduce in the petri dish. The preserved ECM components also provide chemical and biological cues to act as a native-like microenvironment. All these characteristics constitute a complicated combination of biochemical and mechanical cues that may enhance cell attachment, proliferation, survival, and cardiovascular differentiation during later recellularization [57, 58].

2.1 Differences Between Native Myocardial and 2D-Cultured Cardiomyocytes

During human heart development and upon birth, CMs experience a series of intracellular modifications [41]. In the prenatal stage, heart growth is primarily dependent on CM proliferation. More than half of the CMs in the heart withdraw from the cell cycle after birth to become polyploid cells [59, 60] and stop proliferating progressively with a remaining 0.5% turnover [59, 61]. Therefore, the non-proliferative state of postnatal adult CMs has been universally accepted.

Adult CMs have an elongated shape with an anisotropic sarcomere and cell alignment, and they interact with adjacent CMs via intercalated disks, thereby allowing efficient electroconduction and formation of the electromechanically coupled myocardial tissue. Due to the lack of these dynamic, mechanical, and environmental stimuli in 2D cell culture, cultured neonatal CMs are hypothesized to be unable to develop further to reach their physiologically hypertrophic size and build functional structures, as seen in adult CMs [62]. Similarly, hiPSC-CMs differentiated in standard 2D culture do not fully mature and display random alignment, disorganized sarcomeres, no multinucleation, absence of transverse tubules (T-tubules), and vary largely in shape [30]. Understanding the developmental pathways that trigger and stimulate further maturation throughout cardiac development will enhance and improve the generation of adult-like CMs derived from hiPSCs, while insights gained during hiPSC-CM derivation will improve our insights into human myocardial cell development [63]. Here, we outlined the major differences between native adult human CMs and their 2D-cultured counterparts.

2.2 Cell Morphology

As the contractile function of the heart requires a specific coordinated cellular movement, adult CMs need to be anisotropically localized in the myocardium to electromechanically allow efficient coupling and contraction. However, hiPSC-CMs from standard 2D culture lack a defined direction in the petri dish, resulting in poor cellular and sarcomere alignment. At the cellular level, adult CMs have a rod-like shape with a large surface area, ranging from 10,000 to 14,000 μm^2, a length-to-width aspect ratio close to 7: 1 [64, 65], and sarcomere length of about 2.2 μm in diastole [66]. This specific morphology allows the development of long myofibrils and laterally aligned sarcomeres. Compared to adult CMs, the immature hiPSC-CMs are at least tenfold smaller, with a surface area between 1000 and 1300 μm^2 [65, 67], a 3: 1 aspect ratio, and disorganized, shorter sarcomeres (around 1.65 μm) [31, 62, 68]. Sarcomeres are distinct structures in muscle cells and are composed of repeated myofibril subunits [66, 69]. The formation and organization of sarcomeres are dependent on the expression and organization of myofibril proteins like myosin heavy chain (MYH), myosin light chain 2 (MLC2), and troponin I (TNNI), which switch from fetal to adult isoforms during maturation due to transcriptional changes or alternative splicing [70]. MYH7, MLC2V, and TNNI3 are primarily expressed in adult CMs, but MYH6, MLC2a, and TNNI1 are the major isoforms in hiPSC-CMs [71–73]. In addition to changes in protein subtypes, the sarcomere structure of adult CMs and iPSC-CMs differs dramatically. Sarcomeres in highly aligned adult CMs display a Z-disc structure with a uniform width that can be observed histologically, while iPSC-CMs only form sarcomeres with randomly aligned Z-discs and variable width [74]. The shortening of CM sarcomeres results in contraction and force generation. In adult human papillary muscle strips, adult ventricle CMs are electrically quiescent and only beat when stimulated and can produce contractile forces of 40–80 mN/mm^2 [75]. Conversely, hiPSC-CMs beat spontaneously, providing an easy readout for successful CM differentiation, but the contractile force is substantially weaker, varying from 0.22 ± 0.70 to 11.8 ± 4.5 mN/mm^2 [76–79]

2.3 *Electrical Signal Conduction*

Adult CMs are longitudinally interconnected through intercalated discs comprising specific protein structures, such as desmosomes, N-cadherin-based adherent junctions, and gap junctions (containing Connexin-43 (Cx43)) [30, 80, 81]. These protein structures allow ions and low-weight molecules to transduce among CMs, facilitating rapid electrical propagation and preventing cell separation and ripping under stretch. Because of the anisotropic alignment of adult CMs, N-cadherin and Cx43 are predominantly localized in the intercalated discs, instead of the latitudinal edges of CMs. However, in hiPSC-CMs, N-cadherin and Cx43 are distributed circumferentially throughout the cell membrane [82, 83]. The CM action potential (AP) is regulated through voltage-gated ion channels embedded in the CM plasma membrane, but several ion channel proteins show significant differences in expression levels between the adult CMs and hiPSC-CMs [84–88]. These differences, together with the absence of clear intercalated discs, result in an immature electrophysiological phenotype in hiPSC-CMs. For example, the resting membrane potential of mature ventricular CMs is approximately –90 mV, whereas immature hiPSC-CMs are less hyperpolarized, reaching about −60 mV [89, 90]. In addition, the conduction and upstroke velocities in mature ventricular CMs are around 60 cm/s and 150–350 V/s, respectively. Lack of ion channel expression, as well as the less hyperpolarized membrane potential on hiPSC-CMs, leads to slower conduction and upstroke velocities of 10–20 cm/s and 10–50 V/s, respectively [88, 91, 92].

With the aid of T-tubules and a well-developed sarcoplasmic reticulum (SR), the rhythmic beating of adult CMs is regulated by intracellular Ca^{2+} transition and duration. T-tubules are specific invaginations in the CM plasma membrane that facilitate a fast Ca^{2+} transition from the membrane to the SR. In adult CMs, T-tubules and SR are well-organized and allow for Ca^{2+}-induced Ca^{2+} release and rapid excitation-contraction coupling. Two proteins play a critical role in regulating

SR Ca^{2+} release: sarco/endoplasmic reticulum calcium ATPase 2a (SERCA2a) and the ryanodine receptor (RyR2). In hiPSC-CMs, hardly any T-tubules are present, and an underdeveloped SR is generally found with standard 2D differentiation protocols, resulting in poor calcium handling [93–95].

2.4 Metabolism

Cellular metabolism shifts dramatically from the prenatal to the postnatal heart. The uterine environment is hypoxic during the fetal phase; therefore, carbohydrates, particularly lactate, are the primary energy source, and the anaerobic glycolysis produces the energy in CMs [96, 97]. After birth, the oxygen-rich environment progressively stimulates CMs to switch their metabolic substrate from carbohydrates to long-chain fatty acids [98]. β-oxidation of fatty acids becomes the primary energy production pathway in adult postnatal CMs and provides about 80% of total energy [99, 100]. The glycolytic pathway, however, is the primary metabolic pathway for energy production in hiPSC-CMs [101, 102]. In addition to energy generation, the number, morphology, and localization of energy-producing mitochondria vary between adult CMs and hiPSC-CMs. The mitochondria only account for a small fraction of the cell volume in hiPSC-CMs (<5%), whereas 20 to 40% of the cell volume in adult CMs is occupied by mitochondria [103]. Mitochondria in the hiPSC-CMs show an immature morphology with poorly developed cristae. In adult CMs, cristae are dense and show a great extension of their surface area for oxidative respiration [104]. Because of the high energy consumption in adult CMs, mitochondria are explicitly distributed in areas with high energetic loads like the SR and T-tubules, whereas mitochondria are located perinuclearly in iPSC-CMs [105].

Taken together, major differences exist between native cardiac-derived cardiomyocytes and their hiPSC-derived derivatives. CM maturation is a sophisticated and intricate process, involving multiple developmental pathways to allow morphological, functional, and metabolic changes, as detailed and illustrated in Table 1 and Fig. 1.

3 Biophysical Cues for iPSC-CM Maturation

3.1 Matrix Stiffness and Surface Topography in 2D-Culturing

The elastic modulus, also known as Young's modulus E, enables the calculation of material stiffness (unit in Pa) [106]. The stiffness of the myocardium changes during both heart development and various cardiac diseases. For instance, the stiffness of the myocardium in rats ranges between 10 and 15 kPa during cardiogenesis,

Table 1 Comparison between adult cardiomyocytes and monolayer hiPSC-CMs phenotype

		Adult	iPSC-CM
Morphology	Shape	Rod-like	Round
	Alignment	Anisotropic	Random
	Cell area	10,000 to 14,000 μm^2	1000 to 1300 μm^2
	Length-to-width aspect ratio	7:1	3:1
	Sarcomere length in diastole	2.2 μm	1.65 μm
	Sarcomere banding	Z-disc structure with a uniform width	Randomly aligned Z-disc with variable width
	T-tubules presence	Yes	No
Contractility	Myofibril protein subtypes	MYH7	MYH6
		MLC2V	MLC2a
		TNNI3	TNNI1
	Force generation	40–80 mN/mm^2	0.22 ± 0.70 to 11.8 ± 4.5 mN/mm^2
Electrophysiology	Resting membrane potential	−90 mV	−60 mV
	Conduction velocity	60 cm/s	10–20 cm/s
	Upstroke velocity	150–350 V/s	10–50 V/s
Metabolism	Metabolic pathways	Fatty acid based β-oxidation	Glycolysis
	Mitochondria	20 to 40% cell volume	<5% cell volume
		Well-developed cristae	Poorly developed cristae
		Distributed around sarcoplasmic reticulum and T-tubules	Perinuclear

15–20 kPa in healthy adult myocardium, and 35–70 kPa in the infarcted myocardium [107, 108]. Collagens are the key players in determining tissue stiffness due to their ability to form rigid, thick, and long fibrils [109]. The difference in stiffness between the neonatal and adult heart may be attributed to the total amount of collagen, the ratio of collagen type I to collagen type III [110], and the amount of fibril crosslinking.

The effects of substrate stiffness that is comparable to the prenatal myocardium (~ 10 kPa) have been extensively investigated in neonatal rat ventricular myocytes (NRVMs)[48–50, 111–113] and were optimal to induce NRVM maturation, as evidenced by matured cell structures and protein expression. As such, PSC-CM maturation with similar experimental settings has been studied, as highlighted in

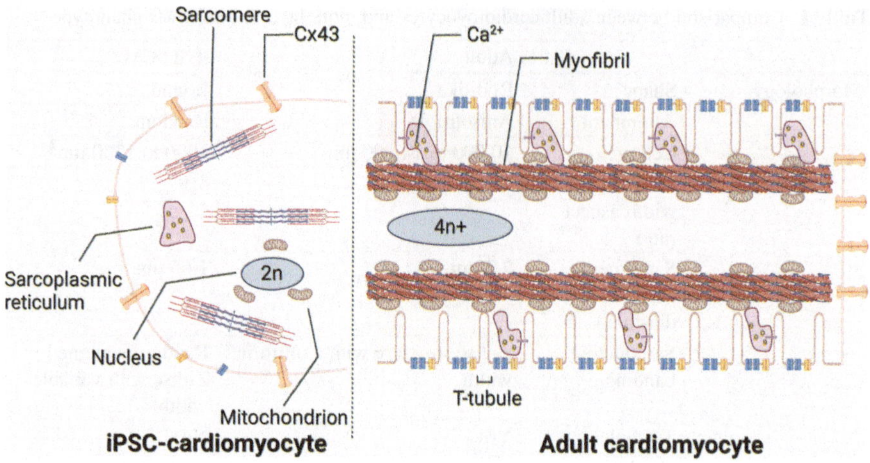

Fig. 1 Cardiomyocyte maturation characteristics. Cardiomyocytes experience a series of changes during maturation in vivo. Adult CMs are rod shaped with well-organized myofibrils. hiPSC-CMs are more rounded in shape and contain disorganized sarcomeres, with no multinucleation, and absence of T-tubules. Because of different energy requirements between hiPSC-CMs and adult CMs, the mitochondrial network is located perinuclearly in iPSC-CMs, and distributed around the SR and T-tubules in adult CMs. This work was created with BioRender.com

Table 2. To determine the impact of substrate stiffness on human PSC-CM maturation, a polyacrylamide (PA) hydrogel with tunable stiffness was used for short-term culture up to 5 days. In comparison to hESC-CMs on a 4 kPa PA hydrogel, 16 kPa PA hydrogels improved intercellular alignment with clearly defined sarcomeres, induced whole-cell distribution of SERCA2a expression, and enhanced calcium handling. Pharmacological inhibition of integrins and associated downstream RhoA/ROCK pathways suggested their involvement in SERCA2 redistribution and the enhancement of calcium handling [114]. Additionally, the contractile stress of iPSC-CMs was positively correlated to the substrate stiffness (4.4 to 99.7 kPa) within 3 days of culture, whereas this functional change was not associated with the improvement of beating rate [115]. However, Körner et al. pointed out that murine iPSC-CMs displayed the best contractility, intercellular connection, and calcium handling when grown on a soft PDMS substrate (1.5 kPa) when studying the stiffness effects ranging from 1.5 kPa to 25 GPa (rigid glass) [123]. These rather different results may be due to species differences. Other studies have used Matrigel to determine substrate stiffness effects on hiPSC-CMs. A thick layer of (0.4- to 0.8-mm) concentrated Matrigel (Matrigel mattress) with 5.8 kPa stiffness promoted iPSC-CM morphological and electrophysiological maturation compared to control 1:60 diluted Matrigel-coated wells (<0.1-mm thick), which gives essentially the same rigidity as the underlying tissue culture plastic, ranging from 0.5 to 1 MPa [30, 116, 124]. Notably, hiPSC-CMs grown on the Matrigel mattress showed comparable contractile properties to freshly isolated rabbit CMs. Likewise, hiPSC-CMs cultured on a soft Matrigel-coated PDMS membrane showed a higher

Table 2 Summary of matrix stiffness and topography effects on PSC-CMs

Mechanical stimulus	Differentiation methods	Max. stimulation duration	Results				References
			Morphology	Contractility	Electrophysiology	Metabolism	
hESC-CMs on 16 kPa PA hydrogel	EB formation	5 days	↑ Sarcomere organization ↑ Myofibril alignment		↑ Calcium handling SERCA2a redistribution		[114]
hPSC-CMs on 4.4 kPa to 99.7 kPa PA hydrogel	Activin A/ BMP-4 induction	3 days	↑ Cell area	↑ Contraction stress			[115]
hiPSC-CMs on 5.8 kPa Matrigel mattress	Wnt-modulation protocol	7 days	Rod-shaped morphology ↑ Sarcomere length ↑ Myofibril alignment	↑ Cell shortening ↑ *TNNI3* expression	↑ Upstroke velocity		[116]
hiPSC-CM on soft Matrigel-coated PDMS membrane	CDI iCell™ human CMs	7 days	↑ Cell area ↑ Binucleation	↑ *cTnI* expression	↑ Upstroke velocity ↑ Conduction velocity ↑ Cx43 expression		[117]
hiPSC-CMs on Fibronectin coated microgroove substrates	CDI iCell™ human CMs	2 weeks	↑ Cellular anisotropy ↑ Myofibril alignment		↑ Calcium handling ↑ SR Ca^{2+} cycling		[118]

(continued)

Table 2 (continued)

Mechanical stimulus	Differentiation methods	Max. stimulation duration	Results				References
			Morphology	Contractility	Electrophysiology	Metabolism	
hiPSC-CMs on fibronectin-coated 20 μm microgrooves	Wnt-modulation protocol	>3 weeks	↑ Myofibril alignment ↑ Sarcomere length ↓ Cell area	↑ Contraction stress		↑ Fatty acid-oxidizing genes	[119]
hiPSC-CMs on Matrigel-coated PDMS 15° chevron micropattern	Wnt-modulation protocol	18 days	↑ Cellular anisotropy ↑ Myofibril alignment	No increase in contraction stress	↑ Conduction velocity		[120]
hPSC-CMs on 2000 μm^2 rectangular Matrigel patterns with optimal 7:1 aspect ratio	Wnt-modulation protocol	9 days	↑ Myofibril alignment No increase in sarcomere length T-tubule-like structures		↑ Upstroke velocity ↓ Resting membrane potential ↑ AP amplitude Anisotropic Ca^{2+} flow		[121]
hiPSC-CMs on 800 nm nanogroove substrates	Activin A/BMP-4 induction	3 weeks	↑ Cell area ↑ Sarcomere length ↑ Cellular anisotropy				[122]

PA polyacrylamide, *ESC-CM* embryonic stem cell-derived cardiomyocytes, *PSC-CM* pluripotent stem cell-derived cardiomyocytes, *PDMS* polydimethylsiloxane

binucleated population and better upstroke and conduction velocity, which might be attributed to increased Cx43 expression [117].

To improve sarcomere alignment in hiPSC-CMs, researchers have employed culture substrates with specific micro- or nano-scaled patterns to anisotropically guide CM growth. Initially, studies using NRVMs grown on micropatterned laminin surfaces [125], micro-groove scaffolds [126], or nanopatterned PEG hydrogels [127] showed improved CM alignment and sarcomere organization. Given these studies, micro- or nano-scale structured culture substrates that resemble native myocardial ECM organization of 100 nm collagen fibers in diameter [128] have been applied to stimulate hiPSC-CM maturation. hiPSC-CMs grown on uniaxial microgroove substrates showed better cell alignment and electrophysiological functionality, such as lower resting membrane potential, increased upstroke velocity, and anisotropic conduction velocity [118, 121, 129]. Of note, it was found that the more mature electrophysiological performance was not associated with changes in related gene expression [121], and Cx43 localization in iPSC-CMs did not shift along with the well-aligned sarcomeres to the short axis of the cells, as observed in adult CMs [118]. When fatty acid was introduced to represent the primary energy source, hiPSC-CMs replated on a micropatterned surface (20 μm grooves) displayed more robust structural and functional improvements in sarcomere arrangement and contractile stress [119]. In addition, when seeded on a 15° chevron micropattern, Napiwocki et al. found that hiPSC-derived cardiac fibroblasts could improve hiPSC-CM performance with faster upstroke velocity and enhanced contraction stress when co-cultured [120]. Like micropatterning methods, nano-topographic substrates with an 800 nm width induced optimal effects on hiPSC-CM maturation by cellular alignment, elongated cell shape, and sarcomere organization [122, 130]. Larger nanogroove substrates performed better to advance iPSC-CM maturation compared to the 100 nm in vivo physiological size, which might be attributed to the limited contact that iPSC-CMs have with their substrates in 2D culture.

Overall, these observations indicate the beneficial effects of matrix stiffness and micro- or nanogroove patterns which simulate the native mechanical properties and architecture, thereby improving hiPSC-CM morphological and functional maturation. The underlying mechanisms for these stiffness-induced effects remain unclear, but integrin-mediated and ROCK signaling pathways, either directly or indirectly [108, 117, 131–133], have been implicated. Although it is challenging to mimic the well-aligned and uniaxial bundles of in vivo fibers, testing different types of substrates provides a useful means to explore the surface topographic role in CM development.

3.2 3D-Culturing via Improved Scaffold Design

Due to the limiting conditions of 2D culture for hiPSC-CM development, in which crucial cell-cell and cell-matrix interactions cannot be recapitulated sufficiently, researchers have shifted to 3D cell culture to study these interactions in a more representative model. The earliest 3D EHT was introduced more than two decades ago utilizing embryonic chick CMs [134] and NRVMs [51]. Since then, 3D cardiac construct development has advanced significantly by using a wide range of ECM-mimicking synthetic or native dECM scaffolds, inducing more adult-like CM morphology and functionality.

Wanjare et al. fabricated microfibrous polycaprolactone (PCL) scaffolds with random or parallel orientation [135], although these anisotropic PCL scaffolds were stiffer (1.81 MPa) than human myocardium (0.02–0.5 MPa) [136–139]. On the parallel fibers, seeded hiPSC-CMs showed enhanced cellular and myofibril alignment with increased length ($\approx$1.6 μm), increased *MYH7* gene expression level, and higher contraction velocity when compared to the randomly aligned scaffolds. Similarly, synthetic scaffolds with defined uniaxial orientation were also discovered to promote CM alignment and intercellular sarcomere organization, both in NVRMs and mouse ESC-CMs [102, 140]. Moreover, advanced printing technology, like melt electrowriting (MEW) and direct laser writing (DLW), has been introduced to fabricate 3D structures. hiPSC-CMs seeded on hexagonal PCL fiber scaffolds printed by MEW displayed enhanced sarcomere density and alignment with upregulated sarcomere, AP, calcium handling, and OXPHOS-related gene expression, as compared to rectangular fiber scaffolds [141], and additionally provided better mechanical properties upon force exposure. The effects of DLW-printed rectangular-shaped micro-scaffolds on single murine iPSC-CMs were evaluated by Silbernagel et al. [5]. Structural analysis revealed reorganized myofibrils with highly parallel alignment and the formation of sarcolemmal T-tubule-like structures on the cell membrane. Structural remodeling in murine iPSC-CMs tremendously impacted Ca^{2+} transient kinetics, along with enhanced clusters of Ca^{2+} handling protein expression (L-type Ca^{2+} channels and ryanodine receptors) [142].

Because of several intrinsic features of synthetic scaffolds, such as the lack of clear cell-adhesion points, the incapability to be remodeled by cells, and the potential activation of immune responses during graft-host integration, many groups have started exploring the use of scaffolds derived from decellularized heart tissue scaffolds. Compared to synthetic scaffolds, dECM scaffolds can largely preserve the native ECM composition and intricate structures [58, 143], providing a natural myocardial environment to support iPSC-CMs and promote their maturation. So far, dECM scaffolds from various mammalian species seeded with hiPSC-CMs have emerged as a promising method for making 3D cardiac tissues and evaluating hiPSC-CM maturation. In 2016, Guyette et al. decellularized a small part of the LV free wall of a donated human heart as a non-perfused cardiac scaffold [143]. The cardiac matrix induced alignment of hiPSC-CMs with an improved striated phenotype and enhanced force generation, when compared to cardiac slices in 2D culture. In addition to human dECM scaffolds, orthologous sources from porcine have also been investigated recently [144, 145]. Circular porcine dECM slices, with 12 mm diameters of 300 μm in thickness, induced elongated hiPSC-CM morphology and enhanced sarcomere organization [144]. When comparing wild-type (WT) and hypertrophic cardiomyopathy (HCM) swine dECM, hiPSC-CMs displayed higher force generation but worsened calcium handling on the stiffer HCM scaffold (± 4.5 kPa vs. WT scaffold 8.05 ± 2 kPa). Moreover, the expression of calcium handling genes was elevated (*ATP2A2*, *CACNA1C*, and *PLN*) in iPSC-CMs grown on the WT scaffold, whereas reduced calcium handling ability was observed on the HCM scaffold [145].

Next to using complete decellularized heart tissue as a scaffold, some studies also investigated the maturation effects of 3D hydrogels made of dECM powder on hiPSC-CMs. Interestingly, dECM derived from tissues of different ages showed different mechanical properties. Adult bovine dECM hydrogel was shown to contain more collagen fibers with increased width and was approximately tenfold stiffer than fetal dECM (67.5 Pa vs. 7.2 Pa) [146]. Comparably, dECM hydrogel from adult mice is more rigid but less elastic than neonatal dECM [147]. Studies on EHTs or cardiac microtissues generated by combining hiPSC-CMs and dECM hydrogels showed significant maturation, as indicated by enhanced cell architecture, contractility, and calcium handling [146–148]. On aged mice ECM it was found that hiPSC-CMs showed accelerated senescence and overgrowth, with sarcomere length above the upper physiological limit [147]. As a result, the age factor for dECM application in the clinic should be considered.

Excitation-contraction coupling is important for heart development and function, and electrical stimulation has become a vital approach to mature CMs in vitro. When PSC-CMs were stimulated on the "Biowire" device with progressively increasing electrical pacing from 1 to 6 Hz over the course of 1 week, indicative maturation was shown with respect to sarcomere reorganization, calcium handling, and a more negative resting membrane potential [33]. Another stimulation scheme (termed as "intensity training"), in which the hiPSC-derived EHTs were trained with a regular fluctuating electrical pacing showed a remarkable morphological maturity with adult-like size and T-tubule presence [34]. Electroconductive materials, such as carbon nanotubes and graphene oxide (GO) [149, 150], were also utilized to couple with dECM to advance the electrophysiological maturation of hiPSC-CMs. iPSC-CMs cultured with adult sheep dECM hydrogel with carbon nanotubes exhibited a higher Cx43 expression level and improved calcium handling [151]. Similarly, hiPSC-CMs cultured on a dECM-reduced GO hydrogel, with stiffness (17.5 ± 0.5 kPa) comparable to native rat myocardium, also displayed dramatically increased twitch force (23.61 ± 2.62 μN), enhanced contraction amplitude and upstroke velocity, and increased gene expression levels related to contractility (*TNNT2*, *TNNI3*, *TTN*, and *N2B*) and electrophysiological function (*CACNA1C*, *ATP2A2*, and *SERCA2*, 166].

Taken together, the maturation effects of synthetic scaffolds, dECM scaffolds, and dECM hydrogel reveal that substrate guidance cues supported by various topographies have an impact on hiPSC-CM morphology and function (Table 3). Particularly, dECM scaffolds from different ages exhibited inconsistent mechanical properties (ECM protein composition, scaffold rigidity, maximum displacement) and different effects on hiPSC-CMs, such as higher beating frequency and better calcium kinetics on young scaffolds, but impaired cardiac function on aged scaffolds [147]. Therefore, generating a standard evaluation for dECM scaffolds is necessary

Table 3 Summary of scaffolds and dECM hydrogels effects on human cardiac tissues

Mechanical stimulus	Differentiation methods	Max. stimulation duration	Results				References
			Morphology	Contractility	Electrophysiology	Metabolism	
Engineered myocardial tissue with parallel-aligned PCL scaffold	Wnt-modulation protocol	12 days	↑ Cellular anisotropy ↑ Sarcomere length	↑ *MYH7* expression ↑ Contraction velocity ↓ Contraction velocity when co-cultured with iPSC-ECs			[135]
MEW hexagonal fiber scaffolds	Wnt-modulation protocol	14 days	↑ Myofibril alignment ↑ Sarcomere length	↑ *MYH7*, *TNNI3*, *MYL2* expression	↑ *Cx43* expression ↑ *SCN5A*, *KCNJ2* expression	↑ *PPARGC1-α* expression	[141]
Human dECM fibers (15 mm length, 2.5 mm diameter)	Wnt-modulation protocol	≥60 days	↑ Cellular anisotropy ↑ Myofibril alignment	↑ Contraction stress			[143]
12-mm diameter, 300-μm thick porcine dECM myocardium slices	Wnt-modulation protocol	>200 days	↑ Cellular anisotropy ↑ Myofibril alignment ↑ Cell length		Anisotropic AP conduction Cx43 along the periphery of the cells		[144]
Swine HCM and WT dECM scaffolds	Wnt-modulation protocol	16 days		↑ Contraction stress on HCM scaffold	↓ Calcium handling on HCM scaffold ↑ *ATP2A2*, *CACNA1C*, and *PLN* on WT scaffold		[145]

Bovine adult and fetal hearts dECM hydrogel	Wnt-modulation protocol	7 days		↑ *MLC2v* expression ↓ Beat rate	↑ Calcium handling genes ↑ Calcium amplitude ↑ Upstroke velocity ↑ Downstroke velocity ↑ *Cx43* expression		[146]
Young (1–3 months), adult (6–9 months), and aged (22–24 months) mice hearts dECM hydrogel	Wnt-modulation protocol	>35 days	↑ Sarcomere length on aged ECM		↑ Calcium handling on young and adult ECM		[147]
Human myocardial dECM hydrogel	Wnt-modulation protocol	28 days	↑ Sarcomere length ↑ Myofibril alignment organized Z-lines, A-bands, and I-bands in sarcomeres	↑ Beating rate ↑ *MYH7*, *MLC2v*, and *TNNI3* expression	↑ Calcium handling		[148]
Adult sheep pericardial matrix-derived electroconductive biohybrid hydrogel	Wnt-modulation protocol	7 days	↑ Cellular anisotropy ↑ Myofibril alignment ↑ Sarcomere length	↑ Contraction amplitude ↑ Contraction speed No arrhythmic contractions	↑ Cx43 expression ↑ Calcium handling		[151]

(continued)

Table 3 (continued)

Mechanical stimulus	Differentiation methods	Max. stimulation duration	Results				References
			Morphology	Contractility	Electrophysiology	Metabolism	
Porcine dECM-rGO hydrogel scaffolds	Wnt-modulation protocol	35 days	↑ Sarcomere length ↑ Z-band width	↑ Contraction stress ↑ Ratio of *TNNI3* to *TNNI1* expression	↑ Calcium amplitude ↑ *CACNA1C*, *ATP2A2* expression ↑ *KCNH2*, *KCNE1* expression ↑ *Cx43* expression ↑ APD_{90} ↑ Conduction velocity		[152]

PCL polycaprolactone, *MEW* melt electrowriting, *DLW* direct laser writing, *rGO* reduced graphene oxide, *Cav1.2* voltage-dependent L-type Ca2+ channel, *ATP2A2* sarcoplasmic/endoplasmic reticulum calcium ATPase 2, *CACNA1C* calcium voltage-gated channel subunit Alpha1 C, *PLN* phospholamban, *KCNH2* potassium voltage-gated channel subfamily H member 2, *KCNE1* potassium voltage-gated channel subfamily E regulatory subunit 1, *KCNJ2* potassium channel, inwardly rectifying subfamily J, member 2, *SCN5A* sodium voltage-gated channel alpha subunit 5

before proceeding with future clinical trials. Alternatively, dECM from different ages could be applied to cardiac disease modeling to reproduce a pathophysiological myocardial condition related to age.

3.3 Mechanical Loading and Stretch

Passive static stress on CMs can be induced by attaching contractile tissues between two posts, where the distance between the two posts is correlated to the force of static stress. Simple static stretching by using two fixed ends [153, 154], flexible stretching with one moveable post [155, 156], and altering the distance between posts [19, 157–159] have all been shown to induce structural, functional, and molecular maturation of 3D hiPSC-CMs-derived tissues (more details are shown in Table 4).

It is noteworthy that when combining electrical pacing with static stretch, 3D cardiac constructs showed an additional increase in contractile stress with upregulated protein expression of RYR2 and SERCA2 compared to tissue maintained at a fixed length, suggesting more mature excitation-contraction coupling [153]. However, when training EHTs with one flexible post to mimic afterload, an upregulation of *GLUT4* expression and downregulation of *CPT1B* expression was shown, indicating more effective glycolysis but decreased fatty acid usage [156]. This outcome suggested that, while structural and electrophysiological maturation were observed with increasing afterload, the afterload was not effective in inducing mature oxidative metabolism. Thus, it is important to also assess the metabolism transition when evaluating other maturation effects. More recently, fibroblasts have been included in 3D cardiac tissues, showing that these cells provide a critical component in maintaining and improving CM function through secreting growth factors [166], aiding in electrical impulse propagation [167], and modulating ECM properties [168]. Engineered human myocardium (EHM), trained by dynamic stretch using flexible silicone holders [158, 169], displayed the maximal contractile forces when coupled with equivalent biopsy-derived fibroblasts under defined, serum-free conditions [19] and advanced hiPSC-CM sarcomere organization, contractility, and cardiac-associated gene expression as compared to parallel monolayer condition. For example, hiPSC-CMs in the EHM showed distinguishable sarcomeric M bands which were hardly visible under standard differentiation and prolonged culture [32]. Functionally, hiPSC-CMs in the EHM not only displayed stronger twitch force (6.2 ± 0.8 mN/mm^2 at 1.5 Hz) than human infants (1 mN/mm^2 at 1 Hz) [170], albeit weaker than adult myocardium (25 mN/mm^2 at 1 Hz) [171], but also exhibited dramatic upregulation of fetal to adult CM gene expression upon global transcriptome profiling.

Providing cyclic stretch with a given frequency resembles the rhythmic blood filling in the ventricles during diastole. Simulating this type of mechanical stimuli on hiPSC-CMs would provide cues more comparable to the native mechanical cardiac environment. Most studies carried out uniaxial cyclic stretching with a frequency of 1 to 2 Hz (60 to 120 cycles/min) and a displacement from 0 to 20% to recapitulate the

Table 4 Summary of the effects of mechanical stretch on PSC-CMs

Mechanical stimulus	Differentiation methods	Max. stimulation duration	Results				References
			Morphology	Contractility	Electrophysiology	Metabolism	
Static and cyclic (1 Hz uniaxial cyclic stress) stretch on 3D cardiac construct	Activin A/ BMP-4 induction	4 days	↑ Myofibril and collagen fiber alignment under static and cyclic stretch ↑ Cell area under cyclic stretch	↑ Beating rate under cyclic stretch ↑ *MYH7*, *cTNT* expression	↑ *CACNA1C*, *RYR2*, *SERCA2A* expression under cyclic stretch	↑ Mitochondria density	[154]
Static stress and electric stimulation on EHT	Activin A/ BMP-4 induction	14 days	↑ Myofibril alignment ↑ CM density ↑ Cellular anisotropy ↑ Cell area	↑ Contraction stress	↑ Calcium handling ↑SERCA2, RYR2 expression		[153]
Engineered cardiac tissue on adjustable afterload platform	Activin A/ BMP-4 induction	3 weeks	↑ Cellular anisotropy ↑ Cell area ↑ Sarcomere length ↑ Z-band width	↑ *MYH7*, *TNNI3*, *MLC2v* expression	↑ AP amplitude	↑ *GLUT4*, expression ↓ *CPT1B*, expression	[156]
Static stretch from 7mm post distance on EHM	Wnt-modulation protocol	50 days	↑ Myofibril alignment	↓ Beating rate ↑ *TNNT2* expression	↑ AP kinetics ↑ *CAV3*, *KCNJ2*, *CACNA1C* expression		[157]
EHM under defined, serum-free condition	Activin A/ BMP-4 and Wnt modulation induction	8 weeks	↑ Cell area rod-shaped cardiomyocytes ↑ Sarcomere length and organization with obvious M bands	↑ Contraction stress ↑ MLC2v expression Positive force-frequency behavior	↑ *Cx43*, *SCN5A* expression Unidirectional conduction ↑ Conduction velocity ↓ Resting membrane potential		[19]

Dynamic EHT model	Activin A/ BMP-4 induction	14 days	↑ Cellular anisotropy ↑ Myofibril alignment ↑ Sarcomere length	↑ Contraction stress ↑ *TNNI3* expression ↑ *DSP*, *PKP2*, and *DSC2* expression	↑ Conduction velocity ↑ Cx43 expression ↓Resting membrane potential	↑ *CPT1B*, *PDK4* expression	[155]
Growing static stretch on bioartificial cardiac tissue	Embryoid bodies formation	21 days	↑ Cellular anisotropy ↑ Myofibril alignment ↑ Sarcomere length	↑ Contraction stress ↑ MLC2V expression			[76]
1 Hz, 15% cyclic stretch on monolayer hiPSC-CMs	CDI iCell™ human CMs	10 days	↑ Cellular anisotropy ↑ Myofibril alignment ↑ Sarcomere length	↑ Contraction stress ↑ MYH7 expression			[160]
1 Hz, 15% cyclic stretch on monolayer hESC-CMs	Wnt-modulation protocol	2 days	↑ Cellular area ↑ Myofibril alignment ↑ Sarcomere length ↑ Cell stiffness	↑ MYH7 mRNA expression ↓ Beating rate	↓ Ca^{2+}-channels, K^+ and Na^+ channels associated genes	↑ROS production	[161]
1 Hz, 15% cyclic stretch on hiPSC-CM spheroids	Wnt-modulation protocol	7 days		↑ Ratio of *MYH7* to *MYH6* expression ↑ cTnI, MLC2v expression	↑ Cx43 expression ↑ Calcium handling		[162]

(continued)

Table 4 (continued)

Mechanical stimulus	Differentiation methods	Max. stimulation duration	Results				References
			Morphology	Contractility	Electrophysiology	Metabolism	
1.25 Hz, 12% cyclic stretch on cardiac tissue constructs	Activin A/ BMP-4 induction	3 days	↑ Gap junctions Well-developed Z bands	↑ Beating rate ↑ *MYH7* expression			[163]
Pulsatile flow (1.48 mL/min) and cyclic strain (5%) on hESC-CMs	Activin A/ BMP-4 induction	20 days	↑ Myofibril alignment ↑ Sarcomere length	↑ *MYH7* expression	↑Cx43 expression ↑*CACNA1C*, *ATP2A2* expression ↑ AP kinetics ↑ SR storage capacity		[164]
Cyclic pulsatile hemodynamic forces on hiPSC-CMs	Wnt-modulation protocol	7 days	↑ Cellular anisotropy ↑ Myofibril alignment ↑ Cellular area	↑ *TNNI3*/ *TNNI1* ratio		↑ Mitochondria density ↑ Mitochondria network	[165]

ROS reactive oxygen species, *GLUT4* glucose transporter type 4, *CPT1B* carnitine palmitoyltransferase 1B, *PDK4* pyruvate dehydrogenase lipoamide kinase isozyme 4, *CAV3* calveolin-3, *DSP* desmoplakin, *PKP2* plakophillin-2, *DSC-2* desmocollin-2

mechanical preload in vivo. Varieties of cellular effects induced by cyclic stretch on NVRMs, both in 2D and 3D culture, have been observed, including highly organized sarcomeres, Cx43 polarization [52, 172], higher twitch force [101, 173, 174], and impacts on focal adhesion kinase (FAK) activation that might involve hypertrophic and adhesive responses [175–177]. At the 2D level, iPSC-CMs seeded on an ECM protein cocktail and cultured under cyclic stretch (1 Hz, 15%) displayed overall enhanced sarcomere alignment, which reoriented them to more striated and parallel patterns compared to hiPSC-CM without cyclic stimulation (0% strain), and approximately doubled force generation with concurrently elevated protein expression of MYH7 [160]. In contrast, Ovchinnikova et al. observed increased cell stiffness on hESC-CMs, less expression of ion-channel genes upon RNA-seq analysis, and essential traits of cardiac hypertrophy [161], when subjected to the same mechanical stimulation. These two studies suggest that the cyclic stretch may induce CM hypertrophy other than maturation. Aside from cyclic stretch, blood flow-induced pulsatile laminar shear stress also plays a critical role in heart development [43]. Despite major effects on endothelial cells, Shen et al. showed that pulsatile flow (1.48 mL/min) with cyclic strain (5%) can influence hESC-CM maturation synergistically [164]. Within 20 days of stimulation, hESC-CMs exhibited improved sarcomere structure and increased contractility and electrophysiology-related gene expression, when compared to the static culture condition. Additionally, a lower β-catenin level was observed in hESC-CMs subjected to shear and cyclic stretch, indicating an involvement and inhibition of the Wnt/β-catenin signaling pathway. In 2020, Kolanowski et al. utilized a microfluidic system, which can provide homodynamic pressure in the physiological range, to study the cyclic pulsatile hemodynamic forces on hiPSC-CMs [165]. Compared to hiPSC-CMs cultured in static conditions for 1 week, the cyclic pulsatile flow clearly enhanced alignment of beating hiPSC-CMs, increased surface area, and facilitated mitochondrial density and network development. These morphological changes were also validated at the molecular level, like higher *TNNI3/TNNI1* ratio and increased expression of mitochondrial marker genes (*MT-CO1, OPA1*). In 3D cardiac constructs, elongated and hypertrophic CMs with enhanced Cx43 protein expression were found, both at 1 Hz [154, 162] and 1.25 Hz cyclic stretch [163], indicating more efficient electroconduction when compared to unstretched conditions. hiPSC-CM within the cardiac tissue also exhibited upregulation of CM maturation-related genes (*MYH7*, *cTnT*, *SERCA2A*, *cTnI*, and *MLC2v*)[162]. In contrast to hiPSC-CMs, hESC-CMs in gelatin-based scaffolds subjected to 1.25 Hz cyclic stretch displayed a faster Ca^{2+} cycle frequency and a lower Ca^{2+} cycle duration period, suggesting more rapid calcium cycling and thereby a higher beating frequency [163]. However, Kensah et al. observed that imposing cyclic stretch (10%, 1 Hz) for a week on murine iPSC-CM 3D aggregates did not improve tissue morphology and maximum active force [76], but beneficial effects could be found through applying stepwise growing static stretch, thereby stimulating prolonged sarcomere length, improved sarcomere organization, and better CM alignment.

4 Underlying Mechanisms of Biomechanical Stimulation

External biophysical cues are translated into intracellular biochemical stimuli by the bridge constructed between the cell cytoskeleton and ECM through focal adhesive contacts and ligand-receptor signaling transduction. CMs can adapt to the substrate stiffness variation and cyclic stretch and exhibit the mechanical force effects by activation or inhibition of a few specific signaling pathways, as illustrated in Fig. 2.

Integrins are transmembrane receptors that transduce mechanical signals to cells and bridge the environmental stimulation and intracellular events [132, 178, 179]. β1 integrin receptors, abundantly expressed in the rat adult heart [180], in combination with following FAK activation [181], have been shown to regulate the structural maturation of iPSC-CMs, like increasing the cTnI protein level [117]. Furthermore, activated PKC located at integrin adhesion sites, along with the downstream Src, FHOD1, and non-muscle myosin activation, has been reported to be involved in the CM rigidity sensing mechanism [112].

The RhoA/ROCK pathway has been found to modulate a wide range of fundamental cellular functions, including contraction, proliferation, and apoptosis [182]. Improper activation of the RhoA/ROCK pathways could likewise contribute

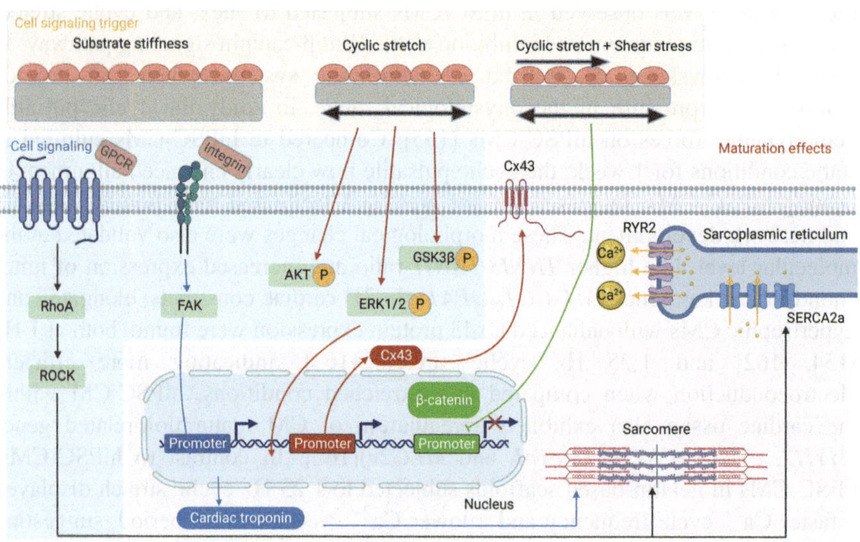

Fig. 2 Underlying molecules and pathways activated upon biophysical stimulation. Substrate stiffness influences CM maturation through RhoA and FAK. RhoA/ROCK pathways are involved to facilitate CM morphology, contraction, and calcium handling [50, 114, 117]. β1 integrin receptor activation, coupled with FAK activation, was reported to regulate sarcomere structure maturation [117]. Cyclic stretch induces phosphorylation of AKT, ERK1/2 and GSK3β, and ERK1/2 pathways can upregulate Cx43 expression [5]. Additionally, ESC-CMs showed a decreased β-catenin level when subjected to cyclic stretch and shear stress, indicating enhanced maturity [164]. This work was created with BioRender.com.

to major cardiovascular disorders [183]. Substrate stiffness has been shown to facilitate NRVM morphology, contraction, and calcium handling, in which the RhoA/ROCK and integrin-mediated pathways are incorporated to modify these mature phenotypes [50, 112, 114, 117]. RhoA/ROCK was discovered to be involved in the adaption of NRVMs to substrate stiffness changes, which led to myofibrils and focal adhesion development and allowed for a positive correlation between contraction force and substrate stiffness, according to Jacot et al. [50]. Also, in hESC-CMs, pharmacological blockade of the RhoA/ROCK pathway was reported to abolish the whole-cell spreading of SERCA2 and impaired the related calcium handling [114].

It is further known that ERK1/2, AKT, and GSK3β are highly involved in myocyte stretch [184]. Salameh et al. observed phosphorylation of these three signal transduction proteins when stimulating NRVMs with cyclic stretch (1 Hz, 10%), showing activation of ERK1/2 and AKT, but inhibition of GSK3β. Additionally, they showed that the ERK1/2 pathways, rather than PKA and PKC, were the main controllers of Cx43 upregulation and polarization in NRVMs [172].

Wnt/β-catenin signaling pathways have been reported to influence cardiogenesis through either activation or inhibition [167]. Wnt/β-catenin signaling is required for cardiac lineage cell development in the early phase, and late-stage CM differentiation is promoted through β-catenin downregulation [185]. When hESC-CMs were subjected to cyclic stretch and shear stress, a lower β-catenin level was detected, compared to cells without dynamic mechanical stimulation, suggesting that stimulated hESC-CMs became more mature [164]. This intriguing phenomenon confirmed the involvement of Wnt/β-catenin signaling pathways, and β-catenin protein level variation could be utilized as a generic marker to reflect the derived CM maturation status.

So far, there is insufficient knowledge about the molecular mechanisms involved in CM maturation, both in primary and PSC-derived cells, as most studies have mainly focused on characterizing maturation-related morphology and functionality. However, understanding the underlying molecular mechanisms contributing to CM maturation is critical for generating more adult-like CMs in vitro, providing researchers with more insights into CM development, and pathophysiological development in cardiac disease modeling.

5 Conclusion

Throughout this chapter, we have described the differences between adult CMs and hiPSC-CMs, as well as the effects of several biophysical strategies (topographies, substrate stiffness, passive and dynamic stretch) on hiPSC-CM morphological and functional maturation. The collective data revealed that hiPSC-CMs can sense microenvironmental biophysical cues and adapt by altering their structure and function. Even though hiPSC-CMs displayed improved maturity over their controls when subjected to mechanical stimulation, it should be highlighted that the level of maturity does not yet fully represent the adult phenotypes.

It is challenging to determine which strategy is the most effective to promote hiPSC-CM maturation because of the inherent restrictions when performing a horizontal comparison between studies with varying experimental setups. Firstly, the stimulation period and starting point of hiPSC-CM age differed from study to study. The duration of hiPSC-CM culture has been reported as a critical factor in maturation [32]. Thus, using aged hiPSC-CMs may have already resulted in the presence of more mature characteristics. Regarding 3D cardiac tissue constructs, different groups have established their own optimal protocols to generate cardiac constructs with different types of non-cardiomyocyte populations and cell inoculation ratios. Because of the heterogeneity in 3D cardiac constructs, directly comparing maturation effects of mechanical stimulation among these studies is difficult. Next, it is acceptable that not all maturation-related features are investigated in each strategy. Most studies focus on morphological, contractile, and electrophysiological changes, but pay less attention to metabolic alterations. Next to maturation, a potential concern is the heterogeneous populations when employing iPSC-CMs as a research platform, since the majority of studies used mixed hiPSC-CM populations, comprising ventricular-like cells, as well as small numbers of pacemaker- and atrial-like cells, thereby posing an obstacle to cell therapy and disease modeling studies. It is possible that mixed myocyte types may influence the outcomes of cardiovascular disease modeling in vitro as well as change the graft performance in vivo. However, progress has been made to define the molecular barriers among various types of cardiac myocytes when differentiating them in vitro [186, 187]. It is critical to establish differentiation strategies that induce the production of each of these cardiomyocyte subtypes in order to treat and simulate illnesses that affect specific parts of the heart. Although biophysical cues are able to induce hiPSC-CM maturation, few studies have moved forward to dig into the underlying mechanisms. Several studies have shown the role of RhoA/ROCK, integrin, ERK1/2, and Wnt/β-catenin pathways in initiating biophysical responses, but they were performed on NRVMs and 2D PSC-CMs. More concrete answers could be found in the native development process. Consequently, it is crucial to understand the role of natural mechanical stimulation and delineate molecular mechanisms in native CMs during growth and maturation, especially in terms of timepoints, intensity, and order. Enhanced knowledge on the underlying mechanisms will allow for generating CMs with more adult-like phenotypes and a better understanding of CM development in a physiological or pathophysiological environment.

From the literature gathered, the more physiological-like the strategy, the more mature traits the hiPSC-CMs could acquire. Nevertheless, CM maturation is an extremely sophisticated process that necessitates well-orchestrated management of multiple signaling pathways in the cells as well as external stimuli. One could imagine that fully matured iPSC-CMs would be achieved if all related stimulations corresponding to native CM development could be ideally simulated in vitro. However, this would be extremely laborious since natural CM development takes at least 10 months. Perhaps it is necessary to better define the minimal maturity requirements of iPSC-CMs for each application. Prior studies on in vivo transplantation showed that injected adult CMs in the injured area did not survive long term [188], but that the in vivo environment can promote transplanted iPSC-CM maturation [189]. Thereby, for the purpose of therapeutic remuscularization after myocardial infarction, it is practical to induce iPSC-CM maturation in the recipient's native environment. On the other hand, determining the "minimal" maturity of iPSC-CMs for disease modelling, drug screening and toxicology research is critical to prevent false positives and negatives and allow predictive modelling. Finally, by elucidating the heart development biophysical molecular mechanisms, e.g. through transcriptionally mapping PSC-CM differentiation and maturation [190], and by using single-cell RNA sequencing (scRNA-seq) technology [191, 192] on CM developmental samples researchers could pinpoint and potentially activate the critical pathways to further induce iPSC-CMs maturation.

Compliance with Ethical Standards This study was funded by the EU-funded project BRAV∃ (H2020, ID:874827). All authors declare they have no conflict of interest. This article does not contain any studies with human participants performed by any of the authors.

References

1. Moran AE, Forouzanfar MH, Roth GA, Mensah GA, Ezzati M, Flaxman A, Murray CJL, Naghavi M (2014) The global burden of ischemic heart disease in 1990 and 2010: the Global Burden of Disease 2010 study. Circulation 129:1493–1501. https://doi.org/10.1161/CIRCULATIONAHA.113.004046
2. Steffens S, Van Linthout S, Sluijter JPG, Tocchetti CG, Thum T, Madonna R (2020) Stimulating pro-reparative immune responses to prevent adverse cardiac remodelling: consensus document from the joint 2019 meeting of the ESC Working Groups of cellular biology of the heart and myocardial function. Cardiovasc Res 116:1850–1862. https://doi.org/10.1093/cvr/cvaa137
3. Bär C, Chatterjee S, Falcão Pires I, Rodrigues P, Sluijter JPG, Boon RA, Nevado RM, Andrés V, Sansonetti M, de Windt L, Ciccarelli M, Hamdani N, Heymans S, Figuinha Videira R, Tocchetti CG, Giacca M, Zacchigna S, Engelhardt S, Dimmeler S, Madonna R, Thum T (2020) Non-coding RNAs: update on mechanisms and therapeutic targets from the ESC Working Groups of Myocardial Function and Cellular Biology of the Heart. Cardiovasc Res 116:1805–1819. https://doi.org/10.1093/cvr/cvaa195
4. Sluijter JPG, Davidson SM, Boulanger CM, Buzás EI, de Kleijn DPV, Engel FB, Giricz Z, Hausenloy DJ, Kishore R, Lecour S, Leor J, Madonna R, Perrino C, Prunier F, Sahoo S, Schiffelers RM, Schulz R, Van Laake LW, Ytrehus K, Ferdinandy P (2018) Extracellular vesicles in diagnostics and therapy of the ischaemic heart: position paper from the working group on cellular biology of the Heart of the European Society of Cardiology. Cardiovasc Res 114:19–34. https://doi.org/10.1093/cvr/cvx211
5. Ruiz-Meana M, Bou-Teen D, Ferdinandy P, Gyongyosi M, Pesce M, Perrino C, Schulz R, Sluijter JPG, Tocchetti CG, Thum T, Madonna R (2020) Cardiomyocyte ageing and cardioprotection: consensus document from the ESC working groups cell biology of the heart and myocardial function. Cardiovasc Res 116:1835–1849. https://doi.org/10.1093/cvr/cvaa132
6. Perrino C, Ferdinandy P, Bøtker HE, Brundel BJJM, Collins P, Davidson SM, den Ruijter HM, Engel FB, Gerdts E, Girao H, Gyöngyösi M, Hausenloy DJ, Lecour S, Madonna R, Marber M, Murphy E, Pesce M, Regitz-Zagrosek V, Sluijter JPG, Steffens S, Gollmann-Tepeköylü C, Van Laake LW, Van Linthout S, Schulz R, Ytrehus K (2021) Improving translational research in sex-specific effects of comorbidities and risk factors in ischaemic

heart disease and cardioprotection: position paper and recommendations of the ESC Working Group on Cellular Biology of the Heart. Cardiovasc Res 117:367–385. https://doi.org/10.1093/cvr/cvaa155

7. Lecour S, Bøtker HE, Condorelli G, Davidson SM, Garcia-Dorado D, Engel FB, Ferdinandy P, Heusch G, Madonna R, Ovize M, Ruiz-Meana M, Schulz R, Sluijter JPG, Van Laake LW, Yellon DM, Hausenloy DJ (2014) ESC working group cellular biology of the heart: position paper: improving the preclinical assessment of novel cardioprotective therapies. Cardiovasc Res 104:399–411. https://doi.org/10.1093/cvr/cvu225
8. Zacchigna S, Paldino A, Falcão-Pires I, Daskalopoulos EP, Dal Ferro M, Vodret S, Lesizza P, Cannatà A, Miranda-Silva D, Lourenço AP, Pinamonti B, Sinagra G, Weinberger F, Eschenhagen T, Carrier L, Kehat I, Tocchetti CG, Russo M, Ghigo A, Cimino J, Hirsch E, Dawson D, Ciccarelli M, Oliveti M, Linke WA, Cuijpers I, Heymans S, Hamdani N, de Boer M, Duncker DJ, Kuster D, van der Velden J, Beauloye C, Bertrand L, Mayr M, Giacca M, Leuschner F, Backs J, Thum T (2021) Towards standardization of echocardiography for the evaluation of left ventricular function in adult rodents: a position paper of the ESC Working Group on Myocardial Function. Cardiovasc Res 117:43–59. https://doi.org/10.1093/cvr/cvaa110
9. Alexander Y, Osto E, Schmidt-Trucksäss A, Shechter M, Trifunovic D, Duncker DJ, Aboyans V, Bäck M, Badimon L, Cosentino F, De Carlo M, Dorobantu M, Harrison DG, Guzik TJ, Hoefer I, Morris PD, Norata GD, Suades R, Taddei S, Vilahur G, Waltenberger J, Weber C, Wilkinson F, Bochaton-Piallat M-L, Evans PC (2021) Endothelial function in cardiovascular medicine: a consensus paper of the European Society of Cardiology Working Groups on Atherosclerosis and Vascular Biology, Aorta and Peripheral Vascular Diseases, Coronary Pathophysiology and Microcirculation, and Thrombosis. Cardiovasc Res 117:29–42. https://doi.org/10.1093/cvr/cvaa085
10. Grigorian-Shamagian L, Sanz-Ruiz R, Climent A, Badimon L, Barile L, Bolli R, Chamuleau S, Grobbee DE, Janssens S, Kastrup J, Kragten-Tabatabaie L, Madonna R, Mathur A, Menasché P, Pompilio G, Prosper F, Sena E, Smart N, Zimmermann W-H, Fernández-Avilés F (2021) Insights into therapeutic products, preclinical research models, and clinical trials in cardiac regenerative and reparative medicine: where are we now and the way ahead. Current opinion paper of the ESC Working Group on Cardiovascular Regenerative and Reparative Medicine. Cardiovasc Res 117:1428–1433. https://doi.org/10.1093/cvr/cvaa337
11. Wang EY, Kuzmanov U, Smith JB, Dou W, Rafatian N, Lai BFL, Lu RXZ, Wu Q, Yazbeck J, Zhang X-O, Sun Y, Gramolini A, Radisic M (2021) An organ-on-a-chip model for pre-clinical drug evaluation in progressive non-genetic cardiomyopathy. J Mol Cell Cardiol 160:97–110. https://doi.org/10.1016/j.yjmcc.2021.06.012
12. Takahashi K, Tanabe K, Ohnuki M, Narita M, Ichisaka T, Tomoda K, Yamanaka S (2007) Induction of pluripotent stem cells from adult human fibroblasts by defined factors. Cell 131:861–872. https://doi.org/10.1016/j.cell.2007.11.019
13. Guo X-M, Zhao Y-S, Chang H-X, Wang C-Y, Zhang X-A, Duan C-M, Dong L-Z, Jiang H, Li J, Song Y, Yang XJ (2006) Creation of engineered cardiac tissue in vitro from mouse embryonic stem cells. Circulation 113:2229–2237. https://doi.org/10.1161/CIRCULATIONAHA.105.583039
14. Wang X, Wei G, Yu W, Zhao Y, Yu X, Ma X (2006) Scalable producing embryoid bodies by rotary cell culture system and constructing engineered cardiac tissue with ES-derived cardiomyocytes in vitro. Biotechnol Prog 22:811–818. https://doi.org/10.1021/bp060018z
15. Querdel E, Reinsch M, Castro L, Köse D, Bähr A, Reich S, Geertz B, Ulmer B, Schulze M, Lemoine MD, Krause T, Lemme M, Sani J, Shibamiya A, Stüdemann T, Köhne M, von Bibra C, Hornaschewitz N, Pecha S, Nejahsie Y, Mannhardt I, Christ T, Reichenspurner H, Hansen A, Klymiuk N, Krane M, Kupatt C, Eschenhagen T, Weinberger F (2021) Human engineered heart tissue patches remuscularize the injured heart in a dose-dependent manner. Circulation 143:1991–2006. https://doi.org/10.1161/CIRCULATIONAHA.120.047904

16. Shiba Y, Gomibuchi T, Seto T, Wada Y, Ichimura H, Tanaka Y, Ogasawara T, Okada K, Shiba N, Sakamoto K, Ido D, Shiina T, Ohkura M, Nakai J, Uno N, Kazuki Y, Oshimura M, Minami I, Ikeda U (2016) Allogeneic transplantation of iPS cell-derived cardiomyocytes regenerates primate hearts. Nature 538:388–391. https://doi.org/10.1038/nature19815
17. Zimmermann W-H (2017) Remuscularization of the failing heart. J Physiol 595:3685–3690. https://doi.org/10.1113/JP273098
18. Gao L, Gregorich ZR, Zhu W, Mattapally S, Oduk Y, Lou X, Kannappan R, Borovjagin AV, Walcott GP, Pollard AE, Fast VG, Hu X, Lloyd SG, Ge Y, Zhang J (2018) Large cardiac muscle patches engineered from human induced-pluripotent stem cell-derived cardiac cells improve recovery from myocardial infarction in swine. Circulation 137:1712–1730. https://doi.org/10.1161/CIRCULATIONAHA.117.030785
19. Tiburcy M, Hudson JE, Balfanz P, Schlick S, Meyer T, Chang Liao M-L, Levent E, Raad F, Zeidler S, Wingender E, Riegler J, Wang M, Gold JD, Kehat I, Wettwer E, Ravens U, Dierickx P, van Laake LW, Goumans MJ, Khadjeh S, Toischer K, Hasenfuss G, Couture LA, Unger A, Linke WA, Araki T, Neel B, Keller G, Gepstein L, Wu JC, Zimmermann W-H (2017) Defined engineered human myocardium with advanced maturation for applications in heart failure modeling and repair. Circulation 135:1832–1847. https://doi.org/10.1161/CIRCULATIONAHA.116.024145
20. Weinberger F, Breckwoldt K, Pecha S, Kelly A, Geertz B, Starbatty J, Yorgan T, Cheng K-H, Lessmann K, Stolen T, Scherrer-Crosbie M, Smith G, Reichenspurner H, Hansen A, Eschenhagen T (2016) Cardiac repair in guinea pigs with human engineered heart tissue from induced pluripotent stem cells. Sci Transl Med 8:363ra148. https://doi.org/10.1126/scitranslmed.aaf8781
21. Doherty KR, Talbert DR, Trusk PB, Moran DM, Shell SA, Bacus S (2015) Structural and functional screening in human induced-pluripotent stem cell-derived cardiomyocytes accurately identifies cardiotoxicity of multiple drug types. Toxicol Appl Pharmacol 285:51–60. https://doi.org/10.1016/j.taap.2015.03.008
22. Mordwinkin NM, Burridge PW, Wu JC (2013) A review of human pluripotent stem cell-derived cardiomyocytes for high-throughput drug discovery, cardiotoxicity screening, and publication standards. J Cardiovasc Transl Res 6:22–30. https://doi.org/10.1007/s12265-012-9423-2
23. Oikonomopoulos A, Kitani T, Wu JC (2018) Pluripotent stem cell-derived cardiomyocytes as a platform for cell therapy applications: progress and hurdles for clinical translation. Mol Ther 26:1624–1634. https://doi.org/10.1016/j.ymthe.2018.02.026
24. Bertero A, Pawlowski M, Ortmann D, Snijders K, Yiangou L, Cardoso de Brito M, Brown S, Bernard WG, Cooper JD, Giacomelli E, Gambardella L, Hannan NRF, Iyer D, Sampaziotis F, Serrano F, Zonneveld MCF, Sinha S, Kotter M, Vallier L (2016) Optimized inducible shRNA and CRISPR/Cas9 platforms for in vitro studies of human development using hPSCs. Development 143:4405–4418. https://doi.org/10.1242/dev.138081
25. Paige SL, Thomas S, Stoick-Cooper CL, Wang H, Maves L, Sandstrom R, Pabon L, Reinecke H, Pratt G, Keller G, Moon RT, Stamatoyannopoulos J, Murry CE (2012) A temporal chromatin signature in human embryonic stem cells identifies regulators of cardiac development. Cell 151:221–232. https://doi.org/10.1016/j.cell.2012.08.027
26. Kattman SJ, Witty AD, Gagliardi M, Dubois NC, Niapour M, Hotta A, Ellis J, Keller G (2011) Stage-specific optimization of activin/nodal and BMP signaling promotes cardiac differentiation of mouse and human pluripotent stem cell lines. Cell Stem Cell 8:228–240. https://doi.org/10.1016/j.stem.2010.12.008
27. Burridge PW, Keller G, Gold JD, Wu JC (2012) Production of de novo cardiomyocytes: human pluripotent stem cell differentiation and direct reprogramming. Cell Stem Cell 10:16–28. https://doi.org/10.1016/j.stem.2011.12.013
28. Burridge PW, Matsa E, Shukla P, Lin ZC, Churko JM, Ebert AD, Lan F, Diecke S, Huber B, Mordwinkin NM, Plews JR, Abilez OJ, Cui B, Gold JD, Wu JC (2014) Chemically defined

generation of human cardiomyocytes. Nat Methods 11:855–860. https://doi.org/10.1038/nmeth.2999

29. Buikema JW, Lee S, Goodyer WR, Maas RG, Chirikian O, Li G, Miao Y, Paige SL, Lee D, Wu H, Paik DT, Rhee S, Tian L, Galdos FX, Puluca N, Beyersdorf B, Hu J, Beck A, Venkamatran S, Swami S, Wijnker P, Schuldt M, Dorsch LM, van Mil A, Red-Horse K, Wu JY, Geisen C, Hesse M, Serpooshan V, Jovinge S, Fleischmann BK, Doevendans PA, van der Velden J, Garcia KC, Wu JC, Sluijter JPG, Wu SM (2020) Wnt activation and reduced cell-cell contact synergistically induce massive expansion of functional human iPSC-derived cardiomyocytes. Cell Stem Cell 27:50–63.e5. https://doi.org/10.1016/j.stem.2020.06.001
30. Karbassi E, Fenix A, Marchiano S, Muraoka N, Nakamura K, Yang X, Murry CE (2020) Cardiomyocyte maturation: advances in knowledge and implications for regenerative medicine. Nat Rev Cardiol 17:341–359. https://doi.org/10.1038/s41569-019-0331-x
31. Yang X, Pabon L, Murry CE (2014) Engineering adolescence: maturation of human pluripotent stem cell-derived cardiomyocytes. Circ Res 114:511–523. https://doi.org/10.1161/CIRCRESAHA.114.300558
32. Kamakura T, Makiyama T, Sasaki K, Yoshida Y, Wuriyanghai Y, Chen J, Hattori T, Ohno S, Kita T, Horie M, Yamanaka S, Kimura T (2013) Ultrastructural maturation of human-induced pluripotent stem cell-derived cardiomyocytes in a long-term culture. Circ J 77:1307–1314. https://doi.org/10.1253/circj.cj-12-0987
33. Nunes SS, Miklas JW, Liu J, Aschar-Sobbi R, Xiao Y, Zhang B, Jiang J, Massé S, Gagliardi M, Hsieh A, Thavandiran N, Laflamme MA, Nanthakumar K, Gross GJ, Backx PH, Keller G, Radisic M (2013) Biowire: a platform for maturation of human pluripotent stem cell-derived cardiomyocytes. Nat Methods 10:781–787. https://doi.org/10.1038/nmeth.2524
34. Ronaldson-Bouchard K, Ma SP, Yeager K, Chen T, Song L, Sirabella D, Morikawa K, Teles D, Yazawa M, Vunjak-Novakovic G (2018) Advanced maturation of human cardiac tissue grown from pluripotent stem cells. Nature 556:239–243. https://doi.org/10.1038/s41586-018-0016-3
35. Correia C, Koshkin A, Duarte P, Hu D, Teixeira A, Domian I, Serra M, Alves PM (2017) Distinct carbon sources affect structural and functional maturation of cardiomyocytes derived from human pluripotent stem cells. Sci Rep 7:8590. https://doi.org/10.1038/s41598-017-08713-4
36. Feyen DAM, McKeithan WL, Bruyneel AAN, Spiering S, Hörmann L, Ulmer B, Zhang H, Briganti F, Schweizer M, Hegyi B, Liao Z, Pölönen R-P, Ginsburg KS, Lam CK, Serrano R, Wahlquist C, Kreymerman A, Vu M, Amatya PL, Behrens CS, Ranjbarvaziri S, Maas RGC, Greenhaw M, Bernstein D, Wu JC, Bers DM, Eschenhagen T, Metallo CM, Mercola M (2020) Metabolic maturation media improve physiological function of human iPSC-derived cardiomyocytes. Cell Rep 32:107925. https://doi.org/10.1016/j.celrep.2020.107925
37. Funakoshi S, Fernandes I, Mastikhina O, Wilkinson D, Tran T, Dhahri W, Mazine A, Yang D, Burnett B, Lee J, Protze S, Bader GD, Nunes SS, Laflamme M, Keller G (2021) Generation of mature compact ventricular cardiomyocytes from human pluripotent stem cells. Nat Commun 12:3155. https://doi.org/10.1038/s41467-021-23329-z
38. Mills RJ, Titmarsh DM, Koenig X, Parker BL, Ryall JG, Quaife-Ryan GA, Voges HK, Hodson MP, Ferguson C, Drowley L, Plowright AT, Needham EJ, Wang Q-D, Gregorevic P, Xin M, Thomas WG, Parton RG, Nielsen LK, Launikonis BS, James DE, Elliott DA, Porrello ER, Hudson JE (2017) Functional screening in human cardiac organoids reveals a metabolic mechanism for cardiomyocyte cell cycle arrest. Proc Natl Acad Sci U S A 114:E8372–E8381. https://doi.org/10.1073/pnas.1707316114
39. Chirico N, Kessler EL, Maas RG, Fang J, Qin J, Dokter I, Daniels M, Šarić T, Neef K, Buikema JW, Lei Z (2022) Small molecule-mediated rapid maturation of human induced pluripotent stem cell-derived cardiomyocytes. Stem Cell Res Ther 13(1):531. https://doi.org/10.1186/s13287-022-03209-z
40. Dunn KK, Reichardt IM, Simmons AD, Jin G, Floy ME, Hoon KM, Palecek SP (2019) Coculture of endothelial cells with human pluripotent stem cell-derived cardiac progenitors

reveals a differentiation stage-specific enhancement of cardiomyocyte maturation. Biotechnol J 14:e1800725. https://doi.org/10.1002/biot.201800725
41. Guo Y, Pu WT (2020) Cardiomyocyte maturation: new phase in development. Circ Res 126: 1086–1106. https://doi.org/10.1161/CIRCRESAHA.119.315862
42. Brade T, Pane LS, Moretti A, Chien KR, Laugwitz K-L (2013) Embryonic heart progenitors and cardiogenesis. Cold Spring Harb Perspect Med 3:a013847. https://doi.org/10.1101/cshperspect.a013847
43. Andrés-Delgado L, Mercader N (2016) Interplay between cardiac function and heart development. Biochim Biophys Acta 1863:1707–1716. https://doi.org/10.1016/j.bbamcr.2016.03.004
44. Gaetani R, Zizzi EA, Deriu MA, Morbiducci U, Pesce M, Messina E (2020) When stiffness matters: mechanosensing in heart development and disease. Front Cell Dev Biol 8:334. https://doi.org/10.3389/fcell.2020.00334
45. Hove JR, Köster RW, Forouhar AS, Acevedo-Bolton G, Fraser SE, Gharib M (2003) Intracardiac fluid forces are an essential epigenetic factor for embryonic cardiogenesis. Nature 421: 172–177. https://doi.org/10.1038/nature01282
46. Taber LA (2001) Biomechanics of cardiovascular development. Annu Rev Biomed Eng 3:1–25. https://doi.org/10.1146/annurev.bioeng.3.1.1
47. Zhu R, Blazeski A, Poon E, Costa KD, Tung L, Boheler KR (2014) Physical developmental cues for the maturation of human pluripotent stem cell-derived cardiomyocytes. Stem Cell Res Ther 5:117. https://doi.org/10.1186/scrt507
48. Bhana B, Iyer RK, Chen WLK, Zhao R, Sider KL, Likhitpanichkul M, Simmons CA, Radisic M (2010) Influence of substrate stiffness on the phenotype of heart cells. Biotechnol Bioeng 105:1148–1160. https://doi.org/10.1002/bit.22647
49. Jacot JG, Kita-Matsuo H, Wei KA, Chen HSV, Omens JH, Mercola M, McCulloch AD (2010) Cardiac myocyte force development during differentiation and maturation. Ann N Y Acad Sci 1188:121–127. https://doi.org/10.1111/j.1749-6632.2009.05091.x
50. Jacot JG, McCulloch AD, Omens JH (2008) Substrate stiffness affects the functional maturation of neonatal rat ventricular myocytes. Biophys J 95:3479–3487. https://doi.org/10.1529/biophysj.107.124545
51. Zimmermann WH, Fink C, Kralisch D, Remmers U, Weil J, Eschenhagen T (2000) Three-dimensional engineered heart tissue from neonatal rat cardiac myocytes. Biotechnol Bioeng 68:106–114
52. Zimmermann W-H, Schneiderbanger K, Schubert P, Didié M, Münzel F, Heubach JF, Kostin S, Neuhuber WL, Eschenhagen T (2002) Tissue engineering of a differentiated cardiac muscle construct. Circ Res 90:223–230. https://doi.org/10.1161/hh0202.103644
53. Li A-H, Liu PP, Villarreal FJ, Garcia RA (2014) Dynamic changes in myocardial matrix and relevance to disease: translational perspectives. Circ Res 114:916–927. https://doi.org/10.1161/CIRCRESAHA.114.302819
54. Kim H, Yoon CS, Kim H, Rah B (1999) Expression of extracellular matrix components fibronectin and laminin in the human fetal heart. Cell Struct Funct 24:19–26. https://doi.org/10.1247/csf.24.19
55. Schwach V, Passier R (2019) Native cardiac environment and its impact on engineering cardiac tissue. Biomater Sci 7:3566–3580. https://doi.org/10.1039/c8bm01348a
56. Hughes CS, Postovit LM, Lajoie GA (2010) Matrigel: a complex protein mixture required for optimal growth of cell culture. Proteomics 10:1886–1890. https://doi.org/10.1002/pmic.200900758
57. Kc P, Hong Y, Zhang G (2019) Cardiac tissue-derived extracellular matrix scaffolds for myocardial repair: advantages and challenges. Regen Biomater 6:185–199. https://doi.org/10.1093/rb/rbz017
58. Ott HC, Matthiesen TS, Goh S-K, Black LD, Kren SM, Netoff TI, Taylor DA (2008) Perfusion-decellularized matrix: using nature's platform to engineer a bioartificial heart. Nat Med 14:213–221. https://doi.org/10.1038/nm1684

59. Bergmann O, Zdunek S, Felker A, Salehpour M, Alkass K, Bernard S, Sjostrom SL, Szewczykowska M, Jackowska T, Dos Remedios C, Malm T, Andrä M, Jashari R, Nyengaard JR, Possnert G, Jovinge S, Druid H, Frisén J (2015) Dynamics of cell generation and turnover in the human heart. Cell 161:1566–1575. https://doi.org/10.1016/j.cell.2015.05.026
60. Mollova M, Bersell K, Walsh S, Savla J, Das LT, Park S-Y, Silberstein LE, dos Remedios CG, Graham D, Colan S, Kühn B (2013) Cardiomyocyte proliferation contributes to heart growth in young humans. Proc Natl Acad Sci USA 110:1446–1451. https://doi.org/10.1073/pnas.1214608110
61. Bergmann O, Bhardwaj RD, Bernard S, Zdunek S, Barnabé-Heider F, Walsh S, Zupicich J, Alkass K, Buchholz BA, Druid H, Jovinge S, Frisén J (2009) Evidence for cardiomyocyte renewal in humans. Science 324:98–102. https://doi.org/10.1126/science.1164680
62. Lundy SD, Zhu W-Z, Regnier M, Laflamme MA (2013) Structural and functional maturation of cardiomyocytes derived from human pluripotent stem cells. Stem Cells Dev 22:1991–2002. https://doi.org/10.1089/scd.2012.0490
63. Meilhac SM, Buckingham ME (2018) The deployment of cell lineages that form the mammalian heart. Nat Rev Cardiol 15:705–724. https://doi.org/10.1038/s41569-018-0086-9
64. McCain ML, Parker KK (2011) Mechanotransduction: the role of mechanical stress, myocyte shape, and cytoskeletal architecture on cardiac function. Pflugers Arch 462:89–104. https://doi.org/10.1007/s00424-011-0951-4
65. Schwartz K, Boheler KR, de la Bastie D, Lompre AM, Mercadier JJ (1992) Switches in cardiac muscle gene expression as a result of pressure and volume overload. Am J Physiol 262: R364–R369. https://doi.org/10.1152/ajpregu.1992.262.3.R364
66. Bird SD, Doevendans PA, van Rooijen MA, Brutel de la Riviere A, Hassink RJ, Passier R, Mummery CL (2003) The human adult cardiomyocyte phenotype. Cardiovasc Res 58:423–434. https://doi.org/10.1016/s0008-6363(03)00253-0
67. Snir M, Kehat I, Gepstein A, Coleman R, Itskovitz-Eldor J, Livne E, Gepstein L (2003) Assessment of the ultrastructural and proliferative properties of human embryonic stem cell-derived cardiomyocytes. Am J Physiol Heart Circ Physiol 285:H2355–H2363. https://doi.org/10.1152/ajpheart.00020.2003
68. Denning C, Borgdorff V, Crutchley J, Firth KSA, George V, Kalra S, Kondrashov A, Hoang MD, Mosqueira D, Patel A, Prodanov L, Rajamohan D, Skarnes WC, Smith JGW, Young LE (2016) Cardiomyocytes from human pluripotent stem cells: from laboratory curiosity to industrial biomedical platform. Biochim Biophys Acta 1863:1728–1748. https://doi.org/10.1016/j.bbamcr.2015.10.014
69. Gregorio CC, Antin PB (2000) To the heart of myofibril assembly. Trends Cell Biol 10:355–362. https://doi.org/10.1016/S0962-8924(00)01793-1
70. Veerman CC, Kosmidis G, Mummery CL, Casini S, Verkerk AO, Bellin M (2015) Immaturity of human stem-cell-derived cardiomyocytes in culture: fatal flaw or soluble problem? Stem Cells Dev 24:1035–1052. https://doi.org/10.1089/scd.2014.0533
71. de Sousa C, Lopes SM, Hassink RJ, Feijen A, van Rooijen MA, Doevendans PA, Tertoolen L, Brutel de la Rivière A, Mummery CL (2006) Patterning the heart, a template for human cardiomyocyte development. Dev Dyn 235:1994–2002. https://doi.org/10.1002/dvdy.20830
72. Mahdavi V, Lompre AM, Chambers AP, Nadal-Ginard B (1984) Cardiac myosin heavy chain isozymic transitions during development and under pathological conditions are regulated at the level of mRNA availability. Eur Heart J 5(Suppl F):181–191. https://doi.org/10.1093/eurheartj/5.suppl_f.181
73. Piccini I, Rao J, Seebohm G, Greber B (2015) Human pluripotent stem cell-derived cardiomyocytes: Genome-wide expression profiling of long-term in vitro maturation in comparison to human heart tissue. Genom Data 4:69–72. https://doi.org/10.1016/j.gdata.2015.03.008
74. Zhang R, Guo T, Han Y, Huang H, Shi J, Hu J, Li H, Wang J, Saleem A, Zhou P, Lan F (2021) Design of synthetic microenvironments to promote the maturation of human pluripotent stem

cell derived cardiomyocytes. J Biomed Mater Res Part B 109:949–960. https://doi.org/10.1002/jbm.b.34759
75. Hasenfuss G, Mulieri LA, Blanchard EM, Holubarsch C, Leavitt BJ, Ittleman F, Alpert NR (1991) Energetics of isometric force development in control and volume-overload human myocardium. Comparison with animal species. Circ Res 68:836–846. https://doi.org/10.1161/01.res.68.3.836
76. Kensah G, Roa Lara A, Dahlmann J, Zweigerdt R, Schwanke K, Hegermann J, Skvorc D, Gawol A, Azizian A, Wagner S, Maier LS, Krause A, Dräger G, Ochs M, Haverich A, Gruh I, Martin U (2013) Murine and human pluripotent stem cell-derived cardiac bodies form contractile myocardial tissue in vitro. Eur Heart J 34:1134–1146. https://doi.org/10.1093/eurheartj/ehs349
77. Kita-Matsuo H, Barcova M, Prigozhina N, Salomonis N, Wei K, Jacot JG, Nelson B, Spiering S, Haverslag R, Kim C, Talantova M, Bajpai R, Calzolari D, Terskikh A, McCulloch AD, Price JH, Conklin BR, Chen HSV, Mercola M (2009) Lentiviral vectors and protocols for creation of stable hESC lines for fluorescent tracking and drug resistance selection of cardiomyocytes. PLoS One 4:e5046. https://doi.org/10.1371/journal.pone.0005046
78. Stoehr A, Neuber C, Baldauf C, Vollert I, Friedrich FW, Flenner F, Carrier L, Eder A, Schaaf S, Hirt MN, Aksehirlioglu B, Tong CW, Moretti A, Eschenhagen T, Hansen A (2014) Automated analysis of contractile force and Ca2+ transients in engineered heart tissue. Am J Physiol Heart Circ Physiol 306:H1353–H1363. https://doi.org/10.1152/ajpheart.00705.2013
79. Zhang D, Shadrin IY, Lam J, Xian H-Q, Snodgrass HR, Bursac N (2013) Tissue-engineered cardiac patch for advanced functional maturation of human ESC-derived cardiomyocytes. Biomaterials 34:5813–5820. https://doi.org/10.1016/j.biomaterials.2013.04.026
80. Angst BD, Khan LU, Severs NJ, Whitely K, Rothery S, Thompson RP, Magee AI, Gourdie RG (1997) Dissociated spatial patterning of gap junctions and cell adhesion junctions during postnatal differentiation of ventricular myocardium. Circ Res 80:88–94. https://doi.org/10.1161/01.res.80.1.88
81. Jansen JA, van Veen TAB, de Bakker JMT, van Rijen HVM (2010) Cardiac connexins and impulse propagation. J Mol Cell Cardiol 48:76–82. https://doi.org/10.1016/j.yjmcc.2009.08.018
82. Chen H-SV, Kim C, Mercola M (2009) Electrophysiological challenges of cell-based myocardial repair. Circulation 120:2496–2508. https://doi.org/10.1161/CIRCULATIONAHA.107.751412
83. Zwi L, Caspi O, Arbel G, Huber I, Gepstein A, Park I-H, Gepstein L (2009) Cardiomyocyte differentiation of human induced pluripotent stem cells. Circulation 120:1513–1523. https://doi.org/10.1161/CIRCULATIONAHA.109.868885
84. Cordeiro JM, Nesterenko VV, Sicouri S, Goodrow RJ, Treat JA, Desai M, Wu Y, Doss MX, Antzelevitch C, Di Diego JM (2013) Identification and characterization of a transient outward K+ current in human induced pluripotent stem cell-derived cardiomyocytes. J Mol Cell Cardiol 60:36–46. https://doi.org/10.1016/j.yjmcc.2013.03.014
85. Davis RP, Casini S, van den Berg CW, Hoekstra M, Remme CA, Dambrot C, Salvatori D, Oostwaard DW, Wilde AAM, Bezzina CR, Verkerk AO, Freund C, Mummery CL (2012) Cardiomyocytes derived from pluripotent stem cells recapitulate electrophysiological characteristics of an overlap syndrome of cardiac sodium channel disease. Circulation 125:3079–3091. https://doi.org/10.1161/CIRCULATIONAHA.111.066092
86. Goversen B, van der Heyden MAG, van Veen TAB, de Boer TP (2018) The immature electrophysiological phenotype of iPSC-CMs still hampers in vitro drug screening: Special focus on IK1. Pharmacol Ther 183:127–136. https://doi.org/10.1016/j.pharmthera.2017.10.001
87. Itzhaki I, Rapoport S, Huber I, Mizrahi I, Zwi-Dantsis L, Arbel G, Schiller J, Gepstein L (2011) Calcium handling in human induced pluripotent stem cell derived cardiomyocytes. PLoS One 6:e18037. https://doi.org/10.1371/journal.pone.0018037

88. Ma J, Guo L, Fiene SJ, Anson BD, Thomson JA, Kamp TJ, Kolaja KL, Swanson BJ, January CT (2011) High purity human-induced pluripotent stem cell-derived cardiomyocytes: electrophysiological properties of action potentials and ionic currents. Am J Physiol Heart Circ Physiol 301:H2006–H2017. https://doi.org/10.1152/ajpheart.00694.2011
89. Koivumäki JT, Naumenko N, Tuomainen T, Takalo J, Oksanen M, Puttonen KA, Lehtonen Š, Kuusisto J, Laakso M, Koistinaho J, Tavi P (2018) Structural immaturity of human iPSC-derived cardiomyocytes: in silico investigation of effects on function and disease modeling. Front Physiol 9:80. https://doi.org/10.3389/fphys.2018.00080
90. Mummery CL, Zhang J, Ng ES, Elliott DA, Elefanty AG, Kamp TJ (2012) Differentiation of human embryonic stem cells and induced pluripotent stem cells to cardiomyocytes: a methods overview. Circ Res 111:344–358. https://doi.org/10.1161/CIRCRESAHA.110.227512
91. Buchanan JW, Saito T, Gettes LS (1985) The effects of antiarrhythmic drugs, stimulation frequency, and potassium-induced resting membrane potential changes on conduction velocity and dV/dtmax in guinea pig myocardium. Circ Res 56:696–703. https://doi.org/10.1161/01.res.56.5.696
92. Lieu DK, Fu J-D, Chiamvimonvat N, Tung KC, McNerney GP, Huser T, Keller G, Kong C-W, Li RA (2013) Mechanism-based facilitated maturation of human pluripotent stem cell-derived cardiomyocytes. Circ Arrhythm Electrophysiol 6:191–201. https://doi.org/10.1161/CIRCEP.111.973420
93. Hwang HS, Kryshtal DO, Feaster TK, Sánchez-Freire V, Zhang J, Kamp TJ, Hong CC, Wu JC, Knollmann BC (2015) Comparable calcium handling of human iPSC-derived cardiomyocytes generated by multiple laboratories. J Mol Cell Cardiol 85:79–88. https://doi.org/10.1016/j.yjmcc.2015.05.003
94. Lieu DK, Liu J, Siu C-W, McNerney GP, Tse H-F, Abu-Khalil A, Huser T, Li RA (2009) Absence of transverse tubules contributes to non-uniform Ca(2+) wavefronts in mouse and human embryonic stem cell-derived cardiomyocytes. Stem Cells Dev 18:1493–1500. https://doi.org/10.1089/scd.2009.0052
95. Satin J, Itzhaki I, Rapoport S, Schroder EA, Izu L, Arbel G, Beyar R, Balke CW, Schiller J, Gepstein L (2008) Calcium handling in human embryonic stem cell-derived cardiomyocytes. Stem Cells 26:1961–1972. https://doi.org/10.1634/stemcells.2007-0591
96. Lopaschuk GD, Spafford MA, Marsh DR (1991) Glycolysis is predominant source of myocardial ATP production immediately after birth. Am J Physiol 261:H1698–H1705. https://doi.org/10.1152/ajpheart.1991.261.6.H1698
97. Werner JC, Sicard RE, Schuler HG (1989) Palmitate oxidation by isolated working fetal and newborn pig hearts. Am J Physiol 256:E315–E321. https://doi.org/10.1152/ajpendo.1989.256.2.E315
98. Bartelds B, Knoester H, Smid GB, Takens J, Visser GH, Penninga L, van der Leij FR, Beaufort-Krol GC, Zijlstra WG, Heymans HS, Kuipers JR (2000) Perinatal changes in myocardial metabolism in lambs. Circulation 102:926–931. https://doi.org/10.1161/01.cir.102.8.926
99. Mootha VK, Arai AE, Balaban RS (1997) Maximum oxidative phosphorylation capacity of the mammalian heart. Am J Physiol 272:H769–H775. https://doi.org/10.1152/ajpheart.1997.272.2.H769
100. Stanley WC, Recchia FA, Lopaschuk GD (2005) Myocardial substrate metabolism in the normal and failing heart. Physiol Rev 85:1093–1129. https://doi.org/10.1152/physrev.00006.2004
101. Kensah G, Gruh I, Viering J, Schumann H, Dahlmann J, Meyer H, Skvorc D, Bär A, Akhyari P, Heisterkamp A, Haverich A, Martin U (2011) A novel miniaturized multimodal bioreactor for continuous in situ assessment of bioartificial cardiac tissue during stimulation and maturation. Tissue Eng Part C Methods 17:463–473. https://doi.org/10.1089/ten.TEC.2010.0405
102. Parrag IC, Zandstra PW, Woodhouse KA (2012) Fiber alignment and coculture with fibroblasts improves the differentiated phenotype of murine embryonic stem cell-derived

cardiomyocytes for cardiac tissue engineering. Biotechnol Bioeng 109:813–822. https://doi.org/10.1002/bit.23353
103. Piquereau J, Caffin F, Novotova M, Lemaire C, Veksler V, Garnier A, Ventura-Clapier R, Joubert F (2013) Mitochondrial dynamics in the adult cardiomyocytes: which roles for a highly specialized cell? Front Physiol 4:102. https://doi.org/10.3389/fphys.2013.00102
104. Hom JR, Quintanilla RA, Hoffman DL, de Mesy Bentley KL, Molkentin JD, Sheu S-S, Porter GA (2011) The permeability transition pore controls cardiac mitochondrial maturation and myocyte differentiation. Dev Cell 21:469–478. https://doi.org/10.1016/j.devcel.2011.08.008
105. Pasqualini FS, Nesmith AP, Horton RE, Sheehy SP, Parker KK (2016) Mechanotransduction and metabolism in cardiomyocyte microdomains. Biomed Res Int 2016:4081638. https://doi.org/10.1155/2016/4081638
106. Zukor LJ (1960) Nature and properties of engineering materials. Z. D. Jastrzebski. Wiley, New York-London, 1959, xvii + 571 pp. $11.00. J Appl Polym Sci 4:372–373. https://doi.org/10.1002/app.1960.070041224
107. Engler AJ, Carag-Krieger C, Johnson CP, Raab M, Tang H-Y, Speicher DW, Sanger JW, Sanger JM, Discher DE (2008) Embryonic cardiomyocytes beat best on a matrix with heart-like elasticity: scar-like rigidity inhibits beating. J Cell Sci 121:3794–3802. https://doi.org/10.1242/jcs.029678
108. Tallawi M, Rai R, Boccaccini AR, Aifantis KE (2015) Effect of substrate mechanics on cardiomyocyte maturation and growth. Tissue Eng Part B Rev 21:157–165. https://doi.org/10.1089/ten.TEB.2014.0383
109. Tang VW (2020) Collagen, stiffness, and adhesion: the evolutionary basis of vertebrate mechanobiology. Mol Biol Cell 31:1823–1834. https://doi.org/10.1091/mbc.E19-12-0709
110. Ward M, Iskratsch T (2020) Mix and (mis-)match - the mechanosensing machinery in the changing environment of the developing, healthy adult and diseased heart. Biochim Biophys Acta Mol Cell Res 1867:118436. https://doi.org/10.1016/j.bbamcr.2019.01.017
111. McCain ML, Yuan H, Pasqualini FS, Campbell PH, Parker KK (2014) Matrix elasticity regulates the optimal cardiac myocyte shape for contractility. Am J Physiol Heart Circ Physiol 306:H1525–H1539. https://doi.org/10.1152/ajpheart.00799.2013
112. Pandey P, Hawkes W, Hu J, Megone WV, Gautrot J, Anilkumar N, Zhang M, Hirvonen L, Cox S, Ehler E, Hone J, Sheetz M, Iskratsch T (2018) Cardiomyocytes sense matrix rigidity through a combination of muscle and non-muscle myosin contractions. Dev Cell 44:326–336.e3. https://doi.org/10.1016/j.devcel.2017.12.024
113. Rodriguez AG, Han SJ, Regnier M, Sniadecki NJ (2011) Substrate stiffness increases twitch power of neonatal cardiomyocytes in correlation with changes in myofibril structure and intracellular calcium. Biophys J 101:2455–2464. https://doi.org/10.1016/j.bpj.2011.09.057
114. Martewicz S, Serena E, Zatti S, Keller G, Elvassore N (2017) Substrate and mechanotransduction influence SERCA2a localization in human pluripotent stem cell-derived cardiomyocytes affecting functional performance. Stem Cell Res 25:107–114. https://doi.org/10.1016/j.scr.2017.10.011
115. Hazeltine LB, Simmons CS, Salick MR, Lian X, Badur MG, Han W, Delgado SM, Wakatsuki T, Crone WC, Pruitt BL, Palecek SP (2012) Effects of substrate mechanics on contractility of cardiomyocytes generated from human pluripotent stem cells. Int J Cell Biol 2012:508294. https://doi.org/10.1155/2012/508294
116. Feaster TK, Cadar AG, Wang L, Williams CH, Chun YW, Hempel JE, Bloodworth N, Merryman WD, Lim CC, Wu JC, Knollmann BC, Hong CC (2015) Matrigel mattress: a method for the generation of single contracting human-induced pluripotent stem cell-derived cardiomyocytes. Circ Res 117:995–1000. https://doi.org/10.1161/CIRCRESAHA.115.307580
117. Herron TJ, Rocha AMD, Campbell KF, Ponce-Balbuena D, Willis BC, Guerrero-Serna G, Liu Q, Klos M, Musa H, Zarzoso M, Bizy A, Furness J, Anumonwo J, Mironov S, Jalife J (2016) Extracellular matrix-mediated maturation of human pluripotent stem cell-derived

cardiac monolayer structure and electrophysiological function. Circ Arrhythm Electrophysiol 9:e003638. https://doi.org/10.1161/CIRCEP.113.003638
118. Rao C, Prodromakis T, Kolker L, Chaudhry UAR, Trantidou T, Sridhar A, Weekes C, Camelliti P, Harding SE, Darzi A, Yacoub MH, Athanasiou T, Terracciano CM (2013) The effect of microgrooved culture substrates on calcium cycling of cardiac myocytes derived from human induced pluripotent stem cells. Biomaterials 34:2399–2411. https://doi.org/10.1016/j.biomaterials.2012.11.055
119. Knight WE, Cao Y, Lin Y-H, Chi C, Bai B, Sparagna GC, Zhao Y, Du Y, Londono P, Reisz JA, Brown BC, Taylor MRG, Ambardekar AV, Cleveland JC, McKinsey TA, Jeong MY, Walker LA, Woulfe KC, D'Alessandro A, Chatfield KC, Xu H, Bristow MR, Buttrick PM, Song K (2021) Maturation of pluripotent stem cell-derived cardiomyocytes enables modeling of human hypertrophic cardiomyopathy. Stem Cell Rep 16:519–533. https://doi.org/10.1016/j.stemcr.2021.01.018
120. Napiwocki BN, Stempien A, Lang D, Kruepke RA, Kim G, Zhang J, Eckhardt LL, Glukhov AV, Kamp TJ, Crone WC (2021) Micropattern platform promotes extracellular matrix remodeling by human PSC-derived cardiac fibroblasts and enhances contractility of co-cultured cardiomyocytes. Physiol Rep 9. https://doi.org/10.14814/phy2.15045
121. Ribeiro AJS, Ang Y-S, Fu J-D, Rivas RN, Mohamed TMA, Higgs GC, Srivastava D, Pruitt BL (2015) Contractility of single cardiomyocytes differentiated from pluripotent stem cells depends on physiological shape and substrate stiffness. Proc Natl Acad Sci U S A 112: 12705–12710. https://doi.org/10.1073/pnas.1508073112
122. Carson D, Hnilova M, Yang X, Nemeth CL, Tsui JH, Smith AST, Jiao A, Regnier M, Murry CE, Tamerler C, Kim D-H (2016) Nanotopography-induced structural anisotropy and sarcomere development in human cardiomyocytes derived from induced pluripotent stem cells. ACS Appl Mater Interfaces 8:21923–21932. https://doi.org/10.1021/acsami.5b11671
123. Körner A, Mosqueira M, Hecker M, Ullrich ND (2021) Substrate stiffness influences structural and functional remodeling in induced pluripotent stem cell-derived cardiomyocytes. Front Physiol 12:710619. https://doi.org/10.3389/fphys.2021.710619
124. Fekete N, Béland AV, Campbell K, Clark SL, Hoesli CA (2018) Bags versus flasks: a comparison of cell culture systems for the production of dendritic cell–based immunotherapies. Transfusion 58:1800–1813. https://doi.org/10.1111/trf.14621
125. McDevitt TC, Angello JC, Whitney ML, Reinecke H, Hauschka SD, Murry CE, Stayton PS (2002) In vitro generation of differentiated cardiac myofibers on micropatterned laminin surfaces. J Biomed Mater Res 60:472–479. https://doi.org/10.1002/jbm.1292
126. Bien H, Yin L, Entcheva E (2003) Cardiac cell networks on elastic microgrooved scaffolds. IEEE Eng Med Biol Mag 22:108–112. https://doi.org/10.1109/memb.2003.1256279
127. Kim D-H, Lipke EA, Kim P, Cheong R, Thompson S, Delannoy M, Suh K-Y, Tung L, Levchenko A (2010) Nanoscale cues regulate the structure and function of macroscopic cardiac tissue constructs. Proc Natl Acad Sci U S A 107:565–570. https://doi.org/10.1073/pnas.0906504107
128. Robinson TF, Factor SM, Capasso JM, Wittenberg BA, Blumenfeld OO, Seifter S (1987) Morphology, composition, and function of struts between cardiac myocytes of rat and hamster. Cell Tissue Res 249:247–255. https://doi.org/10.1007/BF00215507
129. Napiwocki BN, Lang D, Stempien A, Zhang J, Vaidyanathan R, Makielski JC, Eckhardt LL, Glukhov AV, Kamp TJ, Crone WC (2021) Aligned human cardiac syncytium for in vitro analysis of electrical, structural, and mechanical readouts. Biotechnol Bioeng 118:442–452. https://doi.org/10.1002/bit.27582
130. Macadangdang J, Guan X, Smith AST, Lucero R, Czerniecki S, Childers MK, Mack DL, Kim D-H (2015) Nanopatterned human iPSC-based model of a dystrophin-null cardiomyopathic phenotype. Cell Mol Bioeng 8:320–332. https://doi.org/10.1007/s12195-015-0413-8
131. Bayomy AF, Bauer M, Qiu Y, Liao R (2012) Regeneration in heart disease-Is ECM the key? Life Sci 91:823–827. https://doi.org/10.1016/j.lfs.2012.08.034

132. Ross RS, Borg TK (2001) Integrins and the myocardium. Circ Res 88:1112–1119. https://doi.org/10.1161/hh1101.091862
133. Wang P-Y, Yu J, Lin J-H, Tsai W-B (2011) Modulation of alignment, elongation and contraction of cardiomyocytes through a combination of nanotopography and rigidity of substrates. Acta Biomater 7:3285–3293. https://doi.org/10.1016/j.actbio.2011.05.021
134. Eschenhagen T, Fink C, Remmers U, Scholz H, Wattchow J, Weil J, Zimmermann W, Dohmen HH, Schäfer H, Bishopric N, Wakatsuki T, Elson EL (1997) Three-dimensional reconstitution of embryonic cardiomyocytes in a collagen matrix: a new heart muscle model system. FASEB J 11:683–694. https://doi.org/10.1096/fasebj.11.8.9240969
135. Wanjare M, Hou L, Nakayama KH, Kim JJ, Mezak NP, Abilez OJ, Tzatzalos E, Wu JC, Huang NF (2017) Anisotropic microfibrous scaffolds enhance the organization and function of cardiomyocytes derived from induced pluripotent stem cells. Biomater Sci 5:1567–1578. https://doi.org/10.1039/c7bm00323d
136. Nagueh SF, Shah G, Wu Y, Torre-Amione G, King NMP, Lahmers S, Witt CC, Becker K, Labeit S, Granzier HL (2004) Altered titin expression, myocardial stiffness, and left ventricular function in patients with dilated cardiomyopathy. Circulation 110:155–162. https://doi.org/10.1161/01.CIR.0000135591.37759.AF
137. Nakano K, Sugawara M, Ishihara K, Kanazawa S, Corin WJ, Denslow S, Biederman RW, Carabello BA (1990) Myocardial stiffness derived from end-systolic wall stress and logarithm of reciprocal of wall thickness. Contractility index independent of ventricular size. Circulation 82:1352–1361. https://doi.org/10.1161/01.CIR.82.4.1352
138. Venugopal JR, Prabhakaran MP, Mukherjee S, Ravichandran R, Dan K, Ramakrishna S (2012) Biomaterial strategies for alleviation of myocardial infarction. J R Soc Interface 9:1–19. https://doi.org/10.1098/rsif.2011.0301
139. Watanabe S, Shite J, Takaoka H, Shinke T, Imuro Y, Ozawa T, Otake H, Matsumoto D, Ogasawara D, Paredes OL, Yokoyama M (2006) Myocardial stiffness is an important determinant of the plasma brain natriuretic peptide concentration in patients with both diastolic and systolic heart failure. Eur Heart J 27:832–838. https://doi.org/10.1093/eurheartj/ehi772
140. Fleischer S, Miller J, Hurowitz H, Shapira A, Dvir T (2015) Effect of fiber diameter on the assembly of functional 3D cardiac patches. Nanotechnology 26:291002. https://doi.org/10.1088/0957-4484/26/29/291002
141. Castilho M, van Mil A, Maher M, Metz CHG, Hochleitner G, Groll J, Doevendans PA, Ito K, Sluijter JPG, Malda J (2018) Melt electrowriting allows tailored microstructural and mechanical design of scaffolds to advance functional human myocardial tissue formation. Adv Funct Mater 28:1803151. https://doi.org/10.1002/adfm.201803151
142. Silbernagel N, Körner A, Balitzki J, Jaggy M, Bertels S, Richter B, Hippler M, Hellwig A, Hecker M, Bastmeyer M, Ullrich ND (2020) Shaping the heart: Structural and functional maturation of iPSC-cardiomyocytes in 3D-micro-scaffolds. Biomaterials 227:119551. https://doi.org/10.1016/j.biomaterials.2019.119551
143. Guyette JP, Charest JM, Mills RW, Jank BJ, Moser PT, Gilpin SE, Gershlak JR, Okamoto T, Gonzalez G, Milan DJ, Gaudette GR, Ott HC (2016) Bioengineering human myocardium on native extracellular matrix. Circ Res 118:56–72. https://doi.org/10.1161/CIRCRESAHA.115.306874
144. Blazeski A, Lowenthal J, Zhu R, Ewoldt J, Boheler KR, Tung L (2019) Functional properties of engineered heart slices incorporating human induced pluripotent stem cell-derived cardiomyocytes. Stem Cell Rep 12:982–995. https://doi.org/10.1016/j.stemcr.2019.04.002
145. Sewanan LR, Schwan J, Kluger J, Park J, Jacoby DL, Qyang Y, Campbell SG (2019) Extracellular matrix from hypertrophic myocardium provokes impaired twitch dynamics in healthy cardiomyocytes. JACC Basic Transl Sci 4:495–505. https://doi.org/10.1016/j.jacbts.2019.03.004
146. Fong AH, Romero-López M, Heylman CM, Keating M, Tran D, Sobrino A, Tran AQ, Pham HH, Fimbres C, Gershon PD, Botvinick EL, George SC, Hughes CCW (2016) Three-dimensional adult cardiac extracellular matrix promotes maturation of human induced

pluripotent stem cell-derived cardiomyocytes. Tissue Eng Part A 22:1016–1025. https://doi.org/10.1089/ten.tea.2016.0027

147. Ozcebe SG, Bahcecioglu G, Yue XS, Zorlutuna P (2021) Effect of cellular and ECM aging on human iPSC-derived cardiomyocyte performance, maturity and senescence. Biomaterials 268: 120554. https://doi.org/10.1016/j.biomaterials.2020.120554
148. Almeida HV, Tenreiro MF, Louro AF, Abecasis B, Santinha D, Calmeiro T, Fortunato E, Ferreira L, Alves PM, Serra M (2021) Human extracellular-matrix functionalization of 3D hiPSC-based cardiac tissues improves cardiomyocyte maturation. ACS Appl Bio Mater 4: 1888–1899. https://doi.org/10.1021/acsabm.0c01490
149. Castro Neto AH, Guinea F, Peres NMR, Novoselov KS, Geim AK (2009) The electronic properties of graphene. Rev Mod Phys 81:109–162. https://doi.org/10.1103/RevModPhys.81.109
150. Hilding J, Grulke EA, George Zhang Z, Lockwood F (2003) Dispersion of carbon nanotubes in liquids. J Dispers Sci Technol 24:1–41. https://doi.org/10.1081/DIS-120017941
151. Roshanbinfar K, Mohammadi Z, Sheikh-Mahdi Mesgar A, Dehghan MM, Oommen OP, Hilborn J, Engel FB (2019) Carbon nanotube doped pericardial matrix derived electroconductive biohybrid hydrogel for cardiac tissue engineering. Biomater Sci 7:3906–3917. https://doi.org/10.1039/C9BM00434C
152. Tsui JH, Leonard A, Camp ND, Long JT, Nawas ZY, Chavanachat R, Smith AST, Choi JS, Dong Z, Ahn EH, Wolf-Yadlin A, Murry CE, Sniadecki NJ, Kim D-H (2021) Tunable electroconductive decellularized extracellular matrix hydrogels for engineering human cardiac microphysiological systems. Biomaterials 272:120764. https://doi.org/10.1016/j.biomaterials.2021.120764
153. Ruan J-L, Tulloch NL, Razumova MV, Saiget M, Muskheli V, Pabon L, Reinecke H, Regnier M, Murry CE (2016) Mechanical stress conditioning and electrical stimulation promote contractility and force maturation of induced pluripotent stem cell-derived human cardiac tissue. Circulation 134:1557–1567. https://doi.org/10.1161/CIRCULATIONAHA.114.014998
154. Tulloch NL, Muskheli V, Razumova MV, Korte FS, Regnier M, Hauch KD, Pabon L, Reinecke H, Murry CE (2011) Growth of engineered human myocardium with mechanical loading and vascular coculture. Circ Res 109:47–59. https://doi.org/10.1161/CIRCRESAHA.110.237206
155. Bliley JM, Vermeer MCSC, Duffy RM, Batalov I, Kramer D, Tashman JW, Shiwarski DJ, Lee A, Teplenin AS, Volkers L, Coffin B, Hoes MF, Kalmykov A, Palchesko RN, Sun Y, Jongbloed JDH, Bomer N, de Boer RA, Suurmeijer AJH, Pijnappels DA, Bolling MC, van der Meer P, Feinberg AW (2021) Dynamic loading of human engineered heart tissue enhances contractile function and drives a desmosome-linked disease phenotype. Sci Transl Med 13: eabd1817. https://doi.org/10.1126/scitranslmed.abd1817
156. Leonard A, Bertero A, Powers JD, Beussman KM, Bhandari S, Regnier M, Murry CE, Sniadecki NJ (2018) Afterload promotes maturation of human induced pluripotent stem cell derived cardiomyocytes in engineered heart tissues. J Mol Cell Cardiol 118:147–158. https://doi.org/10.1016/j.yjmcc.2018.03.016
157. Abilez OJ, Tzatzalos E, Yang H, Zhao M-T, Jung G, Zöllner AM, Tiburcy M, Riegler J, Matsa E, Shukla P, Zhuge Y, Chour T, Chen VC, Burridge PW, Karakikes I, Kuhl E, Bernstein D, Couture LA, Gold JD, Zimmermann WH, Wu JC (2018) Passive stretch induces structural and functional maturation of engineered heart muscle as predicted by computational modeling. Stem Cells 36:265–277. https://doi.org/10.1002/stem.2732
158. Soong PL, Tiburcy M, Zimmermann W-H (2012) Cardiac differentiation of human embryonic stem cells and their assembly into engineered heart muscle. Curr Protoc Cell Biol 23:Unit23.8. https://doi.org/10.1002/0471143030.cb2308s55
159. Tiburcy M, Meyer T, Soong PL, Zimmermann W-H (2014) Cardiac differentiation of human embryonic stem cells and their assembly into engineered heart muscle. Methods Mol Biol 1181:167–176. https://doi.org/10.1007/978-1-4939-1047-2_15

160. Dou W, Wang L, Malhi M, Liu H, Zhao Q, Plakhotnik J, Xu Z, Huang Z, Simmons CA, Maynes JT, Sun Y (2021) A microdevice platform for characterizing the effect of mechanical strain magnitudes on the maturation of iPSC-Cardiomyocytes. Biosens Bioelectron 175: 112875. https://doi.org/10.1016/j.bios.2020.112875
161. Ovchinnikova E, Hoes M, Ustyantsev K, Bomer N, de Jong TV, van der Mei H, Berezikov E, van der Meer P (2018) Modeling human cardiac hypertrophy in stem cell-derived cardiomyocytes. Stem Cell Rep 10:794–807. https://doi.org/10.1016/j.stemcr.2018.01.016
162. LaBarge W, Mattappally S, Kannappan R, Fast VG, Pretorius D, Berry JL, Zhang J (2019) Maturation of three-dimensional, hiPSC-derived cardiomyocyte spheroids utilizing cyclic, uniaxial stretch and electrical stimulation. PLoS One 14:e0219442. https://doi.org/10.1371/journal.pone.0219442
163. Mihic A, Li J, Miyagi Y, Gagliardi M, Li S-H, Zu J, Weisel RD, Keller G, Li R-K (2014) The effect of cyclic stretch on maturation and 3D tissue formation of human embryonic stem cell-derived cardiomyocytes. Biomaterials 35:2798–2808. https://doi.org/10.1016/j.biomaterials.2013.12.052
164. Shen N, Knopf A, Westendorf C, Kraushaar U, Riedl J, Bauer H, Pöschel S, Layland SL, Holeiter M, Knolle S, Brauchle E, Nsair A, Hinderer S, Schenke-Layland K (2017) Steps toward maturation of embryonic stem cell-derived cardiomyocytes by defined physical signals. Stem Cell Rep 9:122–135. https://doi.org/10.1016/j.stemcr.2017.04.021
165. Kolanowski TJ, Busek M, Schubert M, Dmitrieva A, Binnewerg B, Pöche J, Fisher K, Schmieder F, Grünzner S, Hansen S, Richter A, El-Armouche A, Sonntag F, Guan K (2020) Enhanced structural maturation of human induced pluripotent stem cell-derived cardiomyocytes under a controlled microenvironment in a microfluidic system. Acta Biomater 102:273–286. https://doi.org/10.1016/j.actbio.2019.11.044
166. Ottaviano FG, Yee KO (2011) Communication signals between cardiac fibroblasts and cardiac myocytes. J Cardiovasc Pharmacol 57:513–521. https://doi.org/10.1097/FJC.0b013e31821209ee
167. Gessert S, Kühl M (2010) The multiple phases and faces of wnt signaling during cardiac differentiation and development. Circ Res 107:186–199. https://doi.org/10.1161/CIRCRESAHA.110.221531
168. Zhang P, Su J, Mende U (2012) Cross talk between cardiac myocytes and fibroblasts: from multiscale investigative approaches to mechanisms and functional consequences. Am J Physiol Heart Circ Physiol 303:H1385–H1396. https://doi.org/10.1152/ajpheart.01167.2011
169. Tiburcy M, Meyer T, Soong PL, Zimmermann W-H (2014) Collagen-based engineered heart muscle. Methods Mol Biol 1181:167–176. https://doi.org/10.1007/978-1-4939-1047-2_15
170. Wiegerinck RF, Cojoc A, Zeidenweber CM, Ding G, Shen M, Joyner RW, Fernandez JD, Kanter KR, Kirshbom PM, Kogon BE, Wagner MB (2009) Force frequency relationship of the human ventricle increases during early postnatal development. Pediatr Res 65:414–419. https://doi.org/10.1203/PDR.0b013e318199093c
171. Mulieri LA, Hasenfuss G, Leavitt B, Allen PD, Alpert NR (1992) Altered myocardial force-frequency relation in human heart failure. Circulation 85:1743–1750. https://doi.org/10.1161/01.cir.85.5.1743
172. Salameh A, Wustmann A, Karl S, Blanke K, Apel D, Rojas-Gomez D, Franke H, Mohr FW, Janousek J, Dhein S (2010) Cyclic mechanical stretch induces cardiomyocyte orientation and polarization of the gap junction protein connexin43. Circ Res 106:1592–1602. https://doi.org/10.1161/CIRCRESAHA.109.214429
173. Fink C, Ergün S, Kralisch D, Remmers U, Weil J, Eschenhagen T (2000) Chronic stretch of engineered heart tissue induces hypertrophy and functional improvement. FASEB J 14:669–679. https://doi.org/10.1096/fasebj.14.5.669
174. Naito H, Melnychenko I, Didié M, Schneiderbanger K, Schubert P, Rosenkranz S, Eschenhagen T, Zimmermann W-H (2006) Optimizing engineered heart tissue for therapeutic applications as surrogate heart muscle. Circulation 114:I72–I78. https://doi.org/10.1161/CIRCULATIONAHA.105.001560

175. Crosara-Alberto DP, Inoue RY, Costa CRC (2009) FAK signalling mediates NF-kappaB activation by mechanical stress in cardiac myocytes. Clin Chim Acta 403:81–86. https://doi.org/10.1016/j.cca.2009.01.023
176. Takahashi N, Seko Y, Noiri E, Tobe K, Kadowaki T, Sabe H, Yazaki Y (1999) Vascular endothelial growth factor induces activation and subcellular translocation of focal adhesion kinase (p125FAK) in cultured rat cardiac myocytes. Circ Res 84:1194–1202. https://doi.org/10.1161/01.res.84.10.1194
177. Zentilin L, Puligadda U, Lionetti V, Zacchigna S, Collesi C, Pattarini L, Ruozi G, Camporesi S, Sinagra G, Pepe M, Recchia FA, Giacca M (2010) Cardiomyocyte VEGFR-1 activation by VEGF-B induces compensatory hypertrophy and preserves cardiac function after myocardial infarction. FASEB J 24:1467–1478. https://doi.org/10.1096/fj.09-143180
178. Israeli-Rosenberg S, Chen C, Li R, Deussen DN, Niesman IR, Okada H, Patel HH, Roth DM, Ross RS (2015) Caveolin modulates integrin function and mechanical activation in the cardiomyocyte. FASEB J 29:374–384. https://doi.org/10.1096/fj.13-243139
179. Israeli-Rosenberg S, Manso AM, Okada H, Ross RS (2014) Integrins and integrin-associated proteins in the cardiac myocyte. Circ Res 114:572–586. https://doi.org/10.1161/CIRCRESAHA.114.301275
180. Ross RS, Pham C, Shai SY, Goldhaber JI, Fenczik C, Glembotski CC, Ginsberg MH, Loftus JC (1998) Beta1 integrins participate in the hypertrophic response of rat ventricular myocytes. Circ Res 82:1160–1172. https://doi.org/10.1161/01.res.82.11.1160
181. Cheng Z, DiMichele LA, Rojas M, Vaziri C, Mack CP, Taylor JM (2014) Focal adhesion kinase antagonizes doxorubicin cardiotoxicity via p21(Cip1.). J Mol Cell Cardiol 67:1–11. https://doi.org/10.1016/j.yjmcc.2013.12.002
182. Etienne-Manneville S, Hall A (2002) Rho GTPases in cell biology. Nature 420:629–635. https://doi.org/10.1038/nature01148
183. Loirand G, Guérin P, Pacaud P (2006) Rho kinases in cardiovascular physiology and pathophysiology. Circ Res 98:322–334. https://doi.org/10.1161/01.RES.0000201960.04223.3c
184. Baba HA, Stypmann J, Grabellus F, Kirchhof P, Sokoll A, Schäfers M, Takeda A, Wilhelm MJ, Scheld HH, Takeda N, Breithardt G, Levkau B (2003) Dynamic regulation of MEK/Erks and Akt/GSK-3beta in human end-stage heart failure after left ventricular mechanical support: myocardial mechanotransduction-sensitivity as a possible molecular mechanism. Cardiovasc Res 59:390–399. https://doi.org/10.1016/s0008-6363(03)00393-6
185. Naito AT, Shiojima I, Akazawa H, Hidaka K, Morisaki T, Kikuchi A, Komuro I (2006) Developmental stage-specific biphasic roles of Wnt/beta-catenin signaling in cardiomyogenesis and hematopoiesis. Proc Natl Acad Sci U S A 103:19812–19817. https://doi.org/10.1073/pnas.0605768103
186. Lee JH, Protze SI, Laksman Z, Backx PH, Keller GM (2017) Human pluripotent stem cell-derived atrial and ventricular cardiomyocytes develop from distinct mesoderm populations. Cell Stem Cell 21:179–194.e4. https://doi.org/10.1016/j.stem.2017.07.003
187. Protze SI, Liu J, Nussinovitch U, Ohana L, Backx PH, Gepstein L, Keller GM (2017) Sinoatrial node cardiomyocytes derived from human pluripotent cells function as a biological pacemaker. Nat Biotechnol 35:56–68. https://doi.org/10.1038/nbt.3745
188. Reinecke H, Zhang M, Bartosek T, Murry CE (1999) Survival, integration, and differentiation of cardiomyocyte grafts: a study in normal and injured rat hearts. Circulation 100:193–202. https://doi.org/10.1161/01.cir.100.2.193
189. Nakamura K, Murry CE (2019) Function follows form - a review of cardiac cell therapy. Circ J 83:2399–2412. https://doi.org/10.1253/circj.CJ-19-0567
190. Uosaki H, Cahan P, Lee DI, Wang S, Miyamoto M, Fernandez L, Kass DA, Kwon C (2015) Transcriptional landscape of cardiomyocyte maturation. Cell Rep 13:1705–1716. https://doi.org/10.1016/j.celrep.2015.10.032
191. Grün D, van Oudenaarden A (2015) Design and analysis of single-cell sequencing experiments. Cell 163:799–810. https://doi.org/10.1016/j.cell.2015.10.039

192. Murphy SA, Miyamoto M, Kervadec A, Kannan S, Tampakakis E, Kambhampati S, Lin BL, Paek S, Andersen P, Lee D-I, Zhu R, An SS, Kass DA, Uosaki H, Colas AR, Kwon C (2021) PGC1/PPAR drive cardiomyocyte maturation at single cell level via YAP1 and SF3B2. Nat Commun 12:1648. https://doi.org/10.1038/s41467-021-21957-z

Mechanical Considerations of Myocardial Tissue and Cardiac Regeneration

Ignasi Jorba, Milica Nikolic, and Carlijn V. C. Bouten

1 Current Regenerative Strategies Fail to Restore the Myocardial Mechanical Environment

According to the World Health Organization, one-third of worldwide deaths are caused by cardiovascular diseases, with an increasing trend in the recent years [1]. Specifically, heart failure (HF) following myocardial infarction (MI) is the most fatal cardiovascular disease with a 5-year survival rate of <50% [2]. At the tissue level, HF as a result of MI is characterised by a gradual loss of cardiomyocytes and, therefore, contractile tissue. The biological healing cascades following injury trigger the fibroblasts to change their phenotype and increase extracellular matrix (ECM) production, ultimately leading to the formation of hypo-contractile tissue scar [3]. At the same time, the remaining cardiomyocytes try to compensate for the loss of contractile tissue, initiating pathological cellular responses and further cardiac remodelling. Ultimately, tissue remodelling leads to an increased myocardial wall thickening, followed by dilation and eventually diminished cardiac function [4].

Nowadays, there is no doubt that cells actively respond to their mechanical environment and this interaction is essential in load-bearing and continuously contracting tissues such as the myocardium [5]. For instance, cellular contractility and electrical stability are highly dependent on mechanical environmental cues such as stiffness or cyclic strain [6–9]; and a more physiological, 'healthy' environment – or niche – will elicit beneficial cell behaviour in contrast to the environment present after injury. Still, this insight has been largely ignored in the design of strategies to

I. Jorba · M. Nikolic · C. V. C. Bouten (✉)
Department of Biomedical Engineering, Eindhoven University of Technology, Eindhoven, The Netherlands

Institute for Complex Molecular Systems (ICMS), Eindhoven, The Netherlands
e-mail: i.jorba.masdeu@tue.nl; m.nikolic@tue.nl; c.v.c.bouten@tue.nl

M. Hecker, D. J. Duncker (eds.), *Cardiac Mechanobiology in Physiology and Disease*, Cardiac and Vascular Biology 9,
https://doi.org/10.1007/978-3-031-23965-6_8

repair or regenerate the myocardial tissue. For instance, cell-based therapies aim to restore tissue contractility by injecting new (contractile) cells in the injured area [10]. Despite promising results in in vitro settings, long-term preclinical and clinical studies have shown that these strategies fail to regenerate myocardial tissue and overall cardiac function [11–14]. One of the main reasons is that these studies mainly focus on cell phenotype and function before implantation, disregarding the effects of the highly injured ECM niche with altered mechanical properties that will be found upon transplantation. In an attempt to induce cell adaptation to the diseased environment prior to implantation, some studies exposed cells in vitro to mechanical or biological environmental cues before in vivo injection [15–17]. However, the toxic environment still caused low cell retention, high cell death and low mechanical stability upon injection in the diseased area.

Hence, it is difficult for the cells alone to overcome the highly destabilised mechanical tissue present after cardiac injury. Several studies have shown that cells upon implantation start synthesising and remodelling their own (healthy) ECM [18, 19]. This principle is copied by scientists in the context of in vitro cardiac tissue engineering (cTE) [20]. In vitro cTE seeks to create living myocardial tissue by combining cardiac cells with a temporary 'healthy' niche prior to transplantation, either as a patch or as an injectable gel replacing the damaged area [21]. Multiple biomaterials, from either natural or synthetic sources, have been developed in the form of hydrogels or scaffolds to provide for such niches [22–26]. The use of natural source biomaterials, such as collagen or other ECM components, seems attractive, as they can offer some of the biochemical cues present in the native tissue. Decellularised cardiac tissue – e.g. from animals – can provide scaffolds with appropriate mechanical, topological and biochemical properties [27–29]. On the other hand, synthetic polymers can provide scaffolds with better tuneable mechanical properties compared to the natural processed ECM but offer poor biochemical signals with fewer cell recognition motifs if applied in pristine form. Bioactive modification of such materials is pursued for the design of life-like materials that can mimic both mechanical and biochemical environmental cues [30–33]. Overall, these approaches fulfil the aim to provide new cardiac cells with a relevant micromechanical environment.

cTE constructs, based either on natural or on synthetic biomaterials, are usually implanted on the epicardium near the infarcted region of the myocardium and serve to constrain and mechanically reinforce the ventricular wall in order to prevent or halt pathological tissue remodelling [34]. However, mechanical consequences of patch transplantation or hydrogel injection are still largely overlooked. Proper mechanical integration of the delivered cTE constructs across length scales from local cell-cell and cell-ECM interactions to global tissue contraction is necessary for the success of these strategies. Mechanical mismatch will cause complications, including heart failure and diastolic dysfunction.

More recently, in situ cTE has emerged as a promising alternative to traditional in vitro cTE to achieve tissue regeneration directly at the functional site. In situ cTE is built on the notion that the injection of a biomimetic acellular scaffold into the injured myocardium stimulates endogenous repair processes [35]. However, in situ

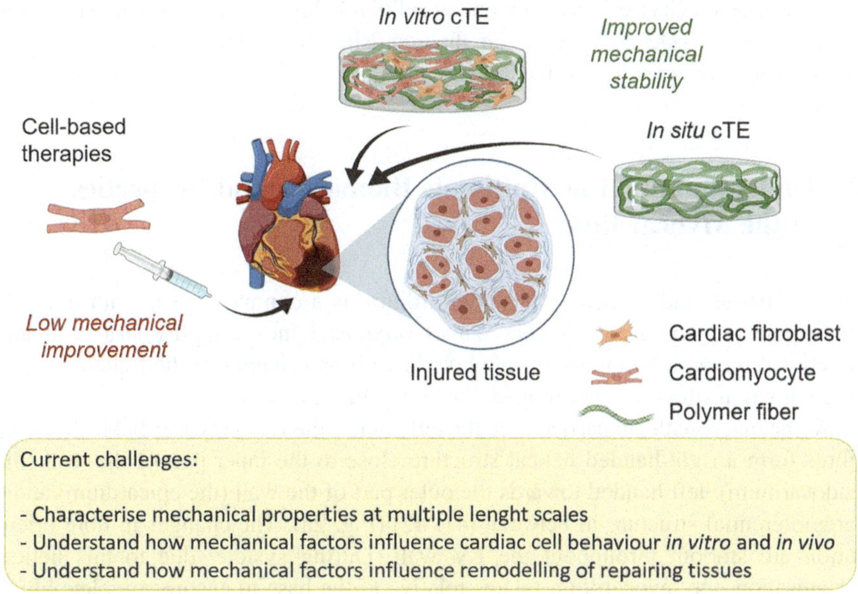

Fig. 1 Cell-based therapies cannot restore the mechanical properties of the injured myocardium. In vitro and in situ tissue engineering strategies using hydrogels and scaffolds reinforce injured myocardium with moderate success. The main challenges to improve the mechanical stability of current cardiac regenerative strategies are highlighted. Partly created with BioRender.com

cTE approaches are still in their infancy, and the governing processes of regeneration are still largely unknown. These processes range from the activated immune response in response to the biomaterial inside the body to the mechanical factors governing de novo ECM formation and organisation. In situ *c*TE has major benefits over in vitro cTE in terms of costs, availability and regulatory complexity, as it uses an acellular scaffold to harness the regenerative potential of the native tissue. In addition, in situ *c*TE does not need to account for the mechanical integration of exogenous cells in the damaged tissue. Nevertheless, the mechanical and electrical coupling and stabilisation at the cellular, tissue and organ level upon biomaterial delivery need to be addressed.

In summary, and with an eye to clinical translation, the design of successful cardiac regenerative strategies should address several aspects regarding the role of biomechanics in myocardial repair or regeneration (Fig. 1). These include questions such as: What are the mechanical properties of native and diseased tissues at the various length scales of the myocardium that will affect therapy outcomes? How will these properties influence cell and tissue behaviour in vitro and upon transplantation in vivo? How do biomechanical factors influence the remodelling of the regenerating myocardium in damaged and remote areas? In this chapter, we first describe the complex cardiac mechanical environment across length scales from macro to micro level and the techniques used for characterising mechanical behaviour at these length scales. Next, we review in vitro and in silico cardiac models to understand the impact of the cardiac mechanical environment on cellular behaviour and tissue regeneration. Finally, we provide an outlook on the requisites to design the next-generation engineering strategies for cardiac regeneration.

2 Understanding the Multiscale Biomechanical Properties of the Myocardium

Cardiac tissue, and especially the myocardium, is a complex tissue with multiple interconnected length scales (Fig. 2a–c) organised into a highly structural and functional hierarchy, ranging from whole heart biomechanics to the functional unit of contraction (the sarcomere) inside the cardiomyocytes [40].

At the macroscale, in particular in the left ventricular (LV) myocardial wall, tissue fibres form a right-handed helical structure close to the inner part of the wall (the endocardium), left-handed towards the outer part of the wall (the epicardium) and a circumferential structure in between [41, 42] (Fig. 2a). The changes in fibre orientation are smooth throughout the LV wall. During systole, due to this helical organisation, the myocardium rotates relative to the base to ensure complete blood ejection from the left ventricle [43]. Nonetheless, this description of LV wall organisation is a simplification because it omits other anatomical structures, including the extensive vascular network across the whole myocardium [44]. At the mesoscale, each of the ventricular fibres consists of an anisotropic array of cardiac cells (cardiomyocytes and fibroblasts) in parallel alignment with the ECM fibres (mainly collagen fibres) to ensure the coordinated contraction of the LV (Fig. 2b). At the microscale, cardiac cells, specifically the cardiomyocytes, form aligned, interconnected bundles that attach to ECM fibres to transduce the cellular contractile forces throughout all myocardium and generate coordinated contraction (Fig. 2c) [36].

The mechanical properties of the myocardium differ across length scales. However, basic mechanical concepts can be identified independently of these scales. These include the deformation and forces present in the tissue (strain, ε, and stress, σ), the elastic stiffness (Young's modulus, E) and the complex G^* modulus accounting for the viscoelastic properties (Table 1). Although they express the same physical concept, the interpretation of the values has to be correlated with tissue physiology and structure at each of the scales addressed.

In clinical practice, there are already established techniques to measure LV macroscopic strain based on ultrasound (echocardiography) and magnetic resonance imaging (MRI) [45–47]. One of the advantages of these techniques is that they are non-invasive, allowing the assessment of cardiac mechanical properties in patients. Echocardiography imaging is established as a gold standard technique, and it is based on analysing the LV wall motion by tracking speckles (natural acoustic markers) in the ultrasonic image. These acoustic markers appear physiologically in the myocardium and can be tracked frame to frame. By post-processing software, the

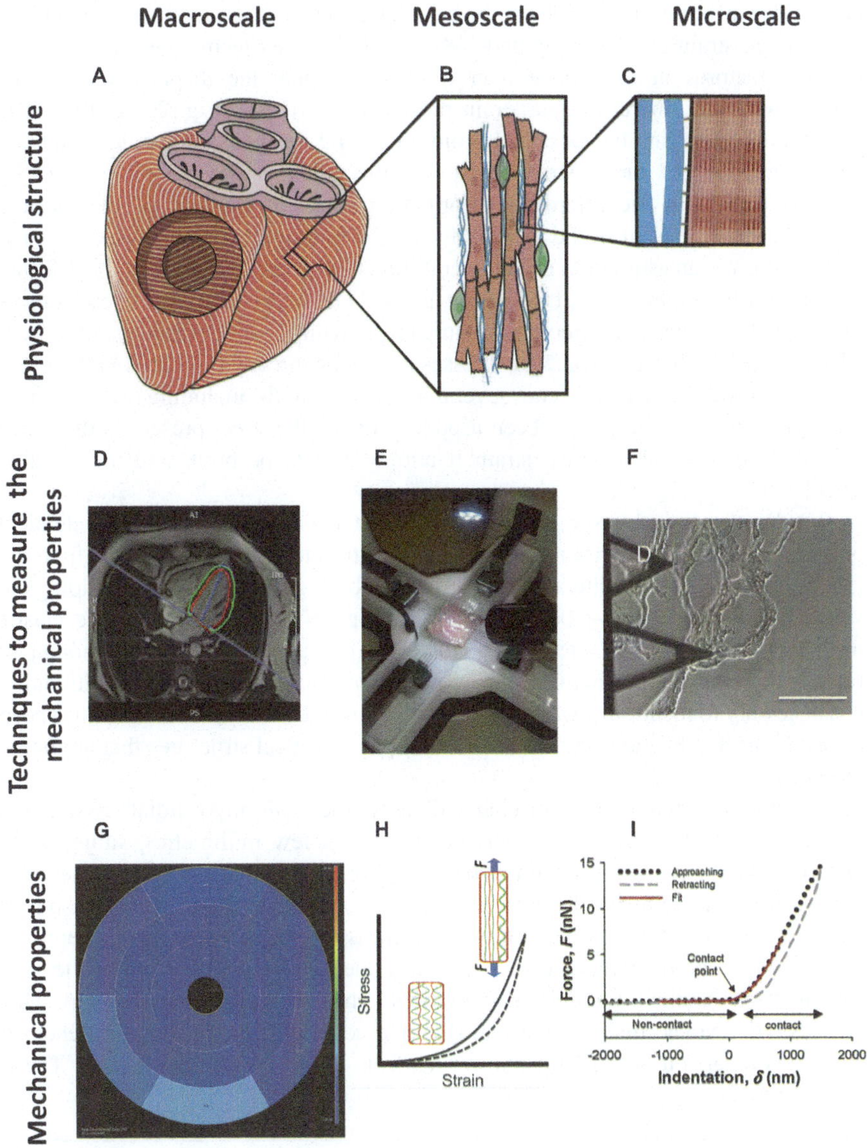

Fig. 2 Physiological structure, techniques to measure mechanical properties of cardiac tissue and mechanical properties at the three relevant scales. (**a**) Differential layer organisation of muscle fibres in the myocardium. (**b**) Cardiac cells (cardiomyocytes, red; fibroblasts, green) are anisotropically oriented along ECM fibres (blue). (**c**) Cardiomyocytes transmit the contraction force to the ECM by cell-ECM attachments (grey). (**d**) MR image of a human thorax. Red and green lines delineate the inner and outer parts of the myocardial wall necessary to compute myocardial strain. (**e**) Biaxial tensile tester setup. A porcine myocardial tissue strip is attached to four lever arms. (**f**) Microscopy phase-contrast images of 12 μm thick decellularised LV mouse heart probed with AFM. (**g**) Circumferential strain bull eye computed by MRI. Values are negative, depicting the contraction of the tissue. Regions closer to the centre correspond to myocardial regions near to the apex of the heart. (**h**) Representative strain-stress (σ-ε) curve measured by tensile testing. Non-linear visco-elastic behaviour is a characteristic of the passive mechanical properties of cardiac tissue. The

geometric shift of each speckle can be calculated to assess tissue movement and, as such, tissue strain can be computed [48]. Similarly, MRI techniques are based on tracking features in the image over time to compute the displacement of the myocardium and, therefore, the strain present in the tissue (Fig. 2d, g) [49]. The information that can be extracted from these 4D images are the longitudinal, circumferential and radial strains. Usually, the strain values are represented by negative values as the reference deformation is set at the end of the diastole [50]. The strain present in the in vivo myocardium accounts for the tissue distensibility and gives an estimation of the overall function. After MI, strain at the affected area diminishes compared to the healthy area [51]. However, detailed knowledge of myocardial wall *stress*, particularly in humans, remains elusive. This lack of knowledge is primarily because forces or stresses cannot be directly measured in the intact myocardial wall [52]. To this end, several indirect methods, including analytical and finite-element modelling, have been used to estimate the stress present in the tissue [53–56]. Based on that, stress-strain relationships can be built, and the E value calculated.

Besides the active tissue contraction present in vivo, the passive mechanical properties of the myocardium significantly contribute to overall cardiac function [57, 58]. Throughout the literature, the E for soft biological tissues has physiological values in the range of 1–100 kPa [5]. The E gives an estimation of the elastic mechanical properties of the tissue. However, just an E value is not enough to fully understand the mechanical behaviour of soft biological tissues. Additionally, it is necessary to mention at which scale and with which technique the value has been measured as this includes the information of physiological structures that are being measured.

At the mesoscale, the mechanical properties of myocardial tissue are characterised from small ex vivo tissue strips of a few millimetres, subjected to uniaxial, biaxial or even triaxial tensile tests (Fig. 2e). The strips are attached to a lever arm controlled by an electromechanical or hydraulic servo-controlled displacement actuator that deforms the strip while it measures the force applied. Therefore, the σ-ε curve can be computed by calculating the relative deformation of the strip and the force applied per unit cross-sectional strip area (Table 1; Fig. 2h). This process is done to normalise deformation and force for the sample size and, hence, to allow a comparison of experimental results from strips with different sizes. The E

Fig. 2 (continued) hysteresis between mechanical loading (solid line) and unloading (dashed line) indicates energy loss. The progressive recruitment of collagen fibres explains the non-linear behaviour. (**i**) Representative force (stress) versus indentation (strain) curve acquired with AFM on a decellularised cardiac tissue slice. Mechanical properties at the microscale are also viscoelastic (the difference between approaching and retracting curves). The appropriate contact model depending on the AFM tip geometry is used to fit the experimental data. (**a–c**, **h**) Adapted from [36] and reprinted under Creative Commons (CC BY) license. (**e**) Adapted from [37] and reprinted under Creative Commons Attribution 4.0 International License. (**f**) Adapted from [38] and reprinted with permission from Elsevier. (**i**) Adapted from [39] and reprinted with permission from Elsevier

Table 1 Basic mechanical concepts of viscoelastic biological tissue characterisation

Mechanical parameter	Symbol	Definition
Engineering strain	ε	Ratio of total deformation to the initial dimension of the tissue
Stress	σ	Normalised force applied per unit area of the tissue
Young's modulus	E	Relationship between σ and ε in the linear region of the tissue
Complex shear modulus	G*	Ratio of σ to ε under vibratory conditions to account for viscoelastic tissue properties Complex G* is decomposed as $G^*(f) = G' + iG''$ being i the imaginary unit. The real part, G′, is the elastic modulus that accounts for the elastic features of the sample, and the imaginary part, G″, is the loss modulus which characterises viscous dissipation
Loss tangent	G″/G′	Ratio of the viscous modulus to elastic modulus in a viscoelastic material. This ratio is an index of solid- or liquid-like behaviour of the sample. For a pure elastic material $G''/G' = 0$. Conversely, a pure viscous material has $G''/G' = \infty$

value of the strip can easily be derived by computing the derivative of the σ-ε curve. However, the σ-ε curve exhibits a more complex mechanical behaviour than just elastic. The hysteresis of the curve (difference between loading and unloading curve) depicts energy loss behaviour, typical of viscoelastic materials. The energy loss results from frictional processes, such as tissue fluid movement, and is commonly observed in soft biological tissues [59]. Moreover, the σ-ε behaviour of living soft tissues is highly non-linear. This behaviour can be mainly explained by the state of the ECM fibres inside the tissue, including the structural organisation of the collagen fibres. At the relaxed state of the tissue (low strains), the collagen fibres are wavy. By increasing the strain present in the tissue, the fibres start to unfold, collagen fibre recruitment increases and the σ-ε curve becomes non-linear (Fig. 2h) [60]. This strain stiffening effect is advantageous for a tissue as it becomes increasingly resistant to extension in order to prevent excessive deformations and tissue damage [61].

The mammalian healthy and diseased mechanical properties of the myocardium at the mesoscale have been the subject of extensive investigation in the past two decades [62–64]. Healthy myocardium shows anisotropic behaviour (e.g. different mechanical properties depending on the measured direction). E of healthy myocardium ranges from 1 to 10 kPa at a physiological tissue strain [65]. On the other hand, the post-MI myocardium shows a stiffer and isotropic behaviour, correlating with the disorganised distribution of collagen fibres found in the post-MI fibrotic scar [66, 67]. These findings support the notion that the ECM fibre organisation dominates the mechanical properties of the scar. In order to measure only the contribution of the ECM in the fibrotic myocardial scar and exclude the cellular effect, some studies have decellularised the samples before testing [68, 69]. Knowledge of the mechanics of decellularised ECM will support the design of novel biomaterials for cardiac tissue engineering.

At the microscale, the mechanical properties are determined by individual ECM fibres and cells. The most suitable technique to measure mechanics at this scale is atomic force microscopy (AFM) [39]. AFM probes micromechanical properties of a thin tissue slice (or a single cell for that matter) by indenting its surface with a microfabricated cantilever ended with a pyramidal or spherical tip (Fig. 2f) [70]. This technique allows the measurement of tip displacement with nanometre resolution and simultaneous measurement of the applied force (Fig. 2i). AFM measurements provide essential mechanical information at the scale at which cells probe their own microenvironment. Micromechanical properties of heart tissue have been studied utilising AFM on fresh (ECM and cells) and decellularised cardiac slices of different species [38, 71–74]. Post-MI tissue mechanical properties showed a dramatic increased E compared to the healthy tissues (approximately three- to fourfold increase). Moreover, the loss tangent of G* (G''/G', see Table 1) showed that the viscous tissue contribution of cardiac tissue at the microscale is around ten times lower than the elastic contribution, indicating that tissues show a solid-like behaviour [38].

Another major contributor to tissue mechanics are the cardiac cells. Characterisation of cardiomyocyte single-cell beating forces, frequency and cellular viscoelasticity has been studied using AFM [75]. A major advantage of AFM to measure cell mechanical parameters over other techniques, such as traction force microscopy or optical tweezers, is that AFM directly measures force and deformation without complex data processing. Different cell types largely used in cTE, such as human embryonic stem cells (hESC)- and induced pluripotent stem cells (iPSC)-derived cardiomyocytes, show different mechanical phenotype [76]. hESC-derived cardiomyocytes showed a beating force twofold lower than iPSC-derived cardiomyocytes, both showing single-cell forces in the range of nanonewton. Moreover, the most interesting fact is that cells showed an increased beating force when cultured in clusters showing that cell-cell connectivity plays a role in overall tissue force [76]. Additionally, cell viscoelasticity has been linked to several diseases [77, 78]. Laminopathies are a family of genetic diseases affecting the cardiomyocyte normal function and are caused by a mutation of the intracellular proteins called lamins. This mutation causes a loss of structural cell integrity, showed by a decreased E, with a lower cell-ECM adhesion affecting force generation and transmission towards surrounding cells and tissues [77]. Also, using primary cardiomyocytes from young and old rats, it was demonstrated that age correlates with a decreased cell shortening, increased relaxation time and increased E. These results indicate that cardiomyocytes from old animals are less deformable and contractile and suggest that cardiomyocyte mechanical changes per se can contribute to age-related diastolic LV dysfunction [79]. Overall, the data demonstrate that active and passive mechanical properties of cardiomyocytes also contribute to the overall tissue mechanics.

Obviously, mechanical data need to be handled with caution when designing novel strategies and biomaterials to mimic the healthy and diseased cardiac cellular niche. Additionally, there is still a lack of knowledge on the relation between cardiac mechanical properties and cardiac mechanical function. Fundamental insights into structure-function properties at all length scales from cell to organ are required and should be integrated to predict the consequences of mechanical changes at the microscale for cardiac function at the macroscale and vice versa. In the following sections, we will review the models used to understand the impact of cardiac mechanical properties on cell and tissue function.

3 Mechanics-Based In Vitro Models to Understand Cardiac Behaviour at the Micro- and Mesoscale

cTE strategies are also designed to study the effect of the mechanical cardiac stimuli on cardiac cells. This section reviews in vitro cTE platforms mimicking the cardiac mechanical properties at the microscale (cells) and mesoscale (microtissues).

3.1 Cardiac ECM Organisation

The healthy myocardium at the micro- and mesoscale shows highly organised, anisotropic ECM fibres in contrast to the highly disorganised, isotropic ECM structure found after injury. Several in vitro methods have been described to mimic this organisation of healthy and diseased myocardium. First, a commonly used approach is microcontact printing that serves to pattern cell culture substrates with various ECM proteins, including fibronectin, laminin, collagen, Matrigel and gelatin [80–87]. Microcontact printing allows recreating the microscale two-dimensional environment by patterning the substrate with a high resolution of just a few micrometres. By changing pattern geometry, cells are forced to adopt specific (dis)organised alignment. A logical consequence of reproducing the highly organised healthy myocardium is that the cardiomyocytes become elongated with an increased contraction force compared to cells on a disorganised pattern [81, 84, 86].

At the mesoscale, the cTE gold standard technique to study cardiac in vitro conditions is the engineered heart tissue (EHT) pioneered at the end of the last century [88]. EHTs are cardiac microtissues composed of cell-laden and ECM mimicking biomaterial, moulded between constraints or stretching posts. Classical EHTs are composed of two stretching posts creating a unidirectional microtissue mimicking ECM anisotropy of the myocardium. More recently, to mimic the chaotic tissue organisation and force distribution after injury, EHTs were fabricated either with uniaxial and biaxial post distribution. To this end, the organisation and mechanical forces have been manipulated from organised to disorganised, respectively [89, 90] (Fig. 3a). This mesoscale model will enable the understanding of the

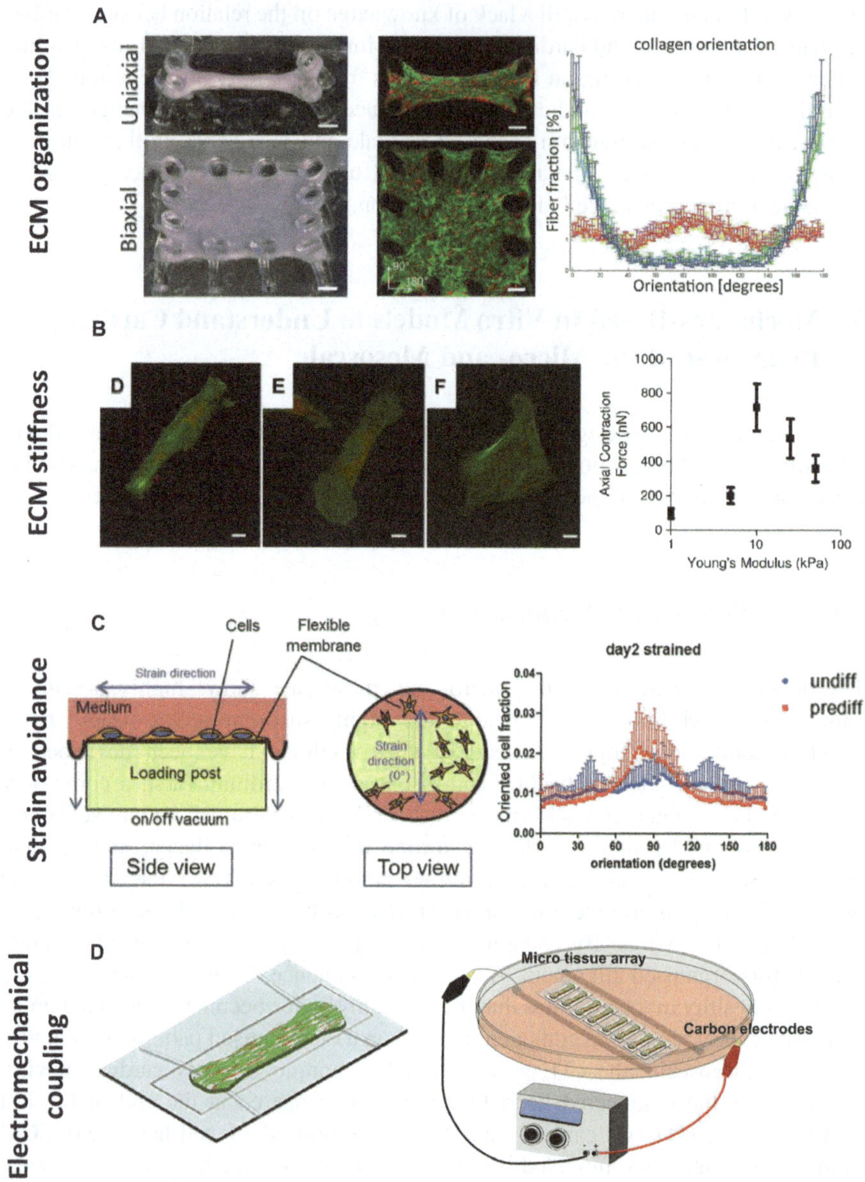

Fig. 3 Mechanical-based microscale and mesoscale in vitro models to understand cardiac cell behaviour. (**a**) Mesoscale microtissue with PDMS uniaxial or biaxial constraints to manipulate tissue organisation. The cells (red) seeded inside the collagen (green)-based hydrogels compacted around the posts. Collagen fibres' orientation shows an (an)isotropic distribution depending on the uniaxially or biaxially constrained tissues. Adapted from [89] and reprinted with permission from Oxford University Press. (**b**) Neonatal rat ventricular cardiomyocytes cultured on top of 1, 10 and 50 kPa PAA gels (left to right). Actin fibres (green) and contraction forces are maximised at physiological ECM stiffness of 10 kPa. Adapted from [91] and reprinted with permission from Elsevier. (**c**) Device based on a flexible membrane with cells seeded on top. The vacuum applied underneath the membrane stretches the flexible membrane which is supported by a loading post.

principles of ECM remodelling after injury and how it can be restored towards an organised ECM.

3.2 ECM Stiffness

A myriad of biomaterials has been developed to simulate the stiffness of healthy and diseased myocardium. Hydrogels from natural (non-cardiac ECM) and synthetic sources such as gelatin methacryloyl, polyethylene glycol (PEG), alginate, polyacrylamide (PAA) and polydimethylsiloxane (PDMS) are regularly used due to tuneability of E [84, 94–100]. It has been demonstrated that culturing cardiomyocytes on substrates with a physiological E maximises their contractility and the number of intracellular actin stress fibres, whereas increased or decreased E disrupts their cytoskeletal structure and reduces their contractile force [84, 91] (Fig. 3b). On the other hand, hydrogels from reconstituted natural cardiac ECM, such as decellularised ECM, induce better biochemical activity and increased remodelling capacity of the encapsulated cells while being able to control their mechanical properties via chemical crosslinking of the gel [29, 101, 102].

Additionally, the viscoelastic properties of some of these materials can be controlled. A recent study developed a nanostructured alginate-based hydrogel allowing control over stress-relaxation properties without changing the E. Using this material, the authors showed that stress relaxation affects cardiomyocyte intracellular contraction [103]. Another study indicated that varying the monomer and crosslinker concentration of PAA hydrogels allows to control the viscoelastic properties [104].

Overall, the described hydrogels can mimic the mechanical properties of healthy and diseased cardiac tissue but cannot capture the dynamics of cardiac remodelling after injury due to the covalent crosslinking between structural polymeric fibres in the gels. A new class of hydrogels offers control over crosslinking dynamics and consequent manipulation of hydrogel dynamic mechanical properties. By incorporating reversible crosslink methods, the properties of such hydrogels can be changed instantly by applying an external trigger, such as temperature, light or a chemical agent [105, 106]. For example, PEG was modified with a photo-sensible and reversible crosslinker that allowed to dynamically tune the viscoelastic properties by the use of blue light. Importantly, these changes were even possible when culturing cells inside the hydrogel [107]. Another recent example addresses a

Fig. 3 (continued) Cardiomyocyte progenitor cells show strain avoidance behaviour depending on their differentiation state. Adapted from [92] and reprinted with permission from Elsevier. (**d**) Schematic representation of the Biowire II platform. Cell-seeded collagen-based hydrogels are attached to uniaxially constrained wires. The system can be electrically paced with carbon electrodes and using an external electrical source. Adapted from [93] and reprinted under ACS AuthorChoice License

pH-sensitive hydrogel that enables changing the crosslinking degree by dynamic pH changes [108].

3.3 Strain on Cells

Cardiac tissue is constantly subjected to static (pre-stress present in the tissue) and cyclic strain (beating), and these strains may change with disease progression. Therefore, it is of utmost importance to investigate the mechanoresponse of cardiac cells to the experienced strain under conditions of health and disease. Most studies investigating strain responses make use of two-dimensional systems to apply (cyclic) uniaxial or equibiaxial strain [109]. Commonly, the cells are seeded on top of a flexible membrane that is stretched, transmitting the deformation of the membrane to the cells [92] (Fig. 3c). Because study designs differ in strain magnitude and frequencies used, comparison of study outcomes is cumbersome [110–112]. However, a common phenomenon has been observed in cardiac fibroblasts in response to cyclic strain. This phenomenon is called *strain avoidance* and refers to the re-orientation of cells perpendicular to the direction of applied cyclic strains [113, 114]. The physiological interpretation of this behaviour is that cells turn away from the stretch direction to experience minimal deformation on their cell body and nucleus. In fibroblasts, strain avoidance has been observed in two-dimensional and three-dimensional in vitro models [115–117]. However, strain avoidance is less clear in cardiomyocytes with several studies indicating that cardiomyocyte strain avoidance depends on the differentiation state of the cell, the strain rate and strain duration [92]. More recently, it was demonstrated that cardiomyocytes derived from a pluripotent stem cell source do not show strain avoidance. However, when co-cultured with cardiac fibroblasts (with a strain avoidance response), the cardiomyocytes did show strain avoidance and rotated along with the fibroblasts [118]. Hence, the effect of strain on cardiac cell behaviour is also influenced by the interplay between different cell types present in the tissue.

Strain generated on cardiac cells also has a significant impact on their phenotype via mechanotransduction pathways. One of the main current challenges regarding cell models in cTE is to obtain highly mature cardiomyocytes from pluripotent stem cell sources resembling adult cardiomyocytes found in vivo. Mechanical factors and, in particular, tissue strain have been shown to play a critical role in maturation process [119]. Numerous studies have been conducted to understand how cyclic mechanical strain affects cardiomyocyte maturation at single-cell and tissue levels [120–124]. Most of these studies showed that cyclic strain increases sarcomere formation, cardiac ion channel expression and contraction force and frequency of cells and tissue. Importantly, the strain magnitude and frequency applied to the tissues are essential to achieve better maturation [125, 126]. Cyclic strains around 10% showed to induce increased cardiomyocyte maturation compared to lower strains of 5% [127]. On the other hand, large strains (mimicking increased afterload) have shown to cause pathological hypertrophy in vitro with larger cardiomyocytes but with decreased contractile function [128, 129].

3.4 Mechanoelectrical Feedback

The highly interconnected cardiomyocyte network controls the coordinated contraction of the myocardium and whole heart rhythm. Cardiomyocytes, at the microscale, depolarise their cellular membrane in the presence of an electrical stimulus. This depolarisation triggers the release of intracellular calcium ions responsible for activating the cell's contractile machinery. In an in vitro setting, generally, the mesoscale tissues containing mature adult cardiomyocytes start beating due to their autorhythmic properties [89]. However, in pathophysiological conditions these tissues often beat non-synchronously due to a lack of proper cardiomyocyte and fibroblast organisation and cell-cell contact inside the hydrogel [130]. To overcome the appearance of arrythmia in cTE tissues, these are generally paced by applying an external electric field during long-term tissue culture. To this end, several approaches have been implemented with notable success [109, 131]. As an example, the Biowire and Biowire II platforms (Fig. 3d) have been demonstrated to improve intracellular calcium handling, contraction force and synchronicity of beating [93, 132].

3.5 Developing Integrated Models for Mechanical Consideration In Vitro

In the context of designing strategies for cardiac regeneration, it is not only necessary to understand how mechanical stimuli influence cell and tissue behaviour, but it is also fundamental to integrate such stimuli across length scales to better recapitulate the in vivo situation. For this purpose, the development and use of bioreactors are pursued. A bioreactor is typically defined as a device that provides tight control of the environmental conditions and external stimuli (biochemical and biomechanical) that influence cell and tissue culture processes [133].

Oxygen tension is of paramount importance in affecting cardiac cellular and tissue behaviour [134, 135]. After MI, there is a loss of perfusion in the scar region that leads to a decreased oxygen level or hypoxia. It has been previously shown that hypoxia enhances the migration and differentiation capacity of pluripotent stem cells derived to cardiomyocytes [136, 137]. Moreover, low oxygen tensions stimulate the ECM-producing phenotype of cardiac fibroblast, maintaining the presence of fibrotic tissue after injury, having a direct impact on cardiac tissue mechanics [138]. Therefore, the use of bioreactors capable to mimic (patho)physiological oxygen tensions is critical to further understand the mechanical implications on the behaviour of cTE strategies.

Biomechanically, the main goal for cTE is to synchronise contraction with appropriately timed mechanical or electrical stimulation to mimic ventricular filling. Independent control over individual input signals further allows for manipulating disease progression, e.g. via changes in the stretch (haemodynamics) and electrical signal patterns (e.g. arrythmias). Various bioreactors have been designed for applying mechanical and electrical stimuli simultaneously [139–141]. For example, an electromechanical bioreactor platform was able to provide static stress to microtissues using a pneumatically driven stretch device. It consisted of a tissue culture chamber where several tissue constructs (20 mm $\times$ 20 mm $\times$ 3 mm) were subjected to frequencies and amplitudes of cyclic stretches and electrical pulses matching the native tissue [142]. Moreover, it is also important to track over time the changes in mechanical properties of the tissues inside bioreactors as a parameter to understand tissue growth and remodelling. A recent study developed a bioreactor to test cardiac tissues under dynamic loading together with an ultrasound system to trace non-invasively the mechanical properties of the tissue over time [143].

4 In Silico Models to Study the Mechanics of Myocardial Remodelling and Regeneration

The rapid development of digital technologies has enabled the development and application of computational models in many fields nowadays. Computational models in bioengineering, commonly referred to as in silico models, enhance the knowledge of various biological tissues' behaviour at different scales. Moreover, a multiscale approach can integrate knowledge from different scales into an overall simulation of the tissue behaviour and thus become more relevant for analysis and predictions.

The use of in silico models for simulating cardiac tissue function has rapidly expanded in the last years, especially for simulating drug testing and considering chemical coupling within the tissue. The computational platforms for testing novel/ existing drugs have achieved widespread approval on different aspects – ethical, toxicological and economic [144]. Besides the platforms for drug testing, the major interest of in silico models lies in capturing cardiac systolic/diastolic functioning and electrical coupling, from micro- to macroscale [145]. These models usually neglect the biomechanics of the heart and/or miss to include the mechanical environment for the cells [146]. Moreover, there is a gap in linking electro-mechanical coupling of the heart to the contraction of the tissue at different length scales. Improvement of in silico models in this area is suggested to result in models that can predict the change of tissue mechanical function in response to tissue remodelling under conditions of health, degeneration and regeneration. A such, these models also have to translate basic research findings on cardiac regeneration into tissue engineering or other

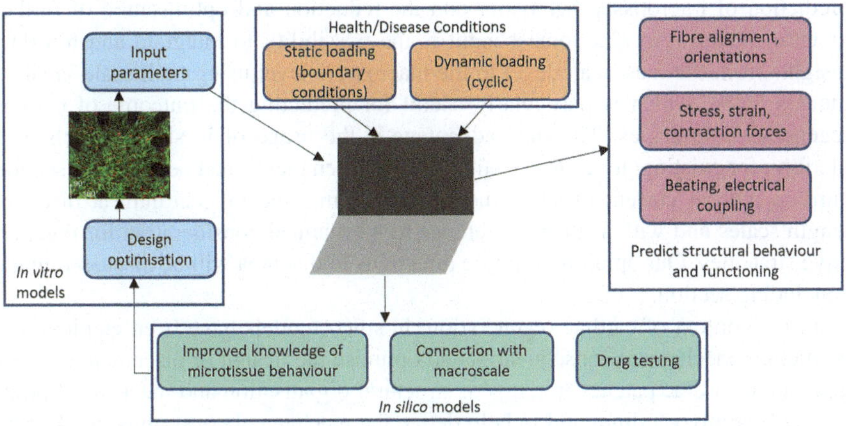

Fig. 4 Scheme of in silico approach in myocardial mesoscale modelling

regenerative strategies under pathological conditions, including their altered hemodynamic, electroconductive and tissue mechanical features [147].

Current regenerative strategies fail to properly restore the cellular microenvironment and aligned structural organisation after cardiac injury, relevant for coordinated contraction and tissue mechanical homeostasis, and this is where in silico models can be very useful in answering the questions about (changing) mechanical impact on cardiac tissue regeneration [148]. They can be developed to replicate certain experimental observations, e.g. at cell and tissue level, and be further extended to in vivo-like conditions at the organ scale. By employing a multiscale approach, they can be also used to investigate the effect of microscale modifications on macroscale function, much easier than experiments can do [149].

The starting point in the development of in silico models is the connection of simulated biological processes with data from in vitro or in vivo experiment(s), or those from the literature, since each computational model needs meaningful input data for robust predictions (Fig. 4). In this first phase of development, model outputs are compared with – or validated against – additional experimental findings so that initially applied boundary conditions of the proposed in silico model are in line with real-life data. As there are now various in silico models in cardiac research, there is a tendency to standardise in silico cardiac models in terms of verification, validation and uncertainty quantification of scientific software [150].

In the scheme depicted in Fig. 4, the input from an in vitro microtissue model is used to feed the computational model at the mesoscale (tissue level). The outputs from the in silico simulations can include, for instance, mechanical properties of the tissue, structural organisation and mechanical contraction in response to the experimental starting conditions. Once validated, the in silico model can be used to understand and predict outputs in response to new boundary/loading conditions, e.g. to mimic healthy/diseased states of the microtissue. The added value of such mesoscale in silico models can be found in the enhanced understanding and prediction of microtissue behaviour and the reduction and optimisation of further in vitro models. An additional benefit is the possibility to integrate and translate insights from the mesoscale level to the macroscale level using multiscale in silico models [149], which is particularly useful for predicting the outcome of cardiac regenerative strategies. The final advantage is the usage of in silico models as a platform for existing/novel drug testing – as has been mentioned before. This section aims to present current in silico models that mimic the myocardium at different length scales and with a special reference to mechanical consideration for regenerative strategies. Our opinion on future directions in this area will be discussed in the concluding section.

In the context of cardiac regeneration, in silico models have been employed to predict the mechanical consequences and optimise the design or placement of next-generation cardiac patches in terms of structural organisation and mechanical properties of the myocardium and to help define the mechanical constraints invoked by such patches that would lead to reversion or inhibition of adverse cardiac remodelling.

Urdeitx and Doweidar developed a mesoscale *finite element* model to simulate cardiac cellular behaviour such as proliferation, migration, maturation and cell-matrix adhesion in response to mechanical and electrical cues from the environment [151]. Cell behaviour was described as a function of cell deformation due to changes in ECM stiffness and/or electrical stimulation. The model predicted that on soft ECM, cell alignment and migration improved with increasing (directional) electrical stimulation. On stiffer ECM, cells showed enhanced maturation and proliferation. However, the mechanical impact considered in the model is limited as it only considers changes in ECM stiffness, without incorporating the cause of ECM stiffness changes or hemodynamic loading.

To better understand the potential of cardiac regeneration, tissue (ECM) production and organisation by resident or newly delivered cells have been studied in vitro. When cultured in 3D environments, such as collagen hydrogels enriched with Matrigel, cardiac cells (both cardiac fibroblasts and cardiomyocytes) contribute to the production and maintenance of the ECM by ECM synthesis, degradation and cell-matrix interactions, including cell traction forces [89]. They also respond to environmental strains by (re)orienting their cell body as well as the ECM fibres around them (see Sect. 3.3). Mesoscale in silico models of the cardiac tissues have been developed to successfully describe the underlying biological phenomena of these processes of healthy tissue remodelling, where fibroblasts align their internal cytoskeleton (stress fibres) and ECM to form an anisotropic tissue [152–156]. In response to cardiac injury, for instance, due to ischemia, the heart's primary response is to create a scar-like tissue and protect the damaged tissue from rupturing by ongoing remodelling. Under these conditions the original cell and tissue anisotropy is lost and fibroblasts differentiate into myofibroblasts, which show a higher production of ECM proteins (mainly collagen), leading to reduced compliance and stiffening of the scar [157]. In silico models that can predict structural (re)-organisation of the cells and collagen can thus predict disease development of the tissue but are also instrumental for testing the effects of mechanical conditions (e.g. constraints by tissue patches) that would allow the transition from an isotropic to an anisotropic organisation. To this end, the models should describe how cardiac fibroblasts remodel the collagenous matrix and incorporate cell-mediated traction forces [158] under both physiological and pathophysiological conditions [154–156]. When successfully validated, the output from an in silico model that describes tissue structural remodelling can also be used to design engineered tissues with optimised and load-bearing collagen organisations [159, 160].

When modelling the mechano-response of cardiomyocytes, in vitro and in silico models have focused on describing cardiomyocyte organisation and alignment at the mesoscale, either or not in the presence of fibroblasts. These models show that cardiomyocytes align with the stress direction and alongside collagen fibres in uniaxially constrained microtissues, but show no alignment in biaxially constrained microtissues [89, 161]. The aligned structure is essential for coordinated and synchronised contraction and proper propagation of electrical signals. Cross-sectional compaction of uniaxial constrained tissues further contributes to alignment and increases with the percentage of fibroblasts present in the tissue. Forceful contraction requires a high percentage of aligned cardiomyocytes and fibroblasts to establish conduction and synchronised beating. Hence, next to structural remodelling, cell ratios and alignment at the mesoscale should be incorporated when mimicking regenerative strategies to reverse or halt remodelling of cardiac scar tissue at the macroscale. Vice versa, the influence of hemodynamic loading conditions as well as active and passive mechanical behaviour at the macroscale [162] will influence cell alignment at the meso- and microscale.

Macroscopically, the myocardium is organised in differently orientated two-dimensional anisotropic sheets that follow a helicoidal shape from the epicardium to the endocardium. Hence, the local coordinate system is represented by three axes: fibre orientation, sheet orientation and normal to the sheet orientation to capture anisotropy of the tissue. There are two approaches to computationally include myocardial organisation within the cardiac geometry at the macroscale: patient-specific and numerically approximated. The inclusion of patient-specific organisation is the ultimate goal of all predictive in silico models, since it takes into account the real structure of the tissue, obtained by image reconstruction of patient-specific MR images. The fibre directions can be measured from MR images leading to the generation of patient-specific structure and geometry, which is a great improvement in numerical modelling despite their time-consuming feature [163]. Sometimes it can be more convenient to avoid image reconstruction for each patient separately. By mapping fibre orientations from the geometry of one patient to the cardiac geometry of another patient, time, money and effort can be reduced, but this approach is less accurate [164]. To simulate generic behaviour of the myocardium at the macroscale, including the influence of tissue organisation, numerical approximations can be used, where myofibril helix angle changes linearly from $-60^{\circ} \pm 10^{\circ}$ at the epicardium to $60^{\circ} \pm 10^{\circ}$ at the endocardium side [165]. Which approach will be selected depends on the purpose of the in silico model.

Cardiac in silico models describing macroscopic mechanical behaviour commonly model left ventricle (LV) behaviour. The LV, representing the largest shape and volume of the four-chambered heart, experiences the highest stresses and strains, and its function is most often and most severely affected by disease. The most valuable output from these LV models is the quantification of myocardial mechanical parameters under in vivo-like hemodynamic loading, in particular LV wall stress, which cannot be measured experimentally but is important in disease prediction since it is indicative of cardiac wall (dis)function. By changing the mechanical properties of the LV wall at certain locations, different pathological conditions can be simulated. For instance, when incorporating differential mechanical properties for scarred tissue, border zone and healthy remote tissue based on patient-specific MR images, these models can be used to predict the severity and functional consequence of myocardial infarction [166]. Yet, the LV models generally do not account for tissue organisation and remodelling at lower length scales, so they cannot simulate the progression of fibrosis and functional consequences at later stages after infarction. A second disadvantage of these models is a simplified representation of electrical conduction and electro-mechanical coupling.

The final goal of in silico models is the integration of insights obtained across length scales (cell, tissue and organ) into multiscale models that can provide more detailed information on cardiac behaviour, predict disease progression and improve existing or establish novel regenerative strategies. So far, multiscale in silico models have mainly focused on describing the effects of pharmacological agents and electrical conduction from cell to organ, while multiscale mechanical interactions have remained largely unexplored. Future directions in the development of in silico myocardial models to support the improvement of existing or creation of novel regenerative strategies lie in predicting the effects of mechanical loading/unloading on tissue remodelling and mechanical functioning at different length scales.

Knowledge obtained from experiments and simulations at cell and tissue scale should be translated into the in vivo situation of the whole heart to ultimately understand how cardiac tissue will remodel under whole-organ hemodynamic loading conditions and depending on (changing/heterogeneous) cardiac wall mechanical properties. The SIMULIA Living Heart Project[1] could offer the next step in the development of more complex multiscale electromechanical cardiac models. This project gathers some of the most prominent researchers in cardiovascular area from different branches, with the purpose to create a multidisciplinary vision in the development of cardiac computational models for clinical use. The project integrates various features relevant for cardiac function, such as the detailed geometry of the whole heart, electrophysiology, active and passive mechanical properties of the cardiac wall, blood flow and the feedback of the circulatory system on cardiac function. Using the available computational methods to model myocardial infarction and tissue remodelling at the mesoscale, the macro model could, for instance, predict

[1]Living Heart Project | SIMULIA™ - Dassault Systèmes® (3ds.com)| SIMULIA™ - Dassault Systèmes® (3ds.com).

how tissue-engineered epicardial patches or biomaterial injection in the cardiac wall would influence tissue remodelling in the infarct area and subsequently how cardiac function as a whole could be improved.

5 Conclusion

Even though the progress in the development of cardiac regenerative strategies can be observed, long-term clinical benefits are still to be expected. Current strategies largely omit the mechanical consequences after implantation of cells or tissue-engineered constructs, suggesting that more attention should be paid to mechanical considerations in the processes of tissue formation and remodelling. The use of in vitro and in silico models, together with a thorough mechanical characterisation of myocardial tissue at different length scales, is required to understand and predict the effects of the mechanical cues (e.g. loading and ECM stiffness) on cells and tissues. In this context, it should be noted that mechanical properties of cardiac tissue are influenced by the scale at which they are measured, highlighting the need to use the appropriate technique at each scale. Microscale and mesoscale in vitro models need to integrate these mechanical properties to better understand the effect on cell and tissue behaviour. Moreover, in silico models can provide significant assistance in simulating a multiscale cardiac response upon improvement of existent or development of novel models. Regarding in silico models that describe mechanical behaviour, the literature mainly reports on models at the meso- and macroscales but does not provide multiscale models that can simulate cardiac tissue remodelling or regeneration based on such processes at cellular or tissue level. The development of multiscale predictive models could illuminate native-like mechanical conditions that can provoke remodelling of fibrotic tissue (e.g. change of mechanical loading and change of stiffness), but more importantly translation of these conditions into the effect of regenerative strategies, such as patches with regionally different stiffness.

Acknowledgements The authors would like to thank Gemma Burcet, MD from Hospital Universitari Vall d'Hebron and Institut de Diagnòstic per la Imatge (Barcelona, Spain), for the supply of magnetic resonance images and strain processing. This work is supported by the partners of 'Regenerative Medicine Crossing Borders' (RegMed XB) and by Health~Holland, Top Sector Life Sciences & Health. We also gratefully acknowledge funding from the Ministry of Education, Culture and Science for the Gravitation Program 024.003.103 'Materials-Driven Regeneration' and Nederlandse Organisatie voor Wetenschappelijk Onderzoek for the NWO Open Competition Domain Science grant, OCENW.XS21.4.146.

Compliance with Ethical Standards **Funding:** This work was funded by 1) the partners of 'Regenerative Medicine Crossing Borders' and Health~Holland, Top Sector Life Sciences & Health (project Cardiac Moonshot), 2) the Ministry of Education, Culture and Science for the Gravitation Program 'Materials-Driven Regeneration' (grant number 024.003.103), and 3) by the Dutch Research Council (OCENW.XS21.4.146).

Conflict of Interest: All authors declare they have no conflict of interest.

Ethical Approval: This chapter does not contain any studies with animals performed by any of the authors. For Fig. 2d, g informed consent was obtained from the patients.

References

1. Mendis S, Puska P, Norrving B, World Health Organization, World Heart Federation, World Stroke Organization (2011) Global atlas on cardiovascular disease prevention and control. World Health Organization, Geneva, p vi, 155
2. Stewart S, MacIntyre K, Hole DJ, Capewell S, McMurray JJV (2001) More "malignant" than cancer? Five-year survival following a first admission for heart failure. Eur J Heart Fail 3:315–322. https://doi.org/10.1016/S1388-9842(00)00141-0
3. Frangogiannis NG (2017) The extracellular matrix in myocardial injury, repair, and remodeling. J Clin Invest 127:1600–1612
4. Prabhu SD, Frangogiannis NG (2016) The biological basis for cardiac repair after myocardial infarction. Circ Res 119:91–112
5. Chaudhuri O, Cooper-White J, Janmey PA, Mooney DJ, Shenoy VB (2020) Effects of extracellular matrix viscoelasticity on cellular behaviour. Nature 584:535–546. https://doi.org/10.1038/s41586-020-2612-2
6. Chin IL, Hool L, Choi YS (2019) A review of in vitro platforms for understanding cardiomyocyte mechanobiology. Front Bioeng Biotechnol 7:133. https://doi.org/10.3389/fbioe.2019.00133
7. Shapira-Schweitzer K, Seliktar D (2007) Matrix stiffness affects spontaneous contraction of cardiomyocytes cultured within a PEGylated fibrinogen biomaterial. Acta Biomater 3:33–41. https://doi.org/10.1016/j.actbio.2006.09.003
8. Boothe SD, Myers JD, Pok S, Sun J, Xi Y, Nieto RM, Cheng J, Jacot JG (2016) The effect of substrate stiffness on cardiomyocyte action potentials. Cell Biochem Biophys 74:527–535. https://doi.org/10.1007/s12013-016-0758-1
9. D'Urso M, Kurniawan NA (2020) Mechanical and physical regulation of fibroblast–myofibroblast transition: from cellular mechanoresponse to tissue pathology. Front Bioeng Biotechnol 8:609653
10. Nakamura K, Murry CE (2019) Function follows form—a review of cardiac cell therapy. Circ J 83:2399–2412
11. Karantalis V, Difede DL, Gerstenblith G, Pham S, Symes J, Zambrano JP, Fishman J, Pattany P, McNiece I, Conte J, Schulman S, Wu K, Shah A, Breton E, Davis-Sproul J, Schwarz R, Feigenbaum G, Mushtaq M, Suncion VY, Lardo AC, Borrello I, Mendizabal A, Karas TZ, Byrnes J, Lowery M, Heldman AW, Hare JM (2014) Autologous mesenchymal stem cells produce concordant improvements in regional function, tissue perfusion, and fibrotic burden when administered to patients undergoing coronary artery bypass grafting: The prospective randomized study of mesenchymal stem cell therapy in patients undergoing cardiac surgery (PROMETHEUS) trial. Circ Res 114:1302–1310. https://doi.org/10.1161/CIRCRESAHA.114.303180
12. Aghajanian H, Kimura T, Rurik JG, Hancock AS, Leibowitz MS, Li L, Scholler J, Monslow J, Lo A, Han W, Wang T, Bedi K, Morley MP, Linares Saldana RA, Bolar NA, McDaid K, Assenmacher CA, Smith CL, Wirth D, June CH, Margulies KB, Jain R, Puré E, Albelda SM, Epstein JA (2019) Targeting cardiac fibrosis with engineered T cells. Nature 573:430–433. https://doi.org/10.1038/s41586-019-1546-z
13. Fisher SA, Brunskill SJ, Doree C, Mathur A, Taggart DP, Martin-Rendon E (2014) Stem cell therapy for chronic ischaemic heart disease and congestive heart failure. Cochrane Database Syst Rev 4:CD007888
14. Tenreiro MF, Louro AF, Alves PM, Serra M (2021) Next generation of heart regenerative therapies: progress and promise of cardiac tissue engineering. NPJ Regen Med 61(6):1–17. https://doi.org/10.1038/s41536-021-00140-4
15. Cassino TR, Drowley L, Okada M, Beckman SA, Keller B, Tobita K, Leduc PRHJ (2012) Mechanical loading of stem cells for improvement of transplantation outcome in a model of acute myocardial infarction: the role of loading history. Tissue Eng Part A 18:1101–1108. https://doi.org/10.1089/TEN.TEA.2011.0285

16. Wu KH, Mo XM, Han ZC, Zhou B (2012) Cardiac cell therapy: pre-conditioning effects in cell-delivery strategies. Cytotherapy 14:260–266. https://doi.org/10.3109/14653249.2011.643780
17. Lee J, Henderson K, Massidda MW, Armenta-Ochoa M, Im BG, Veith A, Lee B-K, Kim M, Maceda P, Yoon E, Samarneh L, Wong M, Dunn AK, Kim J, Baker AB (2021) Mechanobiological conditioning of mesenchymal stem cells for enhanced vascular regeneration. Nat Biomed Eng 51(5):89–102. https://doi.org/10.1038/s41551-020-00674-w
18. Smits AM, van Laake LW, den Ouden K, Schreurs C, Szuhai K, van Echteld CJ, Mummery CL, Doevendans PA, Goumans M-J (2009) Human cardiomyocyte progenitor cell transplantation preserves long-term function of the infarcted mouse myocardium. Cardiovasc Res 83: 527–535. https://doi.org/10.1093/CVR/CVP146
19. Bax NA, van Marion MH, Shah B, Goumans MJ, Bouten CV, van der Schaft DW (2012) Matrix production and remodeling capacity of cardiomyocyte progenitor cells during in vitro differentiation. J Mol Cell Cardiol 53:497–508. https://doi.org/10.1016/J.YJMCC.2012.07.003
20. Montero P, Flandes-Iparraguirre M, Musquiz S, Pérez Araluce M, Plano D, Sanmartín C, Orive G, Gavira JJ, Prosper F, Mazo MM (2020) Cells, materials, and fabrication processes for cardiac tissue engineering. Front Bioeng Biotechnol 8:955
21. Vunjak-Novakovic G, Tandon N, Godier A, Maidhof R, Marsano A, Martens TP, Radisic M (2010) Challenges in cardiac tissue engineering. Tissue Eng Part B Rev 16:169. https://doi.org/10.1089/TEN.TEB.2009.0352
22. Radisic M (2015) Biomaterials for cardiac tissue engineering. Biomed Mater 10:030301. https://doi.org/10.1088/1748-6041/10/3/030301
23. Roshandel M, Dorkoosh F (2021) Cardiac tissue engineering, biomaterial scaffolds, and their fabrication techniques. Polym Adv Technol 32:2290–2305. https://doi.org/10.1002/PAT.5273
24. Majid QA, Fricker ATR, Gregory DA, Davidenko N, Hernandez Cruz O, Jabbour RJ, Owen TJ, Basnett P, Lukasiewicz B, Stevens M, Best S, Cameron R, Sinha S, Harding SE, Roy I (2020) Natural biomaterials for cardiac tissue engineering: a highly biocompatible solution. Front Cardiovasc Med 0:192. https://doi.org/10.3389/FCVM.2020.554597
25. Efraim Y, Sarig H, Cohen Anavy N, Sarig U, de Berardinis E, Chaw SY, Krishnamoorthi M, Kalifa J, Bogireddi H, Duc TV, Kofidis T, Baruch L, Boey FYC, Venkatraman SS, Machluf M (2017) Biohybrid cardiac ECM-based hydrogels improve long term cardiac function post myocardial infarction. Acta Biomater 50:220–233. https://doi.org/10.1016/j.actbio.2016.12.015
26. Li Z, Guan J (2011) Hydrogels for cardiac tissue engineering. Polymers (Basel) 3:740–761
27. Perea-Gil I, Gálvez-Montón C, Prat-Vidal C, Jorba I, Segú-Vergés C, Roura S, Soler-Botija C, Iborra-Egea O, Revuelta-López E, Fernández MA, Farré R, Navajas D, Bayes-Genis A (2018) Head-to-head comparison of two engineered cardiac grafts for myocardial repair: from scaffold characterization to pre-clinical testing. Sci Rep 8. https://doi.org/10.1038/s41598-018-25115-2
28. Prat-Vidal C, Rodríguez-Gómez L, Aylagas M, Nieto-Nicolau N, Gastelurrutia P, Agustí E, Gálvez-Montón C, Jorba I, Teis A, Monguió-Tortajada M, Roura S, Vives J, Torrents-Zapata-S, Coca MI, Reales L, Cámara-Rosell ML, Cediel G, Coll R, Farré R, Navajas D, Vilarrodona A, García-López J, Muñoz-Guijosa C, Querol S, Bayes-Genis A (2020) First-in-human PeriCord cardiac bioimplant: scalability and GMP manufacturing of an allogeneic engineered tissue graft. EBioMedicine 54. https://doi.org/10.1016/J.EBIOM.2020.102729
29. Bejleri D, Davis ME (2019) Decellularized extracellular matrix materials for cardiac repair and regeneration. Adv Healthc Mater 8:1801217. https://doi.org/10.1002/ADHM.201801217
30. Bouten CVC, Dankers PYW, Driessen-Mol A, Pedron S, Brizard AMA, Baaijens FPT (2011) Substrates for cardiovascular tissue engineering. Adv Drug Deliv Rev 63:221–241. https://doi.org/10.1016/J.ADDR.2011.01.007
31. Spaans S, Peter-Paul K, Fransen H, Schotman MJG, van der Wulp R, Lafleur RPM, Kluijtmans SGJM, Dankers PYW (2019) Supramolecular modification of a

sequence-controlled collagen-mimicking polymer. Biomacromolecules 14:46. https://doi.org/10.1021/acs.biomac.9b00353
32. Mol EA, Lei Z, Roefs MT, Bakker MH, Goumans M, Doevendans PA, Dankers PYW, Vader P, Sluijter JPG (2019) Injectable supramolecular ureidopyrimidinone hydrogels provide sustained release of extracellular vesicle therapeutics. Adv Healthc Mater 8:1900847. https://doi.org/10.1002/adhm.201900847
33. Goor OJGM, Hendrikse SIS, Dankers PYW, Meijer EW (2017) From supramolecular polymers to multi-component biomaterials. Chem Soc Rev 46:6621–6637
34. Dwyer KD, Coulombe KLK (2021) Cardiac mechanostructure: Using mechanics and anisotropy as inspiration for developing epicardial therapies in treating myocardial infarction. Bioact Mater 6:2198–2220. https://doi.org/10.1016/J.BIOACTMAT.2020.12.015
35. Jarrell DK, Vanderslice EJ, VeDepo MC, Jacot JG (2020) Engineering myocardium for heart regeneration—advancements, considerations, and future directions. Front Cardiovasc Med 7:586261. https://doi.org/10.3389/fcvm.2020.586261
36. Jorba I, Mostert D, Hermans LHL, van der Pol A, Kurniawan NA, Bouten CVC (2021) In vitro methods to model cardiac mechanobiology in health and disease. Tissue Eng Part C Methods 27:139–151. https://doi.org/10.1089/TEN.TEC.2020.0342
37. Nemavhola F (2017) Biaxial quantification of passive porcine myocardium elastic properties by region. Eng Solid Mech 5:155–166. https://doi.org/10.5267/j.esm.2017.6.003
38. Andreu I, Luque T, Sancho A, Pelacho B, Iglesias-García O, Melo E, Farré R, Prósper F, Elizalde MR, Navajas D (2014) Heterogeneous micromechanical properties of the extracellular matrix in healthy and infarcted hearts. Acta Biomater 10:3235–3242. https://doi.org/10.1016/j.actbio.2014.03.034
39. Alcaraz J, Otero J, Jorba I, Navajas D (2018) Bidirectional mechanobiology between cells and their local extracellular matrix probed by atomic force microscopy. Semin Cell Dev Biol 73:71–81. https://doi.org/10.1016/j.semcdb.2017.07.020
40. Bers DM (2002) Cardiac excitation–contraction coupling. Nature 2002(415):198–205. https://doi.org/10.1038/415198a
41. Streeter D (1979) Gross morphology and fiber geometry of the heart
42. Arts T, Costa KD, Covell JW, McCulloch AD (2001) Relating myocardial laminar architecture to shear strain and muscle fiber orientation. Am J Physiol Heart Circ Physiol 280(5):H2222–H2229. https://doi.org/10.1152/AJPHEART.2001.280.5.H2222
43. Greenbaum RA, Ho SY, Gibson DG, Becker AE, Anderson RH (1981) Left ventricular fibre architecture in man. Br Heart J 45:248–263. https://doi.org/10.1136/hrt.45.3.248
44. Burton RA, Plank G, Schneider JE, Grau V, Ahammer H, Keeling SL, Lee J, Smith NP, Gavaghan D, Trayanova N, Kohl P (2006) Three-dimensional models of individual cardiac histoanatomy: tools and challenges. Ann N Y Acad Sci 1080:301–319. https://doi.org/10.1196/ANNALS.1380.023
45. Phelan D, Collier P, Thavendiranathan P, Popović ZB, Hanna M, Plana JC, Marwick TH, Thomas JD (2012) Relative apical sparing of longitudinal strain using two-dimensional speckle-tracking echocardiography is both sensitive and specific for the diagnosis of cardiac amyloidosis. Heart 98:1442–1448. https://doi.org/10.1136/HEARTJNL-2012-302353
46. Saito M, Okayama H, Yoshii T, Higashi H, Morioka H, Hiasa G, Sumimoto T, Inaba S, Nishimura K, Inoue K, Ogimoto A, Shigematsu Y, Hamada M, Higaki J (2012) Clinical significance of global two-dimensional strain as a surrogate parameter of myocardial fibrosis and cardiac events in patients with hypertrophic cardiomyopathy. Eur Hear J Cardiovasc Imaging 13:617–623. https://doi.org/10.1093/EJECHOCARD/JER318
47. Valente F, Gutierrez L, Rodríguez-Eyras L, Fernandez R, Montano M, Sao-Aviles A, Pineda V, Guala A, Cuéllar H, Evangelista A, Rodríguez-Palomares J (2020) Cardiac magnetic resonance longitudinal strain analysis in acute ST-segment elevation myocardial infarction: A comparison with speckle-tracking echocardiography. IJC Hear Vasc 29:100560. https://doi.org/10.1016/J.IJCHA.2020.100560

48. Dandel M, Lehmkuhl H, Knosalla C, Suramelashvili N, Hetzer R (2009) Strain and strain rate imaging by echocardiography - basic concepts and clinical applicability. Curr Cardiol Rev 5: 133–148. https://doi.org/10.2174/157340309788166642
49. Pedrizzetti G, Claus P, Kilner PJ, Nagel E (2016) Principles of cardiovascular magnetic resonance feature tracking and echocardiographic speckle tracking for informed clinical use. J Cardiovasc Magn Reson 18:1–12. https://doi.org/10.1186/S12968-016-0269-7
50. Scatteia A, Baritussio A, Bucciarelli-Ducci C (2017) Strain imaging using cardiac magnetic resonance. Heart Fail Rev 224(22):465–476. https://doi.org/10.1007/S10741-017-9621-8
51. Mangion K, McComb C, Auger DA, Epstein FH, Berry C (2017) Magnetic resonance imaging of myocardial strain after acute ST-segment-elevation myocardial infarction: a systematic review. Circ Cardiovasc Imaging 10:e006498. https://doi.org/10.1161/CIRCIMAGING.117.006498
52. Huisman RM, Elzinga G, Westerhof N, Sipkema P (1980) Measurement of left ventricular wall stress. Cardiovasc Res 14:142–153. https://doi.org/10.1093/CVR/14.3.142
53. Guccione JM, McCulloch AD, Waldman LK (1991) Passive material properties of intact ventricular myocardium determined from a cylindrical model. J Biomech Eng 113:42–55. https://doi.org/10.1115/1.2894084
54. Feygin J, Hu Q, Swingen C, Zhang J (2008) Relationships between regional myocardial wall stress and bioenergetics in hearts with left ventricular hypertrophy. Am J Physiol Heart Circ Physiol 294:H2313–H2321. https://doi.org/10.1152/AJPHEART.01288.2007
55. Walker JC, Ratcliffe MB, Zhang P, Wallace AW, Hsu EW, Saloner DA, Guccione JM (2008) Magnetic resonance imaging-based finite element stress analysis after linear repair of left ventricular aneurysm. J Thorac Cardiovasc Surg 135:1094–1102.e2. https://doi.org/10.1016/J.JTCVS.2007.11.038
56. Walker JC, Ratcliffe MB, Zhang P, Wallace AW, Fata B, Hsu EW, Saloner D, Guccione JM (2005) MRI-based finite-element analysis of left ventricular aneurysm. Am J Physiol Heart Circ Physiol 289:H692–H700. https://doi.org/10.1152/AJPHEART.01226.2004
57. Humphrey JD (2002) Cardiovascular solid mechanics, 1st edn. Springer, New York, pp 1–758. https://doi.org/10.1007/978-0-387-21576-1
58. Holmes JW, Borg TK, Covell JW (2005) Structure and mechanics of healing myocardial infarcts. Annu Rev Biomed Eng 7:223–253. https://doi.org/10.1146/ANNUREV.BIOENG.7.060804.100453
59. Fung Y-C (1993) Biomechanics, 2nd edn. Springer, New York, pp 1–568. https://doi.org/10.1007/978-1-4757-2257-4
60. Suki B (2014) Assessing the functional mechanical properties of bioengineered organs with emphasis on the lung. J Cell Physiol 229:1134–1140. https://doi.org/10.1002/jcp.24600
61. Storm C, Pastore JJ, MacKintosh FC, Lubensky TC, Janmey PA (2005) Nonlinear elasticity in biological gels. Nature 435:191–194. https://doi.org/10.1038/nature03521
62. Demer LL, Yin FC (1983) Passive biaxial mechanical properties of isolated canine myocardium. J Physiol 339:615. https://doi.org/10.1113/JPHYSIOL.1983.SP014738
63. Novak VP, Yin FCP, Humphrey JD (1994) Regional mechanical properties of passive myocardium. J Biomech 27:403–412. https://doi.org/10.1016/0021-9290(94)90016-7
64. Dokos S, Smaill BH, Young AA, LeGrice IJ (2002) Shear properties of passive ventricular myocardium. Am J Physiol Heart Circ Physiol 283:2650–2659. https://doi.org/10.1152/AJPHEART.00111.2002
65. Engler AJ, Carag-Krieger C, Johnson CP, Raab M, Tang HY, Speicher DW, Sanger JW, Sanger JM, Discher DE (2008) Embryonic cardiomyocytes beat best on a matrix with heart-like elasticity: scar-like rigidity inhibits beating. J Cell Sci 121:3794–3802. https://doi.org/10.1242/jcs.029678
66. Fomovsky GM, Holmes JW (2010) Evolution of scar structure, mechanics, and ventricular function after myocardial infarction in the rat. Am J Physiol Heart Circ Physiol 298. https://doi.org/10.1152/ajpheart.00495.2009

67. Sirry MS, Butler JR, Patnaik SS, Brazile B, Bertucci R, Claude A, Mclaughlin R, Davies NH, Liao J, Franz T (2016) Characterisation of the mechanical properties of infarcted myocardium in the rat under biaxial tension and uniaxial compression. J Mech Behav Biomed Mater 63: 252–264. https://doi.org/10.1016/j.jmbbm.2016.06.029
68. Brazile BL, Butler JR, Patnaik SS, Claude A, Prabhu R, Williams LN, Perez KL, Nguyen KT, Zhang G, Bajona P, Peltz M, Yang Y, Hong Y, Liao J (2021) Biomechanical properties of acellular scar ECM during the acute to chronic stages of myocardial infarction. J Mech Behav Biomed Mater 116:104342. https://doi.org/10.1016/J.JMBBM.2021.104342
69. Farré N, Jorba I, Torres M, Falcones B, Martí-Almor J, Farré R, Almendros I, Navajas D (2018) Passive stiffness of left ventricular myocardial tissue is reduced by ovariectomy in a post-menopause mouse model. Front Physiol 9. https://doi.org/10.3389/fphys.2018.01545
70. Jorba I, Uriarte JJ, Campillo N, Farré R, Navajas D (2017) Probing micromechanical properties of the extracellular matrix of soft tissues by atomic force microscopy. J Cell Physiol 232. https://doi.org/10.1002/jcp.25420
71. Zhang S, Sun A, Ma H, Yao K, Zhou N, Shen L, Zhang C, Zou Y, Ge J (2011) Infarcted myocardium-like stiffness contributes to endothelial progenitor lineage commitment of bone marrow mononuclear cells. J Cell Mol Med 15:2245–2261. https://doi.org/10.1111/J.1582-4934.2010.01217.X
72. Berry MF, Engler AJ, Woo YJ, Pirolli TJ, Bish LT, Jayasankar V, Morine KJ, Gardner TJ, Discher DE, Sweeney HL (2006) Mesenchymal stem cell injection after myocardial infarction improves myocardial compliance. Am J Physiol Heart Circ Physiol. 290:2196–2203. https://doi.org/10.1152/AJPHEART.01017.2005
73. Notari M, Ventura-Rubio A, Bedford-Guaus SJ, Jorba I, Mulero L, Navajas D, Martí M, Raya Á (2018) The local microenvironment limits the regenerative potential of the mouse neonatal heart. Sci Adv 4:eaao5553. https://doi.org/10.1126/sciadv.aao5553
74. Garcia-Puig A, Mosquera JL, Jiménez-Delgado S, García-Pastor C, Jorba I, Navajas D, Canals F, Raya A (2019) Proteomics analysis of extracellular matrix remodeling during zebrafish heart regeneration. Mol Cell Proteomics 18(9):1745–1755. https://doi.org/10.1074/mcp.RA118.001193
75. Borin D, Pecorari I, Pena B, Sbaizero O (2018) Novel insights into cardiomyocytes provided by atomic force microscopy. Semin Cell Dev Biol 73:4–12. https://doi.org/10.1016/J.SEMCDB.2017.07.003
76. Liu J, Sun N, Bruce MA, Wu JC, Butte MJ (2012) Atomic force mechanobiology of pluripotent stem cell-derived cardiomyocytes. PLoS One 7:e37559. https://doi.org/10.1371/JOURNAL.PONE.0037559
77. Lanzicher T, Martinelli V, Long CS, Del FG, Puzzi L, Borelli M, Mestroni L, Taylor MRG, Sbaizero O (2016) AFM single-cell force spectroscopy links altered nuclear and cytoskeletal mechanics to defective cell adhesion in cardiac myocytes with a nuclear lamin mutation. Nucleus 6:394–407. https://doi.org/10.1080/19491034.2015.1084453
78. Benech JC, Benech N, Zambrana AI, Rauschert I, Bervejillo V, Oddone N, Damián JP (2014) Diabetes increases stiffness of live cardiomyocytes measured by atomic force microscopy nanoindentation. Am J Physiol 307:C910–C919. https://doi.org/10.1152/AJPCELL.00192.2013
79. Lieber SC, Aubry N, Pain J, Diaz G, Kim S-J, Vatner SF (2004) Aging increases stiffness of cardiac myocytes measured by atomic force microscopy nanoindentation. Am J Physiol Heart Circ Physiol 287:645–651. https://doi.org/10.1152/AJPHEART.00564.2003
80. Feinberg AW, Ripplinger CM, Van Der Meer P, Sheehy SP, Domian I, Chien KR, Parker KK (2013) Functional differences in engineered myocardium from embryonic stem cell-derived versus neonatal cardiomyocytes. Stem Cell Rep 1:387–396. https://doi.org/10.1016/j.stemcr.2013.10.004
81. Sheehy SP, Grosberg A, Qin P, Behm DJ, Ferrier JP, Eagleson MA, Nesmith AP, Krull D, Falls JG, Campbell PH, McCain ML, Willette RN, Hu E, Parker KK (2017) Toward improved

myocardial maturity in an organ-on-chip platform with immature cardiac myocytes. Exp Biol Med (Maywood) 242:1643–1656. https://doi.org/10.1177/1535370217701006
82. Ariyasinghe NR, Reck CH, Viscio AA, Petersen AP, Lyra-Leite DM, Cho N, McCain ML (2017) Engineering micromyocardium to delineate cellular and extracellular regulation of myocardial tissue contractility. Integr Biol (United Kingdom) 9:730–741. https://doi.org/10.1039/c7ib00081b
83. Pasqualini FS, Agarwal A, O'Connor BB, Liu Q, Sheehy SP, Parker KK (2018) Traction force microscopy of engineered cardiac tissues. PLoS One 13:e0194706. https://doi.org/10.1371/journal.pone.0194706
84. McCain ML, Yuan H, Pasqualini FS, Campbell PH, Parker KK (2014) Matrix elasticity regulates the optimal cardiac myocyte shape for contractility. Am J Physiol Circ Physiol 306:H1525–H1539. https://doi.org/10.1152/ajpheart.00799.2013
85. Gopalan SM, Flaim C, Bhatia SN, Hoshijima M, Knoell R, Chien KR, Omens JH, McCulloch AD (2003) Anisotropic stretch-induced hypertrophy in neonatal ventricular myocytes micropatterned on deformable elastomers. Biotechnol Bioeng 81:578–587. https://doi.org/10.1002/bit.10506
86. Ribeiro AJS, Ang Y-S, Fu J-D, Rivas RN, Mohamed TMA, Higgs GC, Srivastava D, Pruitt BL (2015) Contractility of single cardiomyocytes differentiated from pluripotent stem cells depends on physiological shape and substrate stiffness. Proc Natl Acad Sci 112:12705–12710. https://doi.org/10.1073/pnas.1508073112
87. Ribeiro MC, Tertoolen LG, Guadix JA, Bellin M, Kosmidis G, D'Aniello C, Monshouwer-Kloots J, Goumans M-J, Wang Y, Feinberg AW, Mummery CL, Passier R (2015) Functional maturation of human pluripotent stem cell derived cardiomyocytes in vitro – correlation between contraction force and electrophysiology. Biomaterials 51:138–150. https://doi.org/10.1016/j.biomaterials.2015.01.067
88. Eschenhagen T, Fink C, Remmers U, Scholz H, Wattchow J, Weil J, Zimmermann W, Dohmen HH, Schäfer H, Bishopric N, Wakatsuki T, Elson EL (1997) Three-dimensional reconstitution of embryonic cardiomyocytes in a collagen matrix: a new heart muscle model system. FASEB J 11:683–694. https://doi.org/10.1096/fasebj.11.8.9240969
89. Van Spreeuwel ACC, Bax NAM, Bastiaens AJ, Foolen J, Loerakker S, Borochin M, Van Der Schaft DWJ, Chen CS, Baaijens FPT, Bouten CVC (2014) The influence of matrix (an)-isotropy on cardiomyocyte contraction in engineered cardiac microtissues. Integr Biol (United Kingdom) 6:422–429. https://doi.org/10.1039/c3ib40219c
90. van Spreeuwel ACC, Bax NAM, van Nierop BJ, Aartsma-Rus A, Goumans MJTH, Bouten CVC (2017) Mimicking cardiac fibrosis in a dish: fibroblast density rather than collagen density weakens cardiomyocyte function. J Cardiovasc Transl Res 10:116–127. https://doi.org/10.1007/s12265-017-9737-1
91. Jacot JG, McCulloch AD, Omens JH (2008) Substrate stiffness affects the functional maturation of neonatal rat ventricular myocytes. Biophys J 95:3479–3487. https://doi.org/10.1529/biophysj.107.124545
92. Mauretti A, Bax NAM, Van Marion MH, Goumans MJ, Sahlgren C, Bouten CVC (2016) Cardiomyocyte progenitor cell mechanoresponse unrevealed: Strain avoidance and mechanosome development. Integr Biol (United Kingdom) 8:991–1001. https://doi.org/10.1039/c6ib00117c
93. Wang EY, Rafatian N, Zhao Y, Lee A, Lai BFL, Lu RX, Jekic D, Davenport Huyer L, Knee-Walden EJ, Bhattacharya S, Backx PH, Radisic M (2019) Biowire model of interstitial and focal cardiac fibrosis. ACS Cent Sci 5:1146–1158. https://doi.org/10.1021/ACSCENTSCI.9B00052
94. Bracco Gartner TCL, Deddens JC, Mol EA, Magin Ferrer M, van Laake LW, Bouten CVC, Khademhosseini A, Doevendans PA, Suyker WJL, Sluijter JPG, Hjortnaes J (2019) Anti-fibrotic effects of cardiac progenitor cells in a 3D-model of human cardiac fibrosis. Front Cardiovasc Med 6:52. https://doi.org/10.3389/fcvm.2019.00052

95. Sadeghi AH, Shin SR, Deddens JC, Fratta G, Mandla S, Yazdi IK, Prakash G, Antona S, Demarchi D, Buijsrogge MP, Sluijter JPG, Hjortnaes J, Khademhosseini A (2017) Engineered 3D cardiac fibrotic tissue to study fibrotic remodeling. Adv Healthc Mater 6:1–14. https://doi.org/10.1002/adhm.201601434
96. Crocini C, Walker CJ, Anseth KS, Leinwand LA (2020) Three-dimensional encapsulation of adult mouse cardiomyocytes in hydrogels with tunable stiffness. Prog Biophys Mol Biol 154: 71–79. https://doi.org/10.1016/j.pbiomolbio.2019.04.008
97. Ma Y, Iyer RP, Jung M, Czubryt MP, Lindsey ML (2017) Cardiac fibroblast activation post-myocardial infarction: current knowledge gaps. Trends Pharmacol Sci 38:448–458
98. Lee S, Serpooshan V, Tong X, Venkatraman S, Lee M, Lee J, Chirikian O, Wu JC, Wu SM, Yang F (2017) Contractile force generation by 3D hiPSC-derived cardiac tissues is enhanced by rapid establishment of cellular interconnection in matrix with muscle-mimicking stiffness. Biomaterials 131:111–120. https://doi.org/10.1016/j.biomaterials.2017.03.039
99. Broughton KM, Russell B (2015) Cardiomyocyte subdomain contractility arising from microenvironmental stiffness and topography. Biomech Model Mechanobiol 14:589–602. https://doi.org/10.1007/s10237-014-0624-2
100. Corbin EA, Vite A, Peyster EG, Bhoopalam M, Brandimarto J, Wang X, Bennett AI, Clark AT, Cheng X, Turner KT, Musunuru K, Margulies KB (2019) Tunable and reversible substrate stiffness reveals a dynamic mechanosensitivity of cardiomyocytes. ACS Appl Mater Interfaces 11:20603–20614. https://doi.org/10.1021/acsami.9b02446
101. Spang MT, Christman KL (2018) Extracellular matrix hydrogel therapies: in vivo applications and development. Acta Biomater 68:1–14. https://doi.org/10.1016/J.ACTBIO.2017.12.019
102. Cramer MC, Badylak SF (2019) Extracellular matrix-based biomaterials and their influence upon cell behavior. Ann Biomed Eng 487(48):2132–2153. https://doi.org/10.1007/S10439-019-02408-9
103. Chaudhuri O, Gu L, Klumpers D, Darnell M, Bencherif SA, Weaver JC, Huebsch N, Lee H-P, Lippens E, Duda GN, Mooney DJ (2016) Hydrogels with tunable stress relaxation regulate stem cell fate and activity. Nat Mater 15:326–334. https://doi.org/10.1038/NMAT4489
104. Charrier EE, Pogoda K, Wells RG, Janmey PA (2018) Control of cell morphology and differentiation by substrates with independently tunable elasticity and viscous dissipation. Nat Commun 9:449. https://doi.org/10.1038/s41467-018-02906-9
105. Zhao H, Xu K, Zhu P, Wang C, Chi Q (2019) Smart hydrogels with high tunability of stiffness as a biomimetic cell carrier. Cell Biol Int 43:84–97. https://doi.org/10.1002/cbin.11091
106. Accardo JV, Kalow JA (2018) Reversibly tuning hydrogel stiffness through photocontrolled dynamic covalent crosslinks. Chem Sci 9:5987–5993. https://doi.org/10.1039/c8sc02093k
107. Wu X, Huang W, Wu WH, Xue B, Xiang D, Li Y, Qin M, Sun F, Wang W, Bin ZW, Cao Y (2018) Reversible hydrogels with tunable mechanical properties for optically controlling cell migration. Nano Res 11:5556–5565. https://doi.org/10.1007/s12274-017-1890-y
108. Bastings MMC, Koudstaal S, Kieltyka RE, Nakano Y, Pape ACH, Feyen DAM, van Slochteren FJ, Doevendans PA, Sluijter JPG, Meijer EW, Chamuleau SAJ, Dankers PYW (2014) A fast pH-switchable and self-healing supramolecular hydrogel carrier for guided, local catheter injection in the infarcted myocardium. Adv Healthc Mater 3:70–78. https://doi.org/10.1002/adhm.201300076
109. Stoppel WL, Kaplan DL, Black LD (2016) Electrical and mechanical stimulation of cardiac cells and tissue constructs. Adv Drug Deliv Rev 96:135–155
110. Yamane M, Matsuda T, Ito T, Fujio Y, Takahashi KAJ (2007) Rac1 activity is required for cardiac myocyte alignment in response to mechanical stress. Biochem Biophys Res Commun 353:1023–1027. https://doi.org/10.1016/J.BBRC.2006.12.144
111. Salameh A, Wustmann A, Karl S, Blanke K, Apel D, Rojas-Gomez D, Franke H, Mohr FW, Janousek J, Dhein S (2010) Cyclic mechanical stretch induces cardiomyocyte orientation and polarization of the gap junction protein connexin43. Circ Res 106:1592–1602. https://doi.org/10.1161/CIRCRESAHA.109.214429

112. Debbi L, Drori S, Tzlil S (2018) The influence of the timing of cyclic load application on cardiac cell contraction. Front Physiol 0:917. https://doi.org/10.3389/FPHYS.2018.00917
113. Livne A, Bouchbinder E, Geiger B (2014) Cell reorientation under cyclic stretching. Nat Commun 51(5):1–8. https://doi.org/10.1038/ncomms4938
114. Tondon A, Hsu HJ, Kaunas R (2012) Dependence of cyclic stretch-induced stress fiber reorientation on stretch waveform. J Biomech 45:728–735. https://doi.org/10.1016/J.JBIOMECH.2011.11.012
115. Jungbauer S, Gao H, Spatz JP, Kemkemer R (2008) Two characteristic regimes in frequency-dependent dynamic reorientation of fibroblasts on cyclically stretched substrates. Biophys J 95:3470–3478. https://doi.org/10.1529/BIOPHYSJ.107.128611
116. Boccafoschi F, Bosetti M, Gatti S, Cannas M (2007) Dynamic fibroblast cultures: response to mechanical stretching. Cell Adhes Migr 1:124–128. https://doi.org/10.4161/CAM.1.3.5144
117. Foolen J, Deshpande VS, Kanters FMW, Baaijens FPT (2012) The influence of matrix integrity on stress-fiber remodeling in 3D. Biomaterials 33:7508–7518. https://doi.org/10.1016/j.biomaterials.2012.06.103
118. Mostert D, Klouda L, van Turnhout MC, Kurniawan NA, Bouten CVC (2021) Cardiac fibroblast mechanoresponse guides anisotropic organization of hiPSC-derived cardiomyocytes in vitro. bioRxiv. https://doi.org/10.1101/2021.02.16.431369
119. Carlos-Oliveira M, Lozano-Juan F, Occhetta P, Visone R, Rasponi M (2021) Current strategies of mechanical stimulation for maturation of cardiac microtissues. Biophys Rev 13:717–727. https://doi.org/10.1007/S12551-021-00841-6/FIGURES/4
120. Mihic A, Li J, Miyagi Y, Gagliardi M, Li SH, Zu J, Weisel RD, Keller G, Li RK (2014) The effect of cyclic stretch on maturation and 3D tissue formation of human embryonic stem cell-derived cardiomyocytes. Biomaterials 35:2798–2808. https://doi.org/10.1016/J.BIOMATERIALS.2013.12.052
121. Marsano A, Conficconi C, Lemme M, Occhetta P, Gaudiello E, Votta E, Cerino G, Redaelli A, Rasponi M (2016) Beating heart on a chip: a novel microfluidic platform to generate functional 3D cardiac microtissues. Lab Chip 16:599–610. https://doi.org/10.1039/C5LC01356A
122. Llucià-Valldeperas A, Bragós R, Soler-Botija C, Roura S, Gálvez-Montón C, Prat-Vidal C, Perea-Gil I, Bayes-Genis A (2018) Unravelling the effects of mechanical physiological conditioning on cardiac adipose tissue-derived progenitor cells in vitro and in silico. Sci Rep 81(8):1–12. https://doi.org/10.1038/s41598-017-18799-5
123. LaBarge W, Mattappally S, Kannappan R, Fast VG, Pretorius D, Berry JL, Zhang J (2019) Maturation of three-dimensional, hiPSC-derived cardiomyocyte spheroids utilizing cyclic, uniaxial stretch and electrical stimulation. PLoS One 14:e0219442. https://doi.org/10.1371/JOURNAL.PONE.0219442
124. Kreutzer J, Viehrig M, Pölönen RP, Zhao F, Ojala M, Aalto-Setälä K, Kallio P (2020) Pneumatic unidirectional cell stretching device for mechanobiological studies of cardiomyocytes. Biomech Model Mechanobiol 19:291–303. https://doi.org/10.1007/S10237-019-01211-8/FIGURES/5
125. Dou W, Wang L, Malhi M, Liu H, Zhao Q, Plakhotnik J, Xu Z, Huang Z, Simmons CA, Maynes JT, Sun Y (2021) A microdevice platform for characterizing the effect of mechanical strain magnitudes on the maturation of iPSC-cardiomyocytes. Biosens Bioelectron 175:112875. https://doi.org/10.1016/J.BIOS.2020.112875
126. Lu K, Seidel T, Cao-Ehlker X, Dorn T, Batcha AMN, Schneider CM, Semmler M, Volk T, Moretti A, Dendorfer A, Tomasi R (2021) Progressive stretch enhances growth and maturation of 3D stem-cell-derived myocardium. Theranostics 11:6138–6153. https://doi.org/10.7150/THNO.54999
127. Kroll K, Chabria M, Wang K, Häusermann F, Schuler F, Polonchuk L (2017) Electro-mechanical conditioning of human iPSC-derived cardiomyocytes for translational research. Prog Biophys Mol Biol 130:212–222. https://doi.org/10.1016/J.PBIOMOLBIO.2017.07.003

128. Ovchinnikova E, Hoes M, Ustyantsev K, Bomer N, de Jong TV, van der Mei H, Berezikov E, van der Meer P (2018) Modeling human cardiac hypertrophy in stem cell-derived cardiomyocytes. Stem Cell Rep 10:794–807. https://doi.org/10.1016/J.STEMCR.2018.01.016
129. Viereck J, Bührke A, Foinquinos A, Chatterjee S, Kleeberger JA, Xiao K, Janssen-Peters H, Batkai S, Ramanujam D, Kraft T, Cebotari S, Gueler F, Beyer AM, Schmitz J, Bräsen JH, Schmitto JD, Gyöngyösi M, Löser A, Hirt MN, Eschenhagen T, Engelhardt S, Bär C, Thum T (2020) Targeting muscle-enriched long non-coding RNA H19 reverses pathological cardiac hypertrophy. Eur Heart J 41:3462–3474. https://doi.org/10.1093/EURHEARTJ/EHAA519
130. Nagaraju CK, Dries E, Gilbert G, Abdesselem M, Wang N, Amoni M, Driesen RB, Sipido KR (2019) Myofibroblast modulation of cardiac myocyte structure and function. Sci Rep 9:1–11. https://doi.org/10.1038/s41598-019-45078-2
131. López-Canosa A, Perez-Amodio S, Yanac-Huertas E, Ordoño J, Rodriguez-Trujillo R, Samitier J, Castaño O, Engel E (2021) A microphysiological system combining electrospun fibers and electrical stimulation for the maturation of highly anisotropic cardiac tissue. Biofabrication 13:035047. https://doi.org/10.1088/1758-5090/ABFF12
132. Nunes SS, Miklas JW, Liu J, Aschar-Sobbi R, Xiao Y, Zhang B, Jiang J, Massé S, Gagliardi M, Hsieh A, Thavandiran N, Laflamme MA, Nanthakumar K, Gross GJ, Backx PH, Keller G, Radisic M (2013) Biowire: a platform for maturation of human pluripotent stem cell-derived cardiomyocytes. Nat Methods 10:781–787. https://doi.org/10.1038/nmeth.2524
133. Paez-Mayorga J, Hernández-Vargas G, Ruiz-Esparza GU, Iqbal HMN, Wang X, Zhang YS, Parra-Saldivar R, Khademhosseini A (2019) Bioreactors for cardiac tissue engineering. Adv Healthc Mater 8:1701504. https://doi.org/10.1002/ADHM.201701504
134. Abe H, Semba H, Takeda N (2017) The roles of hypoxia signaling in the pathogenesis of cardiovascular diseases. J Atheroscler Thromb 24:884–894. https://doi.org/10.5551/JAT.RV17009
135. D'Amario D, Migliaro S, Borovac JA, Restivo A, Vergallo R, Galli M, Leone AM, Montone RA, Niccoli G, Aspromonte N, Crea F (2019) Microvascular dysfunction in heart failure with preserved ejection fraction. Front Physiol 10:1347. https://doi.org/10.3389/FPHYS.2019.01347/BIBTEX
136. Niebruegge S, Bauwens CL, Peerani R, Thavandiran N, Masse S, Sevaptisidis E, Nanthakumar K, Woodhouse K, Husain M, Kumacheva E, Zandstra PW (2009) Generation of human embryonic stem cell-derived mesoderm and cardiac cells using size-specified aggregates in an oxygen-controlled bioreactor. Biotechnol Bioeng 102:493–507. https://doi.org/10.1002/BIT.22065
137. Correia C, Serra M, Espinha N, Sousa M, Brito C, Burkert K, Zheng Y, Hescheler J, Carrondo MJT, Šarić T, Alves PM (2014) Combining hypoxia and bioreactor hydrodynamics boosts induced pluripotent stem cell differentiation towards cardiomyocytes. Stem Cell Rev Rep 10:786–801. https://doi.org/10.1007/S12015-014-9533-0/FIGURES/7
138. Wang JH, Zhao L, Pan X, Chen NN, Chen J, Gong QL, Su F, Yan J, Zhang Y, Zhang SH (2016) Hypoxia-stimulated cardiac fibroblast production of IL-6 promotes myocardial fibrosis via the TGF-β1 signaling pathway. Lab Investig 968(96):839–852. https://doi.org/10.1038/labinvest.2016.65
139. Isenberg BC, Tranquillo RT (2003) Long-term cyclic distention enhances the mechanical properties of collagen-based media-equivalents. Ann Biomed Eng 31:937–949. https://doi.org/10.1114/1.1590662
140. Lu L, Mende M, Yang X, Körber HF, Schnittler HJ, Weinert S, Heubach J, Werner C, Ravens U, Lu L, Mende M, Yang X, Körber HF, Schnittler HJ, Weinert S, Heubach J, Werner C, Ravens U (2013) Design and validation of a bioreactor for simulating the cardiac niche: a system incorporating cyclic stretch, electrical stimulation, and constant perfusion. Tissue Eng Part A 19:403–414. https://doi.org/10.1089/TEN.TEA.2012.0135
141. Miklas JW, Nunes SS, Sofla A, Reis LA, Pahnke A, Xiao Y, Laschinger C, Radisic M (2014) Bioreactor for modulation of cardiac microtissue phenotype by combined static stretch and electrical stimulation. Biofabrication 6. https://doi.org/10.1088/1758-5082/6/2/024113

142. Wang B, Wang G, To F, Butler JR, Claude A, McLaughlin RM, Williams LN, de Jongh Curry AL, Liao J (2013) Myocardial scaffold-based cardiac tissue engineering: application of coordinated mechanical and electrical stimulations. Langmuir 29:11109–11117. https://doi.org/10.1021/LA401702W
143. van Kelle MAJ, Oomen PJA, Bulsink JA, Janssen-van den Broek MWJT, Lopata RGP, Rutten MCM, Loerakker S, Bouten CVC (2017) A bioreactor to identify the driving mechanical stimuli of tissue growth and remodeling. Tissue Eng Part C Methods 23:377–387. https://doi.org/10.1089/TEN.TEC.2017.0141
144. Savoji H, Mohammadi MH, Rafatian N, Toroghi MK, Wang EY, Zhao Y, Korolj A, Ahadian S, Radisic M (2019) Cardiovascular disease models: a game changing paradigm in drug discovery and screening. Biomaterials 198:3–26. https://doi.org/10.1016/j.biomaterials.2018.09.036
145. Mijailovich SM, Prodanovic M, Poggesi C, Geeves MA, Regnier M (2021) Multiscale modeling of twitch contractions in cardiac trabeculae. J Gen Physiol 153. https://doi.org/10.1085/JGP.202012604
146. Campbell SG, McCulloch AD (2011) Multi-scale computational models of familial hypertrophic cardiomyopathy: genotype to phenotype. J R Soc Interface 8:1550–1561. https://doi.org/10.1098/RSIF.2011.0184
147. Günter J, Wolint P, Bopp A, Steiger J, Cambria E, Hoerstrup SP, Emmert MY (2016) Microtissues in cardiovascular medicine: regenerative potential based on a 3d microenvironment. Stem Cells Int 2016. https://doi.org/10.1155/2016/9098523
148. Butler DL, Goldstein SA, Guldberg RE, Guo XE, Kamm R, Laurencin CT, McIntire LV, Mow VC, Nerem RM, Sah RL, Soslowsky LJ, Spilker RL, Tranquillo RT (2009) The impact of biomechanics in tissue engineering and regenerative medicine. Tissue Eng Part B Rev 15:477–484
149. Mihalef V, Passerini T, Mansi T (2020) Multi-scale models of the heart for patient-specific simulations. Artif Intell Comput Model Hear 2020:3–42. https://doi.org/10.1016/B978-0-12-817594-1.00011-5
150. Land S, Gurev V, Arens S, Augustin CM, Baron L, Blake R, Bradley C, Castro S, Crozier A, Favino M, Fastl TE, Fritz T, Gao H, Gizzi A, Griffith BE, Hurtado DE, Krause R, Luo X, Nash MP, Pezzuto S, Plank G, Rossi S, Ruprecht D, Seemann G, Smith NP, Sundnes J, Rice JJ, Trayanova N, Wang D, Wang ZJ, Niederer SA (2015) Verification of cardiac mechanics software: benchmark problems and solutions for testing active and passive material behaviour. Proc R Soc A Math Phys Eng Sci 471:20150641. https://doi.org/10.1098/RSPA.2015.0641
151. Urdeitx P, Doweidar MH (2020) A computational model for cardiomyocytes mechano-electric stimulation to enhance cardiac tissue regeneration. Mathematics. 8:1875. https://doi.org/10.3390/MATH8111875
152. Obbink-Huizer C, Foolen J, Oomens CWJ, Borochin M, Chen CS, Bouten CVC, FPT B (2014) Computational and experimental investigation of local stress fiber orientation in uniaxially and biaxially constrained microtissues. Biomech Model Mechanobiol 135(13): 1053–1063. https://doi.org/10.1007/S10237-014-0554-Z
153. Obbink-Huizer C, Oomens CW, Loerakker S, Foolen J, Bouten CV, Baaijens FP (2014) Computational model predicts cell orientation in response to a range of mechanical stimuli. Biomech Model Mechanobiol 13:227–236. https://doi.org/10.1007/S10237-013-0501-4
154. Loerakker S, Obbink-Huizer C, Baaijens FPT (2013) A physically motivated constitutive model for cell-mediated compaction and collagen remodeling in soft tissues. Biomech Model Mechanobiol 135(13):985–1001. https://doi.org/10.1007/S10237-013-0549-1
155. Ristori T, Obbink-Huizer C, Oomens CW, Baaijens FP, Loerakker S (2016) Efficient computational simulation of actin stress fiber remodeling. Comput Methods Biomech Biomed Engin 19:1347–1358. https://doi.org/10.1080/10255842.2016.1140748
156. Ristori T, Notermans TMW, Foolen J, Kurniawan NA, Bouten CVC, Baaijens FPT, Loerakker S (2018) Modelling the combined effects of collagen and cyclic strain on cellular orientation in collagenous tissues. Sci Rep 81(8):1–14. https://doi.org/10.1038/s41598-018-26989-y

157. Davis J, Molkentin JD (2014) Myofibroblasts: trust your heart and let fate decide. J Mol Cell Cardiol 70:9–18. https://doi.org/10.1016/J.YJMCC.2013.10.019
158. Meshel AS, Wei Q, Adelstein RS, Sheetz MP (2005) Basic mechanism of three-dimensional collagen fibre transport by fibroblasts. Nat Cell Biol 72(7):157–164. https://doi.org/10.1038/ncb1216
159. Ristori T, Bouten CVC, Baaijens FPT, Loerakker S (2018) Predicting and understanding collagen remodeling in human native heart valves during early development. Acta Biomater 80:203–216. https://doi.org/10.1016/J.ACTBIO.2018.08.040
160. Emmert MY, Schmitt BA, Loerakker S, Sanders B, Spriestersbach H, Fioretta ES, Bruder L, Brakmann K, Motta SE, Lintas V, Dijkman PE, Frese L, Berger F, Baaijens FPT, Hoerstrup SP (2018) Computational modeling guides tissue-engineered heart valve design for long-term in vivo performance in a translational sheep model. Sci Transl Med 10. https://doi.org/10.1126/SCITRANSLMED.AAN4587
161. Thavandiran N, Dubois N, Mikryukov A, Massé S, Beca B, Simmons CA, Deshpande VS, McGarry JP, Chen CS, Nanthakumar K, Keller GM, Radisic M, Zandstra PW (2013) Design and formulation of functional pluripotent stem cell-derived cardiac microtissues. Proc Natl Acad Sci U S A 110:E4698–E4707. https://doi.org/10.1073/PNAS.1311120110
162. Shim J, Grosberg A, Nawroth JC, Kit Parker K, Bertoldi K (2012) Modeling of cardiac muscle thin films: pre-stretch, passive and active behavior. J Biomech 45:832–841. https://doi.org/10.1016/J.JBIOMECH.2011.11.024
163. Toussaint N, Stoeck CT, Schaeffter T, Kozerke S, Sermesant M, Batchelor PG (2013) In vivo human cardiac fibre architecture estimation using shape-based diffusion tensor processing. Med Image Anal 17:1243–1255. https://doi.org/10.1016/J.MEDIA.2013.02.008
164. Cao Y, Miller MI, Winslow RL, Younes L (2005) Large deformation diffeomorphic metric mapping of vector fields. IEEE Trans Med Imaging 24:1216–1230. https://doi.org/10.1109/TMI.2005.853923
165. Genet M, Lee LC, Kuhl E, Guccione J (2014) Abaqus/Standard-based quantification of human cardiac mechanical properties. In: 2014 SIMULIA Community Conference (SCC2014), May 2014, Providence, Rhode Island, United States
166. Haddad SMH (2016) A novel composite material-based computational model for left ventricle biomechanics simulation. Electronic Thesis and Dissertation Repository

Mechanobiology of Exercise-Induced Cardiac Remodeling in Health and Disease

William K. Cornwell III and Peter Buttrick

Abbreviations

ARVC	Arrhythmogenic right ventricular cardiomyopathy
DCM	Dilated cardiomyopathy
ECG	Electrocardiogram
EICR	Exercise-induced cardiac remodeling
HCM	Hypertrophic cardiomyopathy
LV	Left ventricle
MRI	Magnetic resonance imaging
RBBB	Right bundle branch block
RV	Right ventricle
Qc	Cardiac output
VO_2	Oxygen uptake

1 Introduction

Structural changes in the hearts of athletes were initially described in the late nineteenth century [1, 2]. Using auscultation and a physical examination, Henschen [1] described the heart of Nordic skiers and, around the same time, Darling [2] evaluated collegiate rowers to similarly demonstrate enlarged cardiac dimensions. Since these initial reports, numerous studies over the past century have repetitively demonstrated the impact of repetitive exposure to intense exercise on cardiac structure and function among athletes. With advances in technology, heart size and

W. K. Cornwell III (✉) · P. Buttrick
Department of Medicine-Cardiology, University of Colorado Anschutz Medical Campus, Aurora, CO, USA
e-mail: William.cornwell@cuanschutz.edu

M. Hecker, D. J. Duncker (eds.), *Cardiac Mechanobiology in Physiology and Disease*, Cardiac and Vascular Biology 9,
https://doi.org/10.1007/978-3-031-23965-6_9

function were evaluated in athletes and controls using imaging-based analyses, that while now somewhat outdated (biplane X-ray and echocardiography), at the time, provided great insight into the athletic heart [3, 4]. The process of cardiac remodeling, termed exercise-induced cardiac remodeling (EICR) [5], involves unique physiologic adaptations based on the type of exercise discipline, leading to varying degrees of ventricular myocardial hypertrophy as well as dilatation of the atria and ventricles. This chapter provides an overview of the impact of these different types of exercise on cardiovascular structure and function, as well as the essential characteristics of EICR present among different types of athletes.

2 Cardiovascular Function During Exercise

During exercise, the cardiovascular system delivers oxygenated blood from the lungs to the musculoskeletal system to support the increase in metabolic demand that is necessary to sustain work. The Fick equation describes the relationship between cardiac output (Qc), oxygen extracted from the blood by peripheral tissue, i.e., the a-vO_2 difference (arterial-venous oxygen difference), and pulmonary oxygen uptake (VO_2), according to the following relationship:

$$VO_2 = Qc \times \text{a-}VO_2 \quad \text{difference}$$

Regardless of age and gender, there is an inviolate relationship between Qc and VO_2 such that Qc increases by 5–6 L/min for every 1 L/min increase in VO_2 during exercise, assuming normal hemoglobin concentration and red cell mass [6–8]. Heart rate (HR) augmentation during exercise is in direct proportion to exercise intensity, results from both vagal withdrawal and sympathetic activation, and is responsible for most of the increase in Qc during exercise. During exercise, the arterial baroreceptor operating point is reset to operate at a higher blood pressure (BP) [9]. The degree of resetting is in direct relation to the intensity of exercise and occurs to regulate HR and BP as effectively during exercise as under resting conditions. Resetting of the baroreceptors is mediated by both feedforward (central command) and feedback mechanisms from skeletal muscle afferents (exercise pressor reflex) [10]. Maximal stroke volume of the left ventricle (LV) during exercise occurs at a relatively low level of exercise intensity, typically around 50% of maximal oxygen uptake (VO_2 max) in large part due to decreased peripheral vascular resistance with onset of exercise. Pericardial constraint limits LV end-diastolic volume and prevents further increases in stroke volume (SV) during exercise [11].

While considerable overlap exists, different types of exercise impose differing chronic loads on the heart. Using echocardiography, Morganroth et al. [12] were the first to demonstrate how different types of exercise impact cardiovascular structure and function. Specifically, Morganroth et al. found that athletes participating in isotonic or endurance types of exercise (e.g., running, swimming) had increased LV end-diastolic volume and mass, whereas athletes involved in strength training

Table 1 Classification of sports and exercise (adapted from Mitchell et al. [80])

Static component	Dynamic component Low (<40% VO_2 max)	Medium (40–70% VO_2 max)	High (>70% VO_2 max)
Low	Billiards	Baseball/softball	Badminton
(<20% MVC)	Bowling	Fencing	Cross-county skiing
	Cricket	Table tennis	Field hockey
	Curling	Volleyball	Orienteering
	Golf		Race walking
	Riflery		Running (long distance)
			Soccer
			Tennis
Medium	Archery	American football	Basketball
(20–50% MVC)	Auto racing	Field events (jumping)	Ice hockey
	Diving	Figure skating	Cross-country skiing
	Equestrian	Rodeoing	Lacrosse
	Motorcycling	Rugby	Running (middle-distance)
		Running (sprint)	Swimming
		Surfing	Team handball
		Synchronized swimming	
High	Bobsledding	Body building	Boxing
(>50% MVC)	Field events (throwing)	Downhill skiing	Canoeing/kayaking
	Gymnastics	Skateboarding	Cycling
	Martial Arts	Snowboarding	Decathlon
	Sailing	Wrestling	Rowing
	Sport climbing		Speed-skating
	Water skiing		Triathlon
	Weight lifting		
	Windsurfing		

Classification of sports and exercise, based on Mitchell et al. [80]. This classification scheme provides a general description of different exercises according to the degree of static vs. dynamic exercise incorporated. Herein, specific exercises are stratified according to the relative dynamic component according to degree of oxygen uptake (VO_2) achieved, resulting in an increasing cardiac output, as well as a static component according to degree of maximal voluntary contraction (MVC) achieved, resulting in an increasing blood pressure component

(e.g., shot-putting, wrestling) had normal LV end-diastolic volumes but increased wall thickness and mass [12]. This initial observation led to the concept that isotonic/dynamic exercise leads to eccentric remodeling of the heart, similar to what may be seen in response to conditions of chronic volume overload, while static/isometric types of exercise lead to concentric remodeling, similar to what is observed in response to chronic pressure overload [12, 13]. Building on these initial observations, exercise is generally divided into two broad categories: dynamic and static

(Table 1) [14, 15]. Dynamic exercise (e.g., running, cycling, swimming) involves sustained increases in Qc along with reduced total peripheral resistance and imparts a volume load on the heart. In contrast, static exercise (e.g., weight lifting, strength training) imposes a pressure load on the heart and involves increases in total peripheral resistance. Static exercise may lead to very large increases in BP during exercise. For example, among young (22–28 years old) healthy weight lifters, a leg press with both legs at 95% of one-rep maximum weight led to an average BP of 320/250 mmHg [16]. Even a single arm curl continued to failure led to an average BP of 255/190 mmHg [16].

It is important to emphasize that (1) there is considerable overlap since most types of exercise incorporate varying degrees of both dynamic and static exercise and (2) the hemodynamic response and degree of cardiac remodeling vary and depend on the intensity of endurance/dynamic exercise relative to VO_2 max, as well as the intensity of isometric/static exercise relative to maximal voluntary contraction.

3 The Athletic Heart

3.1 *Structural Remodeling of the Left Ventricle*

LV size and morphology may vary between individuals according to several factors, including overall level of fitness, as well as the type, intensity, and duration of exercise undertaken [17, 18]. Since the initial report by Morganroth et al. [12], numerous other studies have demonstrated that LV hypertrophy is a hallmark of the athletic heart. Two landmark meta-analyses characterized the changes that occur in the heart of different athletes. Fagard found that compared to nonathletic controls, LV mass was 64% greater in cyclists, 48% greater in runners, and 25% greater in strength-trained athletes [19]. Pluim et al. [13] further characterized changes in LV structure by analyzing data from 59 studies, encompassing data from 1451 male athletes engaging in one of three types of exercise patterns: purely dynamic (e.g., running), purely static (e.g., weight lifting), or combined dynamic and static exercise (cycling, rowing) vs. nonathletic controls [13]. LV mass among controls (174 g) was significantly lower than LV mass of endurance-trained athletes (249 g), strength-trained athletes (267 g), and combined endurance/strength-trained athletes (288 g) [13].

Historically, it was postulated that endurance- and strength-based exercise modalities led to one of two distinctly different ventricular phenotypes (Fig. 1). While data indicate that there may be significant overlap [13, 19, 20], the majority of human data support Morganroth's initial findings, and in fact the same physiologic distinctions have been seen in large and small animals subjected to distinct training protocols. However, strength training may not lead to overt concentric hypertrophy. Spence et al. [21] randomly assigned 23 healthy untrained individuals to endurance vs. resistance exercise for 6 months. Participants underwent cardiac magnetic resonance imaging (MRI) prior to and following the intervention.

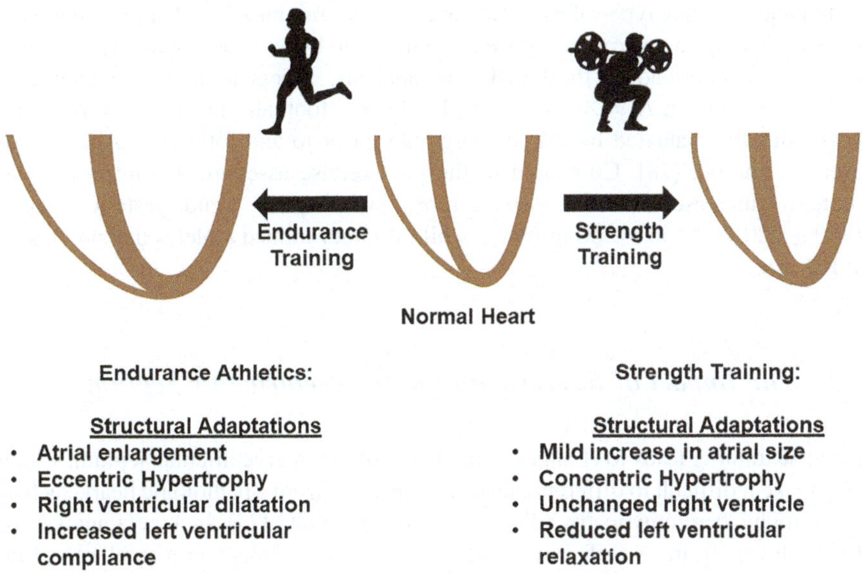

Fig. 1 Sport-specific changes in ventricular structure. While overlap exists between specific types of sport, endurance training generally imposes a volume load on the heart, while strength training imposes a pressure load on the heart. Chronic/repetitive exposure to these types of exercise leads to development of morphologic adaptations characteristic of exercise-induced cardiac remodeling. Modified from Weiner et al. [5]

Endurance-trained individuals showed evidence of eccentric hypertrophy, with increases in LV mass (112.5 ± 7.3 vs. 121.8 ± 6.6 g) and LV end-diastolic volume (by 9.0 ± 5.0 mL), but interventricular wall thickness did not increase following 6 months of resistance training [21].

3.2 *Right Ventricular Remodeling Among Athletes*

In response to endurance exercise, right ventricular (RV) enlargement occurs similarly to LV enlargement, supporting the concept of what has been referred to as "balanced dilatation" [5, 18]. Multiple investigations have found that the RV diameter generally is greater among endurance athletes than nonathletic controls [22–25]. In a study of elite endurance athletes ($N = 127$) participating in either orienteering, cross-country skiing, or middle-distance running, athletes had enlarged RV chamber sizes compared to historical controls [26]. Similar findings were described in an analysis incorporating cardiac MRI in endurance athletes ($N = 21$) and nonathletic controls [22]. In this analysis, RV mass (77 ± 10 vs. 56 ± 8 g) and RV end-diastolic volume (160 ± 26 vs. 128 ± 10 mL) were significantly greater among athletes vs. controls [27].

In contrast, static types of exercise (i.e., strength training) do not appear to impact RV size, perhaps not surprising since the stimulus to hypertrophy is largely reflective of increased afterload on the LV. In one analysis, collegiate athletes engaging in endurance (rowing, $N = 40$) or strength athletes (football players, $N = 40$) were longitudinally evaluated by echocardiography prior to and following 3 months of exercise training [28]. Compared to the pre-exercise assessment, rowers demonstrated increased RV size (pre vs. post end-diastolic area: 1460 ± 220 vs. 1650 ± 200 mm/m^2), while strength-trained athletes had no change in RV size.

3.3 The Impact of Exercise on the Interventricular Septum

Exercise training leads to changes in thickness of the interventricular septum, which may make it difficult to differentiate an athletic heart from a pathologic heart, such as hypertrophic cardiomyopathy (discussed in detail below) [29]. In a meta-analysis of 1451 athletes from 39 different studies, nonathletic controls had a smaller septum than athletes who engage in strength training, endurance training, or combined strength-endurance training [13]. Further, endurance exercise may lead to scar formation within the interventricular septum. Among 40 athletes competing in endurance events of 3–11 h duration, 5 participants displayed evidence of delayed gadolinium enhancement on cardiac MRI, located within the interventricular septum and near the RV attachment points [30]. Interestingly, in this analysis, athletes with evidence of scarring had been competing in endurance events for longer periods of time and also had evidence of RV structural remodeling [30]. The distribution of scarring seen in athletes is similar to descriptions of scarring observed among patients with pulmonary hypertension, suggesting that scarring results, at least in part, from repetitive (in the case of athletes) or chronic (in the case of pulmonary hypertension) exposure to increased pulmonary arterial pressures, increasing RV myocardial wall stress [30–32].

3.4 Atrial Remodeling and the Risk of Atrial Fibrillation

Numerous studies have demonstrated that the left atrium may be enlarged among competitive athletes [22, 33]. In a study of 1777 competitive athletes, 347 (20%) had echocardiographic evidence of left atrial enlargement [34]. On multivariable analysis, left atrial enlargement was explained by a dilated left ventricle as well as participation in endurance sports, such as cycling and rowing [34]. In an analysis of 370 endurance athletes and 245 strength-trained athletes, the left atrial volume index was greater in endurance athletes vs. strength-trained athletes (29.1 ± 9.1 vs. 26.4 ± 8.4 mL/m^2) [35]. Among collegiate endurance athletes (rowers), 90 days of rowing led to a significant increase in left atrial volume index (pre vs. post training: 28.9 ± 5.7 vs. 31.3 ± 6.2 mL/m^2) [28].

Routine exercise reduces the risk of incident atrial fibrillation and also reduces the time spent in atrial fibrillation for individuals with established disease [36–38]. However, there is a U-shaped curve that describes the dose-dependent relationship between habitual exercise and risk of atrial fibrillation [38–42]. While routine physical activity reduces the risk of incident atrial fibrillation [36], among highly trained persons, the risk increases and approximates that of the general population. Multiple physiologic responses to exercise may account for this association between exercise and risk of atrial fibrillation [43]. During exercise, increases in systemic BP and pulmonary arterial pressures predispose to increased intra-atrial pressure [43]. However, consistent with the hemodynamic effect of exercise on the ventricles, chronic exposure to exercise—particularly dynamic exercise—repetitively exposes the atrium to volume overload, leading to atrial dilatation. Interestingly, the lifetime dose of exercise may, in part, determine left atrial size [44]. For example, 60 athletes completing the 2010 Grand Prix of Bern, a 10 mile race, were stratified according to lifetime hours of training into a low (<1500 h), medium (1500–4500 h), or high (>4500 h) amount of lifetime training. The P-wave duration on ECG increased, and left atrial volume increased in concert with hours of lifetime training [44]. Atrial dilatation, a hallmark of the athlete's heart [18, 45], is a well-recognized factor that predisposes to atrial fibrillation [46]. At the same time, inflammation occurs in a dose-dependent fashion with the intensity and duration of exercise [47, 48]. In addition, pro-fibrotic biomarkers, increased among athletes compared to healthy control populations, may promote atrial fibrosis [49–52]. Together, these factors contribute to the development of atrial fibrillation in athletes.

The impact of exercise duration on clinical atrial fibrillation among athletes has been reported. The incidence of atrial fibrillation and stroke was compared with ~209,000 Swedish cross-country skiers competing in the Vasaloppet and a matched sample of ~500,000 non-skier controls. Overall, very well-trained men had a higher incidence of atrial fibrillation than less trained men, but the incidence was comparable to that observed in the general population (HR 0.98, 95% CI 0.93–1.03) [53]. Female skiers had a lower incidence of arrhythmia than female non-skiers (HR 0.55, 95% CI 0.48–0.64), and overall, skiers of either sex had a lower incidence of stroke than non-skiers (HR 0.64, 95% CI 0.60–0.67) [53].

4 Impact of Exercise on Ventricular Function

In response to dynamic exercise, afterload of both the RV and LV increases progressively as workload intensity increases [32, 54]. In addition, volume on both sides of the heart increases acutely, at least until an individual approaches his/her ventilatory threshold [55]. This combination of volume and pressure load applied to the heart, particularly at higher intensities of exercise, creates an environment that is favorable to cardiac remodeling in response to repetitive exposure to this stimulus

[17]. In a study of 12 healthy sedentary individuals, 1 year of endurance exercise training increased LV and RV mass to levels observed in elite endurance athletes [56]. In this study, participants underwent cardiac MRI prior to initiation of exercise training, as well as at 3, 6, 9, and 12 months of training. This unique study design shed new light on the differential responses of the RV and LV to endurance exercise training. Specifically, the RV responded to endurance training with progressive eccentric remodeling throughout all timepoints of training. In contrast, the LV initially responded with concentric remodeling during the first 6–9 months of training, but then dilated to restore the baseline mass/volume ratio at the 12-month follow-up assessment [56]. In a similar study, among healthy middle-aged but previously sedentary individuals, 2 years of dynamic exercise training increased LV end-diastolic volume (measured by echocardiography), along with reductions in LV stiffness constants and a down/rightward shift in the diastolic limb of the pressure-volume loop, indicative of improved LV compliance [57]. In addition, there was an upward shift in the LV Frank-Starling relationship, indicative of increased SV for a given filling pressure [57]. These findings demonstrate that exercise prevents age-related stiffening of the LV and may also protect against the development of heart failure with preserved ejection fraction, a disease characterized by marked LV stiffness compared to the normal heart [58, 59]. The impact of strength training on LV compliance has been less well studied. However, in one study of collegiate strength-trained athletes (American football players), there was imaging evidence of impaired diastolic relaxation, with reductions in echocardiographic indices of early and late LV diastolic tissue velocities [28].

The RV, a thin-walled and highly compliant structure—at least when compared to its left-sided counterpart—has historically been referred to as a "passive conduit" [60] and a "mere bystander" [61], with questionable relevance to overall cardiovascular function. However, in a study of nine healthy adults undergoing invasive cardiopulmonary exercise testing with conductance catheters inserted into the RV to generate real-time pressure-volume analysis, it was found that the RV has substantial contractile reserve, with an approximate fourfold increase in metrics of contractility, as well as metrics of myocardial energy production and utilization, from rest to peak exercise of short-term duration (Fig. 2) [62]. However, prolonged dynamic exercise disproportionately stresses the RV compared to the LV [30, 32, 63–65]. In a study of 39 endurance athletes and 14 nonathlete controls, RV wall stress—defined by the Law of Laplace—was lower among athletes vs. controls under resting conditions, but, during exercise, increased more among athletes vs. controls (125% vs. 14%, $P < 0.001$) [32]. In another study of 40 athletes competing in endurance races of 3–11 h duration, RV volumes increased and metrics of RV systolic function (ejection fraction, tricuspid annular plane systolic excursion, strain) declined following completion of the race compared to baseline, whereas no change in LV volumes or function was observed [30]. Interestingly, a dose-response relationship was observed between exercise duration and degree of reduction in RV systolic function, such that individuals completing ultratriathlon races of 11 h had greater decrements in RV systolic function compared to athletes completing shorter events, such as a marathon (3 h) or endurance triathlon (5.5 h) [30]. These reductions

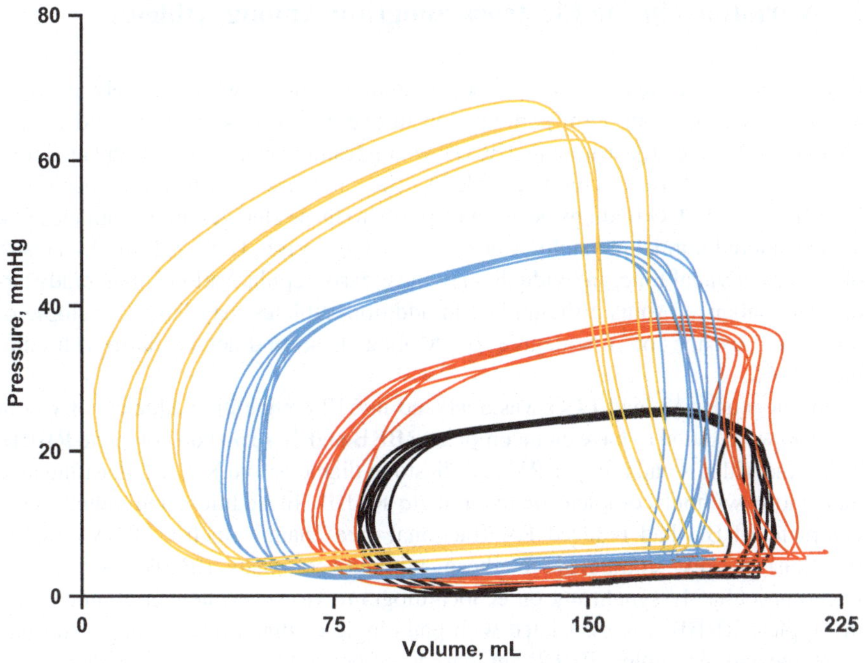

Fig. 2 Pressure volume analysis of right ventricular function during exercise. Real-time right ventricular pressure volume analysis during invasive cardiopulmonary exercise testing on upright cycle ergometry from a healthy 26-year-old man. Black represents resting condition; red and blue represent two levels of submaximal exercise below the ventilatory threshold and gold represents peak exercise. With progressive increases in external work, right ventricular stroke volume increases substantially, as evidenced by a progressive increase in the width of the pressure volume loop. Loops obtained from the corresponding author's laboratory and published in J. Physiology 2020; 598 (13): 2575–2587 [62]

in RV function following completion of endurance events typically normalize after about a week of recovery [30, 64].

Similar results were found in a meta-analysis of 14 studies evaluating longitudinal changes in LV and RV function following at least 90 min of exercise. Following completion of exercise, LV function remained stable, but there were significant reductions in metrics of RV function, including a decline in RV fractional area change by 5.78% (4.46–7.09%), RV ejection fraction by 7.05% (1.8–12.3%), and tricuspid annular plane systolic excursion by 4.77 mm (1.24–8.3 mm) [65].

5 Variations in the Electrocardiogram Among Athletes

Physiological changes to exercise occur among athletes who routinely exercise 4–8 h/week [66]. Long-term participation in exercise at this level leads to unique changes on the electrocardiogram (ECG), which result from an increase in vagal tone [67, 68], as well as structural remodeling (i.e., increased cardiac chamber sizes) [5, 69]. Several ECG findings occur in response to these changes among athletes that are considered normal physiologic adaptations to exercise (Table 2) [66]. As a result of increased vagal tone, individuals may have early repolarization, sinus bradycardia, and enhanced sinus arrhythmias. In addition, athletes may have an ectopic or junctional escape or first-degree or Mobitz I second-degree atrioventricular block [66].

In one study, 12-lead ECG was analyzed in 510 competitive athletes, of whom 44 (9%) were found to have an incomplete RBBB and 13 (3%) had complete RBBB. These individuals had a larger RV end-diastolic diameter and RV end-diastolic area than those without complete or incomplete RBBB. In addition, individuals with complete RBBB had reduced RV fractional area change (metric of RV systolic function) at rest compared to those without complete RBBB, as well as interventricular dyssynchrony on echocardiogram. However, neither complete nor incomplete RBBB was associated with pathologic cardiac disease, suggesting that complete and incomplete RBBB are markers of physiologic cardiac remodeling and reflect a triad of RV dilatation, reduction in resting RV systolic function, and interventricular dyssynchrony [24].

The increase in chamber size and cardiac mass that occurs as a physiologic response to exercise means that many athletes display voltage criteria for left ventricular hypertrophy. In addition, increased right ventricular size may lead to ECG evidence of right ventricular hypertrophy and an incomplete right bundle branch block [70].

Table 2 Changes on the electrocardiogram that are normal variants for an athlete[a]

Increased QRS voltage
Incomplete right bundle branch block
Early repolarization
Black athlete repolarization variant
Juvenile T wave pattern
Sinus bradycardia
Sinus arrhythmia
Ectopic atrial rhythm
Junctional escape rhythm
First-degree atrioventricular block
Mobitz type 1 (Wenckebach) second-degree atrioventricular block

[a]Drezner JA, et al. Br J. Sports Med 2017; 51: 704–731

6 Inherited and Acquired Cardiomyopathies

Considerable overlap exists between structural changes observed with EICR and cardiomyopathies. It is important to determine the underlying cause of these changes, since some cardiomyopathies—specifically, hypertrophic cardiomyopathy, dilated cardiomyopathy, and arrhythmogenic right ventricular dysplasia—have shared features with the athletic heart and, when present, may increase the risk of sudden cardiac death. Considerable attention has been given to identifying parameters that delineate the athletic heart from a true cardiomyopathy [29, 71, 72].

Genetic analysis may be performed in attempt to identify athletes at increased risk of sudden cardiac death. Unfortunately, genetic analysis may be limited by multiple factors [73, 74]. First, there may be high variability in clinical phenotype in patients with the same pathologic variant, even when analyzing patients within the same family tree. Second, penetrance and expressivity may vary [73, 74]. In the Cardiac Arrest Survivors with Preserved Ejection Fraction Registry (CASPER) [75], 375 survivors of cardiac arrest completed genetic testing. Prior syncope and a family history of sudden death were independently associated with the presence of a pathogenic variant on testing [75]. However, pathologic variants were identified in a minority of individuals ($N = 29$, or 17% of the total registry), and the most common variants identified were channelopathy-associated or cardiomyopathy-associated, and variants of unknown significance were identified in 18% of individuals [75]. Thus, the ability to identify individuals at risk for sudden cardiac death from genetic testing may be limited by several factors.

6.1 *Hypertrophic Cardiomyopathy*

Hypertrophic cardiomyopathy (HCM) is the most common cause of sudden cardiac death among individuals in the United States less than 35 years of age, accounting for at least one third of deaths in this demographic [76, 77]. HCM is a common inherited cardiomyopathy, generally affecting about 1 in 500 individuals, and may result from over 1500 mutations in 11 major genes that encode proteins of the cardiac sarcomere, Z disk and calcium handling machinery [71]. While morphologic variations exist, HCM classically involves LV hypertrophy without chamber dilatation. In non-athletes, the upper limit of normal for septal wall thickness is 12 mm by echocardiography. Athletes may have a wall thickness exceeding 12 mm, and a "gray-zone" [29] may exist where it is particularly difficult to distinguish an athletic heart from HCM when the wall thickness is 13–15 mm. Evaluation may require cardiac MRI, exercise testing, genetic testing, and possibly detraining to distinguish etiologies of hypertrophy.

6.2 Dilated Cardiomyopathy

Among endurance athletes, particularly those with dilated LV and a low/reduced ejection fraction, it may be important to rule out an underlying dilated cardiomyopathy. The ejection fraction among trained athletes may decline to as low as 45% [78]. However, patients with dilated cardiomyopathy have limited ability to augment cardiac output during exercise and, as such, will demonstrate reductions in exercise capacity and VO_2 max on cardiopulmonary exercise testing.

6.3 Arrhythmogenic Right Ventricular Cardiomyopathy

For patients with a disproportionately enlarged RV, arrhythmogenic right ventricular cardiomyopathy (ARVC) should be ruled out. Formal criteria exist to appropriately diagnose ARVC [79]. These criteria consist of imaging criteria demonstrating severe enlargement of the RV, as well as tissue diagnosis from endomyocardial biopsy, ECG abnormalities, presence of arrhythmias, and a family history suggesting inherited cardiomyopathy [79].

7 Conclusions

Participation in vigorous sport and exercise, for either professional or recreational purposes, is increasingly popular [5]. Understanding the impact of EICR on cardiovascular morphology and function has multiple clinically relevant applications. First, training programs can be tailored to enhance cardiovascular adaptation and overall performance during sports. In addition, clinicians can more appropriately discriminate between EICR and inherited or acquired cardiomyopathies which, when present, may increase the risk of adverse cardiovascular events. Uncertainty remains as to whether the structural and functional adaptations to exercise—hallmarks of EICR and the athletic heart—are entirely benign or also carry some associated risk, especially in the subgroup of extreme athletes. Additional research is necessary, given the multiple beneficial effects of exercise on the human body for improving cardiovascular health, overall lifespan in general, and quality of life and well-being.

Compliance with Ethical Standards

Funding: Dr. Cornwell and Dr. Buttrick have received research funding from the NIH/NHLBI.

Conflicts of Interest: n/a.

Ethical Approval: Any procedures performed in studies involving human participants were in accordance with the ethical standards of the institutional and/or national research committee and with the 1964 Helsinki declaration and its later amendments or comparable ethical standards.

Informed Consent: Informed consent was obtained from all individual participants included in the study.

References

1. Henschen S, Skidwettlauf S (1899) Eine medizinesche Sportstudie. Mitt Med klin Upsala 2
2. Darling EA (1899) The effects of training. A study of the Harvard University crews. Boston Med Surg J CKLLI:229–233
3. Dickhuth HH, Simon G, Kindermann W, Wildberg A, Keul J (1979) Echocardiographic studies on athletes of various sport-types and non-athletic persons. Z Kardiol 68:449–453
4. Dickhuth HH, Horstmann T, Staiger J, Reindell H, Keul J (1989) The long-term involution of physiological cardiomegaly and cardiac hypertrophy. Med Sci Sports Exerc 21:244–249
5. Weiner RB, Baggish AL (2012) Exercise-induced cardiac remodeling. Prog Cardiovasc Dis 54: 380–386
6. Rowell LB (1986) Human circulation: regulation during physical stress, 1st edn. Oxford University Press, New York
7. Proctor DN, Beck KC, Shen PH, Eikhoff TJ, Halliwill JR, Joyner MJ (1998) Influence of age and gender on cardiac output- Vo2 relationships during submaximal cycle ergometry. J Appl Physiol 84:599–605
8. Gledhill N, Cox D, Jamnik R (1994) Endurance athletes' stroke volume does not plateau: major advantage is diastolic function. Med Sci Sports Exerc 26:1116–1121
9. Raven PB, Young BE, Fadel PJ (2019) Arterial baroreflex resetting during exercise in humans: underlying signaling mechanisms. Exerc Sport Sci Rev 47:129–141
10. Fadel PJ, Raven PB (2012) Human investigations into the arterial and cardiopulmonary baroreflexes during exercise. Exp Physiol 97:39–50
11. Poliner LR, Dehmer GJ, Lewis SE, Parkey RW, Blomqvist G, Willerson JT (1980) Left ventricular performance in normal subjects: a comparison of the responses to exercise in the upright and supine positions. Circulation 62:528–534
12. Morganroth J, Maron BJ, Henry WL, Epstein SE (1975) Comparative left ventricular dimensions in trained athletes. Ann Intern Med 82:802–806
13. Pluim BM, Zwinderman AH, van der Laarse A, Van der Wall EE (1999) The athlete's heart. A meta-analysis of cardiac structure and function. Circulation 100:336–344
14. Mitchell JH, Haskell WL, Raven PB (1994) Classification of sports. J Am Coll Cardiol 24:864–866
15. Levine BD, Baggish AL, Kovacs RJ, Link MS, Maron MS, Mitchell JH, American Heart Association E, Arrhythmias Committee of Council on Clinical Cardiology CoCDiYCoC, Stroke Nursing CoFG, Translational B and American College of C (2015) Eligibility and disqualification recommendations for competitive athletes with cardiovascular abnormalities: Task Force 1: Classification of sports: dynamic, static, and impact: a scientific statement from the American Heart Association and American College of Cardiology. Circulation 132:e262–e266
16. MacDougall JD, Tuxen D, Sale DG, Moroz JR, Sutton JR (1985) Arterial blood pressure response to heavy resistance exercise. J Appl Physiol 58:785–790
17. Beaudry R, Haykowsky MJ, Baggish A, La Gerche A (2016) A modern definition of the athlete's heart-for research and the clinic. Cardiol Clin 34:507–514
18. Prior DL, La Gerche A (2012) The athlete's heart. Heart 98:947–955
19. Fagard RH (1996) Athlete's heart: a meta-analysis of the echocardiographic experience. Int J Sports Med 17(Suppl 3):S140–S144

20. Roy A, Doyon M, Dumesnil JG, Jobin J, Landry F (1988) Endurance vs. strength training – comparison of cardiac structures using normal predicted values. J Appl Physiol 64:2552–2557
21. Spence AL, Naylor LH, Carter HH, Buck CL, Dembo L, Murray CP, Watson P, Oxborough D, George KP, Green DJ (2011) A prospective randomised longitudinal MRI study of left ventricular adaptation to endurance and resistance exercise training in humans. J Physiol 589: 5443–5452
22. Scharhag J, Schneider G, Urhausen A, Rochette V, Kramann B, Kindermann W (2002) Athlete's heart: right and left ventricular mass and function in male endurance athletes and untrained individuals determined by magnetic resonance imaging. J Am Coll Cardiol 40:1856–1863
23. D'Andrea A, Riegler L, Golia E, Cocchia R, Scarafile R, Salerno G, Pezzullo E, Nunziata L, Citro R, Cuomo S, Caso P, Di Salvo G, Cittadini A, Russo MG, Calabro R, Bossone E (2013) Range of right heart measurements in top-level athletes: the training impact. Int J Cardiol 164: 48–57
24. Kim JH, Noseworthy PA, McCarty D, Yared K, Weiner R, Wang F, Wood MJ, Hutter AM, Picard MH, Baggish AL (2011) Significance of electrocardiographic right bundle branch block in trained athletes. Am J Cardiol 107:1083–1089
25. Scharf M, Brem MH, Wilhelm M, Schoepf UJ, Uder M, Lell MM (2010) Cardiac magnetic resonance assessment of left and right ventricular morphologic and functional adaptations in professional soccer players. Am Heart J 159:911–918
26. Henriksen E, Landelius J, Wesslen L, Arnell H, Nystrom-Rosander C, Kangro T, Jonason T, Rolf C, Lidell C, Hammarstrom E, Ringqvist I, Friman G (1996) Echocardiographic right and left ventricular measurements in male elite endurance athletes. Eur Heart J 17:1121–1128
27. Telemedicine in Israel: Israel's health and care system organisation. Momentum. Accessed 5 Mar wt-mei
28. Baggish AL, Wang F, Weiner RB, Elinoff JM, Tournoux F, Boland A, Picard MH, Hutter AM Jr, Wood MJ (1985) Training-specific changes in cardiac structure and function: a prospective and longitudinal assessment of competitive athletes. J Appl Physiol 2008(104):1121–1128
29. Wasfy MM, Weiner RB (2015) Differentiating the athlete's heart from hypertrophic cardiomyopathy. Curr Opin Cardiol 30:500–505
30. La Gerche A, Burns AT, Mooney DJ, Inder WJ, Taylor AJ, Bogaert J, Macisaac AI, Heidbuchel H, Prior DL (2012) Exercise-induced right ventricular dysfunction and structural remodelling in endurance athletes. Eur Heart J 33:998–1006
31. Blyth KG, Groenning BA, Martin TN, Foster JE, Mark PB, Dargie HJ, Peacock AJ (2005) Contrast enhanced-cardiovascular magnetic resonance imaging in patients with pulmonary hypertension. Eur Heart J 26:1993–1999
32. La Gerche A, Heidbuchel H, Burns AT, Mooney DJ, Taylor AJ, Pfluger HB, Inder WJ, Macisaac AI, Prior DL (2011) Disproportionate exercise load and remodeling of the athlete's right ventricle. Med Sci Sports Exerc 43:974–981
33. Scharf M, Brem MH, Wilhelm M, Schoepf UJ, Uder M, Lell MM (2010) Atrial and ventricular functional and structural adaptations of the heart in elite triathletes assessed with cardiac MR imaging. Radiology 257:71–79
34. Pelliccia A, Maron BJ, Di Paolo FM, Biffi A, Quattrini FM, Pisicchio C, Roselli A, Caselli S, Culasso F (2005) Prevalence and clinical significance of left atrial remodeling in competitive athletes. J Am Coll Cardiol 46:690–696
35. D'Andrea A, Riegler L, Cocchia R, Scarafile R, Salerno G, Gravino R, Golia E, Vriz O, Citro R, Limongelli G, Calabro P, Di Salvo G, Caso P, Russo MG, Bossone E, Calabro R (2010) Left atrial volume index in highly trained athletes. Am Heart J 159:1155–1161
36. Jin MN, Yang PS, Song C, Yu HT, Kim TH, Uhm JS, Sung JH, Pak HN, Lee MH, Joung B (2019) Physical activity and risk of atrial fibrillation: a nationwide cohort study in general population. Sci Rep 9:13270

37. Malmo V, Nes BM, Amundsen BH, Tjonna AE, Stoylen A, Rossvoll O, Wisloff U, Loennechen JP (2016) Aerobic interval training reduces the burden of atrial fibrillation in the short term: a randomized trial. Circulation 133:466–473
38. Mozaffarian D, Furberg CD, Psaty BM, Siscovick D (2008) Physical activity and incidence of atrial fibrillation in older adults: the cardiovascular health study. Circulation 118:800–807
39. Aizer A, Giaziano JM, Cook NR, Manson JE, Buring JE, Albert CM (2009) Relation of vigorous exercise risk to risk of atrial fibrillation. Am J Cardiol 103:1572–1577
40. Thelle DS, Selmer R, Gjesdal K, Sakshaug S, Jugessur A, Graff-Iversen G, Tverdal A, Nystad W (2013) Resting heart rate and physical activity as risk factors for lone atrial fibrillation: a prospective study of 309,540 men and women. Heart 99:1713
41. Drca N, Wolk A, Jensen-Urstad M, Larsson SC (2014) Atrial fibrillation is associated with different levels of physical activity levels at different ages in men. Heart 100:1043–1049
42. Andersen K, Farahmand B, Ahlbom A, Held C, Ljunghall S, Michaelsson K, Sundstrom J (2013) Risk of arrhythmias in 52 755 long-distance cross-country skiers: a cohort study. Eur Heart J 34:3624–3631
43. Guasch E, Mont L, Sitges M (2018) Mechanisms of atrial fibrillation in athletes: what we know and what we do not know. Neth Hear J 26:133–145
44. Wilhelm M, Roten L, Tanner H, Wilhelm I, Schmid JP, Saner H (2011) Atrial remodeling, autonomic tone, and lifetime training hours in nonelite athletes. Am J Cardiol 108:580–585
45. Wilhelm M, Roten L, Tanner H, Wilhelm I, Schmied JP, Saner H (2011) Atrial remodeling, autonomic tone, and lifetime training hours in nonelite athletes. Am J Cardiol 108:580–585
46. Vaziri SM, Larson MG, Benjamin EJ, Levy D (1994) Echocardiographic predictors of nonrheumatic atrial fibrillation. The Framingham Heart Study. Circulation 89:724–730
47. Kasapis C, Thompson PD (2005) The effects of physical activity on serum C-reactive protein and inflammatory markers: a systematic review. J Am Coll Cardiol 45:1563–1569
48. La Gerche A, Inder WJ, Roberts TJ, Brosnan MJ, Heidbuchel H, Prior DL (2015) Relationship between inflammatory cytokines and indices of cardiac dysfunction following intense endurance exercise. PLoS One 10:e0130031
49. Hattasch R, Spethmann S, de Boer RA, Ruifrok WP, Schattke S, Wagner M, Schroeckh S, Durmus T, Schimke I, Sanad W, Baumann G, Borges AC, Knebel F (2014) Galectin-3 increase in endurance athletes. Eur J Prev Cardiol 21:1192–1199
50. Roca E, Nescolarde L, Lupon J, Barallat J, Januzzi JL, Liu P, Cruz Pastor M, Bayes-Genis A (2017) The dynamics of cardiovascular biomarkers in non-elite marathon runners. J Cardiovasc Transl Res 10:206–208
51. Baggish AL, Hale A, Weiner RB, Lewis GD, Systrom D, Wang F, Wang TJ, Chan SY (2011) Dynamic regulation of circulating microRNA during acute exhaustive exercise and sustained aerobic exercise training. J Physiol 589:3983–3994
52. Lindsay MM, Dunn FG (2007) Biochemical evidence of myocardial fibrosis in veteran endurance athletes. Br J Sports Med 41:447–452
53. Svedberg N, Sundstrom J, James S, Hallmarker U, Hambraeus K, Andersen K (2019) Long-term incidence of atrial fibrillation and stroke among cross-country skiers. Circulation 140:910–920
54. Claessen G, La Gerche A, Voigt JU, Dymarkowski S, Schnell F, Petit T, Willems R, Claus P, Delcroix M, Heidbuchel H (2016) Accuracy of echocardiography to evaluate pulmonary vascular and RV function during exercise. J Am Coll Cardiol Img 9:532–543
55. La Gerche A, Claessen G, Dymarkowski S, Voigt JU, De Buck F, Vanhees L, Droogne W, Van Cleemput J, Claus P, Heidbuchel H (2015) Exercise-induced right ventricular dysfunction is associated with ventricular arrhythmias in endurance athletes. Eur Heart J 36:1998–2010
56. Arbab-Zadeh A, Perhonen M, Howden E, Peshock RM, Zhang R, Adams-Huet B, Haykowsky MJ, Levine BD (2014) Cardiac remodeling in response to 1 year of intensive endurance training. Circulation 130:2152–2161
57. Howden EJ, Sarma S, Lawley JS, Opondo M, Cornwell W, Stoller D, Urey MA, Adams-Huet B, Levine BD (2018) Reversing the cardiac effects of sedentary aging in middle age – a

randomized controlled trial: implications for heart failure prevention. Circulation 137:1549–1560
58. Borlaug BA, Nishimura RA, Sorajja P, Lam CS, Redfield MM (2010) Exercise hemodynamics enhance diagnosis of early heart failure with preserved ejection fraction. Circ Heart Fail 3:588–595
59. Borlaug BA, Schaff HV, Pochettino A, Pedrotty DM, Asirvatham SJ, Abel MD, Carter RE, Mauermann WJ (2018) Pericardiotomy enhances left ventricular diastolic reserve with volume loading in humans. Circulation 138:2295–2297
60. Dell'Italia LJ (2012) Anatomy and physiology of the right ventricle. Cardiol Clin 30:167–187
61. Voelkel NF, Quaife RA, Leinwand LA, Barst RJ, MD MG, Meldrum DR, Dupuis J, Long CS, Rubin LJ, Smart FW, Suzuki YJ, Gladwin M, Denholm EM, Gail DB, National Heart L, Blood Institute Working Group on C and Molecular Mechanisms of Right Heart F (2006) Right ventricular function and failure: report of a National Heart, Lung, and Blood Institute working group on cellular and molecular mechanisms of right heart failure. Circulation 114:1883–1891
62. Cornwell WK, Tran T, Cerbin L, Coe G, Muralidhar A, Hunter K, Altman N, Ambardekar AV, Tompkins C, Zipse M, Schulte M, O'Gean K, Ostertag M, Hoffman J, Pal JD, Lawley JS, Levine BD, Wolfel E, Kohrt WM, Buttrick P (2020) New insights into resting and exertional right ventricular performance in the healthy heart through real-time pressure-volume analysis. J Physiol 598:2575–2587
63. Oxborough D, Shave R, Warburton D, Williams K, Oxborough A, Charlesworth S, Foulds H, Hoffman MD, Birch K, George K (2011) Dilatation and dysfunction of the right ventricle immediately after ultraendurance exercise: exploratory insights from conventional two-dimensional and speckle tracking echocardiography. Circ Cardiovasc Imaging 4:253–263
64. La Gerche A, Connelly KA, Mooney DJ, MacIsaac AI, Prior DL (2008) Biochemical and functional abnormalities of left and right ventricular function after ultra-endurance exercise. Heart 94:860–866
65. Elliott AD, La Gerche A (2015) The right ventricle following prolonged endurance exercise: are we overlooking the more important side of the heart? A meta-analysis. Br J Sports Med 49:724–729
66. Drezner JA, Sharma S, Baggish A, Papadakis M, Wilson MG, Prutkin JM, Gerche A, Ackerman MJ, Borjesson M, Salerno JC, Asif IM, Owens DS, Chung EH, Emery MS, Froelicher VF, Heidbuchel H, Adamuz C, Asplund CA, Cohen G, Harmon KG, Marek JC, Molossi S, Niebauer J, Pelto HF, Perez MV, Riding NR, Saarel T, Schmied CM, Shipon DM, Stein R, Vetter VL, Pelliccia A, Corrado D (2017) International criteria for electrocardiographic interpretation in athletes: consensus statement. Br J Sports Med 51:704–731
67. Coote JH, White MJ (2015) CrossTalk proposal: bradycardia in the trained athlete is attributable to high vagal tone. J Physiol 593:1745–1747
68. Al-Ani M, Munir SM, White M, Townend J, Coote JH (1996) Changes in R-R variability before and after endurance training measured by power spectral analysis and by the effect of isometric muscle contraction. Eur J Appl Physiol 74
69. Wasfy MM, Weiner RB, Wang F, Berkstresser B, Lewis GD, DeLuca JR, Hutter AM, Picard MH, Baggish AL (2015) Endurance exercise-induced cardiac remodeling: not all sports are created equal. J Am Soc Echocardiogr 28:1434–1440
70. Zaidi A, Ghani S, Sheikh N, Gati S, Bastiaenen R, Madden B, Papadakis M, Raju H, Reed M, Sharma R, Behr ER, Sharma S (2013) Clinical significance of electrocardiographic right ventricular hypertrophy in athletes: comparison with arrhythmogenic right ventricular cardiomyopathy and pulmonary hypertension. Eur Heart J 34:3649–3656
71. Maron BJ, Zipes DP, Kovacs RJ (2015) Eligibility and disqualification recommendations for competitive athletes with cardiovascular abnormalities: preamble, principles, and general considerations: a scientific statement from the American Heart Association and American College of Cardiology. Circulation 132:e256–e261
72. Maron BJ, Pelliccia A (2006) The heart of trained athletes: cardiac remodeling and the risks of sports, including sudden death. Circulation 114:1633–1644

73. Campuzano O, Fernandez-Falgueras A, Sarquella-Brugada G, Cesar S, Arbelo E, Garcia-Alvarez A, Jorda P, Coll M, Fiol V, Iglesias A, Perez-Serra A, Mates J, Del Olmo B, Ferrer C, Alcalde M, Puigmule M, Mademont-Soler I, Pico F, Lopez L, Tiron C, Brugada J, Brugada R (2019) Personalized interpretation and clinical translation of genetic variants associated with cardiomyopathies. Front Genet 10:450
74. Primorac D, Odak L, Peric V, Catic J, Sikic J, Radeljic V, Manola S, Nussbaum R, Vatta M, Aradhya S, Sofrenovic T, Matisic V, Molnar V, Skelin A, Mirat J, Brachmann J (2021) Sudden cardiac death-a new insight into potentially fatal genetic markers. Front Med (Lausanne) 8: 647412
75. Mellor G, Laksman ZWM, Tadros R, Roberts JD, Gerull B, Simpson CS, Klein GJ, Champagne J, Talajic M, Gardner M, Steinberg C, Arbour L, Birnie DH, Angaran P, Leather R, Sanatani S, Chauhan VS, Seifer C, Healey JS, Krahn AD (2017) Genetic testing in the evaluation of unexplained cardiac arrest: from the CASPER (Cardiac Arrest Survivors with Preserved Ejection Fraction Registry). Circ Cardiovasc Genet 10
76. Maron BJ (2003) Sudden death in young athletes. N Engl J Med 349:1064–1075
77. Maron BJ, Doerer JJ, Haas TS, Tierney DM, Mueller FO (2009) Sudden deaths in young competitive athletes: analysis of 1866 deaths in the United States, 1980-2006. Circulation 119: 1085–1092
78. Pelliccia A, Culasso F, Di Paolo FM, Maron BJ (1999) Physiologic left ventricular cavity dilatation in elite athletes. Ann Intern Med 120:23–31
79. Marcus FI, McKenna WJ, Sherrill D, Basso C, Bauce B, Bluemke DA, Calkins H, Corrado D, Cox MG, Daubert JP, Fontaine G, Gear K, Hauer R, Nava A, Picard MH, Protonotarios N, Saffitz JE, Sanborn DM, Steinberg JS, Tandri H, Thiene G, Towbin JA, Tsatsopoulou A, Wichter T, Zareba W (2010) Diagnosis of arrhythmogenic right ventricular cardiomyopathy/ dysplasia: proposed modification of the task force criteria. Circulation 121:1533–1541
80. Mitchell JH, Haskell W, Snell P, Van Camp SP (2005) Task Force 8: Classification of sports. J Am Coll Cardiol 45:1364–1367

[illegible]

Cardiac Microvascular Endothelial Cells and Pressure Overload-Induced Cardiac Fibrosis

Jaafar Al-Hasani and Markus Hecker

Abbreviations

ACE	Angiotensin-converting enzyme
Ang II	Angiotensin II
CTGF	Connective tissue growth factor
DCM	Dilated cardiomyopathy
ECM	Extracellular matrix
EMT	Epithelial-to-mesenchymal transition
EndMT	Endothelial-to-mesenchymal transition
HFpEF	HF with preserved ejection fraction
HFrEF	HF with reduced ejection fraction
MAP kinase	Mitogen-activated protein kinase
MMP	Matrix metalloproteinase
RAAS	Renin-angiotensin-aldosterone system
TGF	Transforming growth factor
TIMP	Tissue inhibitor of metalloproteinases

J. Al-Hasani · M. Hecker (✉)
Department of Cardiovascular Physiology, Heidelberg University, Heidelberg, Germany

German Centre for Cardiovascular Research (DZHK), Partner Site Heidelberg-Mannheim, Berlin, Germany
e-mail: hecker@physiologie.uni-heidelberg.de

M. Hecker, D. J. Duncker (eds.), *Cardiac Mechanobiology in Physiology and Disease*, Cardiac and Vascular Biology 9,
https://doi.org/10.1007/978-3-031-23965-6_10

1 Introduction

To carry out its vital function, the heart itself needs to be properly perfused with blood [1]. For this purpose, it is equipped with larger conduit arteries, i.e., the right and left coronary arteries with their tributaries lined with a monolayer of macrovascular endothelial cells, and a large number of small arteries and arterioles that distribute the blood within the myocardium. Chief function of the subsequent dense network of capillaries is the supply of the cardiomyocytes with nutrients, and oxygen in particular, because their energy metabolism is strictly aerobic. These capillaries are so small that their lumen is encircled by 1–2 microvascular endothelial cells only. Capillary density in the normal myocardium is quite high, with about 3000–4000 capillaries/mm^2 tissue [2]. Recent estimates indicate that microvascular endothelial cells represent the most abundant non-myocyte cell type in the heart, amounting to 60–65% of the non-myocyte cells or about 50% of all cells in the heart [3]. This number seems quite realistic considering that virtually all cardiomyocytes are provided with their individual capillary to ensure their adequate supply of oxygen and nutrients. Another purpose of this dense capillary network is the exchange of small molecules, such as peptides, proteins, micro-vesicles, and microRNAs [4] between endothelial cells and cardiomyocytes.

Heart failure, a pathology among the leading causes of morbidity and death worldwide [5], is characterized by the inability of the heart to pump blood to the body in amounts needed [6]. Symptoms include dyspnoea, peripheral oedema, arrhythmia, fatigue, and stress-induced syncope. The portion of blood present in the ventricle at the end of diastole that is ejected with every heartbeat is termed ejection fraction. Its reduction by about half indicates a clinically significant impairment of systolic force and is commonly referred to as manifest heart failure with reduced ejection fraction (HFrEF). The condition where diastolic filling and relaxation is compromised while ejection fraction and other functional systolic parameters remain largely unchanged is characterized as heart failure with preserved ejection fraction (HFpEF) [7]. HFrEF, which was the dominant variant of heart failure in the past, has now been surpassed by HFpEF as the more prevalent pathological condition [8–11]. This may in part be attributed to better and earlier diagnosis, improved treatment options as well as a shift in ejection fraction thresholds, as can be inferred from the introduction of an intermediate type of heart failure with a mild or moderate reduction of ejection fraction [12]. Maintaining a sufficient stroke volume, especially during exercise [1], where diastole, due to the increase in heart rate, is shortened, requires an increase in filling pressure, i.e., a larger end-diastolic volume in the ventricle. This is achieved through an increase in renal salt and water retention by disinhibiting the sympathetic nervous system as well as by upregulating activity of the renin-angiotensin-aldosterone system (RAAS). This, however, also results in arterial hypertension and, consequently, pressure overload of the left ventricle.

Pressure overload due to arterial hypertension or aortic valve stenosis followed by left ventricular hypertrophy and eventually failure is, after ischemic heart disease, the second-most common cause of heart failure worldwide and especially prevalent among the elderly [13]. Although the exact sequence of events has not been elucidated yet, it is clear that pressure overload causes an initial hypertrophic remodelling of the left ventricle [14]. This has long been viewed as a compensatory mechanism to increase wall thickness and thereby, according to the law of Laplace, reduce wall stress of the left ventricle [15]. Since the cardiomyocytes grow in length, their speed of contraction, referred to as fractional shortening, is significantly impaired, which is viewed as the first step in the transition from left ventricular hypertrophy to heart failure [15]. This initial remodelling process is accompanied by a stiffening of the extracellular matrix (ECM) [16] eventually leading to cardiomyocyte stiffening [17–19], which aggravates diastolic dysfunction in HFpEF [17, 20, 21]. ECM stiffening also causes cardiomyocyte loss due to apoptosis [22] or anoikis [22–24], mainly occurs in HFrEF [25], reinforces replacement fibrosis, and, upon rupture of this scar tissue, leads to dilated cardiomyopathy (DCM) and manifest (left ventricular) heart failure [26]. If left untreated, it will inevitably lead to global heart failure [5]. Risk factors for this condition are age, (arterial) hypertension, (type 2) diabetes, and renal dysfunction, which may be the consequence of hypertension and/or diabetes [8, 27, 28].

2 Pressure Overload

Pressure overload describes a situation in which the left ventricle has an increased afterload to compensate for ejecting the stroke volume, as in arterial hypertension, or as a consequence of aortic valve stenosis, resistance in the outflow tract (proximal aorta) is substantially increased. However, pressure overload initially, and predominantly, leads to left ventricular diastolic rather than systolic dysfunction [29]. When faced with chronic pressure overload, concentric left ventricular hypertrophy ensues resulting in a decrease of left ventricular volume, which is referred to as inward remodelling. In contrast, volume overload, which can also be caused by obesity [30–32] or chronic kidney disease [33], causes eccentric left ventricular hypertrophy that rather corresponds to an outward remodelling process [34]. Persistent pressure overload can thus cause decompensation of the concentrically remodelled left ventricle with HFpEF as the consequence, whereas decompensation of an eccentrically remodelled left ventricle after chronic volume overload rather ends up in HFrEF [29].

Pressure overload not only causes hypertrophy of the left ventricular wall but also provokes chamber dilation [35]. Accordingly, cardiomyocytes throughout the whole heart may switch their phenotype [36], reinforcing cardiac hypertrophy [37, 38]. Moreover, hypertrophied cardiomyocytes are prone to undergo apoptosis, which in turn may have diverse pathological consequences [39–44] among which cardiac fibrosis features rather prominently [39, 45]. Ultimately, persistent pressure overload results in dilated cardiomyopathy and HFrEF [29].

Interestingly, although heart failure develops because of sustained high blood pressure, in some patients one result of this condition can be a drop in blood pressure back to or even below normal. This phenomenon, known as "decapitated hypertension", is a consequence of the impaired pumping capabilities of the heart and hence a reduction in cardiac output. There are several studies in this context suggesting that in patients with heart failure, a higher systolic blood pressure is associated with better survival [46–52].

3 Cardiac Fibrosis

One step in the course from pressure overload to a failing heart is the development of cardiac fibrosis (see Fig. 1). The ECM network surrounding the cells of the heart [53] normally serves as a scaffold for structural support and facilitates force transmission throughout the heart muscle. It is also in close contact with myocardial cells in form of the basement membrane [54], another type of pericellular matrix. In addition, the cardiac ECM serves as a hub for signals and signalling pathways to the cardiomyocytes [55].

As mentioned above, sustained pressure overload leads to concentric remodelling of the left ventricle, i.e., an increase in wall thickness, while ventricular volume is either maintained or decreased. Consequently, the ratio between left ventricular mass and volume increases, too. This development is paralleled by the activation of resident (cardiac) fibroblasts to myofibroblasts and the initiation of matrix synthesis programmes [56, 57], resulting in the production of excessive amounts of ECM. This surplus in ECM increases myocardial stiffness and reduces ventricular compliance. While initially preserving cardiac integrity and function to some extent [58], eventually a dramatic reduction of cardiac function ensues [59].

In contrast, after myocardial infarction/volume overload-induced remodelling, which is accompanied by cardiomyocyte necrosis/apoptosis, disruption and/or degradation of the ECM can be observed, which in part is triggered by the necrotic/apoptotic cardiomyocytes themselves [57, 60]. Since the heart has virtually no repair mechanism and a very limited regenerative capacity, the only way to preserve a certain level of structural and functional integrity is the replacement of necrotic or apoptotic cardiomyocytes by scar tissue, which further exacerbates cardiac fibrosis [57].

Accordingly, two types of fibrosis can be distinguished: *replacement fibrosis*, which normally occurs after myocardial infarction [61] and comprises the replacement of apoptotic cardiomyocytes by connective tissue, and *reactive fibrosis*, which is often further subdivided into an interstitial and a perivascular type [61]. While interstitial fibrosis is characterized by deposition of cross-linked collagen fibres in the interstitium [62, 63], which widens the distance between individual cardiomyocytes, thereby impacting their electrical coupling [64], perivascular fibrosis increases the distance between cardiomyocytes and capillary endothelial cells and thus limits the supply of the cardiomyocytes with oxygen and nutrients [65, 66]. In

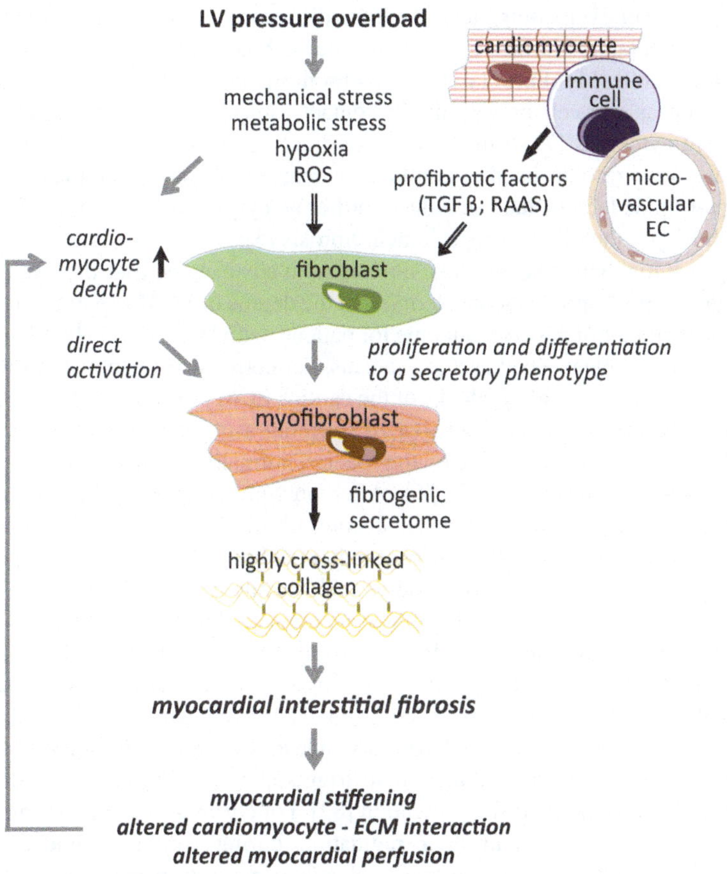

Fig. 1 Direct and indirect intercellular pathways in the heart leading to fibroblast to myofibroblast differentiation and/or myofibroblast activation. For details refer to the text. Abbreviations used: ECM, extracellular matrix; RAAS, renin-angiotensin-aldosterone system; ROS, reactive oxygen species; TGFβ, transforming growth factor β. The grey arrows indicate important functional consequences. The small black arrows point to the release of biologically active factors, and the small double-lined arrows point to the cell type that is affected by these factors

either case, in response to a pathological stimulus, first a release of pro-inflammatory cytokines and pro-fibrotic factors is observed [5], then myofibroblasts secrete collagens and other ECM proteins, such as fibronectin, laminin, and proteoglycans [67].

Although fibrosis is a maladaptive response, cardiac fibrosis after pressure overload is not maladaptive per se, as it initially helps to maintain structural integrity and contractile capacity as well as electrical conductivity of the left ventricular myocardium [64, 68]. But upon its progression, cardiac function dramatically deteriorates due to the increased stiffening of the myocardium, resulting in contractile dysfunction accompanied by a gradual loss of mechano-electrical coupling, and ultimately arrhythmia and death [60]. In the perivascular space, inflammation and fibrosis reduce or even completely block the diffusion of oxygen and nutrients from the microvasculature, further aggravating the pathological situation in the failing myocardium [60]. Myofibroblasts in the fibrotic tissue or scar can be subjected to stimuli such as cell-cell or cell-matrix contacts, tensile stress [69, 70] or growth factors and cytokines [71], which act as survival signals prolonging their lifespan. This, in turn, results in the continuous release of pro-inflammatory and/or pro-hypertrophic factors that can induce cardiomyocyte apoptosis or necrosis, events that both trigger the onset of replacement fibrosis [5].

Despite initial beneficial aspects, fibrosis is an adverse and maladaptive process; this becomes more apparent considering that the degree of ECM expansion itself can serve as a predictor of a worse outcome for patients with HFpEF and HFrEF [72, 73], although it is not clear whether this is a causal or correlative relationship [60].

The predominant building blocks of the cardiac ECM are collagen I, which forms thick, rod-shaped fibres, and collagen III, forming a fibrillary network around the myocardial cells [74]. Depending on the species, normally a collagen I to collagen III ratio of about 8:1 is found [75, 76]. After pressure overload, synthesis of both collagens I and III is upregulated. Other, non-fibrillar collagens, such as collagen IV, which normally is located in the basement membrane, have also been reported to be upregulated after pressure overload [77, 78], while collagen VIII, which is only found at low levels in the adult heart, can be reduced during later stages of disease progression [79]. Interestingly, HFpEF and HFrEF are characterized by different changes in ECM composition, and cardiomyocytes can sense these differences [80]. In HFpEF, large amounts of collagen I are synthesized while at the same time the flexible collagen III is reduced, which leads to cardiac stiffening [81, 82]. In HFrEF, on the other hand, cardiac fibrosis triggered by cardiomyocyte death causes impairment of contractile force generation [83]. For a near all-encompassing review about cardiac collagens and their regulation during cardiac remodelling, see Frangogiannis [60]. These changes in collagen composition are accompanied by alterations of other ECM components such as glycosaminoglycans, fibronectin, glycoproteins, and proteoglycans. In addition, liberation of growth factors and proteases from the ECM could be affected as well [84], and a disturbed protease-protease inhibitor balance leads to degradation of the ECM [85]. Table 1 recapitulates what is currently known about changes in ECM composition in replacement versus reactive cardiac fibrosis.

The main modulators of the ECM are matrix metalloproteinases (MMPs) that degrade ECM proteins, and tissue inhibitors of metalloproteinases (TIMPs), which control activity of the MMPs [100]. In various settings of heart failure, depending on the underlying cause, differential expression of MMPs and TIMPs have been described [101–104]. However, expression levels of these proteases and their inhibitors do not always correlate with activity, and in vitro data do not necessarily reflect the situation in vivo. Nevertheless, the different types of cardiac fibrosis also seem to require a specific profile of MMPs and TIMPs. In interstitial or perivascular fibrosis, mainly occurring in HFpEF or in the later phases of HFrEF, an upregulation of TIMP-1 and TIMP-2 has been reported [105] that seemed to correlate with the degree of interstitial fibrosis. However, in another study [106], an increased

Table 1 Comparison of changes in ECM components as they occur during replacement or reactive fibrosis. Current knowledge on relevant cellular and molecular changes in the different types of fibrosis

	Replacement fibrosis	Reactive fibrosis
Occurrence	Mainly HFrEF, especially following MI	HFpEF and HFrEF
Main trigger	Cardiomyocyte apoptosis or anoikis[a]	TGF-β, pro-inflammatory cytokines[a]
Consequences for		
– Cellular composition	Loss of cardiomyocytes, infiltration by fibroblasts and immune cells[a]	Infiltration of immune cells into the perivascular space[a]
– ECM proteins	Mass production of collagens I and III to replace dead cardiomyocytes[b] Upregulation of collagen III following MI[c]	Increased synthesis, mainly of collagen I[d, e]
– ECM crosslinking	Enhanced crosslinking [f,g]	Enhanced crosslinking, upregulation of lysyl oxidase[d,h,i]
– MMPs/ TIMPs	Upregulation of MMP-2, MMP-9, and TIMP-1[j,k] Upregulation of TIMP[l] following MI (TIMP-1 is pro-fibrotic, TIMP-2, 3, and 4 are anti-fibrotic[m])	
– Matricellular proteins	TSP1—upregulated after MI (in border zone), preventing adverse remodelling after MI and PO[n] Osteopontin—upregulated after both MI and PO, generally pro-fibrotic and pro-hypertrophic, slightly protective following MI[n] Periostin—upregulated after both MI and PO, activates cardiac fibroblasts, protective in acute phase following MI[n]	
– Other mediators	Downregulation of MCP1[a]	Upregulation of MCP1[o] Loss of MCP1 protects from fibrosis[p]
	Integrins α1, 2, and 3 as well as β1 and 3 are upregulated after MI[q]	Integrins α1, 2, and 5 are down-regulated in hypertension but upregulated after PO, β1 is upregulated by both[q]

MI, myocardial infarction; PO, pressure overload
References: [a] [18], [b] [86], [c] [87], [d] [82], [e] [81], [f] [88], [g] [89], [h] [90], [i] [91], [j] [92], [k] [93], [l] [94], [m] [95], [n] [96], [o] [97], [p] [98], [q] [99]

expression of MMPs 1, 2, 3, 9, 13, and 14 has been observed in these settings. This finding must not be counterintuitive per se, as there is not simply an excessive build-up of ECM molecules but also a profound remodelling of the ECM involved. In replacement fibrosis, which is found rather early in HFrEF, e.g., after myocardial ischemia, upregulation of MMP-7 has been associated with a protective effect while that of TIMP-1 rather seemed to be detrimental [107]. Other MMPs and TIMPs have also been found to be differentially expressed [101, 108], but were not associated with the type and degree of fibrosis in the myocardium.

However, counteracting fibrosis by degrading the ECM may not be beneficial per se as ECM fragments are also able to trigger pro-inflammatory pathways and recruit inflammatory cells [109], ultimately leading to cardiomyocyte dysfunction and death. As mentioned before, myofibroblasts surviving in the fibrotic scar for a long time can not only continuously produce pro-inflammatory cytokines, and thus induce cardiomyocyte death, but also constantly stimulate the production of ECM components [110], which would further aggravate cardiac fibrosis.

4 Signalling Pathways in Cardiac Fibrosis

Some of the multiple signalling pathways involved in the onset and course of cardiac fibrosis are well-established (see Fig. 2), while others are still subject to debate. Hitherto, transforming growth factor (TGF)-β1 is the best-known inducer of cardiac fibrosis [111, 112]. The TGF-β family consists of three isoforms encoded by different genes [113], of which isoform 1 is the predominant cytokine. During cardiac fibrosis, levels of TGF-β1 rise [114], activating cardiac fibroblasts to differentiate into myofibroblasts [115] mainly through the Smad2/3 pathway [57, 116, 117]. However, TGF-β1 not only stimulates the production of ECM molecules and other pro-fibrotic factors [118] but also controls ECM degradation by enhancing the expression of TIMPs [119], which inhibit the activity of certain ECM-degrading MMPs. Mouse models using either TGF-β1 [114] or pan-TGF-β neutralizing antibodies [120] or blocking TGF-β1 signalling through inhibition of TGF-β receptor II [121] or vice versa reinforcing [122] TGF-β1 signalling show, consistent with data from humans, less or more fibrosis, respectively. In addition to its effects on fibroblasts and fibrosis, TGF-β1 can also induce cardiomyocyte hypertrophy [123].

Another potent inducer of fibrosis is angiotensin II (Ang II), which has been verified in various animal models by administering the peptide through repeated injections or via implantation of osmotic minipumps [76, 124, 125]. Direct proof for an involvement of Ang II in fibrosis was obtained by pharmaceutical targeting of its main receptor through corresponding type 1 Ang II (AT_1) receptor antagonists such as losartan, which has been shown to potently inhibit cardiac fibrosis [126]. Ang II also plays a major role in the control of blood pressure as part of the RAAS. Moreover, it promotes the conversion of cardiac fibroblasts to myofibroblasts, and induces synthesis of ECM molecules [127] as well as cardiomyocyte hypertrophy [128]. Some of these effects of Ang II, however, were shown to be mediated through TGF-β1 signalling [129]. In fact, Ang II induces the expression of TGF-β1 not only in vitro in cultured cardiomyocytes, cardiac fibroblasts as well as myofibroblasts [130–132] but also in vivo, and this relationship was even found to be causal [133].

Endothelin-1 is known as a powerful vasoconstrictor. However, it can also induce proliferation of cardiac fibroblasts and promote their differentiation into myofibroblasts as well as trigger the synthesis of collagens I and III [134, 135]. Blocking one of its receptors, the ET_A receptor, inhibits TGF-β1-stimulated collagen synthesis by cardiac fibroblasts [136].

Connective tissue growth factor (CTGF) is another cytokine that appears to be involved in fibrosis. It is expressed ubiquitously in the foetal myocardium, whereas

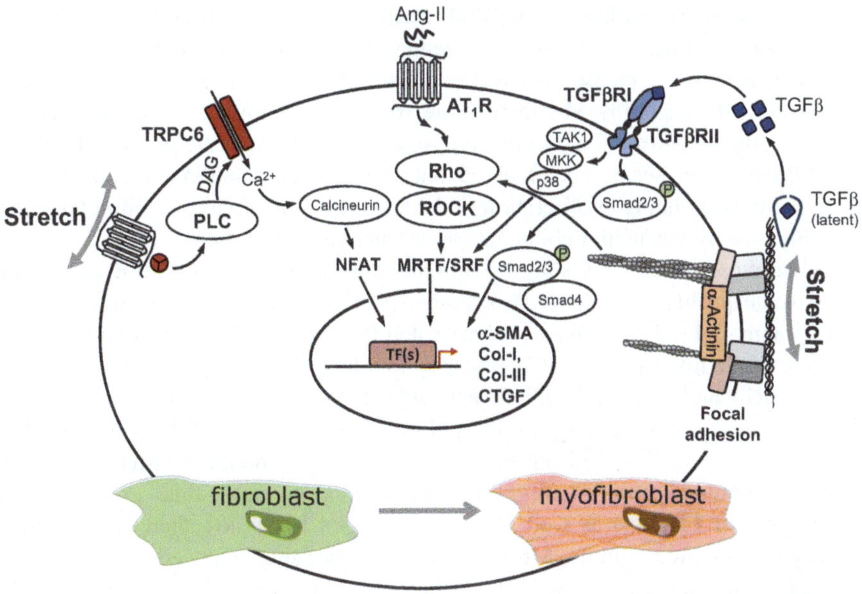

Fig. 2 Major signalling pathways involved in the differentiation of cardiac fibroblasts to myofibroblasts. Pressure overload or mechanical stress is referred to as stretch. TGFβ receptor I (TGFβRI) signals via TGFβ-activated kinase 1 (TAK1), mitogen-activated kinase (MAPK) kinase (MKK) and p38-MAPK to promote nuclear translocation of MRTF-A and serum response factor (SRF) while TGFβ receptor II (TGFβRII) signals via phosphorylation of Smad2/3 leading to recruitment of Smad4 and translocation of the heterodimer to the nucleus. Transient receptor potential 6C (TRPC6) is activated by stretch acting upon a $G_{q/11}$-coupled receptor (without any agonist), which in turn results in phospholipase Cβ-mediated diacylglycerol (DAG) formation. As a consequence of the secondary influx of extracellular calcium, nuclear factor of activated T-cells 1–4 (NFAT) is dephosphorylated by calcineurin, enabling its translocation to the nucleus. Angiotensin II (Ang-II) binding to the AT_1 receptor causes activation of the small G-protein RhoA, which in turn activates Rho kinase (ROCK) that again enables MRTF-A and SRF to translocate to the nucleus. The RhoA-ROCK pathway is also activated by stretch-dependent changes in the cortical actin dynamics

in adults it is restricted to the atria and larger blood vessels [137]. CTGF levels are increased after myocardial infarction and subsequent heart failure [138, 139] where it is expressed predominantly in the fibrotic area. Under these conditions, its expression seems to be induced by TGF-β1 and/or Ang II [140]. However, it is still unclear whether CTGF is required for fibrosis alongside other pro-fibrotic mediators, or whether it can even induce fibrosis on its own [141].

The aforementioned stimuli transmit their signals through various intracellular pathways or factors. In the TGF-β1/Smad pathway, after binding of TGF-β1 to its receptor, different Smad proteins get phosphorylated forming heterodimers, depending on the type of receptor, capable of translocating to the nucleus and act as pro-fibrotic transcription factors [142] stimulating the expression of several genes, mainly those coding for the TGF-β family members and their receptors [143]. The

AMP-activated protein kinase (AMPK)-α pathway is also activated by pro-fibrotic stimuli such as pressure overload, ischemia, TGF-β1 or Ang II. Upregulation of AMPK-α seems to decrease the ensuing pro-fibrotic response, although this finding is still under debate [143]. Members of the Wnt family, known to regulate processes such as body axis formation during embryogenesis, cell proliferation, migration, and fate determination, seem to be involved in the onset and regulation of cardiac fibrosis as well via stimulating proliferation and transformation of cardiac fibroblasts and possibly also by enhancing epithelial-to-mesenchymal transition [143].

Other transcription factors involved in the transmission of pro-fibrotic signals are peroxisome proliferator-activated receptor (PPAR)-γ, activator protein 1, nuclear factor kappa-light-chain-enhancer of activated B cells (NF-κB), activating transcription factor 3, myocardin-related transcription factor (MRTF) A/serum response factor (SRF) and nuclear factor of activated T-cells (NFAT); their effects range from decreasing to increasing fibrosis [143]. Of note, MRTF-A/SRF signalling has been shown to mediate cell adhesion and cell shape-stimulated differentiation of cardiac fibroblasts into myofibroblasts [144], and to upregulate expression of MMP2, 9 and 14 [145] as well as that of collagen Iα2 [146], thereby promoting cardiac fibrosis. Accordingly, MRTF-A deficient mice show reduced cardiac fibrosis after ischemia/Ang II treatment [146], clearly indicating a pro-fibrotic role of MRTF-A/SRF. In cardiac fibroblasts, NFAT can be activated by mechanical stress (stretch) and increased Ca^{2+} levels [143]. It then regulates the expression of collagen III and MRTF [147], and can also induce the differentiation of cardiac fibroblasts to myofibroblasts [148].

5 Cardiac Fibroblasts

Cardiac fibroblasts are responsible for the production of ECM components in the heart. Early estimates suggested that they are the most abundant cell type in the heart, but newer and more precise analyses show that they contribute less than 20% to its cellular composition [3, 149]. Irrespective of their exact number, cardiac fibroblasts play an important physiological role in the homeostasis of the ECM. In certain conditions, such as post-myocardial infarction or pressure overload, or in response to other pro-fibrotic stimuli, such as TGF-βs [57, 150–152], cardiac fibroblasts can be activated to differentiate into myofibroblasts [153, 154]. Several other sources of myofibroblasts have been discussed, including, most obviously, resident fibroblasts but also epicardial epithelial [155–157] or cardiac vascular endothelial cells [158], haematopoietic cells [158, 159] and circulating progenitor cells [158].

Most of these studies, however, have based their conclusions on partially ambiguous markers [160], mainly using Tie1 and fibroblast-specific protein 1 (FSP1) to label endothelial cells and myofibroblasts, respectively. Both markers, however, are known to not exclusively label either cell type, since they can also be expressed by various subpopulations of immune cells [161]. Using Tie2 to trace the lineage of endothelial cells and a collagen1α1-GFP fusion protein as a reporter for myofibroblasts, Moore-Morris et al. [162] concluded that the myofibroblast population formed in the heart upon pro-fibrotic stimulation is derived from two subpopulations of fibroblasts which are already present in the myocardium, where they have accumulated during cardiac development through epithelial-to-mesenchymal transition (EMT) and endothelial-to-mesenchymal transition (EndMT), respectively. However, Tie2 also seems to lack absolute endothelial specificity, and as it is active during early development, other cell types originating from such a common progenitor could also be labelled and hence mistaken for endothelial cells [163–165], leaving the question of the cellular source for the myofibroblasts still not satisfyingly answered.

Myofibroblasts not only produce large amounts of ECM proteins, they also acquire the ability to express contractile proteins such as α-smooth muscle actin [166]. This underscores their role not only in maintaining tissue integrity after a pathological event, but also, to a certain degree, in tissue function, i.e., contractility, when they replace cardiomyocytes as the regular contractile cell type. Myofibroblast activation can also be triggered by altered mechanical stress. Cardiac fibroblasts can sense the increased load [154] which causes activation of surface integrins [167], syndecans [168] and mechanosensitive ion channels [169], each feeding into intracellular signalling cascades such as the RhoA-Rho kinase-MRTF [170], focal adhesion kinase [171] and mitogen-activated protein (MAP) kinase [153] pathways that upregulate TGF-β1 expression [172], and result in myofibroblast formation. As a consequence, the synthesis of integrins [173] and matrix proteins is upregulated, and a general ECM-preserving programme is turned on [56, 57]. Furthermore, pressure overload activates the RAAS. Ang II, through its AT_1 receptor on myofibroblasts, can stimulate these cells to proliferate and synthesize ECM molecules. Synthesis of ECM components can also be exacted via pressure overload-dependent expression of microRNA-21, which in turn disinhibits the extracellular signal regulated kinase (ERK)-MAP kinase pathway in cardiac fibroblasts [174]. However, indirect activation via other mechanosensitive cells in the heart, such as immune cells, vascular cells or cardiomyocytes, which are affected by the increase in mechanical stress, may serve as another mechanism to spur differentiation of cardiac fibroblasts into myofibroblasts [60].

6 Endothelial Cells

The endothelium is an important regulatory element of the vasculature. It not only shields vessels from the flowing blood but also maintains and regulates the permeability of the vessel wall as well as the vascular tone [175]. Because of their special position in the vessel wall, at the interface between the circulating blood and the surrounding tissue, endothelial cells are subjected to unidirectional shear stress exerted on the luminal surface [176, 177]. This biomechanical force, to which solely endothelial cells are constantly exposed, governs the release of nitric oxide (NO) [178, 179], a gaseous mediator with an astonishing array of biological effects, primarily in resistance-sized small arteries and arterioles. In these vascular beds, endothelial cell NO release in response to shear stress counterbalances neurogenic [180] and myogenic [181] vasoconstriction, and thus governs total peripheral resistance [175]. In addition, the shear stress-driven constant release of NO from the endothelial cells ensures that the underlying vascular smooth muscle cells in the *tunica media* are kept in a quiescent contractile state [182, 183], and that the endothelial cells themselves maintain their anti-thrombotic and anti-inflammatory properties [184–186].

Furthermore, the constant exposure to unidirectional shear stress is of paramount importance for continuous expression of endothelial nitric oxide synthase (eNOS) [177, 187–189], the enzyme that produces NO. Any disturbance in laminar flow, hence unidirectional shear stress, as it occurs, e.g., at atherosclerosis predilection sites, i.e., bifurcations or curvatures of the large conduit arteries [177, 190–193], may impact the capacity of endothelial cells to fulfil the aforementioned tasks and give rise to or spur atherosclerosis [194, 195]. This may be further aggravated by enabling, instead of preventing, atherothrombosis as well as the diapedesis of pro-inflammatory leukocytes [196]. Also, endothelial-to-mesenchymal transition may be the consequence of a decline in unidirectional shear stress acting on, and hence NO released from, endothelial cells [197].

The other biomechanical force to which endothelial cells are exposed is tensile stretch [198], primarily affecting the endothelial cell-to-cell contacts [199]. Changes in tensile stretch are typically brought about by (blood) pressure-dependent distension or (active) vasodilation, which both increase circumferential wall tension, and thus strain, on the endothelial cell-to-cell contacts. Osmotic stress leading to a swelling of endothelial cells, on the other hand, does not play a major role in the living organism, whereas an increase in wall tension, namely in the left ventricle of the heart due to pressure overload, will affect the endocardial endothelial cells in a similar way as their vascular counterparts [200, 201]. Typically, enhanced tensile stretch will have a very different impact on mechanosensitive endothelial cell gene expression and phenotype compared to a decrease in unidirectional shear stress (see above), although there may be some overlap. One striking example for this differential control of gene expression in endothelial cells by the aforementioned biomechanical forces is the LIM-domain protein zyxin, which acts as a mechanotransducer and transcription factor in endothelial cells [202–204]. Translocation of zyxin from focal adhesions to the nucleus only occurs when the endothelial cell-to-cell contacts are strained, but not in response to an increase in unidirectional or oscillatory shear stress [202], presumably due to the much smaller magnitude of this particular biomechanical force (see below) which normally needs to be amplified by a mechanosensory complex located at the endothelial cell-to-cell contacts [205] to have an impact on endothelial cell gene expression and hence function.

Typically, unidirectional shear stress ranges from 5 to 20 dyn/cm^2 in resistance-sized small arteries and arterioles in humans under normal physiological conditions, but can rise to 40 dyn/cm^2 or more in larger arteries during increased cardiac output or hypertension [206]. Circumferential tensile stretch in arteries ranges from 5 to 10% under physiological conditions, but can attain up to 20% in hypertension

[207]. Assuming a wall thickness of 0.1 cm, these changes in circumference translate into a change in wall stress of ~600,000 to ~1,200,000 (~2,400,000) dyn/cm^2, i.e., the pressure-induced tensile stretch as a biomechanical force is approximately five orders of magnitude greater than (fluid) shear stress. In addition to its direct effects on the endothelial cell phenotype, an increase in circumferential tensile stretch can sensitize and/or activate the corresponding mechanosensing and transduction mechanisms in endothelial cells [208, 209].

In the heart, the integrity of the vascular endothelial cell monolayer is vital both functionally and mechanically. Disturbances may result mainly in a reduced production/release of NO [25], or in an impairment or even loss of the endothelial barrier function [196]. Consequences of this endothelial cell dysfunction include cardiomyocyte hypertrophy [210], increased myocardial stiffness [211] and/or cardiac fibrosis [212]. In addition, recruitment and diapedesis of circulating pro-inflammatory cells, namely monocytes differentiating into pro-inflammatory macrophages, may occur more often due to the increased expression of chemokines and adhesion molecules by the dysfunctional endothelial cells [213].

6.1 *Macrovascular Versus Microvascular Endothelial Cells*

Arteries in the body range in size from the largest diameter in the aorta to the smallest in pre-capillary arterioles. They are subject to a wide variety of environmental influences. The vascular endothelium has adapted to these specific environment variables expressing distinct and differing characteristics in endothelial cells of large conduit and resistance-sized small arteries and arterioles, i.e., macrovascular and microvascular endothelial cells, respectively. As a result, certain pathologies occur, only or predominantly, at specific sites within the arterial tree (see above). The disturbed blood flow at these sites, which leads to a decline of unidirectional shear stress, hence NO release, in combination with low or no flow zones at the arterial vessel wall allows circulating leukocytes, namely monocytes, to interact with the endothelial cells. This is reinforced by an increased adhesion and activation of circulating platelets via the CD40-CD40 ligand dyad [214–217], which, in turn, release chemokines that greatly facilitate diapedesis of the recruited monocytes.

Furthermore, endothelial cells possess primary cilia that seem to be utilized to sense blood flow and which have been shown to be implicated in TGF-β-signalling [218–220], atherosclerosis and EndMT [221, 222]. However, they seem to be present only in blood vessels with low flow conditions and not in high flow areas [221], adding another layer of distinction between endothelial cells depending on their local environment. Moreover, due to reflections of the pulse wave, endothelial cells are exposed to cyclic tensile stress that causes their enhanced production of reactive oxygen species [223], namely superoxide anions, which, by way of neutralization [224], adds to the much-decreased bioavailability of NO at the aforementioned predilection sites.

Problems like these, arising from the described alterations of these two main biomechanical forces, may be exacerbated by known primary risk factors of atherosclerosis such as hypertension, diabetes, or dyslipidaemia. Cardiac macrovascular and microvascular endothelial cells have been shown to possess different secretory profiles for MMPs and their inhibitors, TIMPs [225], too. When exposed to oxidized low-density lipoprotein, they respond with differential expression of adhesion molecules [226]. Known risk factors for coronary heart disease, such as hypertension or chronic inflammatory diseases, are likely to affect macrovascular and microvascular endothelial cells to a differing degree. One open question among others is whether the epicardial coronary arteries, the bifurcations of which represent important predilection sites for atherosclerosis, and the small arteries and arterioles in the heart contain different subtypes of endothelial cells, or if the same endothelial cell type, depending on its location/environment, reacts differently to a given stimulus, such as disturbed blood flow [227].

6.2 Cardiac Microvascular Endothelial Cells

Cardiac microvascular endothelial cells also have a broad spectrum of functions crucial to the regulation of cardiac activity in response to environmental cues. They possess an elaborate sensing capacity for biomechanical stress, such as tensile stretch, for the metabolic state of the cardiomyocytes affected by the accumulation of metabolites like ADP or carbon dioxide, as well as for changes in local pH and/or pO_2 [189, 228–234]. In addition, they are equipped with receptors for a whole slew of intercellular mediators such as serotonin, prostaglandins, bradykinin, tumour necrosis factor (TNF)-α or vascular endothelial growth factor (VEGF) [235], to name just a few. Cardiac microvascular endothelial cells translate these stimuli through the subsequent formation and/or release of other factors such as NO, Ang II, endothelin-1, or interleukin-6, among many others, into various signals for targeting cardiomyocytes, fibroblasts, or immune cells. This way, they affect cardiomyocyte contractility, growth, death, ECM composition through cardiac fibroblast differentiation into myofibroblasts, or the degree of the primarily innate immune response in the heart [235].

When exposed to pressure overload, cardiac microvascular endothelial cells respond at the transcriptional, metabolic and functional level by upregulating, among many others, TGF-β-induced genes that play a major role in the composition, and hence mechanical properties, of the ECM [36]. They also seem to transition, in a continuous and reversible manner [236], towards a more pro-fibrotic profile, with increased fatty acid oxidation and the synthesis of proline, one of the main building blocks of collagen and pro-collagen I. Consequently, microvascular endothelial cells isolated from left ventricle biopsies of patients with aortic stenosis revealed a lower pro-angiogenic (sprouting) ability than unaffected control cells isolated from the right atrium of these patients. In a nutshell, in conditions of pressure overload, microvascular endothelial cells in the heart seem to switch from an oxygen-

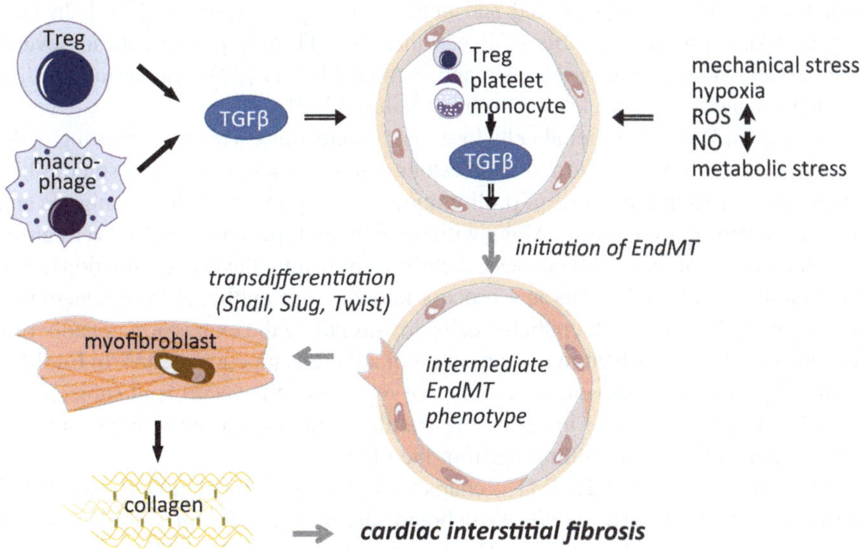

Fig. 3 Cells and stimuli causing and/or promoting endothelial-to-mesenchymal transition (EndMT) of microvascular endothelial cells in the heart. For details refer to the text. Abbreviations used: NO, nitric oxide; ROS, reactive oxygen species; TGFβ, transforming growth factor β; Treg, regulatory T cell

supplying, pro-angiogenic to a pro-fibrotic, collagen I secreting phenotype (see Fig. 3) that is frequently associated with excessive cardiac fibrosis [36].

During the development of the heart, cardiac microvascular endothelial cells undergo a very similar phenotypic switch. In fact, this physiological cellular transdifferentiation process of endothelial-to-mesenchymal transition was first described for embryonic heart valve development [237, 238]. Endothelial cells undergoing EndMT acquire a fibroblast-like mesenchymal profile and subsequently invade the so-called cardiac jelly, where they produce large amounts of ECM molecules, thereby forming the valve tissue [239]. In the adult heart, however, EndMT has not been shown to contribute to heart valve repair or regeneration [240].

The classic toxicology maxim credited to Paracelsus, that *it is the dose that makes the poison*, seems to hold true also in the context of EndMT. As a crucial mechanism in embryonic development, it has to be delicately controlled, since impaired EndMT may lead to the formation of failing heart valves, whereas excessive EndMT may result in valves that are too thick, rendering them too rigid and functionally inadequate [237]. Similarly, EndMT in response to an acute pathological stimulus may initially serve to recruit endothelial cells for cardiac fibrosis, which could be seen as some sort of repair mechanism where fibrosis, e.g., following myocardial infarction, at first delays the loss of structural and functional stability, thereby maintaining an, albeit reduced, myocardial function. Sustained pathological conditions such as pressure overload or a significant loss of cardiomyocytes, on the other hand, could turn EndMT into an exaggerated and hence maladaptive response [237]. In fact, recent studies provide credible evidence that EndMT may play a role in several pathologies, such as atherosclerosis and cardiac fibrosis [241], impairing cardiac mechanics and eventually leading to heart failure [242].

During EndMT, endothelial cells lose their connections with neighbouring cells, change their morphology from a cobblestone-like to a more elongated, spindle-like shape [241], increase their rate of proliferation, and acquire the ability to migrate and invade the surrounding tissue. Along with these phenotypic changes, EndMT causes a certain degree of vascular leakiness, thereby diminishing the barrier function of the endothelium [243, 244]. This is a process known from epithelial-to-mesenchymal transition (EMT), in which epithelial cells downregulate the expression of cadherins and other proteins involved in the formation of adherens junctions [245]. In EndMT, a similar process is observed, indicating that regions of the vessel wall subjected to this phenotype switch lose integrity and stability and become more permeable for (pro-inflammatory) cells infiltrating from the blood.

At the molecular level, EndMT is characterized by downregulation of endothelial markers such as PECAM-1, von Willebrand factor, VE-cadherin, VEGF receptor 2, endothelial NO synthase, Tie-1, Tie-2, and collagen IV [246, 247] while mesenchymal markers such as α-smooth muscle actin, vimentin, fibronectin, collagens I and III, N-cadherin or FSP1 are upregulated [246–248]. However, the transition can manifest itself as a continuum with varying degrees of expression of either marker and, in addition, may be regulated at the epigenetic level [247, 249]. Altogether, this data indicates that notably microvascular endothelial cells contribute to the population of myofibroblasts that form and get activated in response to pressure overload in the left ventricle. One should keep in mind, though, that the aforementioned, apparently contradictory conclusions could be based on the detection of markers, which are characteristic for different time points along the process of transition, and that each study may have captured only part of the spectrum of changes in marker gene expression [243]. Therefore, more research is necessary to elucidate the full extent of the contribution of EndMT to cardiac fibrosis.

6.3 *Effects of Pro-fibrotic Signals on Endothelial Cells*

Regardless of their disputable contribution to cardiac fibrosis via EndMT, the endothelial cell phenotype can be influenced by many factors favouring a pro-fibrotic environment. Thus, TGF-β has not only been shown to activate fibroblasts but endothelial cells as well. However, depending on the composition and structure of the ECM [250], TGF-β seems to affect the phenotype of endothelial cells differently, with effects ranging from pro-inflammatory to anti-inflammatory [251], pro-angiogenic to anti-angiogenic, growth-promoting to anti-proliferative [252], or pro-migratory to anti-migratory [253, 254]. Pro-inflammatory cytokines released by immune cells, mainly tumour necrosis factor α (TNFα) and interleukin-1β (IL-1β), have also been shown to induce EndMT in the brain [255], the vasculature [256], and the heart [257]. Moreover, extravasating monocytes differentiating to pro-inflammatory macrophages can also change the microenvironment to an EndMT-favouring state by releasing the membrane type 1 matrix metalloproteinase (MT1-MMP or MMP14) [257].

There are distinctions to the source of inflammation, and the corresponding tissue change. Local cardiac inflammation caused by either viral, microbial, or non-viral stimuli, such as MI [264], lead to HFrEF because the local inflammatory response primarily results in the infiltration of neutrophils releasing pro-inflammatory cytokines such as TNFα, IL-1β or IL-6 that trigger cardiomyocyte death and degradation of the ECM [265, 266]. The lost cardiomyocytes are replaced by fibrotic material [267, 268], thereby negatively impacting LV compliance and EF. A more general systemic inflammation, by contrast, mostly triggered by diseases such as type 2 diabetes, obesity, or hypertension, is associated with HFpEF (rather than HFrEF) because the pro-inflammatory signals mainly affect the microvascular endothelial cells, which in turn recruit (pro-inflammatory) monocytes from the blood that eventually also stimulate myofibroblast formation and hence fibrosis [269], but without a substantial loss of cardiomyocytes or reduction in EF. Besides TGF-β and the aforementioned pro-inflammatory cytokines, biomechanical stress and other environmental cues are also capable of inducing EndMT and/or endothelial cell-mediated (cardiac) fibrosis. This has been shown for a decline in unidirectional shear stress [196, 197, 258, 259], for an increase in tensile stretch [208, 258, 260] as well as for the composition [261] and stiffness of the surrounding ECM [262, 263].

7 Anti-fibrotic Therapies

So far, there is no cure for heart failure. Several therapeutic approaches exist, and a few of them seem to reduce severity of the disease, but none can prevent or reverse exacerbation of the maladaptive pro-fibrotic remodelling process. A major obstacle is the difficulty to select patients to be enrolled in studies or clinical trials, as early (asymptomatic) stages of heart failure are hard to detect. Therefore, a precise and timely diagnosis would greatly improve the selection and classification of subjects for such studies [270]. This would also help to determine whether certain treatment regimens only had the wrong timing or wrong target(s), or both. To date, strategies either aim at acute fibrosis, myofibroblast activation, EndMT or other underlying mechanisms such as overstimulation of the RAAS. Existing therapeutic approaches and those that are currently being developed have already been comprehensibly summarized in previous review articles [76, 271–273], hence only a very short overview is provided here.

A more generalized approach could aim at passivating the pro-inflammatory environment that strongly reinforces cardiac fibrosis. This could be achieved by providing anti-inflammatory drugs like Saikosaponin A [274], which has been shown to exert cardioprotective effects. As the differentiation of cardiac fibroblasts to myofibroblasts is a pivotal step in the development of cardiac fibrosis, therapeutic approaches mostly target cardiac fibroblasts and/or their activation. Naturally, aiming at TGF-β1 and/or its effectors shows strong anti-fibrotic effects [275]. However, interpretation of the efficacy and primary mechanism of action of anti-TGF-β therapies has been shown to be rather difficult. Most notably, this is due to the ubiquitous nature of expression and maturation of this cytokine as well as its context-depending effects on processes such as the synthesis of ECM components, cell proliferation, apoptosis, angiogenesis, and immune cell function [276, 277]. In addition, there is a rather complex interplay of TGF-β with members of its antagonistic stirps, the bone morphogenetic proteins (BMPs), namely BMP-2 and BMP-9, in the control of fibrosis initiation and progression [278, 279]. Several studies have been carried out (see below), and all encountered strong toxic side effects of whatever kind of drug used at doses sufficiently high to attenuate cardiac fibrosis. Lower doses, on the other hand, that were tolerated better in terms of side effects, had no definite impact on cardiac fibrosis [76].

Components of the TGF-β signalling pathway targeted so far include TGF-β1 itself (via neutralizing antibodies) [114], the type 1 [114, 280] or type 2 TGF-β receptor [121], and downstream elements involved in TGF-β signalling such as various kinases [281], SMAD3 [282] or SMAD4 [283]. They all resulted in high mortality rates in animal models, with the few survivors frequently exhibiting left ventricular dilation [283], widespread pro-inflammatory responses, and eventually multi-organ failure [284, 285]. Therefore, global inhibition of TGF-β signalling and its downstream targets does not seem to be a feasible approach [286]. However, local targeting through nanoparticles was recently shown to present a possible way to reduce endothelial cell-dependent TGF-β signalling [246].

Another possible candidate is angiotensin II (Ang II). As it is a long-standing candidate for anti-hypertensive treatment regimens [287], many therapeutics have been developed that are commonly used. Prototypic are two classes of drugs directed either at the formation of Ang II from Ang I via angiotensin-converting enzyme (ACE) or its effects mediated through the AT_1 receptor. In the treatment of HFrEF, these drugs have proven to be very effective over several decades [287, 288]. Unlike patients with HFrEF, targeting the RAAS had no discernible effects in patients with HFpEF, most likely due to aiming at different Ang II-independent pro-fibrotic pathways [289–292]. Like Ang II, aldosterone causes or reinforces cardiac fibrosis by upregulating pro-fibrotic gene expression [293] or MAP kinases that are relevant in Ang II signalling [294, 295]. Thus, antagonizing the mineralocorticoid receptor has been shown to reduce excessive ECM synthesis and to improve left ventricular function in experimental animals [296–299] but, so far, not in trials with human patients [300, 301].

Atrial and brain natriuretic peptides (BNP) released by cardiomyocytes in response to pressure or volume overload [302, 303] can counteract the development of cardiac fibrosis in experimental animals [304]. In patients with heart failure, however, administration of BNP did not elicit any beneficial effect [305, 306]. This could be due to its short half-life (< 20 min), as natriuretic peptides are metabolized by neprilysin, which is widely expressed in the body [307]. Inhibition of neprilysin itself has been reported to be helpful in patients with HFrEF, especially in combination with antagonizing Ang II [308–310]. In patients with HFpEF, however, a minor effect was observed solely in women [306].

If EndMT could reliably be shown to occur in response to pressure overload or similar stimuli, thereby demonstrating that microvascular endothelial cells are a significant source of myofibroblasts in the heart, attenuating EndMT would be beneficial [311] and an appropriate target. Possible therapeutic approaches would have to be restricted to selectively affecting the non-developmental, excessive variant of EndMT. Alternatively, enhancing or substituting for the cardioprotective effects of endothelial cell-derived NO could be an option, since NO has been shown to exert a strong protective effect in the heart by preventing adverse remodelling [312]. Accordingly, NO donors such as organic nitrates or nitroprusside, administered in combination with hydralazine to prevent side effects such as nitrate tolerance or superoxide anion formation [313], are beneficial in the treatment of cardiac fibrosis [211]. Likewise, enhancing endothelial cell NO formation via stimulation of the type 1 sphingosine-1-phosphate receptor showed cardioprotective effects [314].

NO activates the soluble guanylate cyclase, thereby increasing intracellular cGMP levels and activating cGMP-dependent protein kinases. This can also be achieved by inhibiting the cGMP-specific phosphodiesterase-5 (PDE-5), which blocks the conversion of cGMP to GMP. Maintaining cGMP at an elevated level may lead to an improvement of endothelial cell function [315] and thus attenuate left ventricular remodelling [210]. PDE-5 inhibition may not help, though, if cGMP production in the endothelial cells is already impaired [316].

Therefore, another option would be to activate or stimulate soluble guanylyl cyclase [316]. Initially, this approach was shown to cause strong *hypo*tension, but various stimulators of the enzyme are currently being evaluated in clinical trials to assess their usefulness in this regard [316].

8 Conclusion(s)

Ischemic heart disease, the most important cause of death worldwide, and hypertensive heart disease have in common that they ultimately lead to HFrEF through a maladaptive remodelling process in the myocardium that comprises excessive interstitial fibrosis as a pivotal intermediary step. In response to the loss of cardiomyocytes due to ischemia, triggering replacement fibrosis, or pressure overload, which requires reinforcement of the ECM in addition to cardiomyocyte hypertrophy, cardiac fibroblasts differentiate into myofibroblasts, which synthesize large amounts of ECM molecules. This newly formed scar tissue, however, is prone to rupture. Microvascular endothelial cells that, upon inadequate biomechanical stimulation, i.e., enhanced tensile stretch or exposure to pro-inflammatory and/or pro-fibrotic mediators such as TGF-β signalling, may transdifferentiate into myofibroblasts, therefore pose as another possible source of this ECM-synthesizing cell type in the diseased heart. Apart from HFrEF, excessive cardiac fibrosis also leads to a stiffening of the myocardium, which is thought to play a pivotal role in the development of HFpEF, a variant of heart failure that is as frequent as HFrEF. Systematic unravelling of the underlying signalling mechanisms in the cardiac fibroblasts and microvascular endothelial cells undergoing endothelial-to-mesenchymal transition, which have been reviewed herein, will facilitate the development of effective therapeutic approaches to stop or prevent excessive cardiac fibrosis and thus all-cause heart failure.

Acknowledgements The authors are indebted to Gerd König for critical reading of this manuscript and his help in producing the figures.

Compliance with Ethical Standards **Conflict of Interest**: All authors declare they have no conflict of interest.

Ethical Approval: This article does not contain any studies with human participants or animals performed by any of the authors.

References

1. Duncker DJ, Bache RJ (2008) Regulation of coronary blood flow during exercise. Physiol Rev 88:1009–1086
2. Laughlin MH, Tomanek RJ (1987) Myocardial capillarity and maximal capillary diffusion capacity in exercise-trained dogs. J Appl Physiol 1985(63):1481–1486
3. Pinto AR, Ilinykh A, Ivey MJ, Kuwabara JT, D'Antoni ML, Debuque R, Chandran A, Wang L, Arora K, Rosenthal NA et al (2016) Revisiting cardiac cellular composition. Circ Res 118:400–409
4. Brutsaert DL (2003) Cardiac endothelial-myocardial signaling: its role in cardiac growth, contractile performance, and rhythmicity. Physiol Rev 83:59–115
5. Leask A (2015) Getting to the heart of the matter: new insights into cardiac fibrosis. Circ Res 116:1269–1276
6. Braunwald E, Ross J Jr, Sonnenblick EH (1967) Mechanisms of contraction of the normal and failing heart. N Engl J Med 277:853–863. contd
7. Ponikowski P, Voors AA, Anker SD, Bueno H, Cleland JGF, Coats AJS, Falk V, Gonzalez-Juanatey JR, Harjola VP, Jankowska EA et al (2016) 2016 ESC Guidelines for the diagnosis and treatment of acute and chronic heart failure: the Task Force for the diagnosis and treatment of acute and chronic heart failure of the European Society of Cardiology (ESC)Developed with the special contribution of the Heart Failure Association (HFA) of the ESC. Eur Heart J 37: 2129–2200
8. Dunlay SM, Roger VL, Redfield MM (2017) Epidemiology of heart failure with preserved ejection fraction. Nat Rev Cardiol 14:591–602
9. Groenewegen A, Rutten FH, Mosterd A, Hoes AW (2020) Epidemiology of heart failure. Eur J Heart Fail 22:1342–1356
10. Savarese G, Lund LH (2017) Global public health burden of heart failure. Card Fail Rev 3:7–11
11. Tsao CW, Lyass A, Enserro D, Larson MG, Ho JE, Kizer JR, Gottdiener JS, Psaty BM, Vasan RS (2018) Temporal trends in the incidence of and mortality associated with heart failure with preserved and reduced ejection fraction. JACC Heart Fail 6:678–685
12. Savarese G, Stolfo D, Sinagra G, Lund LH (2022) Heart failure with mid-range or mildly reduced ejection fraction. Nat Rev Cardiol 19:100–116
13. McMurray JJ, Pfeffer MA (2005) Heart failure. Lancet 365:1877–1889

14. Ruwhof C, van der Laarse A (2000) Mechanical stress-induced cardiac hypertrophy: mechanisms and signal transduction pathways. Cardiovasc Res 47:23–37
15. Lorell BH, Carabello BA (2000) Left ventricular hypertrophy: pathogenesis, detection, and prognosis. Circulation 102:470–479
16. Weber KT, Sun Y, Tyagi SC, Cleutjens JP (1994) Collagen network of the myocardium: function, structural remodeling and regulatory mechanisms. J Mol Cell Cardiol 26:279–292
17. Sheng JJ, Feng HZ, Pinto JR, Wei H, Jin JP (2016) Increases of desmin and alpha-actinin in mouse cardiac myofibrils as a response to diastolic dysfunction. J Mol Cell Cardiol 99:218–229
18. Simmonds SJ, Cuijpers I, Heymans S, Jones EAV (2020) Cellular and molecular differences between HFpEF and HFrEF: a step ahead in an improved pathological understanding. Cells 9:242
19. van Heerebeek L, Borbely A, Niessen HW, Bronzwaer JG, van der Velden J, Stienen GJ, Linke WA, Laarman GJ, Paulus WJ (2006) Myocardial structure and function differ in systolic and diastolic heart failure. Circulation 113:1966–1973
20. DeBerge M, Shah SJ, Wilsbacher L, Thorp EB (2019) Macrophages in heart failure with reduced versus preserved ejection fraction. Trends Mol Med 25:328–340
21. Heinzel FR, Hegemann N, Hohendanner F, Primessnig U, Grune J, Blaschke F, de Boer RA, Pieske B, Schiattarella GG, Kuebler WM (2020) Left ventricular dysfunction in heart failure with preserved ejection fraction-molecular mechanisms and impact on right ventricular function. Cardiovasc Diagn Ther 10:1541–1560
22. Nikolova V, Leimena C, McMahon AC, Tan JC, Chandar S, Jogia D, Kesteven SH, Michalicek J, Otway R, Verheyen F et al (2004) Defects in nuclear structure and function promote dilated cardiomyopathy in lamin A/C-deficient mice. J Clin Invest 113:357–369
23. Ding B, Price RL, Goldsmith EC, Borg TK, Yan X, Douglas PS, Weinberg EO, Bartunek J, Thielen T, Didenko VV et al (2000) Left ventricular hypertrophy in ascending aortic stenosis mice: anoikis and the progression to early failure. Circulation 101:2854–2862
24. Kostin S, Pool L, Elsasser A, Hein S, Drexler HC, Arnon E, Hayakawa Y, Zimmermann R, Bauer E, Klovekorn WP et al (2003) Myocytes die by multiple mechanisms in failing human hearts. Circ Res 92:715–724
25. Paulus WJ, Tschope C (2013) A novel paradigm for heart failure with preserved ejection fraction: comorbidities drive myocardial dysfunction and remodeling through coronary microvascular endothelial inflammation. J Am Coll Cardiol 62:263–271
26. Francis GS (1998) Changing the remodeling process in heart failure: basic mechanisms and laboratory results. Curr Opin Cardiol 13:156–161
27. Bekfani T, Pellicori P, Morris DA, Ebner N, Valentova M, Steinbeck L, Wachter R, Elsner S, Sliziuk V, Schefold JC et al (2016) Sarcopenia in patients with heart failure with preserved ejection fraction: impact on muscle strength, exercise capacity and quality of life. Int J Cardiol 222:41–46
28. Streng KW, Nauta JF, Hillege HL, Anker SD, Cleland JG, Dickstein K, Filippatos G, Lang CC, Metra M, Ng LL et al (2018) Non-cardiac comorbidities in heart failure with reduced, mid-range and preserved ejection fraction. Int J Cardiol 271:132–139
29. Messerli FH, Rimoldi SF, Bangalore S (2017) The transition from hypertension to heart failure: contemporary update. JACC Heart Fail 5:543–551
30. Alpert MA, Hashimi MW (1993) Obesity and the heart. Am J Med Sci 306:117–123
31. Aurigemma GP, de Simone G, Fitzgibbons TP (2013) Cardiac remodeling in obesity. Circ Cardiovasc Imaging 6:142–152
32. Vasan RS (2003) Cardiac function and obesity. Heart 89:1127–1129
33. London GM (2002) Left ventricular alterations and end-stage renal disease. Nephrol Dial Transplant 17(Suppl 1):29–36
34. Messerli FH (1982) Cardiovascular effects of obesity and hypertension. Lancet 1:1165–1168
35. Heineke J, Molkentin JD (2006) Regulation of cardiac hypertrophy by intracellular signalling pathways. Nat Rev Mol Cell Biol 7:589–600

36. Trenson S, Hermans H, Craps S, Pokreisz P, de Zeeuw P, Van Wauwe J, Gillijns H, Veltman D, Wei F, Caluwe E et al (2021) Cardiac microvascular endothelial cells in pressure overload-induced heart disease. Circ Heart Fail 14:e006979
37. Dorn GW 2nd, Robbins J, Sugden PH (2003) Phenotyping hypertrophy: eschew obfuscation. Circ Res 92:1171–1175
38. Harvey PA, Leinwand LA (2011) The cell biology of disease: cellular mechanisms of cardiomyopathy. J Cell Biol 194:355–365
39. Bing OH, Ngo HQ, Humphries DE, Robinson KG, Lucey EC, Carver W, Brooks WW, Conrad CH, Hayes JA, Goldstein RH (1997) Localization of alpha1(I) collagen mRNA in myocardium from the spontaneously hypertensive rat during the transition from compensated hypertrophy to failure. J Mol Cell Cardiol 29:2335–2344
40. Chiong M, Wang ZV, Pedrozo Z, Cao DJ, Troncoso R, Ibacache M, Criollo A, Nemchenko A, Hill JA, Lavandero S (2011) Cardiomyocyte death: mechanisms and translational implications. Cell Death Dis 2:e244
41. Conrad CH, Brooks WW, Hayes JA, Sen S, Robinson KG, Bing OH (1995) Myocardial fibrosis and stiffness with hypertrophy and heart failure in the spontaneously hypertensive rat. Circulation 91:161–170
42. Gonzalez A, Fortuno MA, Querejeta R, Ravassa S, Lopez B, Lopez N, Diez J (2003) Cardiomyocyte apoptosis in hypertensive cardiomyopathy. Cardiovasc Res 59:549–562
43. Wencker D, Chandra M, Nguyen K, Miao W, Garantziotis S, Factor SM, Shirani J, Armstrong RC, Kitsis RN (2003) A mechanistic role for cardiac myocyte apoptosis in heart failure. J Clin Invest 111:1497–1504
44. Yussman MG, Toyokawa T, Odley A, Lynch RA, Wu G, Colbert MC, Aronow BJ, Lorenz JN, Dorn GW 2nd (2002) Mitochondrial death protein Nix is induced in cardiac hypertrophy and triggers apoptotic cardiomyopathy. Nat Med 8:725–730
45. Li CJ, Chen CS, Yiang GT, Tsai AP, Liao WT, Wu MY (2019) Advanced evolution of pathogenesis concepts in cardiomyopathies. J Clin Med 8
46. Adamopoulos C, Zannad F, Fay R, Mebazaa A, Cohen-Solal A, Guize L, Juilliere Y, Alla F (2007) Ejection fraction and blood pressure are important and interactive predictors of 4-week mortality in severe acute heart failure. Eur J Heart Fail 9:935–941
47. Bhatia RS, Tu JV, Lee DS, Austin PC, Fang J, Haouzi A, Gong Y, Liu PP (2006) Outcome of heart failure with preserved ejection fraction in a population-based study. N Engl J Med 355: 260–269
48. Fonarow GC, Adams KF Jr, Abraham WT, Yancy CW, Boscardin WJ, ADHERE Scientific Advisory Committee, Study Group, and Investigators (2005) Risk stratification for in-hospital mortality in acutely decompensated heart failure: classification and regression tree analysis. JAMA 293:572–580
49. Gheorghiade M, Abraham WT, Albert NM, Greenberg BH, O'Connor CM, She L, Stough WG, Yancy CW, Young JB, Fonarow GC et al (2006) Systolic blood pressure at admission, clinical characteristics, and outcomes in patients hospitalized with acute heart failure. JAMA 296:2217–2226
50. Lee TT, Chen J, Cohen DJ, Tsao L (2006) The association between blood pressure and mortality in patients with heart failure. Am Heart J 151:76–83
51. Pulignano G, Del Sindaco D, Tavazzi L, Lucci D, Gorini M, Leggio F, Porcu M, Scherillo M, Opasich C, Di Lenarda A et al (2002) Clinical features and outcomes of elderly outpatients with heart failure followed up in hospital cardiology units: data from a large nationwide cardiology database (IN-CHF Registry). Am Heart J 143:45–55
52. Rouleau JL, Roecker EB, Tendera M, Mohacsi P, Krum H, Katus HA, Fowler MB, Coats AJ, Castaigne A, Scherhag A et al (2004) Influence of pretreatment systolic blood pressure on the effect of carvedilol in patients with severe chronic heart failure: the Carvedilol Prospective Randomized Cumulative Survival (COPERNICUS) study. J Am Coll Cardiol 43:1423–1429
53. Theocharis AD, Skandalis SS, Gialeli C, Karamanos NK (2016) Extracellular matrix structure. Adv Drug Deliv Rev 97:4–27

54. Yang H, Borg TK, Wang Z, Ma Z, Gao BZ (2014) Role of the basement membrane in regulation of cardiac electrical properties. Ann Biomed Eng 42:1148–1157
55. Hynes RO (2009) The extracellular matrix: not just pretty fibrils. Science 326:1216–1219
56. Dobaczewski M, Bujak M, Li N, Gonzalez-Quesada C, Mendoza LH, Wang XF, Frangogiannis NG (2010) Smad3 signaling critically regulates fibroblast phenotype and function in healing myocardial infarction. Circ Res 107:418–428
57. Russo I, Cavalera M, Huang SB, Su Y, Hanna A, Chen BJ, Shinde AV, Conway SJ, Graff J, Frangogiannis NG (2019) Protective effects of activated myofibroblasts in the pressure-overloaded myocardium are mediated through smad-dependent activation of a matrix-preserving program. Circ Res 124:1214–1227
58. Sasayama S, Ross J Jr, Franklin D, Bloor CM, Bishop S, Dilley RB (1976) Adaptations of the left ventricle to chronic pressure overload. Circ Res 38:172–178
59. Levy D, Larson MG, Vasan RS, Kannel WB, Ho KK (1996) The progression from hypertension to congestive heart failure. JAMA 275:1557–1562
60. Frangogiannis NG (2019) The extracellular matrix in ischemic and nonischemic heart failure. Circ Res 125:117–146
61. Weber KT, Brilla CG (1991) Pathological hypertrophy and cardiac interstitium. Fibrosis and renin-angiotensin-aldosterone system. Circulation 83:1849–1865
62. Lopez B, Gonzalez A, Hermida N, Valencia F, de Teresa E, Diez J (2010) Role of lysyl oxidase in myocardial fibrosis: from basic science to clinical aspects. Am J Physiol Heart Circ Physiol 299:H1–H9
63. Lopez B, Querejeta R, Gonzalez A, Larman M, Diez J (2012) Collagen cross-linking but not collagen amount associates with elevated filling pressures in hypertensive patients with stage C heart failure: potential role of lysyl oxidase. Hypertension 60:677–683
64. Burstein B, Comtois P, Michael G, Nishida K, Villeneuve L, Yeh YH, Nattel S (2009) Changes in connexin expression and the atrial fibrillation substrate in congestive heart failure. Circ Res 105:1213–1222
65. Dai Z, Aoki T, Fukumoto Y, Shimokawa H (2012) Coronary perivascular fibrosis is associated with impairment of coronary blood flow in patients with non-ischemic heart failure. J Cardiol 60:416–421
66. Ytrehus K, Hulot JS, Perrino C, Schiattarella GG, Madonna R (2018) Perivascular fibrosis and the microvasculature of the heart. Still hidden secrets of pathophysiology? Vascul Pharmacol. https://doi.org/10.1016/j.vph.2018.04.007
67. Gabbiani G (2003) The myofibroblast in wound healing and fibrocontractive diseases. J Pathol 200:500–503
68. Travers JG, Kamal FA, Robbins J, Yutzey KE, Blaxall BC (2016) Cardiac fibrosis: the fibroblast awakens. Circ Res 118:1021–1040
69. Hinz B (2007) Formation and function of the myofibroblast during tissue repair. J Invest Dermatol 127:526–537
70. Hinz B, Lagares D (2020) Evasion of apoptosis by myofibroblasts: a hallmark of fibrotic diseases. Nat Rev Rheumatol 16:11–31
71. Kis K, Liu X, Hagood JS (2011) Myofibroblast differentiation and survival in fibrotic disease. Expert Rev Mol Med 13:e27
72. Aoki T, Fukumoto Y, Sugimura K, Oikawa M, Satoh K, Nakano M, Nakayama M, Shimokawa H (2011) Prognostic impact of myocardial interstitial fibrosis in non-ischemic heart failure. Circ J 75:2605–2613
73. Kato S, Saito N, Kirigaya H, Gyotoku D, Iinuma N, Kusakawa Y, Iguchi K, Nakachi T, Fukui K, Futaki M et al (2015) Prognostic significance of quantitative assessment of focal myocardial fibrosis in patients with heart failure with preserved ejection fraction. Int J Cardiol 191:314–319
74. Bashey RI, Martinez-Hernandez A, Jimenez SA (1992) Isolation, characterization, and localization of cardiac collagen type VI. Associations with other extracellular matrix components. Circ Res 70:1006–1017

75. de Souza RR (2002) Aging of myocardial collagen. Biogerontology 3:325–335
76. Sweeney M, Corden B, Cook SA (2020) Targeting cardiac fibrosis in heart failure with preserved ejection fraction: mirage or miracle? EMBO Mol Med 12:e10865
77. Lencova-Popelova O, Jirkovsky E, Mazurova Y, Lenco J, Adamcova M, Simunek T, Gersl V, Sterba M (2014) Molecular remodeling of left and right ventricular myocardium in chronic anthracycline cardiotoxicity and post-treatment follow up. PLoS One 9:e96055
78. Watanabe T, Kusachi S, Yamanishi A, Kumashiro H, Nunoyama H, Sano I, Nakahama M, Murakami T, Naito I, Ninomiya Y et al (1998) Localization of type IV collagen alpha chain in the myocardium of dilated and hypertrophic cardiomyopathy. Jpn Heart J 39:753–762
79. Skrbic B, Bjornstad JL, Marstein HS, Carlson CR, Sjaastad I, Nygard S, Bjornstad S, Christensen G, Tonnessen T (2013) Differential regulation of extracellular matrix constituents in myocardial remodeling with and without heart failure following pressure overload. Matrix Biol 32:133–142
80. Munch J, Abdelilah-Seyfried S (2021) Sensing and responding of cardiomyocytes to changes of tissue stiffness in the diseased heart. Front Cell Dev Biol 9:642840
81. Echegaray K, Andreu I, Lazkano A, Villanueva I, Saenz A, Elizalde MR, Echeverria T, Lopez B, Garro A, Gonzalez A et al (2017) Role of myocardial collagen in severe aortic stenosis with preserved ejection fraction and symptoms of heart failure. Rev Esp Cardiol (Engl Ed) 70:832–840
82. Kasner M, Westermann D, Lopez B, Gaub R, Escher F, Kuhl U, Schultheiss HP, Tschope C (2011) Diastolic tissue Doppler indexes correlate with the degree of collagen expression and cross-linking in heart failure and normal ejection fraction. J Am Coll Cardiol 57:977–985
83. Baicu CF, Stroud JD, Livesay VA, Hapke E, Holder J, Spinale FG, Zile MR (2003) Changes in extracellular collagen matrix alter myocardial systolic performance. Am J Physiol Heart Circ Physiol 284:H122–H132
84. Barallobre-Barreiro J, Lynch M, Yin X, Mayr M (2016) Systems biology-opportunities and challenges: the application of proteomics to study the cardiovascular extracellular matrix. Cardiovasc Res 112:626–636
85. Iwanaga Y, Aoyama T, Kihara Y, Onozawa Y, Yoneda T, Sasayama S (2002) Excessive activation of matrix metalloproteinases coincides with left ventricular remodeling during transition from hypertrophy to heart failure in hypertensive rats. J Am Coll Cardiol 39: 1384–1391
86. Sun Y, Weber KT (2000) Infarct scar: a dynamic tissue. Cardiovasc Res 46:250–256
87. Mukherjee D, Sen S (1991) Alteration of collagen phenotypes in ischemic cardiomyopathy. J Clin Invest 88:1141–1146
88. McCormick RJ, Musch TI, Bergman BC, Thomas DP (1994) Regional differences in LV collagen accumulation and mature cross-linking after myocardial infarction in rats. Am J Physiol 266:H354–H359
89. Gonzalez-Santamaria J, Villalba M, Busnadiego O, Lopez-Olaneta MM, Sandoval P, Snabel J, Lopez-Cabrera M, Erler JT, Hanemaaijer R, Lara-Pezzi E et al (2016) Matrix cross-linking lysyl oxidases are induced in response to myocardial infarction and promote cardiac dysfunction. Cardiovasc Res 109:67–78
90. Hermida N, Lopez B, Gonzalez A, Dotor J, Lasarte JJ, Sarobe P, Borras-Cuesta F, Diez J (2009) A synthetic peptide from transforming growth factor-beta1 type III receptor prevents myocardial fibrosis in spontaneously hypertensive rats. Cardiovasc Res 81:601–609
91. Herum KM, Lunde IG, Skrbic B, Louch WE, Hasic A, Boye S, Unger A, Brorson SH, Sjaastad I, Tonnessen T et al (2015) Syndecan-4 is a key determinant of collagen cross-linking and passive myocardial stiffness in the pressure-overloaded heart. Cardiovasc Res 106: 217–226
92. Yu M, Wen S, Wang M, Liang W, Li HH, Long Q, Guo HP, Liao YH, Yuan J (2013) TNF-alpha-secreting B cells contribute to myocardial fibrosis in dilated cardiomyopathy. J Clin Immunol 33:1002–1008

93. Rubis P, Wisniowska-Smialek S, Wypasek E, Biernacka-Fijalkowska B, Rudnicka-Sosin L, Dziewiecka E, Faltyn P, Khachatryan L, Karabinowska A, Kozanecki A et al (2016) Fibrosis of extracellular matrix is related to the duration of the disease but is unrelated to the dynamics of collagen metabolism in dilated cardiomyopathy. Inflamm Res 65:941–949
94. Cleutjens JP, Kandala JC, Guarda E, Guntaka RV, Weber KT (1995) Regulation of collagen degradation in the rat myocardium after infarction. J Mol Cell Cardiol 27:1281–1292
95. Fan D, Kassiri Z (2021) Modulation of cardiac fibrosis in and beyond cells. Front Mol Biosci 8:750626
96. Frangogiannis NG (2012) Matricellular proteins in cardiac adaptation and disease. Physiol Rev 92:635–688
97. Collier P, Watson CJ, Voon V, Phelan D, Jan A, Mak G, Martos R, Baugh JA, Ledwidge MT, McDonald KM (2011) Can emerging biomarkers of myocardial remodelling identify asymptomatic hypertensive patients at risk for diastolic dysfunction and diastolic heart failure? Eur J Heart Fail 13:1087–1095
98. Kai H, Kuwahara F, Tokuda K, Imaizumi T (2005) Diastolic dysfunction in hypertensive hearts: roles of perivascular inflammation and reactive myocardial fibrosis. Hypertens Res 28: 483–490
99. Meagher PB, Lee XA, Lee J, Visram A, Friedberg MK, Connelly KA (2021) Cardiac fibrosis: key role of integrins in cardiac homeostasis and remodeling. Cells 10
100. Li YY, McTiernan CF, Feldman AM (2000) Interplay of matrix metalloproteinases, tissue inhibitors of metalloproteinases and their regulators in cardiac matrix remodeling. Cardiovasc Res 46:214–224
101. Kaminski AR, Moore ET, Daseke MJ 2nd, Valerio FM, Flynn ER, Lindsey ML (2020) The compendium of matrix metalloproteinase expression in the left ventricle of mice following myocardial infarction. Am J Physiol Heart Circ Physiol 318:H706–H714
102. Lindsey ML, Iyer RP, Jung M, DeLeon-Pennell KY, Ma Y (2016) Matrix metalloproteinases as input and output signals for post-myocardial infarction remodeling. J Mol Cell Cardiol 91: 134–140
103. Nagatomo Y, Carabello BA, Coker ML, McDermott PJ, Nemoto S, Hamawaki M, Spinale FG (2000) Differential effects of pressure or volume overload on myocardial MMP levels and inhibitory control. Am J Physiol Heart Circ Physiol 278:H151–H161
104. Reinhardt D, Sigusch HH, Hensse J, Tyagi SC, Korfer R, Figulla HR (2002) Cardiac remodelling in end stage heart failure: upregulation of matrix metalloproteinase (MMP) irrespective of the underlying disease, and evidence for a direct inhibitory effect of ACE inhibitors on MMP. Heart 88:525–530
105. Heymans S, Schroen B, Vermeersch P, Milting H, Gao F, Kassner A, Gillijns H, Herijgers P, Flameng W, Carmeliet P et al (2005) Increased cardiac expression of tissue inhibitor of metalloproteinase-1 and tissue inhibitor of metalloproteinase-2 is related to cardiac fibrosis and dysfunction in the chronic pressure-overloaded human heart. Circulation 112:1136–1144
106. Polyakova V, Hein S, Kostin S, Ziegelhoeffer T, Schaper J (2004) Matrix metalloproteinases and their tissue inhibitors in pressure-overloaded human myocardium during heart failure progression. J Am Coll Cardiol 44:1609–1618
107. Kremastiotis G, Handa I, Jackson C, George S, Johnson J (2021) Disparate effects of MMP and TIMP modulation on coronary atherosclerosis and associated myocardial fibrosis. Sci Rep 11:23081
108. Talman V, Ruskoaho H (2016) Cardiac fibrosis in myocardial infarction-from repair and remodeling to regeneration. Cell Tissue Res 365:563–581
109. Silva AC, Pereira C, Fonseca A, Pinto-do OP, Nascimento DS (2020) Bearing my heart: the role of extracellular matrix on cardiac development, homeostasis, and injury response. Front Cell Dev Biol 8:621644
110. Oka T, Xu J, Kaiser RA, Melendez J, Hambleton M, Sargent MA, Lorts A, Brunskill EW, Dorn GW 2nd, Conway SJ et al (2007) Genetic manipulation of periostin expression reveals a role in cardiac hypertrophy and ventricular remodeling. Circ Res 101:313–321

111. Biernacka A, Dobaczewski M, Frangogiannis NG (2011) TGF-beta signaling in fibrosis. Growth Factors 29:196–202
112. Leask A, Abraham DJ (2004) TGF-beta signaling and the fibrotic response. FASEB J 18:816–827
113. Dobaczewski M, Chen W, Frangogiannis NG (2011) Transforming growth factor (TGF)-beta signaling in cardiac remodeling. J Mol Cell Cardiol 51:600–606
114. Kuwahara F, Kai H, Tokuda K, Kai M, Takeshita A, Egashira K, Imaizumi T (2002) Transforming growth factor-beta function blocking prevents myocardial fibrosis and diastolic dysfunction in pressure-overloaded rats. Circulation 106:130–135
115. Desmouliere A, Geinoz A, Gabbiani F, Gabbiani G (1993) Transforming growth factor-beta 1 induces alpha-smooth muscle actin expression in granulation tissue myofibroblasts and in quiescent and growing cultured fibroblasts. J Cell Biol 122:103–111
116. Heldin CH, Miyazono K, ten Dijke P (1997) TGF-beta signalling from cell membrane to nucleus through SMAD proteins. Nature 390:465–471
117. Huang S, Chen B, Su Y, Alex L, Humeres C, Shinde AV, Conway SJ, Frangogiannis NG (2019) Distinct roles of myofibroblast-specific Smad2 and Smad3 signaling in repair and remodeling of the infarcted heart. J Mol Cell Cardiol 132:84–97
118. Leask A (2010) Potential therapeutic targets for cardiac fibrosis: TGFbeta, angiotensin, endothelin, CCN2, and PDGF, partners in fibroblast activation. Circ Res 106:1675–1680
119. Schiller M, Javelaud D, Mauviel A (2004) TGF-beta-induced SMAD signaling and gene regulation: consequences for extracellular matrix remodeling and wound healing. J Dermatol Sci 35:83–92
120. Frantz S, Hu K, Adamek A, Wolf J, Sallam A, Maier SK, Lonning S, Ling H, Ertl G, Bauersachs J (2008) Transforming growth factor beta inhibition increases mortality and left ventricular dilatation after myocardial infarction. Basic Res Cardiol 103:485–492
121. Ikeuchi M, Tsutsui H, Shiomi T, Matsusaka H, Matsushima S, Wen J, Kubota T, Takeshita A (2004) Inhibition of TGF-beta signaling exacerbates early cardiac dysfunction but prevents late remodeling after infarction. Cardiovasc Res 64:526–535
122. Rosenkranz S, Flesch M, Amann K, Haeuseler C, Kilter H, Seeland U, Schluter KD, Bohm M (2002) Alterations of beta-adrenergic signaling and cardiac hypertrophy in transgenic mice overexpressing TGF-beta(1). Am J Physiol Heart Circ Physiol 283:H1253–H1262
123. Lim JY, Park SJ, Hwang HY, Park EJ, Nam JH, Kim J, Park SI (2005) TGF-beta1 induces cardiac hypertrophic responses via PKC-dependent ATF-2 activation. J Mol Cell Cardiol 39: 627–636
124. Regan JA, Mauro AG, Carbone S, Marchetti C, Gill R, Mezzaroma E, Valle Raleigh J, Salloum FN, Van Tassell BW, Abbate A et al (2015) A mouse model of heart failure with preserved ejection fraction due to chronic infusion of a low subpressor dose of angiotensin II. Am J Physiol Heart Circ Physiol 309:H771–H778
125. Sun Y, Ramires FJ, Weber KT (1997) Fibrosis of atria and great vessels in response to angiotensin II or aldosterone infusion. Cardiovasc Res 35:138–147
126. Shibasaki Y, Nishiue T, Masaki H, Tamura K, Matsumoto N, Mori Y, Nishikawa M, Matsubara H, Iwasaka T (2005) Impact of the angiotensin II receptor antagonist, losartan, on myocardial fibrosis in patients with end-stage renal disease: assessment by ultrasonic integrated backscatter and biochemical markers. Hypertens Res 28:787–795
127. Siddesha JM, Valente AJ, Sakamuri SS, Yoshida T, Gardner JD, Somanna N, Takahashi C, Noda M, Chandrasekar B (2013) Angiotensin II stimulates cardiac fibroblast migration via the differential regulation of matrixins and RECK. J Mol Cell Cardiol 65:9–18
128. Sadoshima J, Xu Y, Slayter HS, Izumo S (1993) Autocrine release of angiotensin II mediates stretch-induced hypertrophy of cardiac myocytes in vitro. Cell 75:977–984
129. Rosenkranz S (2004) TGF-beta1 and angiotensin networking in cardiac remodeling. Cardiovasc Res 63:423–432
130. Campbell SE, Katwa LC (1997) Angiotensin II stimulated expression of transforming growth factor-beta1 in cardiac fibroblasts and myofibroblasts. J Mol Cell Cardiol 29:1947–1958

131. Gray MO, Long CS, Kalinyak JE, Li HT, Karliner JS (1998) Angiotensin II stimulates cardiac myocyte hypertrophy via paracrine release of TGF-beta 1 and endothelin-1 from fibroblasts. Cardiovasc Res 40:352–363
132. Wenzel S, Taimor G, Piper HM, Schluter KD (2001) Redox-sensitive intermediates mediate angiotensin II-induced p38 MAP kinase activation, AP-1 binding activity, and TGF-beta expression in adult ventricular cardiomyocytes. FASEB J 15:2291–2293
133. Schultz JEJ, Witt SA, Glascock BJ, Nieman ML, Reiser PJ, Nix SL, Kimball TR, Doetschman T (2002) TGF-beta1 mediates the hypertrophic cardiomyocyte growth induced by angiotensin II. J Clin Invest 109:787–796
134. Piacentini L, Gray M, Honbo NY, Chentoufi J, Bergman M, Karliner JS (2000) Endothelin-1 stimulates cardiac fibroblast proliferation through activation of protein kinase C. J Mol Cell Cardiol 32:565–576
135. Shi-Wen X, Denton CP, Dashwood MR, Holmes AM, Bou-Gharios G, Pearson JD, Black CM, Abraham DJ (2001) Fibroblast matrix gene expression and connective tissue remodeling: role of endothelin-1. J Invest Dermatol 116:417–425
136. Hafizi S, Wharton J, Chester AH, Yacoub MH (2004) Profibrotic effects of endothelin-1 via the ETA receptor in cultured human cardiac fibroblasts. Cell Physiol Biochem 14:285–292
137. Friedrichsen S, Heuer H, Christ S, Cuthill D, Bauer K, Raivich G (2005) Gene expression of connective tissue growth factor in adult mouse. Growth Factors 23:43–53
138. Koitabashi N, Arai M, Kogure S, Niwano K, Watanabe A, Aoki Y, Maeno T, Nishida T, Kubota S, Takigawa M et al (2007) Increased connective tissue growth factor relative to brain natriuretic peptide as a determinant of myocardial fibrosis. Hypertension 49:1120–1127
139. Ohnishi H, Oka T, Kusachi S, Nakanishi T, Takeda K, Nakahama M, Doi M, Murakami T, Ninomiya Y, Takigawa M et al (1998) Increased expression of connective tissue growth factor in the infarct zone of experimentally induced myocardial infarction in rats. J Mol Cell Cardiol 30:2411–2422
140. Zhang J, Chang L, Chen C, Zhang M, Luo Y, Hamblin M, Villacorta L, Xiong JW, Chen YE, Zhang J et al (2011) Rad GTPase inhibits cardiac fibrosis through connective tissue growth factor. Cardiovasc Res 91:90–98
141. Panek AN, Posch MG, Alenina N, Ghadge SK, Erdmann B, Popova E, Perrot A, Geier C, Dietz R, Morano I et al (2009) Connective tissue growth factor overexpression in cardiomyocytes promotes cardiac hypertrophy and protection against pressure overload. PLoS One 4:e6743
142. Zhang Y, Feng XH, Derynck R (1998) Smad3 and Smad4 cooperate with c-Jun/c-Fos to mediate TGF-beta-induced transcription. Nature 394:909–913
143. Ma ZG, Yuan YP, Wu HM, Zhang X, Tang QZ (2018) Cardiac fibrosis: new insights into the pathogenesis. Int J Biol Sci 14:1645–1657
144. O'Connor JW, Gomez EW (2013) Cell adhesion and shape regulate TGF-beta1-induced epithelial-myofibroblast transition via MRTF-A signaling. PLoS One 8:e83188
145. Esnault C, Stewart A, Gualdrini F, East P, Horswell S, Matthews N, Treisman R (2014) Rho-actin signaling to the MRTF coactivators dominates the immediate transcriptional response to serum in fibroblasts. Genes Dev 28:943–958
146. Small EM, Thatcher JE, Sutherland LB, Kinoshita H, Gerard RD, Richardson JA, Dimaio JM, Sadek H, Kuwahara K, Olson EN (2010) Myocardin-related transcription factor-a controls myofibroblast activation and fibrosis in response to myocardial infarction. Circ Res 107:294–304
147. Gonzalez Bosc LV, Layne JJ, Nelson MT, Hill-Eubanks DC (2005) Nuclear factor of activated T cells and serum response factor cooperatively regulate the activity of an alpha-actin intronic enhancer. J Biol Chem 280:26113–26120
148. Molkentin JD, Lu JR, Antos CL, Markham B, Richardson J, Robbins J, Grant SR, Olson EN (1998) A calcineurin-dependent transcriptional pathway for cardiac hypertrophy. Cell 93:215–228

149. Ali SR, Ranjbarvaziri S, Talkhabi M, Zhao P, Subat A, Hojjat A, Kamran P, Muller AM, Volz KS, Tang Z et al (2014) Developmental heterogeneity of cardiac fibroblasts does not predict pathological proliferation and activation. Circ Res 115:625–635
150. Compton LA, Potash DA, Mundell NA, Barnett JV (2006) Transforming growth factor-beta induces loss of epithelial character and smooth muscle cell differentiation in epicardial cells. Dev Dyn 235:82–93
151. Dronkers E, Wauters MMM, Goumans MJ, Smits AM (2020) Epicardial TGFbeta and BMP signaling in cardiac regeneration: what lesson can we learn from the developing heart? Biomolecules 10
152. Vaughan MB, Howard EW, Tomasek JJ (2000) Transforming growth factor-beta1 promotes the morphological and functional differentiation of the myofibroblast. Exp Cell Res 257:180–189
153. Wang J, Chen H, Seth A, McCulloch CA (2003) Mechanical force regulation of myofibroblast differentiation in cardiac fibroblasts. Am J Physiol Heart Circ Physiol 285:H1871–H1881
154. Yong KW, Li Y, Huang G, Lu TJ, Safwani WK, Pingguan-Murphy B, Xu F (2015) Mechanoregulation of cardiac myofibroblast differentiation: implications for cardiac fibrosis and therapy. Am J Physiol Heart Circ Physiol 309:H532–H542
155. Di Gregorio J, Robuffo I, Spalletta S, Giambuzzi G, De Iuliis V, Toniato E, Martinotti S, Conti P, Flati V (2020) The epithelial-to-mesenchymal transition as a possible therapeutic target in fibrotic disorders. Front Cell Dev Biol 8:607483
156. Stone RC, Pastar I, Ojeh N, Chen V, Liu S, Garzon KI, Tomic-Canic M (2016) Epithelial-mesenchymal transition in tissue repair and fibrosis. Cell Tissue Res 365:495–506
157. Zhou B, Honor LB, He H, Ma Q, Oh JH, Butterfield C, Lin RZ, Melero-Martin JM, Dolmatova E, Duffy HS et al (2011) Adult mouse epicardium modulates myocardial injury by secreting paracrine factors. J Clin Invest 121:1894–1904
158. Zeisberg EM, Tarnavski O, Zeisberg M, Dorfman AL, McMullen JR, Gustafsson E, Chandraker A, Yuan X, Pu WT, Roberts AB et al (2007) Endothelial-to-mesenchymal transition contributes to cardiac fibrosis. Nat Med 13:952–961
159. van Amerongen MJ, Bou-Gharios G, Popa E, van Ark J, Petersen AH, van Dam GM, van Luyn MJ, Harmsen MC (2008) Bone marrow-derived myofibroblasts contribute functionally to scar formation after myocardial infarction. J Pathol 214:377–386
160. Moore-Morris T, Tallquist MD, Evans SM (2014) Sorting out where fibroblasts come from. Circ Res 115:602–604
161. Kong P, Christia P, Saxena A, Su Y, Frangogiannis NG (2013) Lack of specificity of fibroblast-specific protein 1 in cardiac remodeling and fibrosis. Am J Physiol Heart Circ Physiol 305:H1363–H1372
162. Moore-Morris T, Guimaraes-Camboa N, Banerjee I, Zambon AC, Kisseleva T, Velayoudon A, Stallcup WB, Gu Y, Dalton ND, Cedenilla M et al (2014) Resident fibroblast lineages mediate pressure overload-induced cardiac fibrosis. J Clin Invest 124:2921–2934
163. Alva JA, Zovein AC, Monvoisin A, Murphy T, Salazar A, Harvey NL, Carmeliet P, Iruela-Arispe ML (2006) VE-Cadherin-Cre-recombinase transgenic mouse: a tool for lineage analysis and gene deletion in endothelial cells. Dev Dyn 235:759–767
164. Kilani B, Gourdou-Latyszenok V, Guy A, Bats ML, Peghaire C, Parrens M, Renault MA, Duplaa C, Villeval JL, Rautou PE et al (2019) Comparison of endothelial promoter efficiency and specificity in mice reveals a subset of Pdgfb-positive hematopoietic cells. J Thromb Haemost 17:827–840
165. Payne S, De Val S, Neal A (2018) Endothelial-specific Cre mouse models. Arterioscler Thromb Vasc Biol 38:2550–2561
166. Leslie KO, Taatjes DJ, Schwarz J, vonTurkovich M, Low RB (1991) Cardiac myofibroblasts express alpha smooth muscle actin during right ventricular pressure overload in the rabbit. Am J Pathol 139:207–216
167. Ross RS, Borg TK (2001) Integrins and the myocardium. Circ Res 88:1112–1119

168. Herum KM, Lunde IG, Skrbic B, Florholmen G, Behmen D, Sjaastad I, Carlson CR, Gomez MF, Christensen G (2013) Syndecan-4 signaling via NFAT regulates extracellular matrix production and cardiac myofibroblast differentiation in response to mechanical stress. J Mol Cell Cardiol 54:73–81
169. Follonier Castella L, Gabbiani G, McCulloch CA, Hinz B (2010) Regulation of myofibroblast activities: calcium pulls some strings behind the scene. Exp Cell Res 316:2390–2401
170. Li Q, Xu Y, Li X, Guo Y, Liu G (2012) Inhibition of Rho-kinase ameliorates myocardial remodeling and fibrosis in pressure overload and myocardial infarction: role of TGF-beta1-TAK1. Toxicol Lett 211:91–97
171. Clemente CF, Tornatore TF, Theizen TH, Deckmann AC, Pereira TC, Lopes-Cendes I, Souza JR, Franchini KG (2007) Targeting focal adhesion kinase with small interfering RNA prevents and reverses load-induced cardiac hypertrophy in mice. Circ Res 101:1339–1348
172. Sarrazy V, Koehler A, Chow ML, Zimina E, Li CX, Kato H, Caldarone CA, Hinz B (2014) Integrins alphavbeta5 and alphavbeta3 promote latent TGF-beta1 activation by human cardiac fibroblast contraction. Cardiovasc Res 102:407–417
173. Kong P, Shinde AV, Su Y, Russo I, Chen B, Saxena A, Conway SJ, Graff JM, Frangogiannis NG (2018) Opposing actions of fibroblast and cardiomyocyte Smad3 signaling in the infarcted myocardium. Circulation 137:707–724
174. Thum T, Gross C, Fiedler J, Fischer T, Kissler S, Bussen M, Galuppo P, Just S, Rottbauer W, Frantz S et al (2008) MicroRNA-21 contributes to myocardial disease by stimulating MAP kinase signalling in fibroblasts. Nature 456:980–984
175. Furchgott RF, Zawadzki JV (1980) The obligatory role of endothelial cells in the relaxation of arterial smooth muscle by acetylcholine. Nature 288:373–376
176. Li YS, Haga JH, Chien S (2005) Molecular basis of the effects of shear stress on vascular endothelial cells. J Biomech 38:1949–1971
177. Topper JN, Cai J, Falb D, Gimbrone MA Jr (1996) Identification of vascular endothelial genes differentially responsive to fluid mechanical stimuli: cyclooxygenase-2, manganese superoxide dismutase, and endothelial cell nitric oxide synthase are selectively up-regulated by steady laminar shear stress. Proc Natl Acad Sci U S A 93:10417–10422
178. Boo YC, Sorescu G, Boyd N, Shiojima I, Walsh K, Du J, Jo H (2002) Shear stress stimulates phosphorylation of endothelial nitric-oxide synthase at Ser1179 by Akt-independent mechanisms: role of protein kinase A. J Biol Chem 277:3388–3396
179. Fisslthaler B, Dimmeler S, Hermann C, Busse R, Fleming I (2000) Phosphorylation and activation of the endothelial nitric oxide synthase by fluid shear stress. Acta Physiol Scand 168:81–88
180. Cellek S, Moncada S (1997) Nitrergic modulation of cholinergic responses in the opossum lower oesophageal sphincter. Br J Pharmacol 122:1043–1046
181. de Wit C, Jahrbeck B, Schafer C, Bolz SS, Pohl U (1998) Nitric oxide opposes myogenic pressure responses predominantly in large arterioles in vivo. Hypertension 31:787–794
182. Sakamoto N, Ohashi T, Sato M (2006) Effect of fluid shear stress on migration of vascular smooth muscle cells in cocultured model. Ann Biomed Eng 34:408–415
183. Wang HQ, Huang LX, Qu MJ, Yan ZQ, Liu B, Shen BR, Jiang ZL (2006) Shear stress protects against endothelial regulation of vascular smooth muscle cell migration in a coculture system. Endothelium 13:171–180
184. Badimon L, Badimon JJ, Penny W, Webster MW, Chesebro JH, Fuster V (1992) Endothelium and atherosclerosis. J Hypertens Suppl 10:S43–S50
185. Blann AD, Lip GY (1998) The endothelium in atherothrombotic disease: assessment of function, mechanisms and clinical implications. Blood Coagul Fibrinolysis 9:297–306
186. Quyyumi AA (1998) Endothelial function in health and disease: new insights into the genesis of cardiovascular disease. Am J Med 105:32S–39S
187. Dimmeler S, Assmus B, Hermann C, Haendeler J, Zeiher AM (1998) Fluid shear stress stimulates phosphorylation of Akt in human endothelial cells: involvement in suppression of apoptosis. Circ Res 83:334–341

188. Dimmeler S, Hermann C, Galle J, Zeiher AM (1999) Upregulation of superoxide dismutase and nitric oxide synthase mediates the apoptosis-suppressive effects of shear stress on endothelial cells. Arterioscler Thromb Vasc Biol 19:656–664
189. Hecker M, Mulsch A, Bassenge E, Forstermann U, Busse R (1994) Subcellular localization and characterization of nitric oxide synthase(s) in endothelial cells: physiological implications. Biochem J 299(Pt 1):247–252
190. Cornhill JF, Roach MR (1976) A quantitative study of the localization of atherosclerotic lesions in the rabbit aorta. Atherosclerosis 23:489–501
191. Karino T, Motomiya M (1983) Flow visualization in isolated transparent natural blood vessels. Biorheology 20:119–127
192. Zarins CK, Giddens DP, Bharadvaj BK, Sottiurai VS, Mabon RF, Glagov S (1983) Carotid bifurcation atherosclerosis. Quantitative correlation of plaque localization with flow velocity profiles and wall shear stress. Circ Res 53:502–514
193. VanderLaan PA, Reardon CA, Getz GS (2004) Site specificity of atherosclerosis: site-selective responses to atherosclerotic modulators. Arterioscler Thromb Vasc Biol 24:12–22
194. Davies PF (2009) Hemodynamic shear stress and the endothelium in cardiovascular pathophysiology. Nat Clin Pract Cardiovasc Med 6:16–26
195. Huo Y, Wischgoll T, Kassab GS (2007) Flow patterns in three-dimensional porcine epicardial coronary arterial tree. Am J Physiol Heart Circ Physiol 293:H2959–H2970
196. Mahmoud MM, Serbanovic-Canic J, Feng S, Souilhol C, Xing R, Hsiao S, Mammoto A, Chen J, Ariaans M, Francis SE et al (2017) Shear stress induces endothelial-to-mesenchymal transition via the transcription factor Snail. Sci Rep 7:3375
197. Mahler GJ, Frendl CM, Cao Q, Butcher JT (2014) Effects of shear stress pattern and magnitude on mesenchymal transformation and invasion of aortic valve endothelial cells. Biotechnol Bioeng 111:2326–2337
198. von Offenberg Sweeney N, Cummins PM, Birney YA, Cullen JP, Redmond EM, Cahill PA (2004) Cyclic strain-mediated regulation of endothelial matrix metalloproteinase-2 expression and activity. Cardiovasc Res 63:625–634
199. Osawa M, Masuda M, Kusano K, Fujiwara K (2002) Evidence for a role of platelet endothelial cell adhesion molecule-1 in endothelial cell mechanosignal transduction: is it a mechanoresponsive molecule? J Cell Biol 158:773–785
200. Kapur NK, Deming CB, Kapur S, Bian C, Champion HC, Donahue JK, Kass DA, Rade JJ (2007) Hemodynamic modulation of endocardial thromboresistance. Circulation 115:67–75
201. Qiu Y, Tarbell JM (2000) Interaction between wall shear stress and circumferential strain affects endothelial cell biochemical production. J Vasc Res 37:147–157
202. Sporkova A, Ghosh S, Al-Hasani J, Hecker M (2021) Lin11-Isl1-Mec3 domain proteins as mechanotransducers in endothelial and vascular smooth muscle cells. Front Physiol 12
203. Suresh Babu S, Wojtowicz A, Freichel M, Birnbaumer L, Hecker M, Cattaruzza M (2012) Mechanism of stretch-induced activation of the mechanotransducer zyxin in vascular cells. Sci Signal 5:ra91
204. Wojtowicz A, Babu SS, Li L, Gretz N, Hecker M, Cattaruzza M (2010) Zyxin mediation of stretch-induced gene expression in human endothelial cells. Circ Res 107:898–902
205. Tzima E, Irani-Tehrani M, Kiosses WB, Dejana E, Schultz DA, Engelhardt B, Cao G, DeLisser H, Schwartz MA (2005) A mechanosensory complex that mediates the endothelial cell response to fluid shear stress. Nature 437:426–431
206. Cunningham KS, Gotlieb AI (2005) The role of shear stress in the pathogenesis of atherosclerosis. Lab Invest 85:9–23
207. Anwar MA, Shalhoub J, Lim CS, Gohel MS, Davies AH (2012) The effect of pressure-induced mechanical stretch on vascular wall differential gene expression. J Vasc Res 49:463–478
208. Clark CB, McKnight NL, Frangos JA (2002) Strain and strain rate activation of G proteins in human endothelial cells. Biochem Biophys Res Commun 299:258–262

209. Islam S, Bostrom KI, Di Carlo D, Simmons CA, Tintut Y, Yao Y, Hsu JJ (2021) The mechanobiology of endothelial-to-mesenchymal transition in cardiovascular disease. Front Physiol 12:734215
210. Takimoto E, Champion HC, Li M, Belardi D, Ren S, Rodriguez ER, Bedja D, Gabrielson KL, Wang Y, Kass DA (2005) Chronic inhibition of cyclic GMP phosphodiesterase 5A prevents and reverses cardiac hypertrophy. Nat Med 11:214–222
211. Bishu K, Hamdani N, Mohammed SF, Kruger M, Ohtani T, Ogut O, Brozovich FV, Burnett JC Jr, Linke WA, Redfield MM (2011) Sildenafil and B-type natriuretic peptide acutely phosphorylate titin and improve diastolic distensibility in vivo. Circulation 124:2882–2891
212. Calderone A, Thaik CM, Takahashi N, Chang DL, Colucci WS (1998) Nitric oxide, atrial natriuretic peptide, and cyclic GMP inhibit the growth-promoting effects of norepinephrine in cardiac myocytes and fibroblasts. J Clin Invest 101:812–818
213. Westermann D, Lindner D, Kasner M, Zietsch C, Savvatis K, Escher F, von Schlippenbach J, Skurk C, Steendijk P, Riad A et al (2011) Cardiac inflammation contributes to changes in the extracellular matrix in patients with heart failure and normal ejection fraction. Circ Heart Fail 4:44–52
214. Moller K, Adolph O, Grunow J, Elrod J, Popa M, Ghosh S, Schwarz M, Schwale C, Grassle S, Huck V et al (2015) Mechanism and functional impact of CD40 ligand-induced von Willebrand factor release from endothelial cells. Thromb Haemost 113:1095–1108
215. Popa M, Tahir S, Elrod J, Kim SH, Leuschner F, Kessler T, Bugert P, Pohl U, Wagner AH, Hecker M (2018) Role of CD40 and ADAMTS13 in von Willebrand factor-mediated endothelial cell-platelet-monocyte interaction. Proc Natl Acad Sci U S A 115:E5556–E5565
216. Sultan CS, Weitnauer M, Turinsky M, Kessler T, Brune M, Gleissner CA, Leuschner F, Wagner AH, Hecker M (2020) Functional association of a CD40 gene single-nucleotide polymorphism with the pathogenesis of coronary heart disease. Cardiovasc Res 116:1214–1225
217. Tahir S, Wagner AH, Dietzel S, Mannell H, Pircher J, Weckbach LT, Hecker M, Pohl U (2020) Endothelial CD40 mediates microvascular von Willebrand factor-dependent platelet adhesion inducing inflammatory venothrombosis in ADAMTS13 knockout mice. Thromb Haemost 120:466–476
218. Egorova AD, Van der Heiden K, Van de Pas S, Vennemann P, Poelma C, DeRuiter MC, Goumans MJ, Gittenberger-de Groot AC, ten Dijke P, Poelmann RE et al (2011) Tgfbeta/Alk5 signaling is required for shear stress induced klf2 expression in embryonic endothelial cells. Dev Dyn 240:1670–1680
219. Ten Dijke P, Egorova AD, Goumans MJ, Poelmann RE, Hierck BP (2012) TGF-beta signaling in endothelial-to-mesenchymal transition: the role of shear stress and primary cilia. Sci Signal 5:pt2
220. Vion AC, Alt S, Klaus-Bergmann A, Szymborska A, Zheng T, Perovic T, Hammoutene A, Oliveira MB, Bartels-Klein E, Hollfinger I et al (2018) Primary cilia sensitize endothelial cells to BMP and prevent excessive vascular regression. J Cell Biol 217:1651–1665
221. Egorova AD, Khedoe PP, Goumans MJ, Yoder BK, Nauli SM, ten Dijke P, Poelmann RE, Hierck BP (2011) Lack of primary cilia primes shear-induced endothelial-to-mesenchymal transition. Circ Res 108:1093–1101
222. Van der Heiden K, Hierck BP, Krams R, de Crom R, Cheng C, Baiker M, Pourquie MJ, Alkemade FE, DeRuiter MC, Gittenberger-de Groot AC et al (2008) Endothelial primary cilia in areas of disturbed flow are at the base of atherosclerosis. Atherosclerosis 196:542–550
223. Howard AB, Alexander RW, Nerem RM, Griendling KK, Taylor WR (1997) Cyclic strain induces an oxidative stress in endothelial cells. Am J Physiol 272:C421–C427
224. Gryglewski RJ, Palmer RM, Moncada S (1986) Superoxide anion is involved in the breakdown of endothelium-derived vascular relaxing factor. Nature 320:454–456
225. Jackson CJ, Nguyen M (1997) Human microvascular endothelial cells differ from macrovascular endothelial cells in their expression of matrix metalloproteinases. Int J Biochem Cell Biol 29:1167–1177

226. Grafe M, Zakrzewicz A, Graf K, Gaehtgens P, Fleck E (1999) [Differential reaction of human cardiac macro- and microvascular endothelial cells with respect to leucocyte adhesion and exposition to atherogenic lipoproteins]. Zeitschrift fur Kardiologie 88:828–837
227. Peng Z, Shu B, Zhang Y, Wang M (2019) Endothelial response to pathophysiological stress. Arterioscler Thromb Vasc Biol 39:e233–e243
228. Dimmeler S, Fleming I, Fisslthaler B, Hermann C, Busse R, Zeiher AM (1999) Activation of nitric oxide synthase in endothelial cells by Akt-dependent phosphorylation. Nature 399:601–605
229. Ley K, Laudanna C, Cybulsky MI, Nourshargh S (2007) Getting to the site of inflammation: the leukocyte adhesion cascade updated. Nat Rev Immunol 7:678–689
230. Mammoto A, Mammoto T, Ingber DE (2012) Mechanosensitive mechanisms in transcriptional regulation. J Cell Sci 125:3061–3073
231. Orr AW, Murphy-Ullrich JE (2004) Regulation of endothelial cell function BY FAK and PYK2. Front Biosci 9:1254–1266
232. True AL, Rahman A, Malik AB (2000) Activation of NF-kappaB induced by H(2)O(2) and TNF-alpha and its effects on ICAM-1 expression in endothelial cells. Am J Physiol Lung Cell Mol Physiol 279:L302–L311
233. Wassmann S, Stumpf M, Strehlow K, Schmid A, Schieffer B, Bohm M, Nickenig G (2004) Interleukin-6 induces oxidative stress and endothelial dysfunction by overexpression of the angiotensin II type 1 receptor. Circ Res 94:534–541
234. Zhang C, Hein TW, Wang W, Ren Y, Shipley RD, Kuo L (2006) Activation of JNK and xanthine oxidase by TNF-alpha impairs nitric oxide-mediated dilation of coronary arterioles. J Mol Cell Cardiol 40:247–257
235. Segers VFM, Brutsaert DL, De Keulenaer GW (2018) Cardiac remodeling: endothelial cells have more to say than just no. Front Physiol 9:382
236. Tombor LS, John D, Glaser SF, Luxan G, Forte E, Furtado M, Rosenthal N, Baumgarten N, Schulz MH, Wittig J et al (2021) Single cell sequencing reveals endothelial plasticity with transient mesenchymal activation after myocardial infarction. Nat Commun 12:681
237. Bischoff J (2019) Endothelial-to-mesenchymal transition. Circ Res 124:1163–1165
238. Eisenberg LM, Markwald RR (1995) Molecular regulation of atrioventricular valvuloseptal morphogenesis. Circ Res 77:1–6
239. Goumans MJ, van Zonneveld AJ, ten Dijke P (2008) Transforming growth factor beta-induced endothelial-to-mesenchymal transition: a switch to cardiac fibrosis? Trends Cardiovasc Med 18:293–298
240. Kim AJ, Alfieri CM, Yutzey KE (2019) Endothelial cell lineage analysis does not provide evidence for EMT in adult valve homeostasis and disease. Anat Rec 302:125–135
241. Sanchez-Duffhues G, Garcia de Vinuesa A, Ten Dijke P (2018) Endothelial-to-mesenchymal transition in cardiovascular diseases: developmental signaling pathways gone awry. Dev Dyn 247:492–508
242. Kovacic JC, Dimmeler S, Harvey RP, Finkel T, Aikawa E, Krenning G, Baker AH (2019) Endothelial to mesenchymal transition in cardiovascular disease: JACC state-of-the-art review. J Am Coll Cardiol 73:190–209
243. Arciniegas E, Frid MG, Douglas IS, Stenmark KR (2007) Perspectives on endothelial-to-mesenchymal transition: potential contribution to vascular remodeling in chronic pulmonary hypertension. Am J Physiol Lung Cell Mol Physiol 293:L1–L8
244. Kovacic JC, Mercader N, Torres M, Boehm M, Fuster V (2012) Epithelial-to-mesenchymal and endothelial-to-mesenchymal transition: from cardiovascular development to disease. Circulation 125:1795–1808
245. Nieto MA, Huang RY, Jackson RA, Thiery JP (2016) Emt: 2016. Cell 166:21–45
246. Chen PY, Schwartz MA, Simons M (2020) Endothelial-to-mesenchymal transition, vascular inflammation, and atherosclerosis. Front Cardiovasc Med 7:53
247. Gong H, Lyu X, Wang Q, Hu M, Zhang X (2017) Endothelial to mesenchymal transition in the cardiovascular system. Life Sci 184:95–102

248. Piera-Velazquez S, Jimenez SA (2019) Endothelial to mesenchymal transition: role in physiology and in the pathogenesis of human diseases. Physiol Rev 99:1281–1324
249. Hulshoff MS, Xu X, Krenning G, Zeisberg EM (2018) Epigenetic regulation of endothelial-to-mesenchymal transition in chronic heart disease. Arterioscler Thromb Vasc Biol 38:1986–1996
250. Madri JA, Pratt BM, Tucker AM (1988) Phenotypic modulation of endothelial cells by transforming growth factor-beta depends upon the composition and organization of the extracellular matrix. J Cell Biol 106:1375–1384
251. Pintavorn P, Ballermann BJ (1997) TGF-beta and the endothelium during immune injury. Kidney Int 51:1401–1412
252. Frater-Schroder M, Muller G, Birchmeier W, Bohlen P (1986) Transforming growth factor-beta inhibits endothelial cell proliferation. Biochem Biophys Res Commun 137:295–302
253. Goumans MJ, Ten Dijke P (2018) TGF-beta signaling in control of cardiovascular function. Cold Spring Harb Perspect Biol 10
254. van Meeteren LA, ten Dijke P (2012) Regulation of endothelial cell plasticity by TGF-beta. Cell Tissue Res 347:177–186
255. Derada Troletti C, Fontijn RD, Gowing E, Charabati M, van Het Hof B, Didouh I, van der Pol SMA, Geerts D, Prat A, van Horssen J et al (2019) Inflammation-induced endothelial to mesenchymal transition promotes brain endothelial cell dysfunction and occurs during multiple sclerosis pathophysiology. Cell Death Dis 10:45
256. Sanchez-Duffhues G, Garcia de Vinuesa A, van de Pol V, Geerts ME, de Vries MR, Janson SG, van Dam H, Lindeman JH, Goumans MJ, Ten Dijke P (2019) Inflammation induces endothelial-to-mesenchymal transition and promotes vascular calcification through downregulation of BMPR2. J Pathol 247:333–346
257. Alonso-Herranz L, Sahun-Espanol A, Paredes A, Gonzalo P, Gkontra P, Nunez V, Clemente C, Cedenilla M, Villalba-Orero M, Inserte J et al (2020) Macrophages promote endothelial-to-mesenchymal transition via MT1-MMP/TGFbeta1 after myocardial infarction. Elife 9
258. Chien S (2007) Mechanotransduction and endothelial cell homeostasis: the wisdom of the cell. Am J Physiol Heart Circ Physiol 292:H1209–H1224
259. Mahmoud M, Mayer M, Cancel LM, Bartosch AM, Mathews R, Tarbell JM (2021) The glycocalyx core protein Glypican 1 protects vessel wall endothelial cells from stiffness-mediated dysfunction and disease. Cardiovasc Res 117:1592–1605
260. Tian Y, Gawlak G, O'Donnell JJ 3rd, Birukova AA, Birukov KG (2016) Activation of vascular endothelial growth factor (VEGF) receptor 2 mediates endothelial permeability caused by cyclic stretch. J Biol Chem 291:10032–10045
261. Dahal S, Huang P, Murray BT, Mahler GJ (2017) Endothelial to mesenchymal transformation is induced by altered extracellular matrix in aortic valve endothelial cells. J Biomed Mater Res A 105:2729–2741
262. Shen J, Xie Y, Liu Z, Zhang S, Wang Y, Jia L, Wang Y, Cai Z, Ma H, Xiang M (2018) Increased myocardial stiffness activates cardiac microvascular endothelial cell via VEGF paracrine signaling in cardiac hypertrophy. J Mol Cell Cardiol 122:140–151
263. Zhong A, Mirzaei Z, Simmons CA (2018) The roles of matrix stiffness and ss-catenin signaling in endothelial-to-mesenchymal transition of aortic valve endothelial cells. Cardiovasc Eng Technol 9:158–167
264. Van Linthout S, Tschope C (2017) Inflammation—cause or consequence of heart failure or both? Curr Heart Fail Rep 14:251–265
265. Dobaczewski M, Gonzalez-Quesada C, Frangogiannis NG (2010) The extracellular matrix as a modulator of the inflammatory and reparative response following myocardial infarction. J Mol Cell Cardiol 48:504–511
266. Satoh M, Minami Y, Takahashi Y, Nakamura M (2008) Immune modulation: role of the inflammatory cytokine cascade in the failing human heart. Curr Heart Fail Rep 5:69–74

267. Lindsey ML, Mann DL, Entman ML, Spinale FG (2003) Extracellular matrix remodeling following myocardial injury. Ann Med 35:316–326
268. van der Laan AM, Nahrendorf M, Piek JJ (2013) Republished: healing and adverse remodelling after acute myocardial infarction: role of the cellular immune response. Postgrad Med J 89:52–58
269. Frangogiannis NG (2008) The immune system and cardiac repair. Pharmacol Res 58:88–111
270. Kelly JP, Mentz RJ, Mebazaa A, Voors AA, Butler J, Roessig L, Fiuzat M, Zannad F, Pitt B, O'Connor CM et al (2015) Patient selection in heart failure with preserved ejection fraction clinical trials. J Am Coll Cardiol 65:1668–1682
271. Hara H, Takeda N, Komuro I (2017) Pathophysiology and therapeutic potential of cardiac fibrosis. Inflamm Regen 37:13
272. Jackson AO, Zhang J, Jiang Z, Yin K (2017) Endothelial-to-mesenchymal transition: a novel therapeutic target for cardiovascular diseases. Trends Cardiovasc Med 27:383–393
273. Lu X, Gong J, Dennery PA, Yao H (2019) Endothelial-to-mesenchymal transition: pathogenesis and therapeutic targets for chronic pulmonary and vascular diseases. Biochem Pharmacol 168:100–107
274. Liu Y, Gao L, Zhao X, Guo S, Liu Y, Li R, Liang C, Li L, Dong J, Li L et al (2018) Saikosaponin A protects from pressure overload-induced cardiac fibrosis via inhibiting fibroblast activation or endothelial cell EndMT. Int J Biol Sci 14:1923–1934
275. Khan R, Sheppard R (2006) Fibrosis in heart disease: understanding the role of transforming growth factor-beta in cardiomyopathy, valvular disease and arrhythmia. Immunology 118:10–24
276. Guo X, Wang XF (2009) Signaling cross-talk between TGF-beta/BMP and other pathways. Cell Res 19:71–88
277. Wu J, Jackson-Weaver O, Xu J (2018) The TGFbeta superfamily in cardiac dysfunction. Acta Biochim Biophys Sin (Shanghai) 50:323–335
278. Morine KJ, Qiao X, York S, Natov PS, Paruchuri V, Zhang Y, Aronovitz MJ, Karas RH, Kapur NK (2018) Bone morphogenetic protein 9 reduces cardiac fibrosis and improves cardiac function in heart failure. Circulation 138:513–526
279. Wang S, Sun A, Li L, Zhao G, Jia J, Wang K, Ge J, Zou Y (2012) Up-regulation of BMP-2 antagonizes TGF-beta1/ROCK-enhanced cardiac fibrotic signalling through activation of Smurf1/Smad6 complex. J Cell Mol Med 16:2301–2310
280. Derangeon M, Montnach J, Cerpa CO, Jagu B, Patin J, Toumaniantz G, Girardeau A, Huang CLH, Colledge WH, Grace AA et al (2017) Transforming growth factor beta receptor inhibition prevents ventricular fibrosis in a mouse model of progressive cardiac conduction disease. Cardiovasc Res 113:464–474
281. Engebretsen KV, Skardal K, Bjornstad S, Marstein HS, Skrbic B, Sjaastad I, Christensen G, Bjornstad JL, Tonnessen T (2014) Attenuated development of cardiac fibrosis in left ventricular pressure overload by SM16, an orally active inhibitor of ALK5. J Mol Cell Cardiol 76: 148–157
282. Divakaran V, Adrogue J, Ishiyama M, Entman ML, Haudek S, Sivasubramanian N, Mann DL (2009) Adaptive and maladaptive effects of SMAD3 signaling in the adult heart after hemodynamic pressure overloading. Circ Heart Fail 2:633–642
283. Umbarkar P, Singh AP, Gupte M, Verma VK, Galindo CL, Guo Y, Zhang Q, McNamara JW, Force T, Lal H (2019) Cardiomyocyte SMAD4-dependent TGF-beta signaling is essential to maintain adult heart homeostasis. JACC Basic Transl Sci 4:41–53
284. Kulkarni AB, Huh CG, Becker D, Geiser A, Lyght M, Flanders KC, Roberts AB, Sporn MB, Ward JM, Karlsson S (1993) Transforming growth factor beta 1 null mutation in mice causes excessive inflammatory response and early death. Proc Natl Acad Sci U S A 90:770–774
285. Shull MM, Ormsby I, Kier AB, Pawlowski S, Diebold RJ, Yin M, Allen R, Sidman C, Proetzel G, Calvin D et al (1992) Targeted disruption of the mouse transforming growth factor-beta 1 gene results in multifocal inflammatory disease. Nature 359:693–699

286. Kotlarz D, Marquardt B, Baroy T, Lee WS, Konnikova L, Hollizeck S, Magg T, Lehle AS, Walz C, Borggraefe I et al (2018) Human TGF-beta1 deficiency causes severe inflammatory bowel disease and encephalopathy. Nat Genet 50:344–348
287. Swedberg K (1987) Effects of enalapril on mortality in severe congestive heart failure. Results of the Cooperative North Scandinavian Enalapril Survival Study (CONSENSUS). N Engl J Med 316:1429–1435
288. Pitt B, Zannad F, Remme WJ, Cody R, Castaigne A, Perez A, Palensky J, Wittes J (1999) The effect of spironolactone on morbidity and mortality in patients with severe heart failure. Randomized Aldactone Evaluation Study Investigators. N Engl J Med 341:709–717
289. Cleland JG, Tendera M, Adamus J, Freemantle N, Polonski L, Taylor J, Investigators P-C (2006) The perindopril in elderly people with chronic heart failure (PEP-CHF) study. Eur Heart J 27:2338–2345
290. Martin N, Manoharan K, Davies C, Lumbers RT (2021) Beta-blockers and inhibitors of the renin-angiotensin aldosterone system for chronic heart failure with preserved ejection fraction. Cochrane Database Syst Rev 5:CD012721
291. Martin N, Manoharan K, Thomas J, Davies C, Lumbers RT (2018) Beta-blockers and inhibitors of the renin-angiotensin aldosterone system for chronic heart failure with preserved ejection fraction. Cochrane Database Syst Rev 6:CD012721
292. Yusuf S, Pfeffer MA, Swedberg K, Granger CB, Held P, McMurray JJ, Michelson EL, Olofsson B, Ostergren J, Investigators C et al (2003) Effects of candesartan in patients with chronic heart failure and preserved left-ventricular ejection fraction: the CHARM-Preserved Trial. Lancet 362:777–781
293. Lemarie CA, Simeone SM, Nikonova A, Ebrahimian T, Deschenes ME, Coffman TM, Paradis P, Schiffrin EL (2009) Aldosterone-induced activation of signaling pathways requires activity of angiotensin type 1a receptors. Circ Res 105:852–859
294. Stockand JD, Meszaros JG (2003) Aldosterone stimulates proliferation of cardiac fibroblasts by activating Ki-RasA and MAPK1/2 signaling. Am J Physiol Heart Circ Physiol 284:H176–H184
295. Tsai CF, Yang SF, Chu HJ, Ueng KC (2013) Cross-talk between mineralocorticoid receptor/angiotensin II type 1 receptor and mitogen-activated protein kinase pathways underlies aldosterone-induced atrial fibrotic responses in HL-1 cardiomyocytes. Int J Cardiol 169:17–28
296. Leader CJ, Moharram M, Coffey S, Sammut IA, Wilkins GW, Walker RJ (2019) Myocardial global longitudinal strain: an early indicator of cardiac interstitial fibrosis modified by spironolactone, in a unique hypertensive rat model. PLoS One 14:e0220837
297. Lother A, Berger S, Gilsbach R, Rosner S, Ecke A, Barreto F, Bauersachs J, Schutz G, Hein L (2011) Ablation of mineralocorticoid receptors in myocytes but not in fibroblasts preserves cardiac function. Hypertension 57:746–754
298. Nishioka T, Suzuki M, Onishi K, Takakura N, Inada H, Yoshida T, Hiroe M, Imanaka-Yoshida K (2007) Eplerenone attenuates myocardial fibrosis in the angiotensin II-induced hypertensive mouse: involvement of tenascin-C induced by aldosterone-mediated inflammation. J Cardiovasc Pharmacol 49:261–268
299. Rickard AJ, Morgan J, Bienvenu LA, Fletcher EK, Cranston GA, Shen JZ, Reichelt ME, Delbridge LM, Young MJ (2012) Cardiomyocyte mineralocorticoid receptors are essential for deoxycorticosterone/salt-mediated inflammation and cardiac fibrosis. Hypertension 60:1443–1450
300. Edelmann F, Wachter R, Schmidt AG, Kraigher-Krainer E, Colantonio C, Kamke W, Duvinage A, Stahrenberg R, Durstewitz K, Loffler M et al (2013) Effect of spironolactone on diastolic function and exercise capacity in patients with heart failure with preserved ejection fraction: the Aldo-DHF randomized controlled trial. JAMA 309:781–791
301. Pitt B, Pfeffer MA, Assmann SF, Boineau R, Anand IS, Claggett B, Clausell N, Desai AS, Diaz R, Fleg JL et al (2014) Spironolactone for heart failure with preserved ejection fraction. N Engl J Med 370:1383–1392

302. Liang F, Gardner DG (1999) Mechanical strain activates BNP gene transcription through a p38/NF-kappaB-dependent mechanism. J Clin Invest 104:1603–1612
303. Pikkarainen S, Tokola H, Kerkela R, Majalahti-Palviainen T, Vuolteenaho O, Ruskoaho H (2003) Endothelin-1-specific activation of B-type natriuretic peptide gene via p38 mitogen-activated protein kinase and nuclear ETS factors. J Biol Chem 278:3969–3975
304. Kerkela R, Ulvila J, Magga J (2015) Natriuretic peptides in the regulation of cardiovascular physiology and metabolic events. J Am Heart Assoc 4:e002423
305. O'Connor CM, Starling RC, Hernandez AF, Armstrong PW, Dickstein K, Hasselblad V, Heizer GM, Komajda M, Massie BM, McMurray JJ et al (2011) Effect of nesiritide in patients with acute decompensated heart failure. N Engl J Med 365:32–43
306. Solomon SD, McMurray JJV, Anand IS, Ge J, Lam CSP, Maggioni AP, Martinez F, Packer M, Pfeffer MA, Pieske B et al (2019) Angiotensin-neprilysin inhibition in heart failure with preserved ejection fraction. N Engl J Med 381:1609–1620
307. Charles CJ, Espiner EA, Nicholls MG, Richards AM, Yandle TG, Protter A, Kosoglou T (1996) Clearance receptors and endopeptidase 24.11: equal role in natriuretic peptide metabolism in conscious sheep. Am J Physiol 271:R373–R380
308. Burke RM, Lighthouse JK, Mickelsen DM, Small EM (2019) Sacubitril/valsartan decreases cardiac fibrosis in left ventricle pressure overload by restoring PKG signaling in cardiac fibroblasts. Circ Heart Fail 12:e005565
309. McMurray JJ, Packer M, Desai AS, Gong J, Lefkowitz MP, Rizkala AR, Rouleau JL, Shi VC, Solomon SD, Swedberg K et al (2014) Angiotensin-neprilysin inhibition versus enalapril in heart failure. N Engl J Med 371:993–1004
310. Suematsu Y, Miura S, Goto M, Matsuo Y, Arimura T, Kuwano T, Imaizumi S, Iwata A, Yahiro E, Saku K (2016) LCZ696, an angiotensin receptor-neprilysin inhibitor, improves cardiac function with the attenuation of fibrosis in heart failure with reduced ejection fraction in streptozotocin-induced diabetic mice. Eur J Heart Fail 18:386–393
311. Zou J, Liu Y, Li B, Zheng Z, Ke X, Hao Y, Li X, Li X, Liu F, Zhang Z (2017) Autophagy attenuates endothelial-to-mesenchymal transition by promoting Snail degradation in human cardiac microvascular endothelial cells. Biosci Rep 37
312. Scherrer-Crosbie M, Ullrich R, Bloch KD, Nakajima H, Nasseri B, Aretz HT, Lindsey ML, Vancon AC, Huang PL, Lee RT et al (2001) Endothelial nitric oxide synthase limits left ventricular remodeling after myocardial infarction in mice. Circulation 104:1286–1291
313. Daiber A, Oelze M, Coldewey M, Kaiser K, Huth C, Schildknecht S, Bachschmid M, Nazirisadeh Y, Ullrich V, Mulsch A et al (2005) Hydralazine is a powerful inhibitor of peroxynitrite formation as a possible explanation for its beneficial effects on prognosis in patients with congestive heart failure. Biochem Biophys Res Commun 338:1865–1874
314. Liu X, Wu J, Zhu C, Liu J, Chen X, Zhuang T, Kuang Y, Wang Y, Hu H, Yu P et al (2020) Endothelial S1pr1 regulates pressure overload-induced cardiac remodelling through AKT-eNOS pathway. J Cell Mol Med 24:2013–2026
315. Katz SD, Balidemaj K, Homma S, Wu H, Wang J, Maybaum S (2000) Acute type 5 phosphodiesterase inhibition with sildenafil enhances flow-mediated vasodilation in patients with chronic heart failure. J Am Coll Cardiol 36:845–851
316. Lim SL, Lam CS, Segers VF, Brutsaert DL, De Keulenaer GW (2015) Cardiac endothelium-myocyte interaction: clinical opportunities for new heart failure therapies regardless of ejection fraction. Eur Heart J 36:2050–2060

Cellular and Subcellular Mechanisms of Ventricular Mechano-Arrhythmogenesis

Breanne A. Cameron, Peter Kohl, and T. Alexander Quinn

1 The Cardiac Mechano-Electric Regulatory Loop

The function of the heart is to provide the driving forces needed for the circulation of blood to meet the metabolic demand of the body. This is achieved through an elegant coordination of cardiac electrical excitation and mechanical pumping action. The body's haemodynamic state is continuously adapting, being modulated, for instance, by breathing, postural changes, exercise or circadian oscillations in blood pressure. To facilitate the matching of blood output from the heart to changes in systemic demand, an auto-regulatory system has evolved that intrinsically tethers electrical and mechanical function of the heart through feed-forward ('excitation-contraction coupling', ECC) [1] and feed-back ('mechano-electric coupling', MEC) pathways [2–7]. Together, ECC and MEC form the cardiac mechano-electric regulatory loop, which allows the heart to acutely sense and respond to physiological changes in its intrinsic and external environment and to maintain adequate pump function [1, 5]. In

B. A. Cameron
Department of Physiology and Biophysics, Dalhousie University, Halifax, Canada

Institute for Experimental Cardiovascular Medicine, University Heart Center Freiburg, Bad Krozingen and Faculty of Medicine, University of Freiburg, Freiburg, Germany

P. Kohl
Institute for Experimental Cardiovascular Medicine, University Heart Center Freiburg, Bad Krozingen and Faculty of Medicine, University of Freiburg, Freiburg, Germany

Faculty of Engineering, University of Freiburg, Freiburg, Germany

T. A. Quinn (✉)
Department of Physiology and Biophysics, Dalhousie University, Halifax, Canada

School of Biomedical Engineering, Dalhousie University, Halifax, Canada
e-mail: alex.quinn@dal.ca

M. Hecker, D. J. Duncker (eds.), *Cardiac Mechanobiology in Physiology and Disease*, Cardiac and Vascular Biology 9,
https://doi.org/10.1007/978-3-031-23965-6_11

physiological settings, MEC can have anti-arrhythmic properties, for instance, by reducing dispersion of repolarisation across the ventricles [8, 9]. However, MEC-mediated response to pathophysiological mechanical perturbations, disease-related alterations in the active or passive mechanical state of the heart or changes in factors driving MEC can destabilise cardiac rhythm and contribute to arrhythmogenesis [7, 10–17].

This chapter considers how pathophysiologically altered MEC, combined with changes in the cardiac mechanical environment, contributes to mechanically induced alterations in ventricular electrophysiology that can lead to arrhythmias (hereafter referred to as 'mechano-arrhythmogenesis'). It then discusses cellular and subcellular components of MEC in ventricular cardiomyocytes (CM) that may contribute to ventricular arrhythmogenesis, including mechano-sensitive ion channels (MSC), biophysical signal transmitters (i.e. microtubules, MT) and mechano-sensitive biochemical signalling pathways (i.e. changes in cytosolic free calcium concentration, $[Ca^{2+}]_i$, or reactive oxygen species, ROS). Finally, it describes clinical manifestations of these cellular and subcellular MEC mechanisms in a local (acute regional myocardial ischaemia) and a global (hypertension) pathophysiological setting. Of note, this chapter will not address mechanical modulation of contractility (mechano-mechanical coupling) [18] or stretch effects on sinus rhythm [19, 20], atrial arrhythmias [21] or tissue conduction [22], as the focus is on MEC in ventricular CM.

2 Clinical Evidence and Experimental Studies of Ventricular Mechano-Arrhythmogenesis

Ventricular mechano-arrhythmogenesis requires a sufficiently large and critically timed mechanical stimulus that directly or indirectly activates MSC, and that may be modulated by mechano-sensitive biophysical transmission or biochemical signals. This can be influenced by pathological changes in ventricular MEC, which alter the likelihood that a given stimulus will overcome the threshold for AP induction in CM [5]. While this chapter is focused on CM, it warrants noting that mechano-arrhythmogenesis also involves hetero-cellular interactions with electrotonically coupled cardiac non-myocytes [23] that also express MSC [24]. The contribution of hetero-cellular coupling will be enhanced upon disease-related changes in non-myocyte levels and phenotypes (e.g. myofibroblast conversion), which increase hetero-cellular interactions (for instance, through enhanced intercellular connectivity by increased fibroblast connexin expression) [25] or affect their MSC expression or function [26, 27].

2.1 Clinical Evidence

One of the most dramatic clinical examples of ventricular mechano-arrhythmogenesis is *Commotio cordis*, during which a non-contusional external impact, usually to the precordium, results in rhythm disturbances of variable nature, including ventricular fibrillation [28]. In fact, mechanically induced ventricular fibrillation is one of the leading causes of sudden cardiac death among young athletes and has been one of the primary reasons for which, in many countries, automated external defibrillators are now increasingly located at sporting facilities where potentially deadly impacts to the chest are most likely to occur [29, 30]. Ventricular mechano-arrhythmogenesis is also common during clinical procedures in which contact of medical instruments (e.g. cardiac catheters) with the myocardium causes local tissue deformations that may lead to ectopic excitation and arrhythmogenesis [31–42]. As with precordial mechanical stimulation [43, 44], such device-tissue contact can also revert pre-existing arrhythmias to normal sinus rhythm [5]. Arrhythmogenic changes in ventricular electrophysiology are additionally seen during acute increases in ventricular mechanical load, such as that occurring with balloon valvuloplasty [45] or during surgical manipulations [46–49].

The risk of an abnormal mechanical stimulus triggering an arrhythmia can be enhanced by cardiovascular disease. In patients with established structural heart disease, acute fluctuations in ventricular volume or pressure are more likely to result in premature excitation and sustained tachyarrhythmias [7, 17, 50]. Indeed, arrhythmia incidence in these patients is affected by blood pressure alterations, including those caused by pharmacological interventions [50], circadian oscillations [51] or day-to-day variations [52], such that acute increases in ventricular load (both preload and afterload) are associated with higher rates of arrhythmogenesis. There is also evidence suggesting that local changes in myocardial mechanics play a role in mechano-arrhythmogenesis, as wall motion abnormalities in patients with ischaemic heart disease are associated with an increased prevalence of premature excitation during acute fluctuations in load [16]. Interestingly, this effect also works in reverse: in patients suffering from chronic ventricular volume overload and tachyarrhythmias, acute ventricular unloading can result in termination of ventricular tachyarrhythmias, but alas, often only for as long as the reduction in load can be maintained [53–57].

2.2 Experimental Studies

Existing clinical evidence of ventricular mechano-arrhythmogenesis has been corroborated in experimental studies that have helped to elucidate potential mechanisms. For instance, in the case of *Commotio cordis*, a pig model of baseball impacts to the chest has confirmed historic observations [58, 59] on critical factors for the induction of ventricular fibrillation (such as pre-cordial location and impact with small and hard projectiles) while adding new information on the link between impact timing and electrophysiological outcomes: while single extra beats can be triggered in diastole, ventricular fibrillation is seen only for impacts in a narrow time-window, ~15–30 ms before the peak T wave of the ECG [60–62]. Isolated heart experiments [63] and computational modelling [64, 65] have further shown that ventricular excitation, induced by a focal mechanical stimulus [66], depends on the degree of myocardial tissue deformation and that it is a local phenomenon (i.e. triggered at the contact site, rather than resulting from the impact-induced intraventricular pressure surge). If a mechanically induced focal excitation overlaps with the trailing wave of the previous regular excitation, it may initiate ventricular fibrillation, highlighting the critical spatio-temporally defined nature of the vulnerable window for mechano-arrhythmogenesis [63].

In addition to elucidating mechanisms of *Commotio cordis*, experimental studies have demonstrated the possibility for acute changes in ventricular load to lead to premature excitation and tachyarrhythmias. Transient increases in intraventricular volume during diastole cause cellular depolarisation, which triggers excitation if supra-threshold [67–84]. Ectopic excitation has also been observed in experiments involving rapid increases in ventricular pressure (due to an acute increase in afterload) [85–87], which has been suggested to be a consequence of enhanced late-systolic myocardial deformation [88]. When an acute ventricular load is applied during systole, it tends to heterogeneously alter repolarisation and refractoriness, furnishing an arrhythmogenic substrate [68–70, 89–111]. In fact, in the isolated heart, an acute increase in intraventricular pressure has been shown to be as arrhythmogenic as the substrate created by the catecholamine-rich milieu and electrical remodeling associated with heart failure [112]. It is important to note that, while changes in intraventricular volume or pressure affect the entire ventricle (whether applied in diastole or systole), the response is generally spatially heterogeneous [113]. This is presumed to be related to variations in myocardial stiffness across the ventricle, such that stretch and the resulting electrophysiological changes are non-uniform [83, 90], with excitation originating from areas of the largest stretch (for instance, in the right ventricular outflow tract) [67, 83, 90].

3 Cellular and Subcellular Mechanisms of Ventricular Mechano-Arrhythmogenesis

Ventricular mechano-arrhythmogenesis in the whole heart arises out of a complex interplay of intra- and intercellular MEC-mediated effects. This next section is focused on the intracellular effects—the cellular and subcellular mechanisms of MEC in ventricular CM, including MSC [24], biophysical signal transmitters [114–116] and biochemical signals [117–121]—and how they shape the effects of MEC at the tissue and organ level.

In ventricular CM, stretch occurring during the resting phase of the cardiac cycle ('electrical diastole') depolarises the sarcolemma. If depolarisation is supra-threshold, this will result in excitation (i.e. it will trigger an action potential, AP). On the other hand, stretch occurring during the AP ('electrical systole') can alter the plateau or affect repolarisation dynamics. As a result, depending on stretch timing, MEC will hasten early phase 3 repolarisation, delay later repolarisation or cause early or delayed afterdepolarisation-like events [68, 69, 89, 90]. Importantly, pathophysiological changes in CM mechanics, electrical activity or factors contributing to MEC may alter this timing dependence while also enabling smaller mechanical stimuli to initiate electrical disturbances [5].

3.1 Mechano-Sensitive Ion Channels

Most acute effects of stretch on ventricular tissue- and CM-level electrophysiology (e.g. diastolic depolarisation, excitation or altered repolarisation) can be explained by MSC [24, 122]. While stretch activated channels (SAC) have been defined exclusively as channels whose activity is directly activated by a mechanical stimulus, MSC also include ion channels whose current is primarily activated by another mechanism (e.g. ligand- or voltage-activated channels), but whose activity can also be mechanically modulated [24, 122]. In fact, it has been suggested that all channels whose opening and closing mechanisms involve changes in their in-plane dimension within the lipid bilayer are MSC and may therefore contribute to MEC [123]. While the exact molecular identity of many MSC in the heart remains a topic of debate, they are generally divided into two groups based on their principal ion permeability. These groups include cation non-specific MSC (MSC_{NS}, which cause depolarisation or altered repolarisation) and potassium (K^+)-selective MSC (MSC_K, which will promote repolarisation [5, 24] and conceivably cause hyperpolarisation in pathological states that are associated with depolarised resting membrane potential, such as ischaemia). There are additional MSC, conducting chloride (Cl^-) ions, which are activated by changes in cell volume (for instance, with osmotic swelling, MSC_{swell}) rather than stretch. They are not believed to contribute to acute MEC-mediated responses, due to a pronounced lag-time in their activation (~ 1 min) and the presumed conservation of cell volume during externally applied deformations of CM [24]. However, in disease states associated with changes in cell volume, such as during cell swelling (e.g. in ischaemia and reperfusion) [124] or CM hypertrophy (where Cl^- channels have been shown to be constitutively active) [125], the contribution of cell volume-sensitive MSC to mechano-arrhythmogenesis may become relevant, for example, by acting as modifiers of the electrophysiological background upon which MEC operates.

3.1.1 MSC_{NS}

With a reversal potential (E_{REV}) between −20 and 0 mV [24], the timing of stretch stimuli (causing MSC_{NS} activation) relative to the AP [122] determines whether MSC_{NS} pass an 'inward' depolarising current (at membrane potentials < E_{REV} of the MSC) or an 'outward' re-/hyperpolarising current (at membrane potentials > E_{REV}) [70, 126, 127]. Accordingly, MSC_{NS} can depolarise resting CM. Even though experimental and computational evidence demonstrates stretch-induced depolarisation and AP-induction [67, 71–74, 122], which can be prevented by pharmacological block of MSC_{NS} [69, 73, 75, 76, 92, 128], no single channel patch clamp recordings of MSC_{NS} have been reported in adult ventricular CM. This is possibly due to localisation of MSC in membrane folds (including caveolae and transverse tubules) or at the Z-disc, locations where their direct measurement by the patch clamp method is difficult [24]. It may also relate to a dependence of MSC activation on cytoskeletal elements, the required deformation of which is not engaged in the available experimental systems (i.e. by negative patch pressure) [129]. While this has hindered identification of MSC_{NS} in adult ventricular CM, two groups of ion channels are thought to make important contributions: Piezo and Transient Receptor Potential (TRP) channels [5, 24].

The discovery of Piezo channels generated a great deal of excitement as a potential key MSC_{NS} in CM [130], because Piezo channel kinetics match macroscopic functional observations [24, 131]. However, while it appears that Piezo channels are expressed in macrophages [132] and cardiac fibroblasts [26, 133], only small amounts of Piezo mRNA have been found in cardiac tissue (from mice) [130], and (at the time of writing) no functional Piezo channels have been shown in ventricular CM from any species. Therefore, there is no direct evidence at present for a role of Piezo in ventricular mechano-arrhythmogenesis.

TRP channels are a ubiquitously expressed group of ion channels comprised of six subfamilies, many of which are found in the heart [27]. These channels have garnered interest regarding their role in the pathological progression of hypertrophy, heart failure and ischaemia, as well as in arrhythmogenesis [134, 135]. As TRP channels can pass inward or outward currents at physiological membrane potentials, and some have been shown to be inherently mechano-sensitive [136] (although this is not uncontested [137]), they may yet be found to play an important role in ventricular mechano-arrhythmogenesis.

3.1.2 MSC_K

The identity of several MSC_K in the heart is well established. TREK-1 and 2 are outwardly rectifying K^+ channels [24]. Being K^+-selective, they pass repolarising currents at all membrane potentials that are naturally experienced by CM. Accordingly, they do not trigger stretch-induced excitation, and their relative contribution to global cardiac MEC phenomena in physiological conditions is generally believed to be low compared to MSC_{NS} [5]. However, in some diseases, additional mechano-sensitive K^+ currents become available for mechanical activation, making contributions of MSC_K more relevant as the 'net reversal potential' of mechanically induced whole cell current will depend on the relative contributions of MSC_{NS} and MSC_K. In ischaemia, for example, mechano- and adenosine triphosphate (ATP)-sensitive K^+ channels (K_{ATP}) are pre-activated by the reduction in ATP (and increase in ADP) and thus alter MEC-mediated effects on cardiac electrophysiology (considered further in Sect. 4.1) [5, 138–140].

3.2 Biophysical Signal Transmitters

Biophysical signal transmitters, such as MT, relay mechanical cues across the cell, affecting mechano-sensitive cellular components, including MSC and elements responsible for the release (e.g. Ca^{2+} from the sarcoplasmic reticulum, SR, via ryanodine receptors, RyR [118], and from mitochondria [141–145]) or production (e.g. ROS by NADPH oxidase 2, NOX2 [120, 121]) of biochemical signals, which elicit and/or modulate electrophysiological responses [114, 116]. In ventricular CM, MT are particularly important for mechanical transmission [114–116, 118, 120, 146, 147] (although sarcomeric proteins and other cytoskeletal elements not discussed here, such as titin, focal adhesion proteins or integrins, will also be involved [148, 149]).

MT form physical links at the Z-disc through interactions with membrane-associated and intermediate proteins (e.g. desmin) [114]. These links create a rigid scaffold, conferring structural integrity to ventricular CM and facilitating their re-lengthening after contraction [116], as illustrated by transient buckling of MT during cell shortening [150]. MT mechano-transmission is enabled by their load-bearing ability, which is enhanced by their lateral reinforcement [151]. MT load-bearing involves interactions between the intermediate protein desmin and detyrosinated MT [150, 152]. Detyrosination is a post-translational modification that confers MT stability, simultaneously preventing MT degradation and promoting the formation of tight junctions with desmin, resulting in a shift from low-energy sliding to energy-costly buckling of MT during cell contraction [150]. Other, less explored post-translational modifications, such as acetylation, also increase the load-bearing capabilities of MT [146, 153]. Acetylation confers resistance to MT breakage during repetitive mechanical stimulation, resulting in more long-lasting ('aged') MT, which are then available for further post-translational modification [147, 154]. Thus, a denser, more detyrosinated (physically anchored) and/or acetylated (aged) MT network would be expected to enhance mechano-transmission in CM. Indeed, increased detyrosination and acetylation have been shown to enhance mechano-dependent biochemical signalling, such as an increase in the stretch-induced release of Ca^{2+} from the SR (i.e. Ca^{2+} sparks) and NOX2-dependent ROS production [115, 155]. Thus, MT network properties (including post-translational modifications) warrant exploration for their role in mechano-arrhythmogenesis (explored in Sect. 4.2).

3.3 Mechano-Sensitive Biochemical Signals

Mechano-sensitive biochemical signals are by-products of intracellular mechano-transmission and modulate the electrophysiological response of CM to a mechanical stimulus [117, 119, 121]. Principal mediators known to be important for ventricular MEC include mechano-sensitive changes in $[Ca^{2+}]_i$, which is determined by (1) trans-sarcolemmal Ca^{2+} in-/efflux [156, 157]; (2) Ca^{2+} release from/re-uptake into the SR [118, 158, 159]; (3) cytosolic Ca^{2+} buffering (such as by dissociation from/binding to troponin C, TnC [160, 161], or release from/re-uptake into mitochondria [142–145]); and (4) mechano-sensitive changes in ROS [120, 121] (although other important factors exist [119]).

These biochemical signals modulate the activity of ion channels, including TRP ankyrin-1 channels (TRPA1) [162, 163] and K^+-selective Ca^{2+}-dependent channels of big conductance (BK_{Ca}) [26, 164] or other trans-sarcolemmal ion flux pathways, such as the sodium/calcium (Na^+/Ca^{2+}) exchanger (NCX) [156, 157]. At the same time, biochemical signals also affect post-translational modification of MT (e.g. acetylation) [165], which will then affect mechano-transmission (see above) [155]. In this way, whether through interactions with MSC, other trans-sarcolemmal ion flux pathways or biophysical signal transmitters, biochemical signals can modulate ventricular MEC.

3.3.1 Mechano-Sensitive Ca^{2+} Handling

In ventricular CM, the total amount of intracellular Ca^{2+} is affected directly by mechano-sensitive trans-sarcolemmal Ca^{2+} flux or secondarily by the effects of trans-sarcolemmal Na^+ flux on NCX activity [156, 157]. 'Free' cytosolic $[Ca^{2+}]_i$ is additionally affected by the mechano-sensitivity of intracellular Ca^{2+} handling. For instance, increases in $[Ca^{2+}]_i$ occur with mechanically induced Ca^{2+} release from mitochondrial stores (which is independent of sarcolemmal MSC or NCX-mediated Ca^{2+} influx) [142, 143, 145], through a process that is dependent on MT integrity. As a result, MT disruption leads to mitochondrial disorganisation, irregular Ca^{2+} propagation and arrhythmogenic Ca^{2+} waves (which may also involve Ca^{2+} release from the SR) [144]. During contraction, $[Ca^{2+}]_i$ is affected by the mechanical modulation of the affinity of TnC for Ca^{2+}, which is increased during stretch (such that more Ca^{2+} binds to myofilaments) [160]. Upon release of stretch, the rapid dissociation of Ca^{2+} from TnC can cause an arrhythmogenic surge of $[Ca^{2+}]_i$ [161]. In diastole, stretch causes an acute increase in Ca^{2+} spark rate through a MT-dependent mechanism (reduced by MT disruption [118, 158]), suggesting that RyR themselves are in fact (non-sarcolemmal) MSC. This stretch-induced increase in Ca^{2+} spark rate would be diminished upon cell shortening during contraction and hence result in a negative-feedback mechanism that aids initiation *and* termination of ECC [166].

In physiological conditions, mechano-sensitive Ca^{2+} handling processes interact, presumably synergistically, and control $[Ca^{2+}]_i$. However, pathological alterations in these processes, such as in RyR open probability (for instance, due to oxidation or nitrosylation) [120, 167–169] or in myofilament Ca^{2+} binding affinity [170], may disturb the control of $[Ca^{2+}]_i$, resulting in Ca^{2+}-mediated arrhythmogenesis. This is particularly relevant for diseases associated with altered mechanics [171] or Ca^{2+} overload, such as acute regional ischaemia [156, 172] and hypertension (considered in Sect. 4) [115, 155].

3.3.2 Mechano-Sensitive ROS Production

The stretch-induced increase in Ca^{2+} spark rate is modulated by another mechano-sensitive biochemical signal: ROS [120]. In the context of ventricular MEC, ROS (and any ROS-mediated Ca^{2+} release) is dependent on MT, as increasing MT stability enhances mechanically induced ROS production [115]. This makes it difficult to distinguish between direct and indirect roles of the MT in mechanical modulation of RyR-mediated Ca^{2+} release [118, 120].

Mechanically induced ROS production is graded by stretch magnitude and is more responsive to cyclic than static stretch, which is in keeping with a role during regular cardiac activity [173]. This contributes to the intracellular tuning of Ca^{2+} signalling in response to stretch [120] and to complex intercellular adjustments of contractility (mechano-mechanical coupling [18]) that has been observed in paired muscle ("duplex") studies [174]. This duplex research showed that stretch effects on cellular Ca^{2+}-balance enable CM to adjust their contractility to changes in external demand. It is conceivable that (at least some of the) MEC-mediated responses are a 'side-effect' of this autoregulation of CM mechanical performance. In any case, physiological (e.g. transmural) or pathophysiologically exacerbated heterogeneity in ventricular electro-mechanics (e.g. during ischaemia) may promote arrhythmogenesis via effects on Ca^{2+}-dynamics [175]. For example, a pathological (including mechanically induced) increase in ROS production could contribute to mechano-arrhythmogenesis by sensitising MSC [162, 163] and promoting RyR Ca^{2+} leak [120] while also stabilising MT (through an increase in acetylation) [165]. The latter may be part of a compensatory response (as stiffer CM will be stretched less), but this would come at a cost (requiring higher forces for contraction).

4 Ventricular Mechano-Arrhythmogenesis in Cardiac Disease

One critical determinant of mechano-arrhythmogenesis, as highlighted in Sect. 2, is the spatial nature of a mechanical disturbance. To illustrate this point, we consider two cardiac disease states with spatially divergent alterations in electro-mechanics and different effects on cellular MEC: (1) acute regional myocardial ischaemia, which is focal in nature and involves acute metabolic, electrophysiological and ionic changes that directly alter MEC, and (2) chronic hypertension, which is global in nature and involves chronic cellular remodelling, with secondary effects on MEC.

4.1 Acute Regional Myocardial Ischaemia

Acute regional myocardial ischaemia is a mismatch between the supply and demand of blood in a localised area of the heart, for example, due to occlusion of a coronary artery. Ischaemia is characterised by three hallmark pathophysiological changes to the cellular milieu that drive pro-arrhythmic electrophysiological changes and MEC-mediated responses: (1) reduced interstitial oxygen content (hypoxia); (2) increased extracellular K^+ concentration (hyperkalemia); and (3) decreased intracellular pH (acidosis). This ischaemic milieu (and reduced or lack of blood flow) causes a cardiometabolic shift, which leads to metabolite accumulation and increased osmotic pressure (Fig. 1) [176]. During progression of acute regional myocardial ischaemia, arrhythmias occur in a bi-modal fashion in periods termed 'phase 1a' (up to ~10 min following artery occlusion) and '1b' (~15–60 min after occlusion).

The arrhythmias occurring in phase 1a appear to be re-entrant in nature and relate to cellular hyper-excitability. This hyper-excitability is driven by diastolic membrane depolarisation secondary to hyperkalemia, combined with stretch of ischaemic tissue (clinically evident as paradoxical segment lengthening) [177], which activates MSC_{NS}, driving further depolarisation and contributing to premature excitation [176, 178]. In phase 1b, excitability is reduced below normal levels in the central ischemic zone [176, 178], which at the same time begins to stiffen and, thus, resist stretch [179–183]. The particularly high incidence of arrhythmias in this phase has been suggested to instead involve systolic stretch of myocardium at the border between the stiffened ischaemic core and healthy contractile tissue, causing MSC_{NS} and K_{ATP} channel activation [179–183]. In low-flow ischaemia, activation of volume-sensitive MSC through cell swelling caused by increased osmotic pressure may also contribute to electrophysiological changes that promote arrhythmogenesis in both phases (Fig. 1) [124, 125].

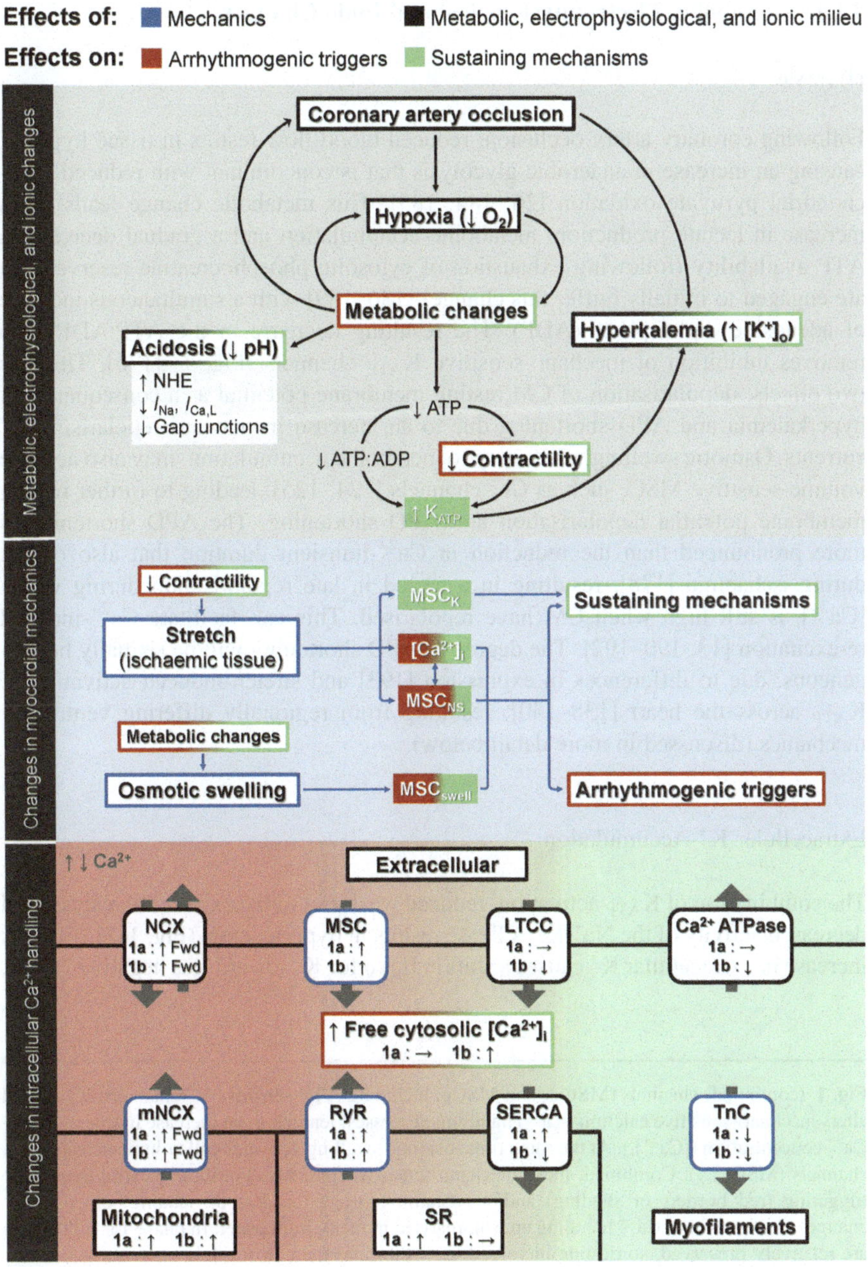

Fig. 1 Cellular and subcellular mechanisms of ventricular mechano-arrhythmogenesis during acute regional myocardial ischaemia. Following coronary artery occlusion, key metabolic, electrophysiological and ionic changes include hypoxia, acidosis and hyperkalemia (black borders, top panel). These changes lead to altered myocardial mechanics through (1) decreased contractility of ischaemic tissue, resulting in stretch of the ischaemic region in phase 1a and of the ischaemic border in phase 1b, and (2) metabolite accumulation, resulting in osmotic swelling (blue borders, middle panel). Stretch activates cation non-specific and potassium (K^+)-selective mechano-sensitive ion

4.1.1 Metabolic, Electrophysiological and Ionic Changes

Hypoxia

Following coronary artery occlusion, reduced blood flow results in tissue hypoxia, causing an increase in anaerobic glycolysis that is concomitant with reduced mitochondrial pyruvate oxidation [26, 184, 185]. This metabolic change leads to an increase in lactate production, metabolite accumulation and a gradual decrease in ATP availability (following exhaustion of cytosolic phosphocreatine reserves that are engaged to initially buffer this change) [186–189] with a simultaneous increase of adenosine diphosphate (ADP). The resulting decrease in the ATP:ADP ratio removes inhibition of mechano-sensitive K_{ATP} channels (Fig. 1) [176]. This has two effects: depolarisation of CM resting membrane potential as a consequence of hyperkalemia and APD shortening due to an increase in outward repolarising K^+ currents. Osmotic swelling, secondary to metabolite accumulation, may also activate volume-sensitive MSC, such as Cl^- channels [124, 125], leading to further resting membrane potential depolarisation and APD shortening. The APD shortening is more pronounced than the reduction in Ca^{2+}-transient duration that also occurs during ischaemia [178], resulting in a period in late repolarisation during which $[Ca^{2+}]_i$ is still high when CM have repolarised. This can facilitate Ca^{2+}-induced re-excitation [13, 190–192]. The degree of APD shortening will be spatially heterogeneous, due to differences in expression [193] and stretch-induced activation of K_{ATP} across the heart [138–140], resulting from regionally differing ventricular mechanics (discussed in more detail below).

Extracellular K^+ Accumulation

The combination of K_{ATP} activation, reduced wash-out of the extracellular space and decreased activity of the Na^+/K^+-ATPase results in hyperkalemia (Fig. 1) [176]. This increase in extracellular K^+ causes a shift in E_{REV} for K^+ toward less negative values,

---

Fig. 1 (continued) channels (MSC_{NS} and MSC_K, including ATP-sensitive K^+ channels, K_{ATP}) and alters mechano-sensitive calcium (Ca^{2+}) handling processes, leading to an increase in free cytosolic Ca^{2+} concentration ($[Ca^{2+}]_i$). At the same time, osmotic swelling activates cell volume-sensitive ion channels (MSC_{swell}). Combined, these mechano-sensitive elements contribute to arrhythmogenic triggering (red borders or shading) and/or sustaining (green borders or shading) effects. Key changes in intracellular Ca^{2+} handling are summarised in the bottom panel. In phase 1a, ATP levels are relatively preserved, so despite increased Ca^{2+} flux into the cytosol, $[Ca^{2+}]_i$ is maintained by normal ATP-dependent extracellular Ca^{2+} extrusion and intra-organelle Ca^{2+} uptake. In phase 1b, ATP levels decrease, so cytosolic Ca^{2+} efflux can no longer balance influx, and there is an increase in $[Ca^{2+}]_i$, which can also trigger and/or sustain arrhythmias. ADP, adenosine diphosphate; $I_{Ca,L}$, L-type Ca^{2+} current; I_{Na}, fast sodium current; $[K^+]_o$, extracellular potassium concentration; LTCC, L-type Ca^{2+} channel; mNCX, mitochondrial NCX; NCX, sodium-Ca^{2+} exchanger; NHE, sodium-hydrogen exchanger; O_2, oxygen; RyR, ryanodine receptors; SERCA, sarcoendoplasmic reticulum ATPase; SR, sarcoplasmic reticulum; TnC, troponin C

resulting in depolarisation of the resting membrane potential of CM. Hyperkalemia also occurs in a bimodal fashion, as in phase 1a, the initial preservation of Na^+/K^+-ATPase activity partially counteracts the K_{ATP}-mediated K^+ efflux. During this period, CM become hyper-excitable, as their slight depolarisation brings the resting membrane potential closer to the threshold for activation of the fast Na^+ channels that underlie AP initiation.

Progressive ATP reduction in phase 1b, however, interferes with Na^+/K^+-ATPase activity, resulting in a greater, secondary rise in extracellular K^+. This leads to further membrane depolarisation, which partially inactivates fast Na^+ channels, thereby reducing cell excitability and prolonging the effective refractory period ('post-repolarisation refractoriness') [178, 194]. In the regionally ischaemic whole heart, the level of extracellular K^+ accumulation differs between the ischaemic core and healthy tissue, due to gradients in extracellular wash-out and in oxygen availability, but in part also due to heterogeneous stretch-induced activation of K_{ATP} in the ischaemic tissue [177]. This gradient results in injury currents that flow from the depolarised ischaemic tissue to the (still electrotonically coupled) healthy tissue, reducing the electrical sink in this region (by shortening the gap between the resting and threshold potential of the electrically coupled healthy myocardium) and contributing to hyper-excitability of tissue at the ischaemic border zone [195, 196].

Intracellular Acidosis

The hypoxia-induced metabolic shift to anaerobic glycolysis, combined with ongoing fatty acid beta-oxidation, enhances proton accumulation and leads to intracellular acidosis. As a result, an initial compensatory efflux of hydrogen through the Na^+/H^+-ATPase (while ATP levels are still sufficient) causes an increase in intracellular Na^+, which reduces Ca^{2+}-extrusion by NCX (or even causes it to operate in reverse mode), leading to intracellular Ca^{2+} accumulation. At the same time, acidosis causes inhibition of fast Na^+ and L-type Ca^{2+} channels [178, 195], along with closure of gap junctions (Fig. 1) [197].

Electrophysiological Changes

At the cellular level, the effects of hypoxia, hyperkalemia and acidosis (as well as osmotic swelling-induced increases in Cl^- currents [124, 125]) manifest as arrhythmogenic changes to the AP: (1) shortened plateau and more rapid repolarisation (resulting in decreased APD and AP triangulation); (2) less negative resting membrane potential; and (3) reduced AP amplitude and upstroke rate [176, 178]. In the regionally ischaemic heart, changes in APD are spatially heterogeneous, with an additional increase in beat-to-beat variability of repolarisation within the peri-infarct region compared to remote, healthy myocardium [198]. This results in dispersion of cell excitability, refractoriness and repolarisation timing, which increases the vulnerability to re-entrant arrhythmias [178, 195, 198].

This arrhythmic risk is further enhanced by increasing the 'excitable gap', i.e. the distance between activation wave-front and wave-end. This occurs though a reduction in APD, and it is enhanced by slowed conduction (due to reduced intercellular coupling and decreased fast Na^+ channel availability). Ventricular conduction may be further slowed by stretch effects, which result in an increase in membrane capacitance [199, 200], caused by the unfolding of membrane invaginations, T-tubule deformation and the incorporation of caveolae into the surface and T-tubule membranes [201, 202].

Alterations in Ca^{2+} Handling

In ischaemia, there is a net increase in cytosolic $[Ca^{2+}]_i$ driven by several mechanisms, including (1) reduced forward-mode NCX activity secondary to the acidosis-induced increase in intracellular Na^+ via the Na^+-H^+ exchanger or stretch-induced (see below) Na^+ entry (and in extreme cases, reverse-mode NCX activity causing Ca^{2+} influx); (2) decreased Ca^{2+} (re-)uptake by the sarco-endoplasmic reticulum ATPase due to reduced ATP levels; (3) an increase in Ca^{2+} leak from the SR, driven by an increase in the open probability of RyR [194]; and (4) stretch effects on Ca^{2+} handling (considered in Sect. 3.3.1). The increase in RyR opening is a result of several effects [176], including Ca^{2+}-induced opening [203], stretch [118, 158] and increased ROS [120]. This overall net gain in intracellular Ca^{2+}leads to an increase in SR Ca^{2+} load. As ischaemia progresses, any compensatory effect of this Ca^{2+} sequestration into the SR is exhausted, further promoting Ca^{2+} leak through RyR and increased $[Ca^{2+}]_i$ (Fig. 1).

4.1.2 Stretch of Ischaemic Myocardium

Altered metabolic, ion channel and Ca^{2+} handling activity in CM during acute regional ischaemia affect myocardial contractility, resulting in regions of diastolic and systolic tissue stretch (Fig. 1).

In phase 1a, there is general stretch of weakened tissue in the central ischaemic region, with additional stretch during mechanical systole. In phase 1b, however, the central ischaemic tissue stiffens, and its stretch is reduced, resulting in stretch of weakened tissue at its border by contraction of adjacent healthy tissue [179–183]. The mechano-arrhythmogenic relevance of ischaemic myocardial stretch is supported by the correlation between wall motion abnormalities and the incidence of ventricular fibrillation seen in patients with coronary artery disease [16]. Experimentally, it has been shown that the onset of tissue stretch and ventricular fibrillation in phase 1a of acute regional ischemia is related [204] and that the degree of tissue stretch is a strong predictor of ventricular fibrillation occurrence [10–12, 177].

In phase 1b, arrhythmia incidence is ventricular load-dependent [13, 14, 80], and aberrant excitation tends to originate at the ischaemic border, where experimental evidence [13, 14] and computational modelling [205] suggest an involvement of stretch-induced depolarisation mediated by MSC_{NS}. Mechanically induced arrhythmias arising from the ischaemic border will be facilitated by heterogeneous slowing of conduction, due to more pronounced gap junction uncoupling within the ischaemic core than at the ischaemic border [197].

Changes in mechano-sensitive biochemical signalling during ischaemia further facilitate mechano-arrhythmogenesis. For instance, the stretch-induced increase in Ca^{2+} spark rate [118] and ROS production [120] that occurs in healthy tissue is enhanced by ischaemia [172], a response potentially caused by an ischaemia-induced increase in SR Ca^{2+} load [194] or RyR sensitivity [120, 168] or a deficit in the antioxidant capacity of CM [206] (for instance, due to a reduction in the key antioxidant glutathione [207]). In fact, enhancement of mechano-sensitive biochemical signalling in regions of stretched myocardium has been shown to cause focal Ca^{2+} waves [208], which, if occurring at the ischaemic border where the injury current tends to reduce the excitation threshold [195], may contribute an additional trigger or serve as a substrate for sustained mechano-arrhythmogenesis.

4.1.3 Tissue-Level Considerations for Mechano-Arrhythmogenesis in Acute Regional Ischaemia

Mechano-arrhythmogenesis in acute regional ischaemia involves the localised interaction of ischaemia- and stretch-induced electrophysiological effects on CM, generating regionally heterogeneous changes in tissue excitability, refractoriness and electrical conduction. As the acute outcome of myocardial stretch is dependent on its magnitude and timing relative to the background electrical and mechanical activity of CM, ischaemic effects that alter CM electrophysiology—for example, shortening of APD—increase the likelihood that stretch will occur during a period of cellular excitability and, thus, provoke an arrhythmia.

In addition to the MEC effects driven by localised tissue stretch during ischaemia described above, there is also evidence for an increase in MEC itself. This includes an increase in mechano-sensitive biochemical signals (Ca^{2+} sparks, ROS) [172], increased RyR sensitivity (and ensuing Ca^{2+} release) [206] and pre-activation of mechano-sensitive K_{ATP} channels [138–140]. Heterogeneous stretch may therefore lead to regionally differing changes that furnish arrhythmogenic triggers and enhance the substrate for arrhythmia sustenance [162, 163, 209]. The heterogeneous expression of mechano-sensitive [139, 140] K_{ATP} channels [193] (or other MSC_K [210]) across the heart will additionally drive dispersion of repolarisation or cause conduction block [211], favouring re-entrant electrical activity. Finally, mechano-arrhythmogenesis in ischaemia may involve heterogeneous changes in the relative dynamics of membrane voltage and Ca^{2+} transients, which facilitate Ca^{2+}-mediated arrhythmias during late repolarisation [13, 190–192].

4.2 *Chronic Hypertension*

Chronic hypertension, defined as a sustained increase in systolic (≥130 mmHg) or diastolic (≥80 mmHg) blood pressure [212], results in increased ventricular afterload (i.e. the load against which ventricular CM must contract) [213]. This increase in afterload results in an increase in systolic intraventricular pressure (which maintains ejection), but may also lead to an increase in ventricular preload (i.e. intraventricular volume) if ejection is reduced. The elevated mechanical load experienced by ventricular myocardium in hypertension results in the stimulation of compensatory CM remodelling (although sustained overload ultimately results in decompensated heart failure) [214, 215], which may enhance MEC and contribute to increased mechano-arrhythmogenesis.

Enhanced MEC, secondary to structural and functional remodelling of CM in hypertension, is thought to contribute to the increased incidence of ectopic ventricular excitation in hypertensive patients [17] that occurs during acute fluctuations in ventricular load, including circadian [51] or pharmacologic [50] modulation of blood pressure. This increase in arrhythmogenic triggers may interact with remodelled myocardium, which acts as an arrhythmia-sustaining substrate, contributing to sustained tachyarrhythmias and/or sudden cardiac death [112, 216–221]. Further, mechano-arrhythmogenesis may be enhanced when hypertensive hypertrophic remodelling is complicated by diffusely distributed ischaemic regions (due to wall thickening and microvasculature remodelling [222]) involving mechanisms described in the previous section.

4.2.1 Structural Remodelling

Tissue-level remodelling in hypertension is characterised by concentric thickening of the ventricular wall (hypertrophy), which counteracts the increase in systolic wall stress caused by an increase in systolic intraventricular pressure with elevated afterload. The benefit of this change in chamber geometry is explained by Laplace's law: ($P \sim [T \times m]/R$). An increased intra-ventricular peak pressure (P) can result either from an increase in CM force production (thus raising tissue tension, T)—which, within normal physiological limits, will be afforded by length-dependent activation of force generation (Frank-Starling law of the heart [223])—or from an increase in ventricular wall thickness (m, hopefully in the absence of an increase in chamber radius, R, which would worsen the situation). Increases in sarcomeric force production are limited in scope, so in pathological settings the increase in ventricular wall thickness accounts for the necessary increase in intra-ventricular pressure (P). This is the result of the parallel addition of sarcomeres in CM, thereby increasing cell diameter, as opposed to their addition in series—which would increase cell length and increase chamber radius (although with prolonged hypertension this may also occur)—or the addition of new cells (CM division is exceedingly uncommon in the adult mammalian heart [224]). The degree of concentric hypertrophy thus scales with the increase in pressure, resulting in a normalisation of wall stress, along with an increase in the ratio of wall thickness to inner chamber diameter [214, 215].

Microtubule Network

In addition to ventricular wall thickening, chronic hypertension is associated with remodelling of the cytoskeleton, characterised by changes in the density (via polymerisation) and stability (via post-translational modifications, altered intermediate protein linkages and changes in expression of MT-associated proteins) of the MT network [225, 226]. These changes to the MT network may work to reduce CM stretch, which would occur if ventricular preload is increased.

While clinical observations [227, 228] and experimental models [229, 230] have demonstrated that an increase in MT density occurs in hypertension, it has remained unclear whether MT proliferation begins during compensatory hypertrophy or is a consequence of the transition to decompensation (discrepancies in published reports may relate to inconsistencies between hypertensive models used, both in terms of species and method to induce hypertension) [231]. Regardless of the moment of onset, the MT network has been consistently shown to become denser and more stable in hypertension [227–230, 232, 233]. MT network remodelling appears to contribute to altered CM shortening and relaxation: hyper-polymerisation of MT in healthy cells decreases contraction and relaxation, while MT de-polymerisation in failing CM results in an improvement [116, 234, 235]. The relative contribution of an increase in the rate of MT polymerisation (which would present as an overall increase in tubulin content) versus an enhanced stability of existing MT (which would manifest as post-translationally modified 'aged' MT) to these effects remains to be clarified [114, 116, 153, 228].

Microtubule Post-Translational Modifications

During contraction, MT buckle at wavelengths corresponding to the distance between adjacent sarcomere units [150]. This process, which also enhances visco-elastic resistance, is dependent on the level of detyrosination, a post-translational modification of α-tubulin that involves the removal of the C-terminal tyrosine, which exposes a glutamate at the newly formed C-terminus [146, 236]. This suggests that the load-bearing (and, by extension, contribution to mechano-transmission) and re-lengthening (which is resisted by viscous forces) [116] of MT are affected by detyrosination, rather than MT density alone. Therefore, pathological increases in the level of MT detyrosination in hypertension may have profound effects on chamber relaxation (which would affect passive ventricular filling during early diastole) [237, 238], ventricular MEC and, as a result, mechano-arrhythmogenesis.

MT detyrosination confers stability to the MT network by preventing the breakdown of existing MT and by facilitating cross-linking with intermediate proteins

[114, 239, 240] (e.g. desmin [150, 152]). This results in a shift from low resistance MT sliding to high-energy buckling during contraction, which modulates the effect of MT on mechano-transmission and segment re-lengthening (by increasing viscoelastic resistance) [116, 150, 228]. In human [228] and murine CM [115], inhibition of detyrosination increases the wavelength and disrupts the organisation of MT buckling (suggesting an attenuation of their resistance to compression and load-bearing capability), which is associated with an increase in the velocity of cellular contraction and relaxation, presumably due to reduced viscoelastic resistance. Conversely, enhancing detyrosination (without a change in MT density) increases both cell stiffness and viscosity, with an associated decrease in contraction and relaxation velocity [115, 150]. This suggests that in ventricular CM subjected to pressure overload, enhanced detyrosination, rather than enhanced MT density, accounts for increased stiffness and impaired contraction and relaxation kinetics. Indeed, in addition to the overall increase in MT detyrosination in failing human CM, there is upregulation of a gene encoding a pre-detyrosinated tubulin [228], as well as of the detyrosination promoting MT-associated protein 4 (MAP4) [241]. As a consequence, suppression of detyrosination in CM from pressure overload patients improves contractile and relaxation function, and the degree of such improvement scales with initial CM stiffness [150, 228].

MT detyrosination is promoted by interactions with MT-associated and intermediate proteins, such as MAP4 [228, 241, 242] and desmin [150, 152]. In pressure overload, there is upregulation of MAP4 (which occurs early in hypertrophy and stabilises MT by preventing their degradation [241]) and desmin, both of which contribute to the pathologic increase in detyrosination levels. Desmin is known to bind specifically to detyrosinated tubulin and facilitate its buckling behaviour [150, 152]. The increase in desmin expression in pressure overload, however, is partially comprised of an isoform prone to misfolding and aggregation [228], which probably explains observations of MT and desmin misalignment [114, 228, 243]. Such MT disorganisation would contribute to decompensated heart failure, including detrimental effects on trafficking of sarcomere precursors, thereby limiting compensatory hypertrophy [114, 244–246].

4.2.2 Proposed Mechanisms of Mechano-Arrhythmogenesis in Hypertension

One function of the MT network is the transmission of mechanical signals across CM [116], which is enhanced by increases in its density or stability (e.g. through detyrosination or acetylation) [114]. Pathophysiological alterations in the MT network may be arrhythmogenic if they modulate MSC activity (which has been demonstrated experimentally by modulation of detyrosination [115] and acetylation [155]), by changing the degree of mechanical stimulation MSC experience (either by reducing the dampening effect of the MT or by enhancing mechano-transmission) [24, 129, 247] or by sensitising MSC [134, 162, 163] (through an increase in MT-dependent biochemical signal production [115, 155]).

In hypertension, remodelling of the CM cytoskeleton results in a laterally reinforced MT network (via interactions between detyrosinated tubulin, desmin and MAP4) [114]. Increased lateral reinforcement facilitates the load-bearing and, thus, biophysical signal transmission capability of the MT network [151]. The remodelled cytoskeleton will thus increase MEC, through an increase in mechano-sensitive biochemical signal production with increased levels of detyrosination [115] and contribute to the constitutive activation of volume-sensitive Cl^- channels following transition to failure [248]. An increase in MEC would be expected to reduce the magnitude of CM stretch needed to cause excitation and trigger an arrhythmia. A role for MT in increased mechano-arrhythmogenesis has been corroborated in several experimental models, including one in which an acute pharmacologically induced rise in MT proliferation and detyrosination in the whole heart (with paclitaxel) was associated with an increased prevalence of acute volume pulse-induced mechano-arrhythmogenesis [81]. However, it should be noted that in this study, potential confounding effects of increased peak intraventricular pressure, associated with volume injections into stiffened ventricles, cannot be excluded (relevant data was not reported).

Enhanced mechanically induced ROS production in hypertension, promoted by elevated levels of detyrosination [115], would be further increased by upregulation of NOX2 [249]. This elevated level of ROS may promote an arrhythmogenic intracellular milieu (i.e. elevated $[Ca^{2+}]_i$ levels) through, for example, increased Ca^{2+} leak from the SR [120, 169]. ROS and Ca^{2+} also modulate the activity of MSC, [162, 163] whose intracellular trafficking and stretch activation may be affected by alterations of the MT network in hypertension [24]. As such, pathological remodelling of biophysical signal transmitters and the resultant effect on ventricular MEC may contribute to formation of arrhythmic triggers and a substrate for mechano-arrhythmogenesis in hypertension.

5 Conclusion

The heart is an electrically controlled mechanical pump with intricate feedback mechanisms that dictate the heart's response to acute changes in its mechanical environment. At the cellular level, these MEC effects involve mechano-sensitive components, including MSC [24], biophysical signal transmitters (e.g. the MT network [114]) and mechano-sensitive biochemical signals [117, 119, 121] (e.g. intracellular ROS and Ca^{2+} [118, 120]). In this chapter, we have discussed how these cellular and subcellular components of MEC in ventricular CM contribute to tissue-level ventricular mechano-arrhythmogenesis.

The potential for a mechanical change to elicit an electrical response depends on the timing of the associated mechanical stimulation relative to the AP and on its magnitude and rate of rise [5]. Therefore, disease states in which regional (e.g. ischaemia [172]) or global (e.g. hypertension [114, 236]) mechanical alterations occur and which are associated with changes in AP dynamics (altering the relative duration of systolic and diastolic periods [13, 134]) or changes in MEC (altering the threshold for stretch-induced excitation [7, 15, 17, 112]) may make the heart more prone to mechano-arrhythmogenesis. In these cases, mechanically induced excitation constitutes an arrhythmogenic trigger, which interacts with a disease-mediated pro-arrhythmic substrate and converts into sustained arrhythmic activity [17, 134].

The nature of metabolic, electrophysiological and ionic changes, as well as MT network remodelling affecting MEC, will vary across cardiac diseases. When considering the relative contribution of cellular and subcellular mechanisms to ventricular mechano-arrhythmogenesis identified in this chapter, it is essential to assess the precipitating factors of the disease being studied, as well their structural and functional manifestations. Comparison of diseases with differences in mechanical loading (e.g. ventricular volume versus pressure overload) or structure-function remodelling (heart failure with preserved versus reduced ejection fraction) will provide critical insight into how disease-induced alterations in cellular MEC drive ventricular mechano-arrhythmogenesis.

For instance, while pressure overload is characterised by densification and increased detyrosination of the MT network, resulting in an increase in MEC, in volume overload there is a reduction in detyrosination-desmin interactions and a disruption of cytoskeletal integrity [243], such that the influence of the MT network on MSC and biochemical signals may be attenuated. Effects of volume overload on MEC may in fact more closely resemble those seen in acute ischaemia (tissue stretch- [177] and osmotic CM swelling-induced [124, 125] MSC activation), but over the entire ventricle (globally), rather than regionally varying (local). In both cases it would be attractive to determine whether a stretch-induced increase in biochemical signalling (i.e. Ca^{2+} sparks and ROS production) involves alterations in post-translational modifications of MT and whether this increases MSC sensitivity.

In failing human CM, both contraction and relaxation are improved by suppression of MT detyrosination [228]. CM from patients with heart failure with preserved ejection fraction, however, show a greater improvement than those with reduced ejection fraction, suggesting there is a greater contribution of detyrosination to the reduced mechanical function in those cells. Thus, it would be informative to compare MEC and mechano-arrhythmogenesis between those two forms of heart failure.

In summary, this chapter has discussed cellular and subcellular mechanisms contributing to mechano-arrhythmogenesis in ventricular CM, with a focus on two disease states that highlighted local (e.g. acute regional ischaemia) and global (e.g. hypertension) pathological manifestation relevant for MEC. The intricate interaction between the various components of ventricular MEC contribute to formation of both triggers and substrate for arrhythmogenesis. Understanding the precise interactions of these components is necessary to facilitate the development of a mechanistic framework for understanding how therapies targeting MEC may contribute to the prevention of ventricular mechano-arrhythmogenesis.

Compliance with Ethical Standards

Funding This work was supported by the Canadian Institutes of Health Research (MOP 342562 to T.A.Q.), the Natural Sciences and Engineering Research Council of Canada (RGPIN-2016-04879 to T.A.Q.) and the Heart and Stroke Foundation of Canada (National New Investigator Award to T. A.Q.). BAC and PK are members of the German Research Foundation Collaborative Research Centre SFB1425 (422681845).

Conflict of Interest All authors declare they have no conflict of interest.

Ethical Approval This article does not contain any studies with human participants or animals performed by any of the authors.

References

1. Bers DM (2002) Cardiac excitation–contraction coupling. Nature 415:198–205. https://doi.org/10.1038/415198a
2. Kohl P, Ravens U (2003) Cardiac mechano-electric feedback: past, present, and prospect. Prog Biophysics Mol Biology 82:3–9. https://doi.org/10.1016/s0079-6107(03)00022-1
3. Kohl P, Sachs F, Franz MR (eds) (2011) Cardiac Mechano-electric coupling and arrhythmias, 2nd edn. Oxford University Press, Oxford
4. Quinn TA, Kohl P, Ravens U (2014) Cardiac mechano-electric coupling research: fifty years of progress and scientific innovation. Prog Biophysics Mol Biology 115:71–75. https://doi.org/10.1016/j.pbiomolbio.2014.06.007
5. Quinn TA, Kohl P (2021) Cardiac mechano-electric coupling: acute effects of mechanical stimulation on heart rate and rhythm. Physiol Rev 101:37–92. https://doi.org/10.1152/physrev.00036.2019
6. Ravens U (2003) Mechano-electric feedback and arrhythmias. Prog Biophysics Mol Biology 82:255–266. https://doi.org/10.1016/s0079-6107(03)00026-9
7. Taggart P, Sutton PMI (1999) Cardiac mechano-electric feedback in man: clinical relevance. Prog Biophysics Mol Biology 71:139–154. https://doi.org/10.1016/s0079-6107(98)00039-x
8. Opthof T, Meijborg VMF, Belterman CNW, Coronel R (2015) Synchronization of repolarization by mechano-electrical coupling in the porcine heart. Cardiovasc Res 108:181–187. https://doi.org/10.1093/cvr/cvv140
9. Quinn TA (2015) Cardiac mechano-electric coupling: a role in regulating normal function of the heart? Cardiovasc Res 108:1–3. https://doi.org/10.1093/cvr/cvv203
10. Barrabés JA, Garcia-Dorado D, González MA et al (1998) Regional expansion during myocardial ischemia predicts ventricular fibrillation and coronary reocclusion. Am J Physiology-Heart Circ Physiol 274:H1767–H1775. https://doi.org/10.1152/ajpheart.1998.274.5.h1767
11. Barrabés JA, Garcia-Dorado D, Padilla F et al (2002) Ventricular fibrillation during acute coronary occlusion is related to the dilation of the ischemic region. Basic Res Cardiol 97:445–451. https://doi.org/10.1007/s003950200051
12. Barrabés JA, Inserte J, Agulló L et al (2015) Effects of the selective stretch-activated channel blocker GsMtx4 on stretch-induced changes in refractoriness in isolated rat hearts and on ventricular premature beats and arrhythmias after coronary occlusion in swine. PLoS One 10: e0125753. https://doi.org/10.1371/journal.pone.0125753
13. Baumeister PA, Lawen T, Rafferty SA et al (2018) Mechanically-induced ventricular arrhythmias during acute regional ischemia. J Mol Cell Cardiol 124:87–88. https://doi.org/10.1016/j.yjmcc.2018.07.021

14. Coronel R, Wilms-Schopman FJG, deGroot JR (2002) Origin of ischemia-induced phase 1b ventricular arrhythmias in pig hearts. J Am Coll Cardiol 39:166–176. https://doi.org/10.1016/s0735-1097(01)01686-2
15. Sideris DA (1993) High blood pressure and ventricular arrhythmias. Eur Heart J 14:1548–1553. https://doi.org/10.1093/eurheartj/14.11.1548
16. Siogas K, Pappas S, Graekas G et al (1998) Segmental wall motion abnormalities alter vulnerability to ventricular ectopic beats associated with acute increases in aortic pressure in patients with underlying coronary artery disease. Heart 79:268. https://doi.org/10.1136/hrt.79.3.268
17. Sutherland GR (2017) Sudden cardiac death: the pro-arrhythmic interaction of an acute loading with an underlying substrate. Eur Heart J 38:2986–2994. https://doi.org/10.1093/eurheartj/ehw449
18. Quinn TA, Kohl P (2016) Rabbit models of cardiac mechano-electric and mechano-mechanical coupling. Prog Biophysics Mol Biology 121:110–122. https://doi.org/10.1016/j.pbiomolbio.2016.05.003
19. Quinn TA, Kohl P (2012) Mechano-sensitivity of cardiac pacemaker function: pathophysiological relevance, experimental implications, and conceptual integration with other mechanisms of rhythmicity. Prog Biophysics Mol Biol 110:257–268. https://doi.org/10.1016/j.pbiomolbio.2012.08.008
20. MacDonald EA, Quinn TA (2021) What keeps us ticking? Sinoatrial node mechano-sensitivity: the grandfather clock of cardiac rhythm. Biophys Rev 13:707–716. https://doi.org/10.1007/s12551-021-00831-8
21. Ravelli F (2003) Mechano-electric feedback and atrial fibrillation. Prog Biophysics Mol Biol 82:137–149. https://doi.org/10.1016/s0079-6107(03)00011-7
22. Pfeiffer ER, Tangney JR, Omens JH, McCulloch AD (2014) Biomechanics of cardiac electromechanical coupling and Mechanoelectric feedback. J Biomech Eng 136:021007. https://doi.org/10.1115/1.4026221
23. Quinn TA, Camelliti P, Rog-Zielinska EA et al (2016) Electrotonic coupling of excitable and nonexcitable cells in the heart revealed by optogenetics. Proc National Acad Sci 113:14852–14857. https://doi.org/10.1073/pnas.1611184114
24. Peyronnet R, Nerbonne JM, Kohl P (2016) Cardiac mechano-gated ion channels and arrhythmias. Circ Res 118:311–329. https://doi.org/10.1161/circresaha.115.305043
25. Baudino TA, Borg TK (2011) The origin of fibroblasts, extracellular matrix, and potential contributions to cardiac mechano-electric coupling. In: Kohl P, Sachs F, Franz MR (eds) Cardiac Mechano-electric coupling and arrhythmias, 2nd edn. Oxford University Press, Oxford, pp 138–142
26. Jakob D, Klesen A, Allegrini B et al (2021) Piezo1 and BKCa channels in human atrial fibroblasts: interplay and remodelling in atrial fibrillation. J Mol Cell Cardiol 158:49–62. https://doi.org/10.1016/j.yjmcc.2021.05.002
27. Hof T, Chaigne S, Récalde A et al (2019) Transient receptor potential channels in cardiac health and disease. Nat Rev Cardiol 16:344–360. https://doi.org/10.1038/s41569-018-0145-2
28. Kohl P, Nesbitt AD, Cooper PJ, Lei M (2001) Sudden cardiac death by commotio cordis: role of mechano—electric feedback. Cardiovasc Res 50:280–289. https://doi.org/10.1016/s0008-6363(01)00194-8
29. Maron BJ, Doerer JJ, Haas TS et al (2009) Sudden deaths in young competitive athletes. Circulation 119:1085–1092. https://doi.org/10.1161/circulationaha.108.804617
30. Maron BJ, Estes NAM (2010) Commotio cordis. New Engl J Medicine 362:917–927. https://doi.org/10.1056/nejmra0910111
31. Bõhm A, Pintér A, Préda I (2017) Ventricular tachycardia induced by a pacemaker lead. Acta Cardiol 57:23–24. https://doi.org/10.2143/ac.57.1.2005375
32. Damen J (1985) Ventricular arrhythmias during insertion and removal of pulmonary artery catheters. Chest 88:190–193. https://doi.org/10.1378/chest.88.2.190

33. Elliott CG, Zimmerman GA, Clemmer TP (1979) Complications of pulmonary artery catheterization in the care of critically iii patients a prospective study. Chest 76:647–652. https://doi.org/10.1378/chest.76.6.647
34. Fiaccadori E, Gonzi G, Zambrelli P et al (1996) Cardiac arrhythmias during central venous catheter procedures in acute renal failure: a prospective study. J Am Soc Nephrol 7:1079–1084. https://doi.org/10.1681/asn.v771079
35. Iberti TJ, Benjamin E, Gruppi L, Raskin JM (1985) Ventricular arrhythmias during pulmonary artery catheterization in the intensive care unit prospective study. Am J Medicine 78:451–454. https://doi.org/10.1016/0002-9343(85)90337-7
36. Kusminsky RE (2007) Complications of central venous catheterization. J Am Coll Surgeons 204:681–696. https://doi.org/10.1016/j.jamcollsurg.2007.01.039
37. Lee JC, Epstein LM, Huffer LL et al (2009) ICD lead proarrhythmia cured by lead extraction. Heart Rhythm 6:613–618. https://doi.org/10.1016/j.hrthm.2009.01.039
38. Lee T-Y, Sung C-S, Chu Y-C et al (1996) Incidence and risk factors of guidewire-induced arrhythmia during internal jugular venous catheterization: comparison of marked and plain J-wires. J Clin Anesth 8:348–351. https://doi.org/10.1016/0952-8180(96)00083-9
39. Lindsay AC, Wong T, Segal O et al (2006) An unusual twist: ventricular tachycardia induced by a loop in a right ventricular pacing wire. Qjm Int J Medicine 99:347–348. https://doi.org/10.1093/qjmed/hcl043
40. Michel J, Johnson AD, Bridges WC et al (1950) Arrhythmias during intracardiac catheterization. Circulation 2:240–244. https://doi.org/10.1161/01.cir.2.2.240
41. Sprung CL, Pozen RG, Rozanski JJ et al (1982) Advanced ventricular arrhythmias during bedside pulmonary artery catheterization. Am J Medicine 72:203–208. https://doi.org/10.1016/0002-9343(82)90811-7
42. Stuart R, Shikora S, Akerman P et al (1990) Incidence of arrhythmia with central venous catheter insertion and exchange. Jpen-parenter Enter 14:152–155. https://doi.org/10.1177/0148607190014002152
43. Haman L, Parizek P, Vojacek J (2009) Precordial thump efficacy in termination of induced ventricular arrhythmias. Resuscitation 80:14–16. https://doi.org/10.1016/j.resuscitation.2008.07.022
44. Pennington JE, Taylor J, Lown B (1970) Chest thump for reverting ventricular tachycardia. New Engl J Medicine 283:1192–1195. https://doi.org/10.1056/nejm197011262832204
45. Levine JH, Guarnieri T, Kadish AH et al (1988) Changes in myocardial repolarization in patients undergoing balloon valvuloplasty for congenital pulmonary stenosis: evidence for contraction-excitation feedback in humans. Circulation 77:70–77. https://doi.org/10.1161/01.cir.77.1.70
46. Orini M, Taggart P, Bhuva A et al (2021) Direct in-vivo assessment of global and regional mechano-electric feedback in the intact human heart. Heart Rhythm 18:1406. https://doi.org/10.1016/j.hrthm.2021.04.026
47. Taggart P, Sutton P, Lab M et al (1992) Effect of abrupt changes in ventricular loading on repolarization induced by transient aortic occlusion in humans. Am J Physiology-Heart Circ Physiol 263:H816–H823. https://doi.org/10.1152/ajpheart.1992.263.3.h816
48. Taggart P, Sutton P, John R et al (1992) Monophasic action potential recordings during acute changes in ventricular loading induced by the Valsalva manoeuvre. Brit Heart J 67:221. https://doi.org/10.1136/hrt.67.3.221
49. Taggart P, Sutton P (2011) Load dependence of ventricular repolarization. In: Kohl P, Sachs F, Franz MR (eds) Cardiac Mechano-electric coupling and arrhythmias, 2nd edn. Oxford University Press, Oxford, pp 269–273
50. Sideris DA, Kontoyannis DA, Michalis L et al (1987) Acute changes in blood pressure as a cause of cardiac arrhythmias. Eur Heart J 8:45–52. https://doi.org/10.1093/oxfordjournals.eurheartj.a062158
51. Muller JE, Tofler GH, Stone PH (1989) Circadian variation and triggers of onset of acute cardiovascular disease. Circulation 79:733–743. https://doi.org/10.1161/01.cir.79.4.733

52. Reiter MJ, Stromberg KD, Whitman TA et al (2013) Influence of intracardiac pressure on spontaneous ventricular arrhythmias in patients with systolic heart failure. Circulation Arrhythmia Electrophysiol 6:272–278. https://doi.org/10.1161/circep.113.000223
53. Ambrosi P, Habib G, Kreitmann B et al (1995) Valsalva manoeuvre for supraventricular tachycardia in transplanted heart recipient. Lancet 346:713. https://doi.org/10.1016/s0140-6736(95)92331-4
54. Hwang E-S, Pak H-N (2012) Mid-septal hypertrophy and apical ballooning; potential mechanism of ventricular tachycardia storm in patients with hypertrophic cardiomyopathy. Yonsei Med J 53:221–223. https://doi.org/10.3349/ymj.2012.53.1.221
55. Waxman MB, Wald RW, Finley JP et al (1980) Valsalva termination of ventricular tachycardia. Circulation 62:843–851. https://doi.org/10.1161/01.cir.62.4.843
56. Perticone F, Ceravolo R, Maio R et al (1993) Mechano-electric feedback and ventricular arrhythmias in heart failure. The possible role of permanent cardiac stimulation in preventing ventricular tachycardia. Cardiol Rome Italy 38:247–252
57. Wei JY, Greene HL, Weisfeldt ML (1980) Cough-facilitated conversion of ventricular tachycardia. Am J Cardiol 45:174–176. https://doi.org/10.1016/0002-9149(80)90235-0
58. Nesbitt AD, Cooper PJ, Kohl P (2001) Rediscovering commotio cordis. Lancet 357:1195–1197. https://doi.org/10.1016/s0140-6736(00)04338-5
59. Schlomka G (1934) Ergeb Inn Med Kinderheilkd 47:1–91. https://doi.org/10.1007/978-3-642-90672-5_1
60. Link MS, Wang PJ, Pandian NG et al (1998) An experimental model of sudden death due to low-energy chest-wall impact (commotio cordis). New Engl J Medicine 338:1805–1811. https://doi.org/10.1056/nejm199806183382504
61. Link MS, Maron BJ, VanderBrink BA et al (2001) Impact directly over the cardiac silhouette is necessary to produce ventricular fibrillation in an experimental model of commotio cordis. J Am Coll Cardiol 37:649–654. https://doi.org/10.1016/s0735-1097(00)01142-6
62. Link MS, Maron BJ, Wang PJ et al (2003) Upper and lower limits of vulnerability to sudden arrhythmic death with chest-wall impact (commotio cordis). J Am Coll Cardiol 41:99–104. https://doi.org/10.1016/s0735-1097(02)02669-4
63. Quinn TA, Jin H, Lee P, Kohl P (2017) Mechanically induced ectopy via stretch-activated cation-nonselective channels is caused by local tissue deformation and results in ventricular fibrillation if triggered on the repolarization wave edge (commotio cordis). Circ Arrhythmia Electrophysiol 10:e004777. https://doi.org/10.1161/circep.116.004777
64. Garny A, Kohl P (2004) Mechanical induction of arrhythmias during ventricular repolarization: modeling cellular mechanisms and their interaction in two dimensions. Ann N Y Acad Sci 1015:133–143. https://doi.org/10.1196/annals.1302.011
65. Li W, Kohl P, Trayanova N (2004) Induction of ventricular arrhythmias following mechanical impact: a simulation study in 3D. J Mol Histol 35:679–686. https://doi.org/10.1007/s10735-004-2666-8
66. Alsheikh-Ali AA, Akelman C, Madias C et al (2008) Endocardial mapping of ventricular fibrillation in commotio cordis. Heart Rhythm 5:1355–1356. https://doi.org/10.1016/j.hrthm.2008.03.009
67. Franz MR, Cima R, Wang D et al (1992) Electrophysiological effects of myocardial stretch and mechanical determinants of stretch-activated arrhythmias. Circulation 86:968–978. https://doi.org/10.1161/01.cir.86.3.968
68. Bode F, Franz MR, Wilke I et al (2006) Ventricular fibrillation induced by stretch pulse: implications for sudden death due to commotio cordis. J Cardiovasc Electr 17:1011–1017. https://doi.org/10.1111/j.1540-8167.2006.00547.x
69. Eckardt L, Kirchhof P, Mönnig G et al (2000) Modification of stretch-induced shortening of repolarization by streptomycin in the isolated rabbit heart. J Cardiovasc Pharmacol 36:711–721. https://doi.org/10.1097/00005344-200012000-00005

70. Zabel M, Koller BS, Sachs F et al (1996) Stretch-induced voltage changes in the isolated beating heart: importance of the timing of stretch and implications for stretch-activated ion channels. Cardiovasc Res 32:120–130. https://doi.org/10.1016/s0008-6363(96)00089-2
71. Franz MR, Burkhoff D, Yue DT et al (1989) Mechanically induced action potential changes and arrhythmia in isolated and in situ canine hearts. Cardiovasc Res 23:213–223. https://doi.org/10.1093/cvr/23.3.213
72. Hansen DE, Craig CS, Hondeghem LM (1990) Stretch-induced arrhythmias in the isolated canine ventricle. Evidence for the importance of mechanoelectrical feedback. Circulation 81:1094–1105. https://doi.org/10.1161/01.cir.81.3.1094
73. Huang H, Wei H, Liu P et al (2009) A simple automated stimulator of mechanically induced arrhythmias in the isolated rat heart. Exp Physiol 94:1054–1061. https://doi.org/10.1113/expphysiol.2009.048660
74. Kim DY, White E, Saint DA (2012) Increased mechanically-induced ectopy in the hypertrophied heart. Prog Biophysics Mol Biol 110:331–339. https://doi.org/10.1016/j.pbiomolbio.2012.07.004
75. Dhein S, Englert C, Riethdorf S et al (2014) Arrhythmogenic effects by local left ventricular stretch: effects of flecainide and streptomycin. N-S Arch Pharmacol 387:763–775. https://doi.org/10.1007/s00210-014-0988-y
76. Wei H, Zhang Z-F, Huang H-X, Niu W-Z (2008) [Arrhythmia triggered by stretching rabbit left ventricles and the block effect of streptomycin] Zhongguo Ying Yong Sheng Li Xue Za Zhi Zhongguo Yingyong Shenglixue Zazhi Chin. J Appl Physiol 24:286–289
77. Dick DJ, Lab MJ (1998) Mechanical modulation of stretch-induced premature ventricular beats: induction of a mechanoelectric adaptation period. Cardiovasc Res 38:181–191. https://doi.org/10.1016/s0008-6363(97)00314-3
78. Hansen DE, Borganelli M, Stacy GP Jr et al (1991) Dose-dependent inhibition of stretch-induced arrhythmias by gadolinium in isolated canine ventricles. Evidence for a unique mode of antiarrhythmic action. Circ Res 69:820–831. https://doi.org/10.1161/01.res.69.3.820
79. Nazir SA, Lab MJ (1996) Mechanoelectric feedback in the atrium of the isolated Guinea-pig heart. Cardiovasc Res 32:112–119
80. Parker KK, Lavelle JA, Taylor LK et al (2004) Stretch-induced ventricular arrhythmias during acute ischemia and reperfusion. J Appl Physiol 97:377–383. https://doi.org/10.1152/japplphysiol.01235.2001
81. Parker KK, Taylor LK, Atkinson JB et al (2001) The effects of tubulin-binding agents on stretch-induced ventricular arrhythmias. Eur J Pharmacol 417:131–140. https://doi.org/10.1016/s0014-2999(01)00856-1
82. Reiter MJ, Synhorst DP, Mann DE (1988) Electrophysiological effects of acute ventricular dilatation in the isolated rabbit heart. Circ Res 62:554–562. https://doi.org/10.1161/01.res.62.3.554
83. Seo K, Inagaki M, Nishimura S et al (2010) Structural heterogeneity in the ventricular wall plays a significant role in the initiation of stretch-induced arrhythmias in perfused rabbit right ventricular tissues and whole heart preparations. Circ Res 106:176–184. https://doi.org/10.1161/circresaha.109.203828
84. Stacy GP, Jobe RL, Taylor LK et al (1992) Stretch-induced depolarizations as a trigger of arrhythmias in isolated canine left ventricles. Am J Physiology-Heart Circ Physiol 263:H613–H621. https://doi.org/10.1152/ajpheart.1992.263.2.h613
85. Sideris DA, Chrysos DN, Maliaras GK et al (1988) Effect of acute hypertension on the cardiac rhythm. Experimental observations J Electrocardiol 21:183–191. https://doi.org/10.1016/s0022-0736(88)80015-3
86. Sideris DA, Toumanidis ST, Kostis EB et al (1989) Arrhythmogenic effect of high blood pressure: some observations on its mechanism. Cardiovasc Res 23:983–992. https://doi.org/10.1093/cvr/23.11.983
87. Sideris DA, Toumanidis ST, Kostis EB et al (1991) Effect of adrenergic blockade on pressure-related ventricular arrhythmias. Acta Cardiol 46:215–225

88. Haemers P, Sutherland G, Cikes M et al (2015) Further insights into blood pressure induced premature beats: transient depolarizations are associated with fast myocardial deformation upon pressure decline. Heart Rhythm 12:2305–2315. https://doi.org/10.1016/j.hrthm.2015.06.037
89. Calkins H, Maughan WL, Weisman HF et al (1989) Effect of acute volume load on refractoriness and arrhythmia development in isolated, chronically infarcted canine hearts. Circulation 79:687–697. https://doi.org/10.1161/01.cir.79.3.687
90. Chen RL, Penny DJ, Greve G et al (2004) Stretch-induced regional mechanoelectric dispersion and arrhythmia in the right ventricle of anesthetized lambs. Am J Physiology Hear Circ Physiol 286:H1008–H1014. https://doi.org/10.1152/ajpheart.00724.2003
91. Zabel M, Portnoy S, Franz MR (1996) Effect of sustained load on dispersion of ventricular repolarization and conduction time in the isolated intact rabbit heart. J Cardiovasc Electr 7:9–16. https://doi.org/10.1111/j.1540-8167.1996.tb00455.x
92. Belus A, White E (2003) Streptomycin and intracellular calcium modulate the response of single Guinea-pig ventricular myocytes to axial stretch. J Physiol 546:501–509. https://doi.org/10.1113/jphysiol.2002.027573
93. Benditt DG, Kriett JM, Tobler HG et al (1985) Electrophysiological effects of transient aortic occlusion in intact canine heart. Am J Physiology-Heart Circ Physiol 249:H1017–H1023. https://doi.org/10.1152/ajpheart.1985.249.5.h1017
94. Burton FL, Cobbe SM (1998) Effect of sustained stretch on dispersion of ventricular fibrillation intervals in normal rabbit hearts. Cardiovasc Res 39:351–359. https://doi.org/10.1016/s0008-6363(98)00092-3
95. Calkins H, Levine JH, Kass DA (1991) Electrophysiological effect of varied rate and extent of acute in vivo left ventricular load increase. Cardiovasc Res 25:637–644. https://doi.org/10.1093/cvr/25.8.637
96. Coulshed DS, Cowan JC (1991) Contraction-excitation feedback in an ejecting whole heart model – dependence of action potential duration on left ventricular diastolic and systolic pressures. Cardiovasc Res 25:343–352. https://doi.org/10.1093/cvr/25.4.343
97. Coulshed DS, Cowan JC, Drinkhill MJ et al (1992) The effects of ventricular end-diastolic and systolic pressures on action potential and duration in anaesthetized dogs. J Physiol 457:75–91. https://doi.org/10.1113/jphysiol.1992.sp019365
98. Coulshed DS, Hainsworth R, Cowan JC (1994) The influence of myocardial systolic shortening on action potential duration following changes in left ventricular end-diastolic pressure. J Cardiovasc Electr 5:919–932. https://doi.org/10.1111/j.1540-8167.1994.tb01132.x
99. Dean JW, Lab MJ (1989) Effect of changes in load on monophasic action potential and segment length of pig heart in situ. Cardiovasc Res 23:887–887. https://doi.org/10.1093/cvr/23.10.887
100. Dean JW, Lab MJ (1990) Regional changes in ventricular excitability during load manipulation of the in situ pig heart. J Physiology 429:387–400. https://doi.org/10.1113/jphysiol.1990.sp018263
101. Greve G, Lab MJ, Chen R et al (2001) Right ventricular distension alters monophasic action potential duration during pulmonary arterial occlusion in anaesthetised lambs: evidence for arrhythmogenic right ventricular mechanoelectrical feedback. Exp Physiol 86:651–657. https://doi.org/10.1113/eph8602225
102. Halperin BD, Adler SW, Mann DE et al (1993) Mechanical correlates of contraction-excitation feedback during acute ventricular dilatation. Cardiovasc Res 27:1084–1087. https://doi.org/10.1093/cvr/27.6.1084
103. Horner SM, Dick DJ, Murphy CF et al (1996) Cycle length dependence of the electrophysiological effects of increased load on the myocardium. Circulation 94:1131–1136. https://doi.org/10.1161/01.cir.94.5.1131
104. Lab MJ (1980) Transient depolarisation and action potential alterations following mechanical changes in isolated myocardium. Cardiovasc Res 14:624–637. https://doi.org/10.1093/cvr/14.11.624

105. Lerman BB, Burkhoff D, Yue DT et al (1985) Mechanoelectrical feedback: independent role of preload and contractility in modulation of canine ventricular excitability. J Clin Invest 76: 1843–1850. https://doi.org/10.1172/jci112177
106. Reiter MJ, Zetelaki Z, Kirchhof CJ et al (1994) Interaction of acute ventricular dilatation and d-sotalol during sustained reentrant ventricular tachycardia around a fixed obstacle. Circulation 89:423–431. https://doi.org/10.1161/01.cir.89.1.423
107. Reiter MJ, Landers M, Zetelaki Z et al (1997) Electrophysiological effects of acute dilatation in the isolated rabbit heart: cycle length–dependent effects on ventricular refractoriness and conduction velocity. Circulation 96:4050–4056. https://doi.org/10.1161/01.cir.96.11.4050
108. Sung D, Mills RW, Schettler J et al (2003) Ventricular filling slows epicardial conduction and increases action potential duration in an optical mapping study of the isolated rabbit heart. J Cardiovasc Electr 14:739–749. https://doi.org/10.1046/j.1540-8167.2003.03072.x
109. Wang K, Terrar D, Gavaghan DJ et al (2014) Living cardiac tissue slices: an organotypic pseudo two-dimensional model for cardiac biophysics research. Prog Biophysics Mol Biology 115:314–327. https://doi.org/10.1016/j.pbiomolbio.2014.08.006
110. Wang X-X, Cheng L-X, Chen J-Z et al (2003) Dependence of ventricular wall stress-induced refractoriness changes on pacing cycle lengths and its mechanism. Sheng Li Xue Bao Acta Physiologica Sinica 55:336–338
111. Werdich AA, Brzezinski A, Jeyaraj D et al (2012) The zebrafish as a novel animal model to study the molecular mechanisms of mechano-electrical feedback in the heart. Prog Biophysics Mol Biology 110:154–165. https://doi.org/10.1016/j.pbiomolbio.2012.07.006
112. Quintanilla JG, Moreno J, Archondo T et al (2015) Increased intraventricular pressures are as harmful as the electrophysiological substrate of heart failure in favoring sustained reentry in the swine heart. Heart Rhythm 12:2172–2183. https://doi.org/10.1016/j.hrthm.2015.05.017
113. Quinn TA (2014) The importance of non-uniformities in mechano-electric coupling for ventricular arrhythmias. J Interv Card Electr 39:25–35. https://doi.org/10.1007/s10840-013-9852-0
114. Caporizzo MA, Chen CY, Prosser BL (2019) Cardiac microtubules in health and heart disease. Exp Biol Med 244:1255–1272. https://doi.org/10.1177/1535370219868960
115. Kerr JP, Robison P, Shi G et al (2015) Detyrosinated microtubules modulate mechanotransduction in heart and skeletal muscle. Nat Commun 6:8526. https://doi.org/10.1038/ncomms9526
116. Robison P, Prosser BL (2017) Microtubule mechanics in the working myocyte. J Physiol 595: 3931–3937. https://doi.org/10.1113/jp273046
117. Chen-Izu Y, Izu LT (2017) Mechano-chemo-transduction in cardiac myocytes. J Physiol 595: 3949–3958. https://doi.org/10.1113/jp273101
118. Iribe G, Ward CW, Camelliti P et al (2009) Axial stretch of rat single ventricular cardiomyocytes causes an acute and transient increase in Ca^{2+} spark rate. Circ Res 104:787–795. https://doi.org/10.1161/circresaha.108.193334
119. Izu LT, Kohl P, Boyden PA et al (2020) Mechano-electric and mechano-chemo-transduction in cardiomyocytes. J Physiol 598:1285–1305. https://doi.org/10.1113/jp276494
120. Prosser BL, Ward CW, Lederer WJ (2011) X-ROS signaling: rapid mechano-chemo transduction in heart. Science 333:1440–1445. https://doi.org/10.1126/science.1202768
121. Prosser BL, Ward CW (2014) Mechano-chemo transduction tunes the heartstrings. Sci Signal 7:pe7. https://doi.org/10.1126/scisignal.2005214
122. Kohl P, Bollensdorff C, Garny A (2006) Effects of mechanosensitive ion channels on ventricular electrophysiology: experimental and theoretical models. Exp Physiol 91:307–321. https://doi.org/10.1113/expphysiol.2005.031062
123. Morris CE (2011) Pacemaker, potassium, calcium, sodium: stretch modulation of the voltage-gated channels. In: Kohl P, Sachs F, Franz MR (eds) Cardiac Mechano-electric coupling and arrhythmias, 2nd edn. Oxford University Press, Oxford, pp 42–49
124. Vandenberg JI, Rees SA, Wright AR et al (1996) Cell swelling and ion transport pathways in cardiac myocytes. Cardiovasc Res 32:85–97. https://doi.org/10.1016/s0008-6363(96)00048-x

125. Baumgarten CM, Clemo HF (2003) Swelling-activated chloride channels in cardiac physiology and pathophysiology. Prog Biophysics Mol Biology 82:25–42. https://doi.org/10.1016/s0079-6107(03)00003-8
126. Craelius W (1993) Stretch-activation of rat cardiac myocytes. Exp Physiol 78:411–423. https://doi.org/10.1113/expphysiol.1993.sp003695
127. Riemer TL, Tung L (2003) Stretch-induced excitation and action potential changes of single cardiac cells. Prog Biophysics Mol Biology 82:97–110. https://doi.org/10.1016/s0079-6107(03)00008-7
128. Gannier F, White E, Lacampagne A et al (1994) Streptomycin reverses a large stretch induced increase in $[Ca^{2+}]_i$ in isolated Guinea pig ventricular myocytes. Cardiovasc Res 28:1193–1198. https://doi.org/10.1093/cvr/28.8.1193
129. Smani T, Dionisio N, López JJ et al (2014) Cytoskeletal and scaffolding proteins as structural and functional determinants of TRP channels. Biochimica Et Biophysica Acta Bba - Biomembr 1838:658–664. https://doi.org/10.1016/j.bbamem.2013.01.009
130. Coste B, Mathur J, Schmidt M et al (2010) Piezo1 and Piezo2 are essential components of distinct mechanically activated cation channels. Science 330:55–60. https://doi.org/10.1126/science.1193270
131. Volkers L, Mechioukhi Y, Coste B (2015) Piezo channels: from structure to function. Pflügers Archiv - European J Physiol 467:95–99. https://doi.org/10.1007/s00424-014-1578-z
132. Solis AG, Bielecki P, Steach HR et al (2019) Mechanosensation of cyclical force by PIEZO1 is essential for innate immunity. Nature 573:69–74. https://doi.org/10.1038/s41586-019-1485-8
133. Blythe NM, Muraki K, Ludlow MJ et al (2019) Mechanically activated Piezo1 channels of cardiac fibroblasts stimulate p38 mitogen-activated protein kinase activity and interleukin-6 secretion. J Biol Chem 294:17395–17408. https://doi.org/10.1074/jbc.ra119.009167
134. Cameron BA, Stoyek MR, Bak JJ et al (n.d.) TRPA1 channels are a source of calcium-driven cardiac mechano-arrhythmogenicity. bioRxiv. https://doi.org/10.1101/2020.10.01.321638
135. Watanabe H, Murakami M, Ohba T et al (2008) TRP channel and cardiovascular disease. Pharmacol Therapeut 118:337–351. https://doi.org/10.1016/j.pharmthera.2008.03.008
136. Inoue R, Jian Z, Kawarabayashi Y (2009) Mechanosensitive TRP channels in cardiovascular pathophysiology. Pharmacol Therapeut 123:371–385. https://doi.org/10.1016/j.pharmthera.2009.05.009
137. Nikolaev YA, Cox CD, Ridone P et al (2019) Mammalian TRP ion channels are insensitive to membrane stretch. J Cell Sci 132:jcs238360. https://doi.org/10.1242/jcs.238360
138. Li W, Kohl P, Trayanova N (2006) Myocardial ischemia lowers precordial thump efficacy: an inquiry into mechanisms using three-dimensional simulations. Heart Rhythm 3:179–186. https://doi.org/10.1016/j.hrthm.2005.10.033
139. Wagoner DRV, Lamorgese M (1994) Ischemia potentiates the mechanosensitive modulation of atrial ATP-sensitive potassium channels. Ann N Y Acad Sci 723:392–395. https://doi.org/10.1111/j.1749-6632.1994.tb36755.x
140. Wagoner DRV (1993) Mechanosensitive gating of atrial ATP-sensitive potassium channels. Circ Res 72:973–983. https://doi.org/10.1161/01.res.72.5.973
141. Rog-Zielinska EA, O'Toole ET, Hoenger A et al (2019) Mitochondrial deformation during the cardiac mechanical cycle. Anatomical Rec 302:146–152. https://doi.org/10.1002/ar.23917
142. Belmonte S, Morad M (2008) 'Pressure–flow'-triggered intracellular Ca^{2+} transients in rat cardiac myocytes: possible mechanisms and role of mitochondria. J Physiol 586:1379–1397. https://doi.org/10.1113/jphysiol.2007.149294
143. Belmonte S, Morad M (2008) Shear fluid-induced Ca^{2+} release and the role of mitochondria in rat cardiac myocytes. Ann N Y Acad Sci 1123:58–63. https://doi.org/10.1196/annals.1420.007
144. Miragoli M, Sanchez-Alonso JL, Bhargava A et al (2015) Microtubule-dependent mitochondria alignment regulates calcium release in response to nanomechanical stimulus in heart myocytes. Cell Rep 14:140–151. https://doi.org/10.1016/j.celrep.2015.12.014

145. Morad M, Javaheri A, Risius T et al (2005) Multimodality of Ca^{2+} signaling in rat atrial myocytes. Ann N Y Acad Sci 1047:112–121. https://doi.org/10.1196/annals.1341.010
146. Janke C, Bulinski JC (2011) Post-translational regulation of the microtubule cytoskeleton: mechanisms and functions. Nat Rev Mol Cell Bio 12:773–786. https://doi.org/10.1038/nrm3227
147. Portran D, Schaedel L, Xu Z et al (2017) Tubulin acetylation protects long-lived microtubules against mechanical ageing. Nat Cell Biol 19:391–398. https://doi.org/10.1038/ncb3481
148. Hirata H, Tatsumi H, Hayakawa K et al (2015) Non-channel mechanosensors working at focal adhesion-stress fiber complex. Pflügers Archiv - European J Physiol 467:141–155. https://doi.org/10.1007/s00424-014-1558-3
149. Israeli-Rosenberg S, Chen C, Li R et al (2015) Caveolin modulates integrin function and mechanical activation in the cardiomyocyte. FASEB J 29:374–384. https://doi.org/10.1096/fj.13-243139
150. Robison P, Caporizzo MA, Ahmadzadeh H et al (2016) Detyrosinated microtubules buckle and bear load in contracting cardiomyocytes. Science 352:aaf0659. https://doi.org/10.1126/science.aaf0659
151. Brangwynne CP, MacKintosh FC, Kumar S et al (2006) Microtubules can bear enhanced compressive loads in living cells because of lateral reinforcement. J Cell Biol 173:733–741. https://doi.org/10.1083/jcb.200601060
152. Salomon AK, Okami N, Heffler J et al (2021) Desmin intermediate filaments and tubulin detyrosination stabilize growing microtubules in the cardiomyocyte bioRxiv 2021.05.26.445641. https://doi.org/10.1101/2021.05.26.445641
153. Belmadani S, Poüs C, Ventura-Clapier R et al (2002) Post-translational modifications of cardiac tubulin during chronic heart failure in the rat. Mol Cell Biochem 237:39–46. https://doi.org/10.1023/a:1016554104209
154. Xu Z, Schaedel L, Portran D et al (2017) Microtubules acquire resistance from mechanical breakage through intralumenal acetylation. Science 356:328–332. https://doi.org/10.1126/science.aai8764
155. Coleman AK, Joca HC, Shi G et al (2021) Tubulin acetylation increases cytoskeletal stiffness to regulate mechanotransduction in striated muscle. J Gen Physiol 153:e202012743. https://doi.org/10.1085/jgp.202012743
156. ter Keurs HEDJ, Boyden PA (2007) Calcium and arrhythmogenesis. Physiol Rev 87:457–506. https://doi.org/10.1152/physrev.00011.2006
157. Landstrom AP, Dobrev D, Wehrens XHT (2017) Calcium signaling and cardiac arrhythmias. Circ Res 120:1969–1993. https://doi.org/10.1161/circresaha.117.310083
158. Iribe G, Kohl P (2008) Axial stretch enhances sarcoplasmic reticulum Ca^{2+} leak and cellular Ca^{2+} reuptake in Guinea pig ventricular myocytes: experiments and models. Prog Biophysics Mol Biology 97:298–311. https://doi.org/10.1016/j.pbiomolbio.2008.02.012
159. Schönleitner P, Schotten U, Antoons G (2017) Mechanosensitivity of microdomain calcium signalling in the heart. Prog Biophysics Mol Biology 130:288–301. https://doi.org/10.1016/j.pbiomolbio.2017.06.013
160. Allen DG, Kentish JC (1988) Calcium concentration in the myoplasm of skinned ferret ventricular muscle following changes in muscle length. J Physiol 407:489–503. https://doi.org/10.1113/jphysiol.1988.sp017427
161. Wakayama Y, Miura M, Stuyvers BD et al (2005) Spatial nonuniformity of excitation–contraction coupling causes arrhythmogenic Ca^{2+}; waves in rat cardiac muscle. Circ Res 96:1266–1273. https://doi.org/10.1161/01.res.0000172544.56818.54
162. Andersson DA, Gentry C, Moss S et al (2008) Transient receptor potential A1 is a sensory receptor for multiple products of oxidative stress. J Neurosci 28:2485–2494. https://doi.org/10.1523/jneurosci.5369-07.2008
163. Zurborg S, Yurgionas B, Jira JA et al (2007) Direct activation of the ion channel TRPA1 by Ca^{2+}. Nat Neurosci 10:277–279. https://doi.org/10.1038/nn1843

164. Iribe G, Jin H, Kaihara K et al (2010) Effects of axial stretch on sarcolemmal BKCa channels in post-hatch chick ventricular myocytes. Exp Physiol 95:699–711. https://doi.org/10.1113/expphysiol.2009.051896
165. Goldblum RR, McClellan M, White K et al (2021) Oxidative stress pathogenically remodels the cardiac myocyte cytoskeleton via structural alterations to the microtubule lattice. Dev Cell 56:2252. https://doi.org/10.1016/j.devcel.2021.07.004
166. Cannell MB (2009) Pulling on the heart strings: a new mechanism within Starling's law of the heart? Circ Res 104:715–716. https://doi.org/10.1161/circresaha.109.195511
167. Belevych AE, Terentyev D, Viatchenko-Karpinski S et al (2009) Redox modification of ryanodine receptors underlies calcium alternans in a canine model of sudden cardiac death. Cardiovasc Res 84:387–395. https://doi.org/10.1093/cvr/cvp246
168. Prosser BL, Ward CW, Lederer WJ (2010) Subcellular Ca^{2+} signaling in the heart: the role of ryanodine receptor sensitivity. J Gen Physiology 136:135–142. https://doi.org/10.1085/jgp.201010406
169. Prosser BL, Khairallah RJ, Ziman AP et al (2013) X-ROS signaling in the heart and skeletal muscle: stretch-dependent local ROS regulates $[Ca^{2+}]_i$. J Mol Cell Cardiol 58:172–181. https://doi.org/10.1016/j.yjmcc.2012.11.011
170. Huke S, Knollmann BC (2010) Increased myofilament Ca^{2+}-sensitivity and arrhythmia susceptibility. J Mol Cell Cardiol 48:824–833. https://doi.org/10.1016/j.yjmcc.2010.01.011
171. ter Keurs HEDJ, Shinozaki T, Zhang YM et al (2008) Sarcomere mechanics in uniform and non-uniform cardiac muscle: a link between pump function and arrhythmias. Prog Biophysics Mol Biology 97:312–331. https://doi.org/10.1016/j.pbiomolbio.2008.02.013
172. Cameron BA, Kai H, Kaihara K et al (2020) Ischemia enhances the acute stretch-induced increase in calcium spark rate in ventricular myocytes. Front Physiol 11:289. https://doi.org/10.3389/fphys.2020.00289
173. Prosser BL, Ward CW, Lederer WJ (2013) X-ROS signalling is enhanced and graded by cyclic cardiomyocyte stretch. Cardiovasc Res 98:307–314. https://doi.org/10.1093/cvr/cvt066
174. Solovyova O, Katsnelson LB, Konovalov PV et al (2014) The cardiac muscle duplex as a method to study myocardial heterogeneity. Prog Biophysics Mol Biology 115:115–128. https://doi.org/10.1016/j.pbiomolbio.2014.07.010
175. Tyberg JV, Parmley WW, Sonnenblick EH (1969) In-vitro studies of myocardial asynchrony and regional hypoxia. Circ Res 25:569–579. https://doi.org/10.1161/01.res.25.5.569
176. Carmeliet E (1999) Cardiac ionic currents and acute ischemia: from channels to arrhythmias. Physiol Rev 79:917–1017. https://doi.org/10.1152/physrev.1999.79.3.917
177. Bollensdorff C, Lab M (2011) Stretch effects on potassium accumulation and alternans in pathological myocardium. In: Kohl P, Sachs F, Franz MR (eds) Cardiac Mechano-electric coupling and arrhythmias, 2nd edn. Oxford University Press, Oxford, pp 173–179
178. Janse MJ, Wit AL (1989) Electrophysiological mechanisms of ventricular arrhythmias resulting from myocardial ischemia and infarction. Physiol Rev 69:1049–1169. https://doi.org/10.1152/physrev.1989.69.4.1049
179. Gallagher KP, Gerren RA, Choy M et al (1987) Subendocardial segment length shortening at lateral margins of ischemic myocardium in dogs. Am J Physiology-Heart Circ Physiol 253: H826–H837. https://doi.org/10.1152/ajpheart.1987.253.4.h826
180. Leuven SLV, Waldman LK, McCulloch AD et al (1994) Gradients of epicardial strain across the perfusion boundary during acute myocardial ischemia. Am J Physiology-Heart Circ Physiol 267:H2348–H2362. https://doi.org/10.1152/ajpheart.1994.267.6.h2348
181. Prinzen FW, Arts T, Hoeks APG et al (1989) Discrepancies between myocardial blood flow and fiber shortening in the ischemic border zone as assessed with video mapping of epicardial deformation. Pflugers Arch 415:220–229. https://doi.org/10.1007/bf00370596
182. Sakai K, Watanabe K, Millard RW (1985) Defining the mechanical border zone: a study in the pig heart. Am J Physiology-Heart Circ Physiol 249:H88–H94. https://doi.org/10.1152/ajpheart.1985.249.1.h88

183. Theroux P, Franklin D, Ross J et al (1974) Regional myocardial function during acute coronary artery occlusion and its modification by pharmacologic agents in the dog. Circ Res 35:896–908. https://doi.org/10.1161/01.res.35.6.896
184. Lopaschuk GD, Ussher JR, Folmes CDL et al (2010) Myocardial fatty acid metabolism in health and disease. Physiol Rev 90:207–258. https://doi.org/10.1152/physrev.00015.2009
185. Stanley WC (2001) Changes in cardiac metabolism: a critical step from stable angina to ischaemic cardiomyopathy. Eur Heart J Suppl 3:O2–O7. https://doi.org/10.1016/s1520-765x(01)90147-6
186. Bittl JA, Weisfeldt ML, Jacobus WE (1985) Creatine kinase of heart mitochondria. The progressive loss of enzyme activity during in vivo ischemia and its correlation to depressed myocardial function. J Biol Chem 260:208–214
187. Califf RM, Abdelmeguid AE, Kuntz RE et al (1998) Myonecrosis after revascularization procedures. J Am Coll Cardiol 31:241–251. https://doi.org/10.1016/s0735-1097(97)00506-8
188. Cao F, Zervou S, Lygate CA (2018) The creatine kinase system as a therapeutic target for myocardial ischaemia–reperfusion injury. Biochem Soc T 46:1119–1127. https://doi.org/10.1042/bst20170504
189. Lygate CA, Bohl S, ten Hove M et al (2012) Moderate elevation of intracellular creatine by targeting the creatine transporter protects mice from acute myocardial infarction. Cardiovasc Res 96:466–475. https://doi.org/10.1093/cvr/cvs272
190. Takahashi M, Yokoshiki H, Mitsuyama H et al (2021) SK channel blockade prevents hypoxia-induced ventricular arrhythmias through inhibition of Ca^{2+}/voltage uncoupling in hypertrophied hearts. Am J Physiology-Heart Circ Physiol 320:H1456–H1469. https://doi.org/10.1152/ajpheart.00777.2020
191. Lang D, Holzem K, Kang C et al (2015) Arrhythmogenic remodeling of β2 versus β1 adrenergic signaling in the human failing heart. Circ Arrhythmia Electrophysiol 8:409–419. https://doi.org/10.1161/circep.114.002065
192. Tang L, Joung B, Ogawa M et al (2012) Intracellular calcium dynamics, shortened action potential duration, and late-phase 3 early afterdepolarization in Langendorff-perfused rabbit ventricles. J Cardiovasc Electr 23:1364–1371. https://doi.org/10.1111/j.1540-8167.2012.02400.x
193. Michailova A, Lorentz W, McCulloch A (2007) Modeling transmural heterogeneity of K_{ATP} current in rabbit ventricular myocytes. Am J Physiol-Cell Ph 293:C542–C557. https://doi.org/10.1152/ajpcell.00148.2006
194. Baumeister P, Quinn TA (2016) Altered calcium handling and ventricular arrhythmias in acute ischemia. Clin Medicine Insights Cardiol 10s1:61–69. https://doi.org/10.4137/cmc.s39706
195. Coronel R, Fiolet JW, Wilms-Schopman FJ et al (1988) Distribution of extracellular potassium and its relation to electrophysiologic changes during acute myocardial ischemia in the isolated perfused porcine heart. Circulation 77:1125–1138. https://doi.org/10.1161/01.cir.77.5.1125
196. Coronel R, Wilms-Schopman FJ, Opthof T et al (1991) Injury current and gradients of diastolic stimulation threshold, TQ potential, and extracellular potassium concentration during acute regional ischemia in the isolated perfused pig heart. Circ Res 68:1241–1249. https://doi.org/10.1161/01.res.68.5.1241
197. de Groot JR, Coronel R (2004) Acute ischemia-induced gap junctional uncoupling and arrhythmogenesis. Cardiovasc Res 62:323–334. https://doi.org/10.1016/j.cardiores.2004.01.033
198. Dries E, Amoni M, Vandenberk B et al (2020) Altered adrenergic response in myocytes bordering a chronic myocardial infarction underlies in vivo triggered activity and repolarization instability. J Physiology 598:2875–2895. https://doi.org/10.1113/jp278839
199. Mills RW, Narayan SM, McCulloch AD (2008) Mechanisms of conduction slowing during myocardial stretch by ventricular volume loading in the rabbit. Am J Physiology-Heart Circ Physiol 295:H1270–H1278. https://doi.org/10.1152/ajpheart.00350.2008

200. de Oliveira BL, Pfeiffer ER, Sundnes J et al (2015) Increased cell membrane capacitance is the dominant mechanism of stretch-dependent conduction slowing in the rabbit heart: a computational study. Cell Mol Bioeng 8:237–246. https://doi.org/10.1007/s12195-015-0384-9
201. Kohl P, Cooper PJ, Holloway H (2003) Effects of acute ventricular volume manipulation on in situ cardiomyocyte cell membrane configuration. Prog Biophysics Mol Biology 82:221–227. https://doi.org/10.1016/s0079-6107(03)00024-5
202. Rog-Zielinska EA, Scardigli M, Peyronnet R et al (2021) Beat-by-beat cardiomyocyte T-tubule deformation drives tubular content exchange. Circ Res 128:203–215. https://doi.org/10.1161/circresaha.120.317266
203. Cannell M, Cheng H, Lederer W (1995) The control of calcium release in heart muscle. Science 268:1045–1049. https://doi.org/10.1126/science.7754384
204. Hj H, Hoeher M, Risse JH (1987) Cardiac energetics, basic mechanisms and clinical implications. Basic Res Cardiol 82(Suppl 2):301–310. https://doi.org/10.1007/978-3-662-11289-2_29
205. Jie X, Gurev V, Trayanova N (2010) Mechanisms of mechanically induced spontaneous arrhythmias in acute regional ischemia. Circ Res 106:185–192. https://doi.org/10.1161/circresaha.109.210864
206. Limbu S, Hoang-Trong TM, Prosser BL et al (2015) Modeling local X-ROS and calcium signaling in the heart. Biophys J 109:2037–2050. https://doi.org/10.1016/j.bpj.2015.09.031
207. Ferrari R, Ceconi C, Curello S et al (1985) Oxygen-mediated myocardial damage during ischaemia and reperfusion: role of the cellular defences against oxygen toxicity. J Mol Cell Cardiol 17:937–945. https://doi.org/10.1016/s0022-2828(85)80074-2
208. Miura M, Taguchi Y, Handoh T et al (2018) Regional increase in ROS within stretched region exacerbates arrhythmias in rat trabeculae with nonuniform contraction. Pflügers Archiv - European J Physiol 470:1349–1357. https://doi.org/10.1007/s00424-018-2152-x
209. Meents JE, Fischer MJM, McNaughton PA (2016) Agonist-induced sensitisation of the irritant receptor ion channel TRPA1. J Physiology 594:6643–6660. https://doi.org/10.1113/jp272237
210. Tan JHC, Liu W, Saint DA (2004) Differential expression of the mechanosensitive potassium channel TREK-1 in epicardial and endocardial myocytes in rat ventricle. Exp Physiol 89:237–242. https://doi.org/10.1113/expphysiol.2003.027052
211. Coronel R, Wilms-Schopman FJG, Opthof T et al (2009) Dispersion of repolarization and arrhythmogenesis. Heart Rhythm 6:537–543. https://doi.org/10.1016/j.hrthm.2009.01.013
212. Whelton PK, Carey RM, Aronow WS et al (2018) 2017 ACC/AHA/AAPA/ABC/ACPM/AGS/APhA/ASH/ASPC/NMA/PCNA guideline for the prevention, detection, evaluation, and management of high blood pressure in adults. Hypertension 71:e13–e115. https://doi.org/10.1161/hyp.0000000000000065
213. Frohlich ED, Susic D (2012) Pressure overload. Heart Fail Clin 8:21–32. https://doi.org/10.1016/j.hfc.2011.08.005
214. Konstam MA, Kramer DG, Patel AR et al (2011) Left ventricular remodeling in heart failure current concepts in clinical significance and assessment. JACC Cardiovasc Imaging 4:98–108. https://doi.org/10.1016/j.jcmg.2010.10.008
215. Lorell BH, Carabello BA (2000) Left ventricular hypertrophy. Circulation 102:470–479. https://doi.org/10.1161/01.cir.102.4.470
216. Kahan T, Bergfeldt L (2005) Left ventricular hypertrophy in hypertension: its arrhythmogenic potential. Heart 91:250. https://doi.org/10.1136/hrt.2004.042473
217. McLenachan JM, Henderson E, Morris KI et al (1987) Ventricular arrhythmias in patients with hypertensive left ventricular hypertrophy. New Engl J Medicine 317:787–792. https://doi.org/10.1056/nejm198709243171302
218. Rapsomaniki E, Timmis A, George J et al (2014) Blood pressure and incidence of twelve cardiovascular diseases: lifetime risks, healthy life-years lost, and age-specific associations in 1·25 million people. Lancet 383:1899–1911. https://doi.org/10.1016/s0140-6736(14)60685-1
219. Shenasa M, Shenasa H (2017) Hypertension, left ventricular hypertrophy, and sudden cardiac death. Int J Cardiol 237:60–63. https://doi.org/10.1016/j.ijcard.2017.03.002

220. Tereshchenko LG, Soliman EZ, Davis BR et al (2017) Risk stratification of sudden cardiac death in hypertension. J Electrocardiol 50:798–801. https://doi.org/10.1016/j.jelectrocard.2017.08.012
221. Yiu K-H, Tse H-F (2008) Hypertension and cardiac arrhythmias: a review of the epidemiology, pathophysiology and clinical implications. J Hum Hypertens 22:380–388. https://doi.org/10.1038/jhh.2008.10
222. Dunn FG, Pringle SD (1987) Left ventricular hypertrophy and myocardial ischemia in systemic hypertension. Am J Cardiol 60:19–22. https://doi.org/10.1016/0002-9149(87)90454-1
223. Katz AM (2002) Ernest Henry Starling, his predecessors, and the "Law of the Heart.". Circulation 106:2986–2992. https://doi.org/10.1161/01.cir.0000040594.96123.55
224. Maliken BD, Molkentin JD (2018) Undeniable evidence that the adult mammalian heart lacks an endogenous regenerative stem cell. Circulation 138:806–808. https://doi.org/10.1161/circulationaha.118.035186
225. Cooper G (2000) Cardiocyte cytoskeleton in hypertrophied myocardium. Heart Fail Rev 5:187–201. https://doi.org/10.1023/a:1009836918377
226. ter Keurs HEDJ (1998) Microtubules in cardiac hypertrophy. Circ Res 82:828–831. https://doi.org/10.1161/01.res.82.7.828
227. Caporizzo MA, Chen CY, Bedi K et al (2020) Microtubules increase diastolic stiffness in failing human cardiomyocytes and myocardium. Circulation 141:902–915. https://doi.org/10.1161/circulationaha.119.043930
228. Chen CY, Caporizzo MA, Bedi K et al (2018) Suppression of detyrosinated microtubules improves cardiomyocyte function in human heart failure. Nat Med 24:1225–1233. https://doi.org/10.1038/s41591-018-0046-2
229. Tagawa H, Wang N, Narishige T et al (1997) Cytoskeletal mechanics in pressure-overload cardiac hypertrophy. Circ Res 80:281–289. https://doi.org/10.1161/01.res.80.2.281
230. Yamamoto S, Tsutsui H, Takahashi M et al (1998) Role of microtubules in the viscoelastic properties of isolated cardiac muscle. J Mol Cell Cardiol 30:1841–1853. https://doi.org/10.1006/jmcc.1998.0747
231. Tagawa H, Koide M, Sato H et al (1998) Cytoskeletal role in the transition from compensated to decompensated hypertrophy during adult canine left ventricular pressure overloading. Circ Res 82:751–761. https://doi.org/10.1161/01.res.82.7.751
232. Nishimura S, Nagai S, Katoh M et al (2006) Microtubules modulate the stiffness of cardiomyocytes against shear stress. Circ Res 98:81–87. https://doi.org/10.1161/01.res.0000197785.51819.e8
233. Zile MR, Richardson K, Cowles MK et al (1998) Constitutive properties of adult mammalian cardiac muscle cells. Circulation 98:567–579. https://doi.org/10.1161/01.cir.98.6.567
234. Koide M, Hamawaki M, Narishige T et al (2000) Microtubule depolymerization normalizes in vivo myocardial contractile function in dogs with pressure-overload left ventricular hypertrophy. Circulation 102:1045–1052. https://doi.org/10.1161/01.cir.102.9.1045
235. Tsutsui H, Ishihara K, Cooper G (1993) Cytoskeletal role in the contractile dysfunction of hypertrophied myocardium. Science 260:682–687. https://doi.org/10.1126/science.8097594
236. Sato H, Nagai T, Kuppuswamy D et al (1997) Microtubule stabilization in pressure overload cardiac hypertrophy. J Cell Biol 139:963–973. https://doi.org/10.1083/jcb.139.4.963
237. Nishimura RA, Housmans PR, Hatle LK et al (1989) Assessment of diastolic function of the heart: background and current applications of doppler echocardiography. Part i. physiologic and pathophysiologic features. Mayo Clin Proc 64:71–81. https://doi.org/10.1016/s0025-6196(12)65305-1
238. Nishimura RA, Abel MD, Hatle LK et al (1989) Assessment of diastolic function of the heart: background and current applications of doppler echocardiography. Part ii. Clinical studies. Mayo Clin Proc 64:181–204. https://doi.org/10.1016/s0025-6196(12)65673-0

239. Kreitzer G, Liao G, Gundersen GG (1999) Detyrosination of tubulin regulates the interaction of intermediate filaments with microtubules in vivo via a kinesin-dependent mechanism. Mol Biol Cell 10:1105–1118. https://doi.org/10.1091/mbc.10.4.1105
240. Liao G, Gundersen GG (1998) Kinesin is a candidate for cross-bridging microtubules and intermediate filaments. J Biol Chem 273:9797–9803. https://doi.org/10.1074/jbc.273.16.9797
241. Cheng G, Takahashi M, Shunmugavel A et al (2010) Basis for MAP4 dephosphorylation-related microtubule network densification in pressure overload cardiac hypertrophy. J Biol Chem 285:38125–38140. https://doi.org/10.1074/jbc.m110.148650
242. Takahashi M, Shiraishi H, Ishibashi Y et al (2003) Phenotypic consequences of β 1 -tubulin expression and MAP4 decoration of microtubules in adult cardiocytes. Am J Physiology-Heart Circ Physiol 285:H2072–H2083. https://doi.org/10.1152/ajpheart.00396.2003
243. Guichard JL, Rogowski M, Agnetti G et al (2017) Desmin loss and mitochondrial damage precede left ventricular systolic failure in volume overload heart failure. Am J Physiology-Heart Circ Physiol 313:H32–H45. https://doi.org/10.1152/ajpheart.00027.2017
244. Lewis YE, Moskovitz A, Mutlak M et al (2018) Localization of transcripts, translation, and degradation for spatiotemporal sarcomere maintenance. J Mol Cell Cardiol 116:16–28. https://doi.org/10.1016/j.yjmcc.2018.01.012
245. Scarborough EA, Uchida K, Vogel M et al (2021) Microtubules orchestrate local translation to enable cardiac growth. Nat Commun 12:1547. https://doi.org/10.1038/s41467-021-21685-4
246. Scholz D, Baicu CF, Tuxworth WJ et al (2008) Microtubule-dependent distribution of mRNA in adult cardiocytes. Am J Physiology-Heart Circ Physiol 294:H1135–H1144. https://doi.org/10.1152/ajpheart.01275.2007
247. Goswami C, Hucho T (2008) Submembraneous microtubule cytoskeleton: biochemical and functional interplay of TRP channels with the cytoskeleton. FEBS J 275:4684–4699. https://doi.org/10.1111/j.1742-4658.2008.06617.x
248. Clemo HF, Stambler BS, Baumgarten CM (1999) Swelling-activated chloride current is persistently activated in ventricular myocytes from dogs with tachycardia-induced congestive heart failure. Circ Res 84:157–165. https://doi.org/10.1161/01.res.84.2.157
249. Kuroda J, Sadoshima J (2010) NADPH oxidase and cardiac failure. J Cardiovasc Transl 3: 314–320. https://doi.org/10.1007/s12265-010-9184-8

Computational Biomechanics of Ventricular Dyssynchrony and Resynchronization Therapy

Angela W. C. Lee, Marina Strocchi, Cristobal Rodero, Caroline Mendonca Costa, and Steven A. Niederer

1 Introduction

Heart failure (HF) is a complex disease causing enormous social and economic burden worldwide, with some of the most recent estimations pointing to a prevalence of 64.34 million cases and costing 346.17 billion dollars in the United States alone [1]. This disease results from structural or functional abnormalities impairing the ability of the heart to pump blood and deliver sufficient oxygen and nutrients to meet the body's metabolic needs. As HF progresses, the heart changes or "remodels," affecting cardiac function over time. These changes are both macroscopic, with changes in mass and shape, and microscopic, with changes in the cellular structure and the distribution and density of the ionic channels. It is also possible that the heart reverts to a less pathological phenotype over time, a process known as "reverse remodeling." This positive progression is usually inferred clinically using surrogate markers such as a reduction in ventricular volume and mass, as well as improvements in the contractile function of the heart. In Yu et al. [2], the authors showed the existence of a positive correlation between reverse remodeling and better therapeutic outcomes for patients.

In a healthy heart, an electrical wave spreads throughout the atria and passes slowly through the atrioventricular (AV) node, before rapidly spreading through the bundle of His, down through the right and left bundle branches and into the Purkinje fibers to synchronously activate the normal ventricular myocardium. In left bundle branch block (LBBB), as per the name, an electrical blockage occurs along the left bundle branch, causing a delay in the activation of the left ventricle and consequently

A. W. C. Lee · M. Strocchi · C. Rodero · C. Mendonca Costa · S. A. Niederer (✉)
Cardiac Electro-Mechanics Research Group, School of Biomedical Engineering and Imaging Sciences, King's College London, London, UK
e-mail: steven.niederer@kcl.ac.uk

M. Hecker, D. J. Duncker (eds.), *Cardiac Mechanobiology in Physiology and Disease*, Cardiac and Vascular Biology 9,
https://doi.org/10.1007/978-3-031-23965-6_12

dyssynchronous activation of the ventricles. This is shown in an ECG with a prolonged QRS duration. In such cases, some of the largest clinical trials (REVERSE, MADIT-CRT, or RAFT) have shown how cardiac resynchronization therapy (CRT) can be an effective treatment and promote reverse remodeling for patients with mild to severe HF [3–5].

In CRT, the heart is artificially paced in the right atrium (RA), right ventricle (RV), and left ventricle (LV) to resynchronize the electrical activation and subsequent mechanical contraction of the heart. The time delay between activation at the atria and at the ventricles (AVD) as well as the time delay between the ventricles (VVD) can be set by the clinicians. Optimization of the AVD can improve the left atrial contribution to LV filling, stroke volume, and cardiac output and thus is an important factor in CRT response [4, 6, 7]. Despite the benefits of CRT, within the cohort of HF patients that are indicated for an implant, there remains 40–50% of patients that do not respond positively to (conventional epicardial) CRT [8, 9]. To address this issue, novel alternative methods of CRT delivery have been explored in recent years, such as multisite pacing (MSP), endocardial pacing, His bundle pacing, or left bundle branch pacing.

MSP stimulates multiple LV sites via leads placed in several CS tributary veins (multiple leads) or with a multipolar LV pacing lead (single lead with two or more electrodes). Although preliminary studies have shown MSP can improve CRT response in both the acute [10–15] and chronic timescales [16, 17], more recent work in porcine models [18] and human patients [19] found that MSP does not significantly improve acute CRT response. For a more definitive answer as to the benefits of MSP, one would have to wait for the results of larger randomized trials such as the SMART multisite pacing sub-study [20].

One of the main disadvantages of conventional epicardial CRT and MSP is that potential LV pacing locations are restricted, as leads are implanted within the coronary sinus veins while avoiding phrenic nerve stimulation. In contrast, endocardial pacing is unconstrained by venous anatomy, allowing for targeted pacing at non-scarred, latest activation locations. Initial studies of endocardial pacing have reported that 47% of patients that do not respond to conventional CRT showed a positive response to endocardial CRT [21]. Novel wireless endocardial pacing systems have also been shown to reliably deliver pacing and improved responses in patients that do not respond to conventional therapy [22].

As opposed to conventional CRT and MSP, where electrical activation spreads from the LV epicardium to the endocardium, His bundle pacing and left bundle branch pacing aim to restore synchronous activation prior to the block by pacing the ventricular fast conducting system [23, 24]. These methods have been shown to be superior or comparable to standard CRT in resynchronizing the electrical activation of the ventricles [25] [26], although left bundle pacing required AV delay optimization to prevent delayed activation of the RV [27, 28].

In any of its modalities, the goal of CRT is to stop or reverse the progression of HF by resynchronizing the electrical activation and mechanical contraction of the ventricles leading to a functional improvement in the pumping of blood throughout the circulatory system. Previous studies have found that one of the main reasons for the high number of non-responders is that the device settings or the lead positions were suboptimal [29]. Computer models of the heart have thus been used to identify the optimal lead location and pacing settings of the heart to predict the response of the heart to CRT, as well as to better understand the underlying pathologies that give rise to cardiac dyssynchrony [28, 30–36]. Moreover, advances in cardiac computational modelling now provide platforms for performing in silico clinical studies [37, 38].

Cardiac models can be used to simulate the electrical activity of the heart from the cellular level through to the tissue level (electrophysiology models [39]), giving rise to contraction of the ventricles (mechanical models [40]), as well as the pumping of the blood throughout the cardiac system (circulatory models [41]). These modeling methods have been used to provide insight into the mechanisms behind LBBB dyssynchrony and reverse remodeling due to CRT, to optimize lead location and device settings, to investigate the efficacy of new pacing technologies, as well as to predict clinical outcomes for patients. Shape analysis can also be used to link specific parts of the anatomy (whether from images directly or using 3D models) directly to clinical outcomes, thereby skipping the simulation step required in electrophysiology, mechanical, or circulatory models [42].

The relationship between cardiac electromechanics models and their inputs and outputs is shown in Fig. 1. Medical images and prior knowledge are used as inputs to generate anatomical models of the heart that are patient- or subject-specific. Patient-specific anatomical models can be used to study pathological changes in shape and electromechanics of the heart. If available, electrical measurements can be used to parameterize and validate the electrophysiology models of the heart, while functional measurements are used as inputs to the circulatory and mechanical models. Prior knowledge from literature such as fiber characteristics are used to constrain the models to simulate physiologically plausible conditions. The outcomes from the models include changes in the mechanical, electrical, hemodynamic, and anatomical responses to LBBB and CRT. In this review, we will focus on the contribution of electrophysiology, mechanics, and circulatory computer models of the heart to understanding mechanisms underlying electrical and mechanical dyssynchrony and CRT response.

2 Modelling Anatomy

2.1 Geometry

Idealized shape models of the heart were first used to investigate the electromechanics of the heart under LBBB and CRT conditions [43]. Generic shape models are still regularly used as they can provide mechanistic insight into CRT response. These models allow for the investigation of the underlying substrate [44], the importance of scar location [45] and total scar burden [46, 47], pacing device settings, and pacing lead locations [34]. In more recent years, there has been a

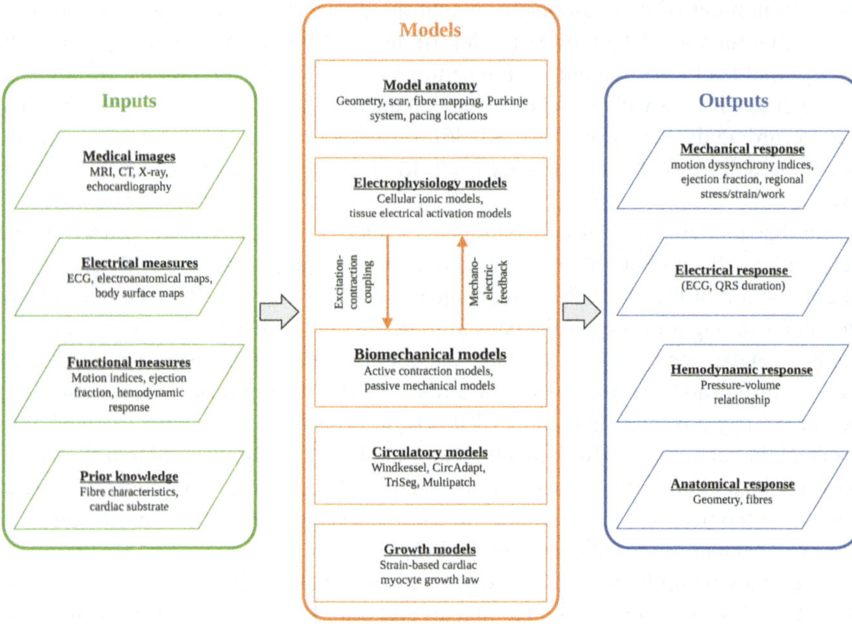

Fig. 1 Schematic of the modelling pipeline for biophysical simulations of LBBB and CRT response. Patient data and prior knowledge can be used to create and parameterize the anatomical, electrophysiological, biomechanical, circulatory, or growth models of the heart. The model outputs can be used to infer and integrate information regarding the diseased heart, generate hypotheses, and predict responses to LBBB and CRT. The current state of the art models does not encompass this entire theoretical framework, but rather only aspects of it depending on the availability of data, computational resources, and the clinical question of interest

shift towards personalized shape models for in silico simulations of the electromechanics of the heart. Segmentations of noninvasive anatomical medical images (CT, MR, X-ray, and echo) of the heart have been used to build subject-specific canine models [48, 49] or patient-specific models [19, 33, 50–52]. Computational meshes built from segmentations of the heart can be used for shape analysis alone [53] or electrophysiological and mechanical simulations [30].

Warriner et al. [53] used statistical shape analysis on pre-implant LV meshes of patients undergoing CRT. This technique allowed the authors to create an "average" heart from 50 patients and find the anatomical remodeling pattern that best discriminates between responders and non-responders. They found that not simply size, but an asymmetric thickening in the pre-implant LV shape, is an independent predictor of a positive response to CRT.

Lee et al. [54] used the same CRT cohort as in Warriner 2018 and over 1,000 healthy subjects to assess the impact of LV size on QRS duration disaggregating by sex. Using electrophysiology simulations and statistical shape models, the authors found that LV size is a mechanistic explanation for sex differences on CRT response.

Moreover, the simulations revealed that lower QRS duration thresholds for females in CRT guidelines could benefit diagnosis and response.

Patient-specific models simulating the electrical and mechanical activation of the heart have typically been focused on the ventricles [55–57]. Recently, there has been increased interest in the role that the atria and ventricular filling play in CRT response optimization [58]. Therefore, in future studies, the importance of atria and other cardiac structures, such as the pericardium [40, 59, 60], valves, and blood vessels, in simulating CRT and LBBB will need to be determined.

2.2 *Fibers*

The mechanical and electrical properties of the heart are highly dependent on the orientation of the fibers within the cardiac tissue. Studies have found that electrical activation spreads two to seven times faster along the direction of the fibers rather than across them [61–63]. Mechanical stiffness is also affected by the fiber orientation, with biaxial tests showing an increase in stiffness along the fiber directions of cardiac tissues [64], while the force from active contraction acts primarily along the fiber directions. For these reasons, a physiological representation of the local microstructure organization of the myocardium is fundamental.

Heart biomechanics models [31, 33, 43, 44, 51, 52, 65–71] have typically used rule-based methods [65] based on diffusion tensor MRI (DT-MRI) or histological studies to define fiber orientation. Personalization of the fiber orientation through in vivo [72, 73] and ex vivo DT-MRI [74, 75] have also been investigated.

In vivo DT-MRI has been successfully performed on human volunteers [76–78] and patients with hypertrophic cardiomyopathy [79], requiring multiple breath holds to acquire an averaged signal for each image slice. While DT-MRI shows promise as a method for subject-specific fiber orientation personalization, technical challenges remain to be overcome before this method can be applied outside of the research space. Low signal to noise ratio, the bulk motion of the heart obfuscating the fiber orientation measurements, as well as multiple breath holds to acquire averaged imaging data remain as challenges to be overcome before this method can be used widely to incorporate personalized fiber orientations in patient-specific computer models of the heart. Studies have found that the electrical activation in models using rule-based methods is comparable to DT-MRI-based fiber orientations in both rat [80] and canine ventricular models [65]. Therefore, to date rule-based methods for determining the fiber orientation remain the norm in the cardiac ventricular models. Rule-based methods are also available to define fiber orientation on atrial meshes. Using the universal atrial coordinates (UACs), it is now possible to map atrial fiber measurements from ex vivo DTMRI onto atrial meshes [81].

2.3 Scar

Scar has been found to have a detrimental effect on CRT response, with both the pacing within scar regions' location [82] and the total scar burden [83] being correlated with negative CRT response. Scarred regions have been found to be associated with fiber orientation disorder, which can lead to both an increase in mechanical stiffness and a reduction in the electrical conductivity of the cardiac tissue [84, 85]. In models simulating HF and LBBB, scar tissue has been represented with reduced anisotropy in the material laws, increasing the passive stiffness of the cardiac tissue, altering the active tension models with a decrease in the contraction force, as well as reducing the electrical conductivity to slow the spread of electrical activation within scarred regions [47, 68, 71, 86]. In silico studies have found that the location of scarred regions can also influence the optimal pacing site [45]. In more recent studies, electrophysiology models incorporating scar were used to investigate the risk of ventricular tachycardia associated with pacing in proximity to scar during epicardial [32] and endocardial pacing [87]; it was found that epicardial pacing in proximity to scar increased VT risk, in agreement with clinical studies [88].

Scar regions are generally identified using signal intensity thresholding techniques on contrast-enhanced cardiac MRI. Typically, the ventricular wall is segmented, and the grayscale intensity values within the myocardium are used to differentiate between scar and normal tissue. Full-width-half-maximum (FWHM) [89] and standard deviation [90] methods are the most widely used. FWHM identifies scars as regions in the image where the signal intensity is greater than 50% of the maximum intensity of the pixels that lie within the wall. In the standard deviation method, the user manually delineates a region of healthy myocardium, from which mean and standard deviation values are computed, and scar tissue is then defined as the voxels where signal intensities are beyond 2 or 3 standard deviations above the mean value. While there is a lack of consensus on the optimal strategy for segmenting scar from contrast-enhanced MRI [91], FWHM method has been shown to be more robust to subjective bias as it does not require user selection of regions of healthy tissue [92].

Contrast-enhanced MRI typically has low out-of-plane resolution (6–20 mm), which may give rise to uncertainties in the scar segmentation. In Ukwatta et al. [93] an interpolation method to reconstruct high-resolution scar geometry from low-resolution MRI datasets was proposed to deal with this issue. Different scar segmentation methods yield differences in scar size, shape, and distribution [91], which could affect CRT modelling studies. The location of the scarred regions has an important effect on the optimal pacing location as studies have shown that pacing within or proximal to scar has a detrimental effect on response to CRT [94]. Accurate scar localization may therefore be important in accurate optimal pacing site selection. The quantitative differences between scar segmentation methods on simulation outcomes are yet unknown, as a comparison study has not yet been done.

An outline of a typical workflow for generating the anatomical models used in electrophysiology and electromechanical cardiac simulations is shown in Fig. 2.

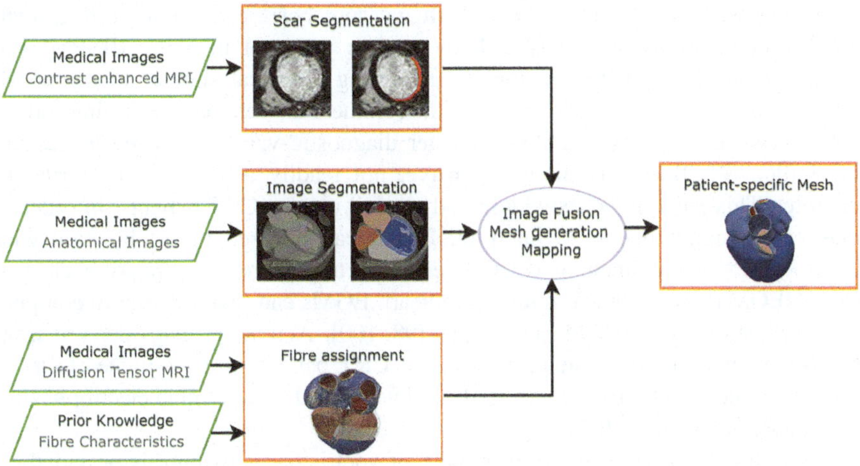

Fig. 2 A schematic of a workflow for the generation of a patient-specific anatomical model of the heart from medical images. Segmentations of the anatomical image are used to generate patient-specific meshes. Scar segmentations (from contrast-enhanced MR) and fiber information (from rule-based or image-based methods) can then be mapped onto the mesh

Subject-specific geometry of the cardiac structures (such as the atria and ventricles) is identified from image segmentations of anatomical medical images (such as MR or CT). Anatomical meshes are generated from the segmentations and can then be used for electrophysiological or mechanical modelling. Scarred regions can be segmented from contrast-enhanced MRI and incorporated into the models via image registration between the anatomical and contrast-enhanced MR images. Fiber orientations in the meshes can be assigned typically with rule-based methods or potentially from DT-MRI.

3 Clinical Measurements

3.1 *Electrical Measurements*

In order to provide useful information in a clinical setting, electrophysiology models need to replicate the activation of the patient's heart and how it changes in response to CRT. Routinely, heart dyssynchrony is diagnosed by looking at the QRS duration and morphology on 12-lead electrocardiograms (ECG), which are then used to indicate whether the patient would benefit from CRT. However, because the 12 leads are located on the torso far away from the heart, ECGs have limited capability of capturing local activation of the heart. Nevertheless, QRS duration is often used to approximate ventricular total activation time and to personalize the electrophysiology model by fitting the conduction velocity [31, 33, 35, 66–68, 95].

As opposed to 12-lead ECGs, which measure heart activation from a small number of locations, noninvasive body surface potential mapping (BSPM) can provide more detailed information by measuring electrical signals on the whole torso with a multi-electrode vest (with commercial versions recording up to 256 ECGs). Despite its potentially greater diagnostic value compared to 12-lead ECG [96, 97], BSPM is more expensive, not readily available, and harder to interpret. This has limited its widespread clinical use for CRT patients. In order to make torso signals easier to interpret, BSPM can be used in combination with electrocardiographic imaging (ECGi) reconstruction to recover unipolar electrograms (EGMs) on the epicardium of the heart. EGMs can then be used to compute local activation and repolarization times [98–100]. Activation maps derived from ECGi have been used on small numbers of CRT patients to investigate electrical synchrony induced by pacing [25, 101] and were used to determine electrophysiology model parameters [102].

More direct and accurate measurement of ventricular activation is provided by invasive electroanatomical maps (EAM) of the LV endocardium and the coronary veins. However, EAM can only be used intra-procedurally, meaning that only acute response can be quantified, with no long-term information available. At present, QRS duration and/or EAM are still the most widely used clinical data to fit electrophysiology models for CRT [19, 31, 33, 35, 52, 66, 68, 95, 103], with only one research group using BSPM to determine electrophysiology parameters [19, 31, 33, 35, 52, 66, 68, 95, 103, 104].

3.2 Functional Measurements

Because electromechanical models simulate heart deformation and pressure transients as well as electrical activation, they also need to reflect the functional changes due to LBBB and CRT response. Large deformations that the heart undergoes throughout a cardiac cycle can be measured with noninvasive imaging (cine MRI or echocardiography). Global metrics such as LV volume transient, end-diastolic and systolic volumes, ejection fraction, and LV mass and LV motion dyssynchrony indices and local metrics such as regional strains can be derived from imaging. Furthermore, tagged MRI can provide wall motion quantification, which can be used to validate deformations predicted by the model [66, 105].

Both invasive intra-procedural LV pressure and noninvasive arterial pressure measurements can be used together with volume transients to adapt the geometry parameters in the circulatory models [105–107] and to determine the active and passive material parameters in the biomechanical models [31, 33, 52, 66, 67, 71] of the heart.

4 Electrophysiology Models

Patient-specific anatomical models may be combined with electrophysiological models to simulate activation sequences and study a patient's electrical response to CRT. A typical workflow for electrophysiology personalization is shown in Fig. 3. Different types of electrophysiology models have been employed in CRT studies, such as the monodomain and bidomain models, the eikonal model, cellular automaton models, and surface source models. Clinical electrical measurements (ECG, EAM, or BSPM) are typically used to (1) personalize the parameters within the model and/or (2) assess the accuracy of the model.

4.1 Models of Cardiac Electrophysiology

Two of the most widespread multi-scale models of cardiac electrophysiology are the monodomain and bidomain models. Both models act on two scales: at the cellular scale, they describe the electrical activity, where action potentials are generated, and at the tissue scale, they describe how action potentials propagate from one cell to another. The main difference between the two is that the bidomain model includes

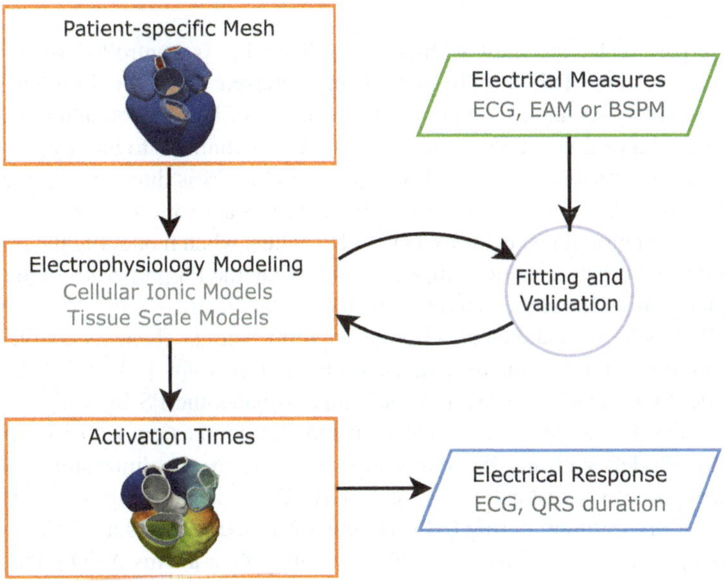

Fig. 3 A schematic representation of the electrophysiology models is shown. Personalized shape models of the heart and electrical measures, such as the ECG, electroanatomical maps (EAM), or body surface potential maps (BSPM), are used as inputs into multi-scale electrophysiology models of the heart to simulate the electrical activation times across the ventricles. The model outcomes can be evaluated in terms of clinical measures such as ECG traces and QRS duration

the extracellular domain, while the monodomain model does not (only describes propagation in the intracellular space). Most CRT studies employ the monodomain model [44, 68, 70, 95, 108] since it is computationally cheaper.

The eikonal model has been developed to make simulations faster, since solving the monodomain model still requires high spatial and temporal resolutions. This model describes the location of the activation upstroke without resolving the complex action potential kinetics, rendering it computationally very efficient, and has also been used in CRT modelling studies [37, 109] to simulate activation sequences.

With a similar goal, cellular automaton models [110] have also been employed in CRT studies [111–113]. In this model, instead of explicitly solving the complex current flow interaction between the intracellular and extracellular domains, action potentials are pre-calculated based on ionic current equations. These are then stored and applied as a set of rules allowing fast computation of ventricular activation and repolarization sequences.

A type of surface source model called the equivalent dipole layer model, which computes body surface potentials from the transmembrane potential on the myocardial surface (source) based on the bidomain formulation [114, 115], has also been employed in CRT studies [116].

4.2 *Purkinje System or Fast Endocardial Conduction?*

At the organ scale, the cardiac bioelectric behavior is controlled by a cardiac conduction system. This system is typically referred to as the Purkinje system (PS) in the ventricles and is composed of the bundle of His, the left and right bundle branches, and an extensive Purkinje network, which connects to the myocardium at the Purkinje-ventricular junctions. Propagation in the PS is three to six times faster than in the myocardium, allowing for rapid electrical activation at the endocardium and ensuring ventricular synchrony [117, 118]. Thus, when modelling the activation sequence at the organ scale, it is important to include the PS properties in the model, as it significantly affects the activation pattern in the heart.

The PS has been modelled as a tree of one-dimensional elements coupled to the myocardium in CRT models of realistic ventricular anatomy [70, 109, 119] or as a fast-conducting endocardial layer, which approximates the PS by assigning tissue conductivities to the endocardium that match the conduction velocity in the PS [44, 108]. Models with the PS represented as a tree of one-dimensional elements have been used to investigate the effect of AVD on the distribution of activation times in the myocardium during CRT and demonstrated that a 30 ms AVD yields an activation pattern more similar to healthy activation than a 0 ms AVD [109]. In this study, the atria were not included in the model, so the AVD was approximated as a timed delay between activation at the AV node and the ventricular pacing locations. In a recent study, a fractal tree representation of the His-Purkinje tree [120] has been used to show that left bundle pacing requires AVD optimization to ensure short RV activation times, leading to a comparable response to His bundle pacing [28].

A similar approach was used to investigate the role of electrical conduction in the PS during CRT pacing in failing hearts [70]. Motivated by experimental evidence of retrograde activation in the PS [121, 122], Romero et al. [70] compared a realistic PS represented by a tree of one-dimensional elements with a fast endocardial layer. Their results show that retrograde conduction in the PS is key to accurately estimate VVD and that a fast endocardial layer cannot accurately capture this effect [70]. However, a fast endocardial layer was employed in a biventricular (BiV) model to study the effect of endocardial versus epicardial pacing during CRT and showed good agreement with experimental results [44]. Their results demonstrated that early access to fast-conducting endocardial tissue reduces ventricular activation time during endocardial pacing, providing a physiological explanation for the observed benefit of endocardial pacing compared with epicardial pacing [123, 124]. In a more recent study, fast endocardial layer was also found to be the most important factor for recovering epicardial activation patterns from QRSd and anatomy for CRT patients [54].

The effect of CRT on electrical activation in the presence of ECG characteristic of LBBB has been studied using in silico models of the heart [95]. LBBB can be caused by conduction block in the left branch of the His bundle due to damage to the His fibers or due to myocardial uncoupling caused by reduced expression of connexins (gap junctions) in its vicinity. This study showed that a LBBB ECG pattern can be replicated in the models with myocardial uncoupling. In addition, they showed that CRT improves ventricular activation in the presence of LBBB but not in the case of myocardial uncoupling mimicking LBBB.

4.3 Lead Optimization

Electrophysiology models also offer the opportunity to carry out in silico automatic CRT optimization. Briefly, the approach consists of simulating electrical activation in a BiV anatomy for different AVD and VVD as well as several different lead locations. Then the error between the obtained activation sequences for each case and a simulated synchronous activation is minimized to determine which option yields the best acute response to CRT [112, 113, 125, 126]. Such models have demonstrated that patient-specific optimization of lead location and AVD and VVD can improve CRT efficacy and impact treatment success [69, 113, 126] and that the use of body surface potential maps can further improve in silico CRT optimization [127]. This technique has also been used recently to establish the optimal electrode configuration in quadripolar leads taking into account how this configuration changes with reverse remodeling [37].

4.4 ECG Simulations

In terms of ECG simulation, ECGSIM is a program to simulate ECG based on the equivalent dipole layer model [128]. This program was used by van Huysduynen et al. [116] to investigate the effects of CRT on repolarization. The model incorporated heart and thorax geometry (based on MRI), conduction heterogeneity, and transmural dispersion of repolarization (TDR). In this study, the authors investigated if BiV pacing in CRT increases TDR in comparison with conventional RV pacing. Their results show that TDR during BiV pacing is not significantly larger than during RV pacing. The main consequence is that increased TDR does not explain the possible proarrhythmic effects of CRT.

Computational models of electrophysiology involve a high number of parameters that need to be fitted, a complex task to be done so they represent accurately a patient's electrophysiology. Different parameters have a different impact on the models, so in some cases, the parameter fitting step can be reduced to the most influential parameters. For instance, in Sanchez et al. [108], the authors investigated the role of myocardial properties in the activation sequence on the LV endocardial surface and the ECG morphology in HF patients. They showed that the effects of changes in tissue properties vary between ECG leads, whereas ionic changes entail similar effects in all ECG leads.

5 Mechanical Models

When the electrical activation of the heart is dyssynchronous, mechanical dyssynchrony emerges. Clinical studies have demonstrated that improvements in electrical synchrony do not imply improved mechanical synchrony [129–131]. Therefore, it is desirable for the model to account for mechanical deformation as well. Figure 4 shows a typical workflow for an electromechanical model of the heart. Mechanical deformation of the heart can be simulated by solving continuum mechanics on idealized or personalized shape models of the heart with mapped fiber orientation (from rule-based or image-based methods). Structural and functional heterogeneities (from regions such as scar) can be included to account for differences in local mechanical properties of the tissue [132].

A multi-scale model for cardiac electromechanics needs to have the following three elements, in addition to the anatomical mesh and boundary conditions: (1) a passive constitutive law to represent how the myocardium reacts to external loads; (2) an active contraction model to compute active tension along myofibers generated by the myocytes once they are activated; and (3) a coupling method between the electrophysiology simulation and the mechanics simulation.

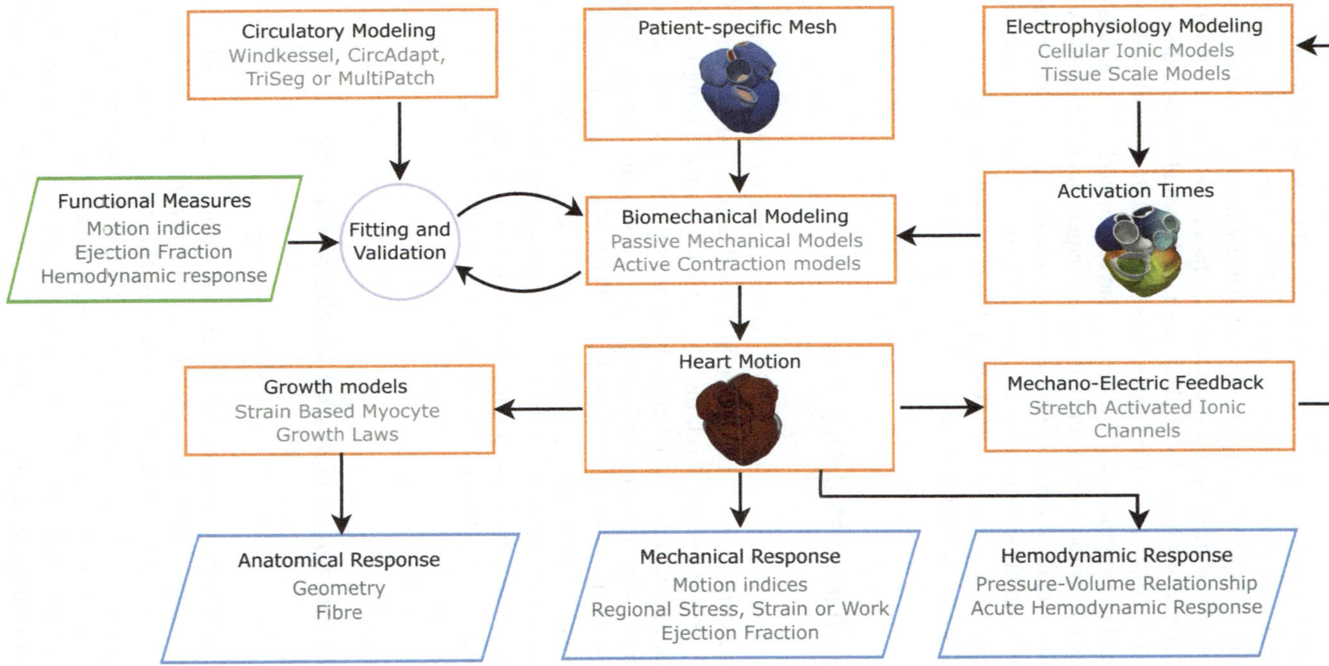

Fig. 4 A schematic representation of the electromechanical workflow is shown. Personalized shape models of the heart, electrophysiology models, and circulatory models are used as inputs into the biomechanical models of the heart which use finite elasticity equations to model dynamic cardiac deformations. Mechano-electric feedback via stretch-activated ionic channels feedbacks into the electrophysiology modeling of the electrical activation time input into the mechanical model simulations. The model outcomes can be evaluated in terms of the mechanical and hemodynamic responses and potentially the anatomical response (mediated with growth models)

5.1 Passive Mechanical Models

The internal stresses developed by the myocardium depending on its deformation are typically represented with a hyperelastic formulation [132]. A strain energy function is defined to quantify the internal energy of the material as a function of local strains. Shear tests on samples of mammalian myocardium showed that the myocardium presents different stiffnesses in the local fiber, sheet, and normal to sheet direction [133]. Therefore, orthotropic [134–136] rather than simplified transversely isotropic [137–141] material laws are more suitable for cardiac electromechanics models. In patient-specific models, parameters of the constitutive equation are fitted and validated using functional clinical measurements, such as the passive pressure-volume relationship during ventricular filling [68]. Because orthotropic models require more parameters to be fitted, leading to a more complex optimization problem, transversely isotropic laws are still widely used.

5.2 Electromechanics Modeling

Excitation-contraction coupling describes the process from which the electrical activation of the myocardium gives rise to the mechanical contraction and relaxation of the heart [142]. At the cellular level, the myocytes depolarize, giving rise to a calcium signal that activates the sarcomeres; this allows the myocytes to generate tension and contract, which then translates to the macro-scale cardiac motion pumping the blood in the heart around the body. While the electrical activation affects the mechanical contraction of the heart, there is also feedback from the mechanical contraction of the heart to the electrophysiology of the cardiac cells. This process, known as mechano-electric feedback (MEF) [143], occurs due to the presence of elements such as stretch-activated ion channels [144, 145] changing the ionic currents in the cell.

5.3 Active Contraction Models and Mechano-Electric Feedback

Active tension describes the contractile force generated along the myofibers from cellular excitation during systole. Active tension is typically incorporated into the mechanics model as an additional tension component along the fiber orientations. There are three ways to model ECC in electromechanics models: phenomenological, weakly, and strongly coupled methods. Phenomenological coupling [68, 146] and weakly coupled models [146, 147] are based on the common assumption that electrophysiology affects the mechanical contraction of the heart through ECC, but that MEF is unimportant and can be ignored. In contrast, limited work has been done with strongly coupled models which take into account the MEF [148–150].

To the best of our knowledge, computer models simulating CRT and LBBB have been implemented with weakly coupled or phenomenological models. The lack of studies incorporating strongly coupled electromechanical models is likely due to the large computational costs of implementing strongly coupled models. In personalized models, clinical measurements of the systolic cardiac function (such as systolic pressure, volume transients, or cardiac motion) are used to define the parameters used to describe the active tension transient. The simulated mechanical deformation of the heart models can then be further analyzed to determine mechanical and hemodynamic response outcomes.

5.4 Cardiac Myocyte Growth Models

Cardiac myocyte growth models can be used to predict the pattern of cardiac growth due to changes in mechanical loading conditions. For a broader perspective on cardiac myocyte growth laws, we refer the reader to an article by Witzenburg and Holmes [151], a comparison study of eight published phenomenological cardiac myocyte growth laws. Nonetheless, these models are not usually incorporated into LBBB or CRT models, therefore ignoring the effects of remodeling, whether pathological or reverse. In a study by Kerckhoffs et al. [152], the effects of myocyte shape changes in response to the strain on the cell were investigated in a canine model of LBBB. A later study by Gurev et al. [153] investigated the effect of growth models on a human model with myocardial infarction. If growth models are not incorporated, a typical approach is to model the effects of remodeling, for instance, with an increase in LV mass and LV dilation or a reduction in EF. However, cardiac remodeling also encompasses changes to the electrics, mechanics, and function of the heart, as well as volumetric changes, and these should also be investigated in future studies. In a recent study, the chronic effect of reverse remodeling on quadripolar leads was investigated by modeling reverse remodeled hearts using healthy (asymptomatic) subjects [37]. However, reverse remodeling effects associated with CRT still need to be determined to model the chronic effects of CRT response.

5.5 Acute Hemodynamic Response

Acute hemodynamic response (AHR), given as the maximal rate of increase in the rise of systolic LV pressure, has been shown to correlate with long-term CRT response, and therefore the optimization of the AHR is one strategy for optimizing CRT response [154, 155]. Electromechanical models of the dyssynchronous failing heart have been used to investigate AHR optimization. However, prior to achieving the lofty goal of translating the biomechanical models of the heart to the CRT clinic, models need to demonstrate their ability in, accurately and robustly, simulating clinical measurements. Extensive and rich clinical datasets have been used to develop patient-specific biomechanical models of the heart under LBBB and CRT conditions. These models were used to predict the AHR with varied pacing protocols with the predictions showing good agreement with clinical measurements of the AHR [19, 30, 31, 33, 67, 71, 156]. While these results are promising, low patient numbers (1–9 subjects) mean that any conclusions drawn from these studies are limited. The requirement for extensive data including, but not limited to, invasive LV pressure recording and EAM, in addition to the large computational costs in creating and parameterizing personalized biomechanical models of the heart has restricted these modeling studies to the research domain, and clinical translation remains unrealistic. The parameterization of personalized biophysically based cardiac models requires multiple simulations with different input parameters to be run. Supercomputing resources allow for the simulation of cardiac electrophysiology at clinical timescales, allowing for the simulation of a single heartbeat within seconds [157] or minutes [158]. However, the computational costs of these simulations are high, requiring up to 1.6 million cores to run each simulation, limiting this method to the research field. As an alternative, techniques such as Gaussian process emulators can accelerate the process of parameter fitting, with the use of surrogate models calibrated with a large number of simulations, as applied in LV models of aortic-banded rats [159] and healthy human volunteers [160]. Nonetheless, this technique has yet to be used in models for CRT patients.

5.6 Lead Location Optimization

CRT response is influenced by LV pacing lead location with suboptimal LV lead placement identified as a cause in 21% of CRT non-responders [161, 162]. Optimizing the LV pacing lead location has been one of the goals of CRT electromechanical modeling with studies using canine [34, 163] and patient-specific [31, 66] models to identify the optimal LV pacing site. In these electromechanical modeling studies of CRT, the optimal LV pacing location was found to be in the lateral LV free wall in agreement with experimental canine [162, 164] and clinical [161, 165] studies. The optimization metric for LV pacing lead location placement to improve CRT response has been varied across studies, including AHR maximization [31, 34, 66], stroke work maximization [34], electromechanical delay minimization [163], or LV electrical activation time reduction [66].

5.7 Device Setting Optimization

Device setting optimization has been shown to improve CRT response, with acute measures of the hemodynamic response, cardiac output, and echocardiographic indexes showing improvement with AVD/VVD optimization in the clinical setting [29, 166–169]. Optimization of the device settings using electromechanical models has focused on measures of mechanical synchrony [31, 156]. The intrinsic electrical activation of the heart passes from the atria to the ventricles via the atrio-ventricular node, before traveling through the Purkinje system to activate the ventricles. In ventricular models where the atria were not explicitly modelled, AVD can be simulated by altering the timings between the intrinsic activation that is set off by atrial pacing and the activation at the ventricular lead pacing locations.

In agreement with clinical studies [167, 170], in silico electromechanical simulations have found that the optimal AVD/VVD is highly personalized [66, 69, 156] and dynamic, with chronic CRT requiring a different set of optimal AVD/VVD in comparison to the acute timepoint [31]. Although AVD/VVD optimization can improve response to CRT, in real-world practice, patients may be left with suboptimal device settings post-CRT due to a lack of time and qualified staff [29, 171]. In silico personalized biomechanical models of the heart potentially allow for a systemic evaluation of the optimal AVD/VVD on a patient-specific basis outside of the clinic.

5.8 Multisite Pacing

In addition to using models to optimize CRT response to existing treatment methods, computer models of the heart have also been used to evaluate emerging technologies for delivering pacing to the heart [19, 31, 156]. MSP is one such novel technology, where multiple sites on the LV are able to be stimulated. In Lee et al. [31], personalized biophysical models of three MSP CRT patients were created at both acute and chronic timepoints. These models were used to assess how the optimal pacing site would change with reverse remodeling. It was found that optimal device and lead location settings remained consistent with chronic CRT. While this implies there is diminished benefit in the post-implant optimization of the pacing location that MSP allows, it also emphasized the importance of the acute optimization of the LV lead pacing location which may potentially be easier with MSP in contrast to conventional CRT.

While many preliminary studies have shown an improved acute CRT response with MSP [10–15], a more recent study found that MSP had no improvement in contrast to CRT. These clinical observations were confirmed using idealized LV electromechanical models [19].

Patients with ischemia have an improved response to MSP CRT in comparison to conventional CRT, according to computational electromechanical studies

[156]. MSP CRT is therefore a promising new technology for improving response rates to CRT since this subgroup of patients is one with the poorest response rate to conventional CRT [172].

5.9 Endocardial Pacing

Endocardial pacing has also emerged as a potential alternative for patients who do not respond to conventional epicardial CRT [173, 174]. Experimental canine studies have found that endocardial pacing has a higher efficacy for CRT in comparison to conventional epicardial CRT [175]. The mechanisms for this difference in epicardial vs endocardial pacing have been explored using computer models of the heart. Hyde et al. [44] found that endocardial pacing recruits the faster conducting endocardial layer early, allowing for a faster total activation time of the ventricles. Similar to epicardial pacing studies, biophysical models have also been used to identify target endocardial pacing location at the latest activating regions [176] while avoiding scarred locations in endocardial pacing studies [87].

5.10 His and Left Bundle Pacing

His bundle pacing and left bundle pacing are also promising novel CRT delivery methods. Computational electrophysiology models have been used in the past to simulate electrical activation on 24 patient-specific geometries [28] and a rabbit heart inclusive of PS [177] during His bundle and left bundle branch pacing. These models showed a relation between the effectiveness of left bundle branch pacing and AVD optimization.

5.11 Limitations of Electromechanical Models

In the heart, the time delay between the atria and ventricular electrical activation allows for the atrial contribution to the ventricular preload. However, most of the electromechanical models looking at AVD settings do not include atrial structures [31, 156]. Some studies on canine heart modeling included the atrial contribution to the ventricular preload via alterations of a lumped parameter model of the circulatory system [178]. Even in models like this one, the anatomical effects of the atria on the activation pattern are still disregarded.

The change in the atrial contribution with AVD optimization is currently not evaluated in the human models, and the consequence of neglecting the effects of the atria will need to be investigated in future studies. Four-chamber heart models

[40, 59, 179], containing the atria and ventricles, have the potential to address this problem.

Most published electromechanical models predicting CRT response have used the AHR as an outcome measure. AHR can be measured acutely and is predictive of reverse remodeling [180]. However, earlier clinical studies have shown poor reproducibility of AHR due to biological variability, especially when only measured once [181]. In the study by Whinnett et al. [181], it was suggested that repeated measurements (six or more) and relative measurements improve the reliability of AHR measurements. Increasingly, models have adopted the prediction of the more reliable relative measure of AHR; however, few clinical groups are making recordings with six repeats due to the inherent constraints in the clinic [182, 183]. Other measures of acute improvement such as diastolic parameters and pressure-volume loop have also been proposed. However, unlike AHR, which has been shown to predict reverse remodeling 6 months post-CRT [180], they have yet to be linked to long-term clinical response [184, 185].

Ventricular models have tended to focus on the simulation of a single beat with boundary conditions being represented by Windkessel models. Contraction models connected to the circulatory models [163, 186, 187] have linked organ-scale models to closed-loop cardiovascular models. As the heart operates as two pumps in series, the blood leaving the RV must equal the blood entering the LV with each beat. To account for these hemodynamic effects requires modeling not only the ventricles but also the closed-loop circulatory system within which the heart operates. In a later study, additional growth models were further incorporated into the models described in [186] to investigate the long-term effects of LBBB [152]. The changes observed in the model were in line with experimental observations of the pathological remodeling effects of the progression of LBBB, such as LV dilation, reduction in LV ejection fraction, and occurrence of septal flash [188].

6 Circulatory Models

Three-dimensional electromechanics models offer a detailed representation of complex multi-scale mechanisms, ranging from sub-cellular to the whole-organ scale. The complexity of these models leads to high computational costs, therefore limiting their clinical translation. Furthermore, simulations are often restricted to the cardiac chambers and one single heartbeat, even though the interaction between the atria, the ventricles, and the circulatory system is likely to play a role in the afterload and the preload of the ventricles, which are important aspects for CRT response and optimization.

One way to solve this issue is to use zero-dimensional models, discarding or significantly simplifying the patient's anatomy. This approach has the benefit of fast simulations, with computational times reduced from hours to seconds/minutes, allowing to simulate the whole circulatory system represented as a closed loop for several cardiac cycles [189–191].

6.1 CircAdapt Model

As mentioned above, zero-dimensional models do not include anatomical information in favor of faster simulations. To make up for the loss of spatial information that three-dimensional anatomy provides, additional parameters need to be estimated. The CircAdapt model [106] partially addresses this issue by adapting the geometry parameters based on known physiological adaptation rules to match target values of circulatory pressure and flow. A schematic representation of CircAdapt is shown in Fig. 5. The model simulates the atria and ventricles coupled to the systemic and pulmonary circulations through cardiac valves. CircAdapt provides fast simulations, requires fewer parameters to be estimated, and allows for quantification of anatomical remodeling and global indices for cardiac function, such as cardiac output and ejection fraction [106].

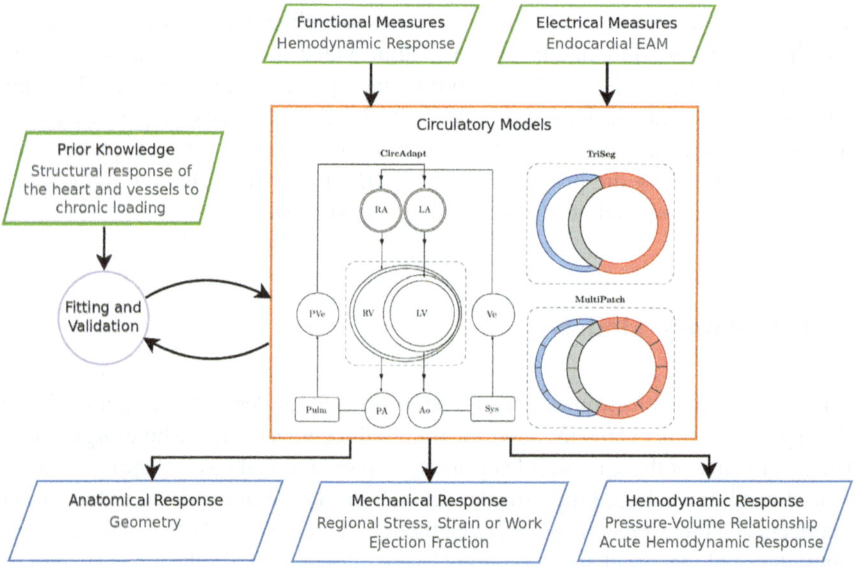

Fig. 5 A schematic representation of the CircAdapt model, together with the TriSeg and the MultiPatch models as its extensions are given in the orange box. Known physiological adaptation rules are used to adapt geometry parameters with the aim to match clinical measurements for mean systemic pressure and flow given in input to the model. The model gives in output hemodynamic and anatomical responses in the form of geometry parameters as a consequence of adaptation and pressure-volume relationships. When extended to the MultiPatch model, local electrical activation is needed as an additional input to provide information about stress, strain, and work within ventricular walls

6.2 TriSeg Model

Although CircAdapt is a very flexible framework, it cannot account for interventricular interaction, because the ventricles are modelled as two shells one inside the other, but other than that they act independently. However, interaction mechanisms between the right and the left ventricles influence CRT and HF, as the function of one chamber directly affects the other [192–195]. CircAdapt was therefore extended to the TriSeg model to capture the heterogeneity of activation times and wall tension of the left and right ventricular free walls and intraventricular septum [107], as shown in Fig. 5. Thanks to this improvement compared to CircAdapt, the TriSeg model allowed to investigate how ventricular function is affected by LBBB [196]. The observations reported by this computational study are in agreement with clinical observations on dyssynchronous hearts and provide valuable information about LBBB consequences on the circulatory system. For instance, the model was able to associate mitral valve regurgitation with LBBB, again consistently with clinical observations [197].

In some patients, LBBB has been found to be associated with abnormal septal motion (septal flash), whereby the mechanical dyssynchrony in the heart manifests in a pre-ejection shortening of the interventricular septal wall followed by a rebound stretch; septal flash has been found to be a predictor of CRT response [198]. The TriSeg model was used by [199] to investigate different types of septal deformation patterns in CRT and to understand the underlying mechanisms of these patterns. Clinical measurements in this study indicated that in addition to septal flash, there were other patterns of motion linked with LBBB patients, with a distinct pattern of motion associated with patients with LV scar. Computer models were able to simulate a septal flash pattern with the imposition of a time difference between the septum and the LV free wall. Imposing a reduction of contractility in this model recreated the pattern of motion for patients with scar, consistent with the understanding of the effect of scar on cardiac tissue.

6.3 MultiPatch Model

The MultiPatch model represents a further improvement based on the TriSeg model, to account for the heterogeneity of electrical activation within one ventricular segment and to provide information about local stresses and work [105]. To this aim, the LV free wall, the RV free wall, and the septum were divided into different sections (or patches), assigned with different activation times provided by EAM measurements. Additional to the whole circulatory system, the model then solves for changes in local stress distribution induced by different activation patterns. Thanks to its more comprehensive representation, the MultiPatch model is more suitable for CRT simulations compared to simpler CircAdapt and TriSeg models. Indeed, Lumens et al. used this framework to show that, despite the differences in activation achieved with BiV and LV only pacing, these methods result in similar improved cardiac function [200]. This was due to an equivalent increase in myocardial work and its linear relation with LV systolic function.

As already mentioned, the MultiPatch model allows accounting for heterogeneity in electrical activation within one wall segment. This heterogeneity can be extended to mechanical properties as well, to simulate areas with increased stiffness and decreased contractility, such as scar [45]. Using this framework, Huntjens et al. were able to show that, in ischemic hearts with LBBB, the LV lead should be placed as far away from the car as possible, in agreement with previous clinical studies [45, 174, 201]. The MultiPatch model has been used to investigate more sensitive indices to classify responders and non-responders to CRT based on different electrical and mechanical substrates [202]. Combined with clinical input and animal measurements, the MultiPatch model was also able to assess how both ventricles respond to different pacing delays [7, 36, 203] and to quantify contributions of inter- and intraventricular dyssynchrony in response to CRT [204]. In Jones et al. [7], it was found that the AVD is an important mechanism for CRT response. In a more recent study, it was found that optimizing AVD has a positive impact on the LV preload and leads to a reduction in diastolic mitral regurgitation, thus improving the hemodynamic CRT response [203].

Although the CircAdapt model, the TriSeg model, and the MultiPatch model have been used to provide mechanistic insight into LBBB and CRT, these frameworks still lack a representation of microscopic mechanisms, although these might be important for CRT and HF. Thanks to its modular structure, the CircAdapt model is suitable for coupling with more complex three-dimensional electromechanics models [60, 205] and sarcomere contraction [206]. Kuijpers et al. coupled the TriSeg model with cellular models for myocyte electrical excitation and sarcomere contraction and used this extended framework to analyze how microscopic mechanisms affect macroscopic cardiac function. The model was run for several beats, thanks to its efficiency, providing insight into both acute and sustained mechanisms because of LBBB, together with CRT delivery [206].

To conclude, the CircAdapt model and its extensions provide an efficient alternative to more complex and computationally expensive three-dimensional electromechanics models. Their modular structure facilitates the coupling with more detailed models that can be used to represent specific dynamics, depending on the clinical question. The lack of representation of local mechanisms in the ventricles is partially overcome by TriSeg and MultiPatch, making them more suitable for LBBB and CRT simulations. The efficient simulations provided by CircAdapt have led to faster clinical translation compared to three-dimensional models for cardiac electromechanics. In future work, the coupling between three-dimensional electromechanics and CircAdapt together with clinical data could provide an invaluable tool to improve our understanding of changes induced by LBBB and reverse remodeling due to CRT in HF patients.

7 Challenges, Opportunities, and Future Directions

The electromechanical models discussed in this review require extensive clinical information, oftentimes including invasive measurements. Future work requires the development of pipelines that allow for the autogeneration of models and parameters from noninvasive data, as well as methods to reduce the computational costs or improvements in the parallel scalability of the simulations.

Patient-specific models for CRT are often very complex with many free parameters. At the same time, the large patient variability in the CRT patient populations motivates personalizing model parameters to each patient. However, efficiently and uniquely constraining model parameters remains a significant challenge in model creation. The inability to uniquely constrain parameters using available clinical data will, in some cases, limit the predictive ability of the model. However, in biological systems, not all parameters will have an equal impact on all model predictions [207], allowing some parameters to be set at representative values, without compromising the model's predictive capacity. Further, when models fail to make accurate predictions, we can identify important parameters that were not well constrained by available measurements. We can then use this information to identify measurements that need to be made to achieve better predictions for future patients. Finally, as models move from research techniques to clinical tools, there will need to be a greater emphasis placed on uncertainty quantification so that the effect of unknown or poorly constrained parameters on model predictions can be included in estimated confidence intervals that can in turn guide a clinician in the reliability of the model predictions.

This is especially true in light of the heterogeneous patient population within a standard CRT cohort, where questions remain unanswered regarding the accuracy and reliability of model predictions. The uncertainties inherent in the models due to reliability and robustness of data measurements (such as AHR) and modeling assumptions need to be addressed upfront when presenting biophysical models of the heart to clinicians. Computational modeling offers an additional tool to make clinical decisions; however, diagnostic and treatment decisions based on model simulations need to be made with the full knowledge of the errors and confidence in the models.

The use of machine learning (ML) methods to predict CRT response in patients has been increasingly explored in recent years [208]. Cardiovascular disease is a complex problem with a multitude of factors determining the best clinical outcome for patients, while ML methods allow for the analysis of different features to identify which ones have the most impact. ML algorithms to predict CRT response have considered clinical data from ECGs (QRSd, 12 wave ECG morphology) [209, 210], medical images (anatomical, strain analysis, and scar) [209, 211, 212], demographics (NYHA stage, age, sex) [213], and other comorbidities (AF, ischemia) [210, 212] to predict CRT response. In future studies, information from biophysical models of the heart could also be implemented in ML methods to predict CRT response.

As discussed in this chapter, one of the strengths of biophysical models is the ability to predict outcomes and states of cardiac tissue with in silico simulations. In Sánchez et al. [214], biophysically based computer models of the cardiac substrate were used to train and validate a ML classifier for fibrotic regions in atrial fibrillation patients, while in Shade et al. [215] EP simulations of personalized LA models were used in a ML classifier to predict atrial fibrillation recurrence post-procedurally. ML classifiers have similarly been applied to personalized ventricular models to predict the risk of sudden cardiac death in patients with cardiac sarcoidosis [216]. For a more in-depth review on the application of ML in cardiac electrophysiology and arrhythmia to date, we refer the reader to Trayanova et al. [217]. In future work, a similar approach could be made where biophysical cardiac models for CRT patients could be combined with ML techniques to improve classification methods.

The current state of the computational modeling of LBBB and CRT patients relies on expertise in image processing, numerical analysis, mesh generation, cardiac electrophysiology, mechanics, and circulatory response which presents another barrier to clinical translation. The development of user-friendly tools that allow the non-expert to model the electrical, functional, and anatomical response of the patient to CRT within a clinically useful timeframe remains a challenge for the community. The development of these tools requires close collaboration and feedback from clinical stakeholders.

8 Conclusion

In this book chapter, we have reviewed some of the most relevant cardiac computer models used to investigate LBBB and CRT. Electrophysiology, electromechanical, and circulatory models have been used to identify optimal pacing location and intracardiac timings in CRT. These models have also proved to be useful to investigate the cardiac substrate (endocardial layers or scar burden or location) to provide insight into the pathophysiology causing electrical and mechanical dyssynchrony in LBBB and response to CRT. Lastly, computational models can also be used to study novel technologies such as MSP, endocardial LV pacing, or conduction system pacing.

Several lines of future work remain open. These include using in silico models to examine interventions, aiding clinicians with prognosis and diagnosis of new treatments, predicting which patients will most benefit from CRT (thus reducing the need for unnecessary procedures), or providing a noninvasive testing ground for emerging technologies.

Funding

This work was supported by the European Research Council (grant number 864055), Engineering and Physical Sciences Research Council (grant number EP/M012492/1), and the British Heart

Foundation (grant PG/15/91/31812 and RG/20/4/34803) to SAN.
Conflict of Interest Authors declare that they have no conflict of interest.

Ethical Approval This article does not contain any studies with human participants or animals performed by any of the authors.

References

1. Lippi G, Sanchis-Gomar F (2020) Global epidemiology and future trends of heart failure. AME Med J 5
2. Yu C-M, Bleeker GB, Fung JW-H, Schalij MJ, Zhang Q, van der Wall EE, Chan Y-S, Kong S-L, Bax JJ (2005) Left ventricular reverse remodeling but not clinical improvement predicts long-term survival after cardiac resynchronization therapy. Circulation 112(11):1580–1586
3. Linde C, Abraham WT, Gold MR, St John Sutton M, Ghio S, Daubert C, R. S. Group (2008) Randomized trial of cardiac resynchronization in mildly symptomatic heart failure patients and in asymptomatic patients with left ventricular dysfunction and previous heart failure symptoms. J Am Coll Cardiol 52(23):1834–1843
4. Solomon SD, Foster E, Bourgoun M, Shah A, Viloria E, Brown MW, Hall WJ, Pfeffer MA, Moss AJ (2010) Effect of cardiac resynchronization therapy on reverse remodeling and relation to outcome: multicenter automatic defibrillator implantation trial: cardiac resynchronization therapy. Circulation 122(10):985–992
5. Tang ASL, Wells GA, Talajic M, Arnold MO, Sheldon R, Connolly S, Hohnloser SH, Nichol G, Birnie DH, Sapp JL, Yee R, Healey JS, Rouleau JL (2010) Cardiac-resynchronization therapy for mild-to-moderate heart failure. N Engl J Med 363(25): 2385–2395
6. Ghio S, Freemantle N, Scelsi L, Serio A, Magrini G, Pasotti M, Shankar A, Cleland JGF, Tavazzi L (2009) Long-term left ventricular reverse remodelling with cardiac resynchronization therapy: results from the CARE-HF trial. Eur J Heart Fail 11(5):480–488
7. Jones S, Lumens J, Sohaib SA, Finegold JA, Kanagaratnam P, Tanner M, Duncan E, Moore P, Leyva F, Frenneaux M (2017) Cardiac resynchronization therapy: mechanisms of action and scope for further improvement in cardiac function. Europace 19(7):1178–1186
8. Auricchio A, Prinzen FW (2011) Non-responders to cardiac resynchronization therapy. Circ J 75(3):521–527
9. Birnie DH, Tang AS (2006) The problem of non-response to cardiac resynchronization therapy. Curr Opin Cardiol 21(1):20–26
10. Pappone C, Ćalović Ž, Vicedomini G, Cuko A, McSpadden LC, Ryu K, Romano E, Saviano M, Baldi M, Pappone A (2014) Multipoint left ventricular pacing improves acute hemodynamic response assessed with pressure-volume loops in cardiac resynchronization therapy patients. Heart Rhythm 11(3):394–401
11. Rinaldi CA, Burri H, Thibault B, Curnis A, Rao A, Gras D, Sperzel J, Singh JP, Biffi M, Bordachar P, Leclercq C (2015) A review of multisite pacing to achieve cardiac resynchronization therapy. Europace 17(1):7–17
12. Rinaldi CA, Kranig W, Leclercq C, Kacet S, Betts T, Bordachar P, Gutleben K, Keel A, Ryu K, Farazi TG, Simon M, Naqvi T (2012) Multisite left ventricular pacing improves acute mechanical Dyssynchrony in heart failure patients. J Am Coll Cardiol 59(13s1):E918
13. Rinaldi CA, Leclercq C, Kranig W, Kacet S, Betts T, Bordachar P, Gutleben K-J, Shetty A, Donal E, Keel A, Ryu K, Farazi TG, Simon M, Naqvi TZ (2014) Improvement in acute contractility and hemodynamics with multipoint pacing via a left ventricular quadripolar pacing lead. J Interv Card Electrophysiol 40(1):75–80

14. Thibault B, Dubuc M, Khairy P, Guerra PG, Macle L, Rivard L, Roy D, Talajic M, Karst E, Ryu K (2013) Acute haemodynamic comparison of multisite and biventricular pacing with a quadripolar left ventricular lead. Europace 15(7):984–991
15. Zanon F, Baracca E, Pastore G, Marcantoni L, Fraccaro C, Lanza D, Picariello C, Aggio S, Roncon L, Dell'Avvocata F, Rigatelli G, Pacetta D, Noventa F, Prinzen FW (2015) Multipoint pacing by a left ventricular quadripolar lead improves the acute hemodynamic response to CRT compared with conventional biventricular pacing at any site. Heart Rhythm 12(5): 975–981
16. Pappone C, Ćalović Ž, Vicedomini G, Cuko A, McSpadden LC, Ryu K, Jordan CD, Romano E, Baldi M, Saviano M, Pappone A, Vitale R, Catalano C, Ciaccio C, Giannelli L, Ionescu B, Petretta A, Fragakis N, Fundaliotis A, Tavazzi L, Santinelli V (2015) Improving cardiac resynchronization therapy response with multipoint left ventricular pacing: twelve-month follow-up study. Heart Rhythm 12(6)
17. Zanon F, Marcantoni L, Baracca E, Pastore G, Lanza D, Aggio S, Roncon L, Conte L, Picariello C, Carraro M, Noventa F, Prinzen FW (2016) Acute optimization of left ventricular pacing site plus multipoint pacing improve remodeling and clinical response of Crt at one year follow up. J Am Coll Cardiol 67(13):1280
18. Heckman LIB, Kuiper M, Anselme F, Ziglio F, Shan N, Jung M, Zeemering S, Vernooy K, Prinzen FW (2020) Evaluating multisite pacing strategies in cardiac resynchronization therapy in the preclinical setting. Heart Rhythm O2 1(2):111–119
19. Dell'Era G, Gravellone M, Scacchi S, Franzone PC, Pavarino LF, Boggio E, Prenna E, De Vecchi F, Occhetta E, Devecchi C, Patti G (2021) A clinical-in silico study on the effectiveness of multipoint bicathodic and cathodic-anodal pacing in cardiac resynchronization therapy. Comput Biol Med 136:104661
20. Gardner RS, D'Onofrio A, Mark G, Gras D, Hu Y, Veraghtert S, Garcia-Bolao I (2021) Real-world outcomes in cardiac resynchronization therapy patients: design and baseline demographics of the SMART-registry. ESC Heart Fail 8(2):1675–1680
21. Biffi M, Defaye P, Jaïs P, Ruffa F, Leclercq C, Gras D, Yang Z, Gerritse B, Ziacchi M, Morgan JM (2018) Benefits of left ventricular endocardial pacing comparing failed implants and prior non-responders to conventional cardiac resynchronization therapy: a subanalysis from the ALSYNC study. Int J Cardiol 259:88–93
22. Sidhu BS, Porter B, Gould J, Sieniewicz B, Elliott M, Mehta V, Delnoy P, Deharo JC, Butter C, Seifert M, Boersma LVA, Riahi S, James S, Turley AJ, Auricchio A, Betts TR, Niederer S, Sanders P, Rinaldi CA (2020) Leadless left ventricular endocardial pacing in nonresponders to conventional cardiac resynchronization therapy. Pacing Clin Electrophysiol 43(9):966–973
23. Ali N, Keene D, Arnold A, Shun-Shin M, Whinnett ZI, Afzal Sohaib SM (2018) His bundle pacing: a new frontier in the treatment of heart failure. Arrhythmia Electrophysiol Rev 7(2): 103–110
24. Singh R, Devabhaktuni S, Ezzeddine F, Simon J, Khaira K, Dandamudi G (2020) His-bundle pacing: a novel treatment for left bundle branch block-mediated cardiomyopathy. J Cardiovasc Electrophysiol 31(10):2730–2736
25. Arnold AD, Shun-Shin MJ, Keene D, Howard JP, Sohaib SMA, Wright IJ, Cole GD, Qureshi NA, Lefroy DC, Koa-Wing M, Linton NWF, Lim PB, Peters NS, Davies DW, Muthumala A, Tanner M, Ellenbogen KA, Kanagaratnam P, Francis DP, Whinnett ZI (2018) His resynchronization versus biventricular pacing in patients with heart failure and left bundle branch block. J Am Coll Cardiol 72(24):3112–3122
26. Lustgarten DL, Crespo EM, Arkhipova-Jenkins I, Lobel R, Winget J, Koehler J, Liberman E, Sheldon T (2015) His-bundle pacing versus biventricular pacing in cardiac resynchronization therapy patients: a crossover design comparison. Heart Rhythm 12(7):1548–1557
27. Lin J, Dai Y, Wang H, Li Y, Chen K, Zhang S (2020) A comparison of left bundle branch pacing with his bundle pacing in a patient with heart failure and left bundle branch block. HeartRhythm Case Rep 6(6):293–296

28. Strocchi M, Lee AWC, Neic A, Bouyssier J, Gillette K, Plank G, Elliott MK, Gould J, Behar JM, Sidhu B, Mehta V, Bishop MJ, Vigmond EJ, Rinaldi CA, Niederer SA (2020) His-bundle and left bundle pacing with optimized atrioventricular delay achieve superior electrical synchrony over endocardial and epicardial pacing in left bundle branch block patients. Heart Rhythm 17(11):1922–1929
29. Mullens W, Grimm RA, Verga T, Dresing T, Starling RC, Wilkoff BL, Tang WHW (2009) Insights from a cardiac resynchronization optimization clinic as part of a heart failure disease management program. J Am Coll Cardiol 53(9):765–773
30. Isotani A, Yoneda K, Iwamura T, Watanabe M, Okada JI, Washio T, Sugiura S, Hisada T, Ando K (2020) Patient-specific heart simulation can identify non-responders to cardiac resynchronization therapy. Heart Vessels 35(8):1135–1147
31. Lee AW, Crozier A, Hyde ER, Lamata P, Truong M, Sohal M, Jackson T, Behar JM, Claridge S, Shetty A, Sammut E, Plank G, Rinaldi CA, Niederer S (2017) Biophysical modeling to determine the optimization of left ventricular pacing site and AV/VV delays in the acute and chronic phase of cardiac resynchronization therapy. J Cardiovasc Electrophysiol 28(2):208–215
32. Mendonca Costa C, Neic A, Kerfoot E, Porter B, Sieniewicz B, Gould J, Sidhu B, Chen Z, Plank G, Rinaldi CA, Bishop MJ, Niederer SA (2019) Pacing in proximity to scar during cardiac resynchronization therapy increases local dispersion of repolarization and susceptibility to ventricular arrhythmogenesis. Heart Rhythm 16(10):1475–1483
33. Okada JI, Washio T, Nakagawa M, Watanabe M, Kadooka Y, Kariya T, Yamashita H, Yamada Y, Momomura SI, Nagai R, Hisada T, Sugiura S (2017) Multi-scale, tailor-made heart simulation can predict the effect of cardiac resynchronization therapy. J Mol Cell Cardiol 108:17–23
34. Pluijmert M, Bovendeerd PH, Lumens J, Vernooy K, Prinzen FW, Delhaas T (2016) New insights from a computational model on the relation between pacing site and CRT response. Europace 18(Suppl 4):iv94–iv103
35. Villongco CT, Krummen DE, Omens JH, McCulloch AD (2016) Non-invasive, model-based measures of ventricular electrical dyssynchrony for predicting CRT outcomes. Europace 18 (Suppl 4):iv104–iv112
36. Willemen E, Schreurs R, Huntjens PR, Strik M, Plank G, Vigmond E, Walmsley J, Vernooy K, Delhaas T, Prinzen FW, Lumens J (2019) The left and right ventricles respond differently to variation of pacing delays in cardiac resynchronization therapy: a combined experimental- computational approach. Front Physiol 10:17
37. Rodero C, Strocchi M, Lee AWC, Rinaldi CA, Vigmond EJ, Plank G, Lamata P, Niederer SA (2022) Impact of anatomical reverse remodelling in the design of optimal quadripolar pacing leads: a computational study. Comput Biol Med 140:105073
38. Viceconti M, Juarez MA, Curreli C, Pennisi M, Russo G, Pappalardo F (2020) Credibility of in silico trial technologies-a theoretical framing. IEEE J Biomed Health Inform 24(1):4–13
39. Campos FO, Orini M, Arnold R, Whitaker J, O'Neill M, Razavi R, Plank G, Hanson B, Porter B, Rinaldi CA, Gill J, Lambiase PD, Taggart P, Bishop MJ (2021) Assessing the ability of substrate mapping techniques to guide ventricular tachycardia ablation using computational modelling. Comput Biol Med 130:104214
40. Strocchi M, Gsell MAF, Augustin CM, Razeghi O, Roney CH, Prassl AJ, Vigmond EJ, Behar JM, Gould JS, Rinaldi CA, Bishop MJ, Plank G, Niederer SA (2020) Simulating ventricular systolic motion in a four-chamber heart model with spatially varying robin boundary conditions to model the effect of the pericardium. J Biomech 101:109645
41. Karabelas E, Gsell MAF, Augustin CM, Marx L, Neic A, Prassl AJ, Goubergrits L, Kuehne T, Plank G (2018) Towards a computational framework for modeling the impact of aortic coarctations upon left ventricular load. Front Physiol 9(538)
42. Rodero C, Strocchi M, Marciniak M, Longobardi S, Whitaker J, O'Neill MD, Gillette K, Augustin C, Plank G, Vigmond EJ, Lamata P, Niederer SA (2021) Linking statistical shape

models and simulated function in the healthy adult human heart. PLoS Comput Biol 17(4): e1008851
43. Usyk TP, McCulloch AD (2003) Electromechanical model of cardiac resynchronization in the dilated failing heart with left bundle branch block. J Electrocardiol 36:57–61
44. Hyde ER, Behar JM, Claridge S, Jackson T, Lee AW, Remme EW, Sohal M, Plank G, Razavi R, Rinaldi CA, Niederer SA (2015) Beneficial effect on cardiac resynchronization from left ventricular endocardial pacing is mediated by early access to high conduction velocity tissue: electrophysiological simulation study. Circ Arrhythm Electrophysiol 8(5): 1164–1172
45. Huntjens PR, Walmsley J, Ploux S, Bordachar P, Prinzen FW, Delhaas T, Lumens J (2014) Influence of left ventricular lead position relative to scar location on response to cardiac resynchronization therapy: a model study. Europace 16(Suppl 4):iv62–iv68
46. Claus P, Bijnens B, Breithardt O-A, Sutherland GR (2003) Why ischemic hearts respond less to cardiac resynchronisation therapy. A modeling study. Lect Notes Comput Sci:287–294
47. Kerckhoffs RC, McCulloch AD, Omens JH, Mulligan LJ (2009) Effects of biventricular pacing and scar size in a computational model of the failing heart with left bundle branch block. Med Image Anal 13(2):362–369
48. Gurev V, Lee T, Constantino J, Arevalo H, Trayanova NA (2011) Models of cardiac electromechanics based on individual hearts imaging data. Biomech Model Mechanobiol 10(3):295–306
49. Panthee N, Okada J, Washio T, Mochizuki Y, Suzuki R, Koyama H, Ono M, Hisada T, Sugiura S (2016) Tailor-made heart simulation predicts the effect of cardiac resynchronization therapy in a canine model of heart failure. Med Image Anal 31:46–62
50. Aguado-Sierra J, Krishnamurthy A, Villongco C, Chuang J, Howard E, Gonzales MJ, Omens J, Krummen DE, Narayan S, Kerckhoffs RC (2011) Patient-specific modeling of dyssynchronous heart failure: a case study. Prog Biophys Mol Biol 107(1):147–155
51. Crozier A, Augustin CM, Neic A, Prassl AJ, Holler M, Fastl TE, Hennemuth A, Bredies K, Kuehne T, Bishop MJ, Niederer SA, Plank G (2015) Image-based personalization of cardiac anatomy for coupled electromechanical Modeling. Ann Biomed Eng
52. Tobon-Gomez C, Duchateau N, Sebastian R, Marchesseau S, Camara O, Donal E, De Craene M, Pashaei A, Relan J, Steghofer M (2013) Understanding the mechanisms amenable to CRT response: from pre-operative multimodal image data to patient-specific computational models. Med Biol Eng Comput 51(11):1235–1250
53. Warriner DR, Jackson T, Zacur E, Sammut E, Sheridan P, Hose DR, Lawford P, Razavi R, Niederer SA, Rinaldi CA, Lamata P (2018) An asymmetric wall-thickening pattern predicts response to cardiac resynchronization therapy. JACC Cardiovasc Imaging 11(10):1545–1546
54. Lee AW, Nguyen UC, Razeghi O, Gould J, Sidhu B, Sieniewicz B, Behar J, Mafi-Rad M, Plank G, Prinzen FW (2019) A rule-based method for predicting the electrical activation of the heart with cardiac resynchronization therapy from non-invasive clinical data. Med Image Anal 57:197–213
55. Campos JO, Sundnes J, Dos Santos RW, Rocha BM (2020) Uncertainty quantification and sensitivity analysis of left ventricular function during the full cardiac cycle. Philos Trans A Math Phys Eng Sci 378(2173):20190381
56. Lopez-Perez A, Sebastian R, Izquierdo M, Ruiz R, Bishop M, Ferrero JM (2019) Personalized cardiac computational models: from clinical data to simulation of infarct-related ventricular tachycardia. Front Physiol 10:580
57. Margara F, Wang ZJ, Levrero-Florencio F, Santiago A, Vázquez M, Bueno-Orovio A, Rodriguez B (2021) In-silico human electro-mechanical ventricular modelling and simulation for drug-induced pro-arrhythmia and inotropic risk assessment. Prog Biophys Mol Biol 159: 58–74
58. Arnold A, Shun-Shin M, Ali N, Howard J, Keene D, Chow J, Qureshi N, Lefroy D, Koa-Wing M, Linton N, Lim P, Peters N, Kanagaratnam P, Francis D, Whinnett Z (2021)

Atrioventricular shortening is the dominant mechanism of benefit of biventricular pacing in left bundle branch block. Europace 23(Supplement_3)
59. Pfaller MR, Hörmann JM, Weigl M, Nagler A, Chabiniok R, Bertoglio C, Wall WA (2019) The importance of the pericardium for cardiac biomechanics: from physiology to computational modeling. Biomech Model Mechanobiol 18(2):503–529
60. Strocchi M, Augustin CM, Gsell MAF, Karabelas E, Neic A, Gillette K, Roney CH, Razeghi O, Behar JM, Rinaldi CA, Vigmond EJ, Bishop MJ, Plank G, Niederer SA (2021) The effect of ventricular myofibre orientation on atrial dynamics. Springer, Cham
61. Clerc L (1976) Directional differences of impulse spread in trabecular muscle from mammalian heart. J Physiol 255(2):335–346
62. Roberts DE, Hersh LT, Scher AM (1979) Influence of cardiac fiber orientation on wavefront voltage, conduction velocity, and tissue resistivity in the dog. Circ Res 44(5):701–712
63. Roth BJ (1997) Electrical conductivity values used with the bidomain model of cardiac tissue. IEEE Trans Biomed Eng 44(4):326–328
64. Yin FC, Strumpf RK, Chew PH, Zeger SL (1987) Quantification of the mechanical properties of noncontracting canine myocardium under simultaneous biaxial loading. J Biomech 20(6): 577–589
65. Bayer JD, Blake RC, Plank G, Trayanova NA (2012) A novel rule-based algorithm for assigning myocardial fiber orientation to computational heart models. Ann Biomed Eng 40(10):2243–2254
66. Crozier A, Blazevic B, Lamata P, Plank G, Ginks M, Duckett S, Sohal M, Shetty A, Rinaldi CA, Razavi R, Smith NP, Niederer SA (2015) The relative role of patient physiology and device optimisation in cardiac resynchronisation therapy: a computational modelling study. J Mol Cell Cardiol 96:93–100
67. Kayvanpour E, Mansi T, Sedaghat-Hamedani F, Amr A, Neumann D, Georgescu B, Seegerer P, Kamen A, Haas J, Frese KS (2015) Towards personalized cardiology: multi-scale modeling of the failing heart. PLoS One 10(7):e0134869
68. Niederer SA, Plank G, Chinchapatnam P, Ginks M, Lamata P, Rhode KS, Rinaldi CA, Razavi R, Smith NP (2011) Length-dependent tension in the failing heart and the efficacy of cardiac resynchronization therapy. Cardiovasc Res 89(2):336–343
69. Reumann M, Farina D, Miri R, Lurz S, Osswald B, Dossel O (2007) Computer model for the optimization of AV and VV delay in cardiac resynchronization therapy. Med Biol Eng Comput 45(9):845–854
70. Romero D, Sebastian R, Bijnens BH, Zimmerman V, Boyle PM, Vigmond EJ, Frangi AF (2010) Effects of the purkinje system and cardiac geometry on biventricular pacing: a model study. Ann Biomed Eng 38(4):1388–1398
71. Sermesant M, Chabiniok R, Chinchapatnam P, Mansi T, Billet F, Moireau P, Peyrat JM, Wong K, Relan J, Rhode K, Ginks M, Lambiase P, Delingette H, Sorine M, Rinaldi CA, Chapelle D, Razavi R, Ayache N (2012) Patient-specific electromechanical models of the heart for the prediction of pacing acute effects in CRT: a preliminary clinical validation. Med Image Anal 16(1):201–215
72. Nguyen C, Fan Z, Sharif B, He Y, Dharmakumar R, Berman DS, Li D (2014) In vivo three-dimensional high resolution cardiac diffusion-weighted MRI: a motion compensated diffusion-prepared balanced steady-state free precession approach. Magn Reson Med 72(5): 1257–1267
73. Wei H, Viallon M, Delattre BM, Moulin K, Yang F, Croisille P, Zhu Y (2015) Free-breathing diffusion tensor imaging and tractography of the human heart in healthy volunteers using wavelet-based image fusion. IEEE Trans Med Imaging 34(1):306–316
74. Helm PA, Tseng HJ, Younes L, McVeigh ER, Winslow RL (2005) Ex vivo 3D diffusion tensor imaging and quantification of cardiac laminar structure. Magn Reson Med 54(4): 850–859

75. Wang VY, Casta C, Zhu YM, Cowan BR, Croisille P, Young AA, Clarysse P, Nash MP (2016) Image-based investigation of human in vivo myofibre strain. IEEE Trans Med Imaging 35(11):2486–2496
76. Lau AZ, Tunnicliffe EM, Frost R, Koopmans PJ, Tyler DJ, Robson MD (2015) Accelerated human cardiac diffusion tensor imaging using simultaneous multislice imaging. Magn Reson Med 73(3):995–1004
77. Mekkaoui C, Reese TG, Jackowski MP, Bhat H, Kostis WJ, Sosnovik DE (2014) In vivo fiber tractography of the right and left ventricles using diffusion tensor MRI of the entire human heart. J Cardiovasc Magn Reson 16(1):P17
78. Nielles-Vallespin S, Mekkaoui C, Gatehouse P, Reese TG, Keegan J, Ferreira PF, Collins S, Speier P, Feiweier T, de Silva R, Jackowski MP, Pennell DJ, Sosnovik DE, Firmin D (2013) In vivo diffusion tensor MRI of the human heart: reproducibility of breath-hold and navigator-based approaches. Magn Reson Med 70(2):454–465
79. McGill L-A, Ismail TF, Nielles-Vallespin S, Ferreira P, Scott AD, Roughton M, Kilner PJ, Ho SY, McCarthy KP, Gatehouse PD, de Silva R, Speier P, Feiweier T, Mekkaoui C, Sosnovik DE, Prasad SK, Firmin DN, Pennell DJ (2012) Reproducibility of in-vivo diffusion tensor cardiovascular magnetic resonance in hypertrophic cardiomyopathy. J Cardiovasc Magn Reson 14(1):86
80. Bishop MJ, Hales P, Plank G, Gavaghan DJ, Scheider J, Grau V (2009) Comparison of rule-based and DTMRI-derived fibre architecture in a whole rat ventricular computational model. International conference on functional imaging and modeling of the heart
81. Roney CH, Pashaei A, Meo M, Dubois R, Boyle PM, Trayanova NA, Cochet H, Niederer SA, Vigmond EJ (2019) Universal atrial coordinates applied to visualisation, registration and construction of patient specific meshes. Med Image Anal 55:65–75
82. Daoulah A, Alsheikh-Ali AA, Al-Faifi SM, Ocheltree SR, Haq E, Asrar FM, Fathey A, Haneef AA, Al Mousily F, El-Sayed O (2015) Cardiac resynchronization therapy in patients with postero-lateral scar by cardiac magnetic resonance: a systematic review and meta-analysis. J Electrocardiol 48(5):783–790
83. White JA, Yee R, Yuan X, Krahn A, Skanes A, Parker M, Klein G, Drangova M (2006) Delayed enhancement magnetic resonance imaging predicts response to cardiac resynchronization therapy in patients with intraventricular dyssynchrony. J Am Coll Cardiol 48(10):1953–1960
84. Holmes JW, Laksman Z, Gepstein L (2016) Making better scar: emerging approaches for modifying mechanical and electrical properties following infarction and ablation. Prog Biophys Mol Biol 120(1):134–148
85. Rog-Zielinska EA, Norris RA, Kohl P, Markwald R (2016) The living scar–cardiac fibroblasts and the injured heart. Trends Mol Med 22(2):99–114
86. Kovacheva E, Gerach T, Schuler S, Ochs M, Dössel O, Loewe A (2021) Causes of altered ventricular mechanics in hypertrophic cardiomyopathy: an in-silico study. Biomed Eng Online 20(1):69
87. Mendonca Costa C, Neic A, Gillette K, Porter B, Gould J, Sidhu B, Chen Z, Elliott M, Mehta V, Plank G, Rinaldi CA, Bishop MJ, Niederer SA (2020) Left ventricular endocardial pacing is less arrhythmogenic than conventional epicardial pacing when pacing in proximity to scar. Heart Rhythm 17(8):1262–1270
88. Nayak HM, Verdino RJ, Russo AM, Gerstenfeld EP, Hsia HH, Lin D, Dixit S, Cooper JM, Callans DJ, Marchlinski FE (2008) Ventricular tachycardia storm after initiation of biventricular pacing: incidence, clinical characteristics, management, and outcome. J Cardiovasc Electrophysiol 19(7):708–715
89. Amado LC, Gerber BL, Gupta SN, Rettmann DW, Szarf G, Schock R, Nasir K, Kraitchman DL, Lima JOAC (2004) Accurate and objective infarct sizing by contrast-enhanced magnetic resonance imaging in a canine myocardial infarction model. J Am Coll Cardiol 44(12): 2383–2389

90. Kim RJ, Fieno DS, Parrish TB, Harris K, Chen EL, Simonetti O, Bundy J, Finn JP, Klocke FJ, Judd RM (1999) Relationship of MRI delayed contrast enhancement to irreversible injury, infarct age, and contractile function. Circulation 100(19):1992–2002
91. Karim R, Housden RJ, Balasubramaniam M, Chen Z, Perry D, Uddin A, Al-Beyatti Y, Palkhi E, Acheampong P, Obom S, Hennemuth A, Lu Y, Bai W, Shi W, Gao Y, Peitgen H-O, Radau P, Razavi R, Tannenbaum A, Rueckert D, Cates J, Schaeffter T, Peters D, MacLeod R, Rhode K (2013) Evaluation of current algorithms for segmentation of scar tissue from late gadolinium enhancement cardiovascular magnetic resonance of the left atrium: an open-access grand challenge. J Cardiovasc Magn Reson 15(15):105
92. Flett AS, Hasleton J, Cook C, Hausenloy D, Quarta G, Ariti C, Muthurangu V, Moon JC (2011) Evaluation of techniques for the quantification of myocardial scar of differing etiology using cardiac magnetic resonance. JACC Cardiovasc Imaging 4(2):150–156
93. Ukwatta E, Rajchl M, White J, Pashakhanloo F, Herzka D, McVeigh E, Lardo A, Trayanova N, Vadakkumpadan F (2015) Image-based reconstruction of 3D myocardial infarct geometry for patient specific applications, SPIE
94. Bose A, Kandala J, Upadhyay GA, Riedl L, Ahmado I, Padmanabhan R, Gewirtz H, Mulligan LJ, Singh JP (2014) Impact of myocardial viability and left ventricular lead location on clinical outcome in cardiac resynchronization therapy recipients with ischemic cardiomyopathy. J Cardiovasc Electrophysiol 25(5):507–513
95. Potse M, Krause D, Bacharova L, Krause R, Prinzen FW, Auricchio A (2012) Similarities and differences between electrocardiogram signs of left bundle-branch block and left-ventricular uncoupling. Europace 14(Suppl 5):33–39
96. McClelland AJJ, Owens CG, Menown IBA, Lown M, Adgey AAJ (2003) Comparison of the 80-lead body surface map to physician and to 12-lead electrocardiogram in detection of acute myocardial infarction. Am J Cardiol 92(3):252–257
97. Owens C, McClelland A, Walsh S, Smith B, Adgey J (2008) Comparison of value of leads from body surface maps to 12-lead electrocardiogram for diagnosis of acute myocardial infarction. Am J Cardiol 102(3):257–265
98. Bear LR, Huntjens PR, Walton RD, Bernus O, Coronel R, Dubois R (2018) Cardiac electrical dyssynchrony is accurately detected by noninvasive electrocardiographic imaging. Heart Rhythm 15(7):1058–1069
99. Bear LR, LeGrice IJ, Sands GB, Lever NA, Loiselle DS, Paterson DJ, Cheng LK, Smaill BH (2018) How accurate is inverse electrocardiographic mapping? Circ Arrhythm Electrophysiol 11(5):e006108
100. Cluitmans M, Brooks DH, MacLeod R, Dössel O, Guillem MS, van Dam PM, Svehlikova J, He B, Sapp J, Wang L, Bear L (2018) Validation and opportunities of electrocardiographic imaging: from technical achievements to clinical applications. Front Physiol 9:1305
101. Elliott MK, Mehta V, Sidhu BS, Niederer S, Rinaldi CA (2020) Electrocardiographic imaging of His bundle, left bundle branch, epicardial, and endocardial left ventricular pacing to achieve cardiac resynchronization therapy. HeartRhythm Case Rep 6(7):460–463
102. Cardone-Noott L, Bueno-Orovio A, Mincholé A, Zemzemi N, Rodriguez B (2016) Human ventricular activation sequence and the simulation of the electrocardiographic QRS complex and its variability in healthy and intraventricular block conditions. Europace 18(suppl 4):iv4–iv15
103. Delingette H, Billet F, Wong KC, Sermesant M, Rhode K, Ginks M, Rinaldi CA, Razavi R, Ayache N (2012) Personalization of cardiac motion and contractility from images using variational data assimilation. IEEE Trans Biomed Eng 59(1):20–24
104. Giffard-Roisin S, Jackson T, Fovargue L, Lee J, Delingette H, Razavi R, Ayache N, Sermesant M (2017) Noninvasive personalization of a cardiac electrophysiology model from body surface potential mapping. IEEE Trans Biomed Eng 64(9):2206–2218
105. Walmsley J, Arts T, Derval N, Bordachar P, Cochet H, Ploux S, Prinzen FW, Delhaas T, Lumens J (2015) Fast simulation of mechanical heterogeneity in the electrically asynchronous heart using the multipatch module. PLoS Comput Biol 11(7):e1004284

106. Arts T, Delhaas T, Bovendeerd P, Verbeek X, Prinzen FW (2005) Adaptation to mechanical load determines shape and properties of heart and circulation: the CircAdapt model. Am J Physiol Heart Circ Physiol 288(4):H1943–H1954
107. Lumens J, Delhaas T, Kirn B, Arts T (2009) Three-wall segment (TriSeg) model describing mechanics and hemodynamics of ventricular interaction. Ann Biomed Eng 37(11):2234–2255
108. Sanchez C, D'Ambrosio G, Maffessanti F, Caiani EG, Prinzen FW, Krause R, Auricchio A, Potse M (2017) Sensitivity analysis of ventricular activation and electrocardiogram in tailored models of heart-failure patients. Med Biol Eng Comput:1–14
109. Pashaei A, Romero D, Sebastian R, Camara O, Frangi AF (2011) Fast multiscale modeling of cardiac electrophysiology including Purkinje system. IEEE Trans Biomed Eng 58(10 Part 2): 2956–2960
110. Werner CD, Sachse FB, Dossel O (2000) Electrical excitation propagation in the human heart. Int J Electromagn
111. Miri R, Graf IM, Dossel O (2009) Efficiency of timing delays and electrode positions in optimization of biventricular pacing: a simulation study. IEEE Trans Biomed Eng 56(11): 2573–2582
112. Miri R, Reumann M, Keller DUJ, Farina D, Dossel O (2008) Comparison of the electrophysiologically based optimization methods with different pacing parameters in patient undergoing resynchronization treatment. Engineering in Medicine and Biology Society, 2008. EMBS 2008. 30th annual international conference of the IEEE 2008, pp 1741–1744
113. Reumann M, Osswald B, Doessel O (2007) Noninvasive, automatic optimization strategy in cardiac resynchronization therapy. Anadolu Kardiyol Derg 7(Suppl 1(6)):209–212
114. Van Dam PM, Oostendorp TF, Linnenbank AC, Van Oosterom A (2009) Non-invasive imaging of cardiac activation and recovery. Ann Biomed Eng 37(9):1739–1756
115. Van Oosterom A (2001) Genesis of the T wave as based on an equivalent surface source model. J Electrocardiol 34(4):217–227
116. van Huysduynen BH, Swenne CA, Bax JJ, Bleeker GB, Draisma HHM, van Erven L, Molhoek SG, van de Vooren H, van der Wall EE, Schalij MJ (2005) Dispersion of repolarization in cardiac resynchronization therapy. Heart Rhythm 2(12):1286–1293
117. Draper MH, Mya-Tu M (1959) A comparison of the conduction velocity in cardiac tissues of various mammals. Q J Exp Physiol Cogn Med Sci 44(1):91–109
118. Myerburg RJ, Gelband H, Nilsson K, Castellanos A, Morales AR, Bassett AL (1978) The role of canine superficial ventricular muscle fibers in endocardial impulse distribution. Circ Res 42(1):27–35
119. Barber F, Langfield P, Lozano M, García-Fernández I, Duchateau J, Hocini M, Haïssaguerre M, Vigmond E, Sebastian R (2021) Estimation of personalized minimal Purkinje systems from human electro-anatomical maps. IEEE Trans Med Imaging 40(8):2182–2194
120. Sahli Costabal F, Hurtado DE, Kuhl E (2016) Generating Purkinje networks in the human heart. J Biomech 49(12):2455–2465
121. Durrer D, van Dam RT, Freud GE, Janse MJ, Meijler FL, Arzbaecher RC (1970) Total excitation of the isolated human heart. Circulation 41(6):899–912
122. Ramanathan C, Jia P, Ghanem R, Ryu K, Rudy Y (2006) Activation and repolarization of the normal human heart under complete physiological conditions. Proc Natl Acad Sci 103(16): 6309–6314
123. Behar JM, Jackson T, Hyde E, Claridge S, Gill J, Bostock J, Sohal M, Porter B, O'Neill M, Razavi R, Niederer S, Rinaldi CA (2016) Optimized left ventricular endocardial stimulation is superior to optimized epicardial stimulation in ischemic patients with poor response to cardiac resynchronization therapy: a combined magnetic resonance imaging, electroanatomic contact mapping, and hemodynamic study to target endocardial Lead placement. JACC Clin Electrophysiol 2(7):799–809
124. Bordachar P, Grenz N, Jais P, Ritter P, Leclercq C, Morgan JM, Gras D, Yang P (2012) Left ventricular endocardial or triventricular pacing to optimize cardiac resynchronization therapy

in a chronic canine model of ischemic heart failure. Am J Physiol Heart Circ Physiol 303(2): H207–H215
125. Miri R, Dossel O (2009) Computerized optimization of biventricular pacing using body surface potential map. Proceedings of the 31st annual international conference of the IEEE engineering in medicine and biology society: engineering the future of biomedicine, EMBC 2009, pp 2815–2818
126. Miri R, Reumann M, Farina D, Dossel O (2009) Concurrent optimization of timing delays and electrode positioning in biventricular pacing based on a computer heart model assuming 17 left ventricular segments. Biomed Tech 54(2):55–65
127. Miri R, Graf IM, Bayarri JV, Dossel O (2010) Applicability of body surface potential map in computerized optimization of biventricular pacing. Ann Biomed Eng 38(3):865–875
128. van Oosterom A, Oostendorp TF (2004) ECGSIM: an interactive tool for studying the genesis of QRST waveforms. Heart 90(2):165–168
129. Auricchio A, Yu CM (2004) Beyond the measurement of QRS complex toward mechanical dyssynchrony: cardiac resynchronisation therapy in heart failure patients with a normal QRS duration. Heart 90(5):479–481
130. Derval N, Bordachar P, Lim HS, Sacher F, Ploux S, Laborderie J, Steendijk P, Deplagne A, Ritter P, Garrigue S (2014) Impact of pacing site on QRS duration and its relationship to hemodynamic response in cardiac resynchronization therapy for congestive heart failure. J Cardiovasc Electrophysiol 25(9):1012–1020
131. Kass DA (2003) Predicting cardiac resynchronization response by QRS duration. J Am Coll Cardiol
132. Nordsletten DA, Niederer SA, Nash MP, Hunter PJ, Smith NP (2011) Coupling multi-physics models to cardiac mechanics. Prog Biophys Mol Biol 104(1–3):77–88
133. Dokos S, Smaill BH, Young AA, LeGrice IJ (2002) Shear properties of passive ventricular myocardium. Am J Physiol Heart Circ Physiol 283(6):H2650–H2659
134. Holzapfel GA, Ogden RW (2009) Constitutive modelling of passive myocardium: a structurally based framework for material characterization. Philos Trans R Soc A Math Phys Eng Sci 367(1902):3445–3475
135. Kohl P, Noble D, Hunter PJ, Costa KD, Holmes JW, Mcculloch AD (2001) Modelling cardiac mechanical properties in three dimensions. Philos Trans R Soc London Ser A 359(1783): 1233–1250
136. Usyk TP, Mazhari R, McCulloch AD (2000) Effect of laminar orthotropic myofiber architecture on regional stress and strain in the canine left ventricle. J Elasticity Phys Sci Solids 61(1): 143–164
137. Demer LL, Yin FC (1983) Passive biaxial mechanical properties of isolated canine myocardium. J Physiol 339:615–630
138. Guccione JM, McCulloch AD, Waldman LK (1991) Passive material properties of intact ventricular myocardium determined from a cylindrical model. J Biomech Eng 113(1):42–55
139. Humphrey JD, Strumpf RK, Yin FC (1990) Determination of a constitutive relation for passive myocardium: I. A new functional form. J Biomech Eng 112(3):333–339
140. Humphrey JD, Yin FC (1987) On constitutive relations and finite deformations of passive cardiac tissue: I. A pseudostrain-energy function. J Biomech Eng 109(4):298–304
141. Humphrey JD, Yin FC (1989) Constitutive relations and finite deformations of passive cardiac tissue II: stress analysis in the left ventricle. Circ Res 65(3):805–817
142. Bers DM (2002) Cardiac excitation-contraction coupling. Nature 415(6868):198–205
143. Kohl P, Sachs F (2001) Mechanoelectric feedback in cardiac cells. Philos Trans R Soc London Ser A 359(1783):1173–1185
144. Craelius W, Chen V, el-Sherif N (1988) Stretch activated ion channels in ventricular myocytes. Biosci Rep 8(5):407–414
145. Kuijpers NH, ten Eikelder HM, Bovendeerd PH, Verheule S, Arts T, Hilbers PA (2007) Mechanoelectric feedback leads to conduction slowing and block in acutely dilated atria: a

modeling study of cardiac electromechanics. Am J Physiol Heart Circ Physiol 292(6):H2832–H2853
146. Kerckhoffs RCP, Bovendeerd PHM, Prinzen FW, Smits K, Arts T (2003) Intra- and interventricular asynchrony of electromechanics in the ventricularly paced heart. J Eng Math 47(3):201–216
147. Trayanova NA (2011) Whole-heart modeling: applications to cardiac electrophysiology and electromechanics. Circ Res 108(1):113–128
148. Berberoğlu E, Solmaz HO, Göktepe S (2014) Computational modeling of coupled cardiac electromechanics incorporating cardiac dysfunctions. Eur J Mech A/Solids 48:60–73
149. Göktepe S, Kuhl E (2009) Electromechanics of the heart: a unified approach to the strongly coupled excitation–contraction problem. Comput Mech 45(2–3):227–243
150. Salvador M, Regazzoni F, Pagani S, Dede L, Trayanova N, Quarteroni A (2022) The role of mechano-electric feedbacks and hemodynamic coupling in scar-related ventricular tachycardia. Comput Biol Med 142:105203
151. Witzenburg CM, Holmes JW (2017) A comparison of phenomenologic growth laws for myocardial hypertrophy. J Elast 129(1):257–281
152. Kerckhoffs RC, Omens JH, McCulloch AD (2012) Mechanical discoordination increases continuously after the onset of left bundle branch block despite constant electrical dyssynchrony in a computational model of cardiac electromechanics and growth. Europace 14(Suppl 5):v65–v72
153. Lee LC, Sundnes J, Genet M, Wenk JF, Wall ST (2016) An integrated electromechanical-growth heart model for simulating cardiac therapies. Biomech Model Mechanobiol 15(4): 791–803
154. Gold MR, Auricchio A, Hummel JD, Giudici MC, Ding J, Tockman B, Spinelli J (2005) Comparison of stimulation sites within left ventricular veins on the acute hemodynamic effects of cardiac resynchronization therapy. Heart Rhythm 2(4):376–381
155. Zanon F, Baracca E, Pastore G, Fraccaro C, Roncon L, Aggio S, Noventa F, Mazza A, Prinzen F (2014) Determination of the longest intrapatient left ventricular electrical delay may predict acute hemodynamic improvement in patients after cardiac resynchronization therapy. Circ Arrhythm Electrophysiol 7(3):377–383
156. Niederer SA, Shetty AK, Plank G, Bostock J, Razavi R, Smith NP, Rinaldi CA (2012) Biophysical modeling to simulate the response to multisite left ventricular stimulation using a quadripolar pacing lead. Pacing Clin Electrophysiol 35(2):204–214
157. Richards DF, Glosli JN, Draeger EW, Mirin AA, Chan B, Fattebert JL, Krauss WD, Oppelstrup T, Butler CJ, Gunnels JA (2013) Towards real-time simulation of cardiac electrophysiology in a human heart at high resolution. Comput Methods Biomech Biomed Eng 16(7): 802–805
158. Niederer S, Mitchell L, Smith N, Plank G (2011) Simulating human cardiac electrophysiology on clinical time-scales. Front Physiol 2(April):14
159. Longobardi S, Lewalle A, Coveney S, Sjaastad I, Espe EKS, Louch WE, Musante CJ, Sher A, Niederer SA (2020) Predicting left ventricular contractile function via Gaussian process emulation in aortic-banded rats. Philos Trans A Math Phys Eng Sci 378(2173):20190334
160. Noè U, Lazarus A, Gao H, Davies V, Macdonald B, Mangion K, Berry C, Luo X, Husmeier D (2019) Gaussian process emulation to accelerate parameter estimation in a mechanical model of the left ventricle: a critical step towards clinical end-user relevance. J R Soc Interface 16(156):20190114
161. Butter C, Auricchio A, Stellbrink C, Fleck E, Ding J, Yu Y, Huvelle E, Spinelli J (2001) Effect of resynchronization therapy stimulation site on the systolic function of heart failure patients. Circulation 104(25):3026–2029
162. Helm RH, Byrne M, Helm PA, Daya SK, Osman NF, Tunin R, Halperin HR, Berger RD, Kass DA, Lardo AC (2007) Three-dimensional mapping of optimal left ventricular pacing site for cardiac resynchronization. Circulation 115(8):953–961

163. Constantino J, Hu Y, Trayanova NA (2012) A computational approach to understanding the cardiac electromechanical activation sequence in the normal and failing heart, with translation to the clinical practice of CRT. Prog Biophys Mol Biol 110(2):372–379
164. Strik M, van Deursen CJ, van Middendorp LB, van Hunnik A, Kuiper M, Auricchio A, Prinzen FW (2013) Transseptal conduction as an important determinant for cardiac resynchronization therapy, as revealed by extensive electrical mapping in the dyssynchronous canine heart. Circ Arrhythm Electrophysiol 6(4):682–689
165. Vardas PE, Auricchio A, Blanc J-J, Daubert J-C, Drexler H, Ector H, Gasparini M, Linde C, Morgado FB, Oto A, Sutton R, Trusz-Gluza M (2007) Guidelines for cardiac pacing and cardiac resynchronization therapy. The task force for cardiac pacing and cardiac resynchronization therapy of the European Society of Cardiology. Developed in collaboration with the European Heart Rhythm Association. Europace 9(10):959–998
166. Bogaard MD, Meine M, Tuinenburg AE, Maskara B, Loh P, Doevendans PA (2012) Cardiac resynchronization therapy beyond nominal settings: who needs individual programming of the atrioventricular and interventricular delay? Europace 14(12):1746–1753
167. Boriani G, Biffi M, Müller CP, Seidl KH, Grove R, Vogt J, Danschel W, Schuchert A, Deharo JC, Becker T, Boulogne E, Trappe HJ (2009) A prospective randomized evaluation of VV delay optimization in CRT-D recipients: echocardiographic observations from the RHYTHM II ICD study. Pacing Clin Electrophysiol 32(s1):S120–S125
168. Porciani MC, Dondina C, Macioce R, Demarchi G, Pieragnoli P, Musilli N, Colella A, Ricciardi G, Michelucci A, Padeletti L (2005) Echocardiographic examination of atrioventricular and interventricular delay optimization in cardiac resynchronization therapy. Am J Cardiol 95(9):1108–1110
169. Riedlbauchova L, Kautzner J, Fridl P (2005) Influence of different atrioventricular and interventricular delays on cardiac output during cardiac resynchronization therapy. Pacing Clin Electrophysiol 28(s1):S19–S23
170. Porciani MC, Dondina C, Macioce R, Demarchi G, Cappelli F, Lilli A, Pappone A, Ricciardi G, Colombo PC, Padeletti M, Jelic S, Padeletti L (2006) Temporal variation in optimal atrioventricular and interventricular delay during cardiac resynchronization therapy. J Card Fail 12(9):715–719
171. Gras D, Gupta MS, Boulogne E, Guzzo L, Abraham WT (2009) Optimization of AV and VV delays in the real-world CRT patient population: an international survey on current clinical practice. PACE – Pacing Clin Electrophysiol 32(Suppl 1):236–239
172. Ypenburg C, Schalij MJ, Bleeker GB, Steendijk P, Boersma E, Dibbets-Schneider P, Stokkel MPM, van der Wall EE, Bax JJ (2007) Impact of viability and scar tissue on response to cardiac resynchronization therapy in ischaemic heart failure patients. Eur Heart J 28(1):33–41
173. Ginks MR, Shetty AK, Lambiase PD, Duckett SG, Bostock J, Peacock JL, Rhode KS, Bucknall C, Gill J, Taggart P, Leclercq C, Carr-White GS, Razavi R, Rinaldi CA (2012) Benefits of endocardial and multisite pacing are dependent on the type of left ventricular electric activation pattern and presence of ischemic heart disease: insights from electroanatomic mapping. Circ Arrhythm Electrophysiol 5(5):889–897
174. Spragg DD, Dong J, Fetics BJ, Helm R, Marine JE, Cheng A, Henrikson CA, Kass DA, Berger RD (2010) Optimal left ventricular endocardial pacing sites for cardiac resynchronization therapy in patients with ischemic cardiomyopathy. J Am Coll Cardiol 56(10):774–781
175. Strik M, Rademakers LM, van Deursen CJ, van Hunnik A, Kuiper M, Klersy C, Auricchio A, Prinzen FW (2012) Endocardial left ventricular pacing improves cardiac resynchronization therapy in chronic asynchronous infarction and heart failure models. Circ Arrhythm Electrophysiol 5(1):191–200
176. Sidhu BS, Lee AWC, Haberland U, Rajani R, Niederer S, Rinaldi CA (2020) Combined computed tomographic perfusion and mechanics with predicted activation pattern can successfully guide implantation of a wireless endocardial pacing system. Europace 22(2):298
177. Zhu H, Jin L, Huang Y, Wu X (2021) A computer simulation research of two types of cardiac physiological pacing. Appl Sci 11(1):449

178. Hu Y, Gurev V, Constantino J, Trayanova N (2013) Efficient preloading of the ventricles by a properly timed atrial contraction underlies stroke work improvement in the acute response to cardiac resynchronization therapy. Heart Rhythm 10(12):1800–1806
179. Augustin CM, Neic A, Liebmann M, Prassl AJ, Niederer SA, Haase G, Plank G (2016) Anatomically accurate high resolution modeling of human whole heart electromechanics: a strongly scalable algebraic multigrid solver method for nonlinear deformation. J Comput Phys 305:622–646
180. Sohal M, Hamid S, Perego G, Della Bella P, Adhya S, Paisey J, Betts T, Kamdar R, Lambiase P, Leyva F, McComb JM, Behar J, Jackson T, Claridge S, Mehta V, Elliott M, Niederer S, Razavi R, Rinaldi CA (2021) A multicentre prospective randomised controlled trial of cardiac resynchronization therapy guided by invasive dP/dt. Heart Rhythm O2
181. Whinnett ZI, Francis DP, Denis A, Willson K, Pascale P, van Geldorp I, De Guillebon M, Ploux S, Ellenbogen K, Haïssaguerre M, Ritter P, Bordachar P (2013) Comparison of different invasive hemodynamic methods for AV delay optimization in patients with cardiac resynchronization therapy: implications for clinical trial design and clinical practice. Int J Cardiol 168(3):2228–2237
182. Duckett SG, Ginks M, Shetty AK, Bostock J, Gill JS, Hamid S, Kapetanakis S, Cunliffe E, Razavi R, Carr-White G, Rinaldi CA (2011) Invasive acute hemodynamic response to guide left ventricular lead implantation predicts chronic Remodeling in patients undergoing cardiac resynchronization therapy. J Am Coll Cardiol 58(11):1128–1136
183. Suzuki H, Shimano M, Yoshida Y, Inden Y, Muramatsu T, Tsuji Y, Tsuboi N, Hirayama H, Shibata R, Murohara T (2010) Maximum derivative of left ventricular pressure predicts cardiac mortality after cardiac resynchronization therapy. Clin Cardiol 33(12):E18–E23
184. Chung ES, Leon AR, Tavazzi L, Sun J-P, Nihoyannopoulos P, Merlino J, Abraham WT, Ghio S, Leclercq C, Bax JJ, Yu C-M, Gorcsan J, St John Sutton M, De Sutter J, Murillo J (2008) Results of the predictors of response to CRT (PROSPECT) trial. Circulation 117(20): 2608–2616
185. Steendijk P, Tulner SA, Bax JJ, Oemrawsingh PV, Bleeker GB, van Erven L, Putter H, Verwey HF, van der Wall EE, Schalij MJ (2006) Hemodynamic effects of long-term cardiac resynchronization therapy: analysis by pressure-volume loops. Circulation 113(10): 1295–1304
186. Kerckhoffs RCP, Neal ML, Gu Q, Bassingthwaighte JB, Omens JH, McCulloch AD (2007) Coupling of a 3D finite element model of cardiac ventricular mechanics to lumped systems models of the systemic and pulmonic circulation. Ann Biomed Eng 35(1):1–18
187. Land S, Niederer SA (2017) Influence of atrial contraction dynamics on cardiac function. Int J Numer Method Biomed Eng
188. Vernooy K, Verbeek XA, Peschar M, Crijns HJ, Arts T, Cornelussen RN, Prinzen FW (2004) Left bundle branch block induces ventricular remodelling and functional septal hypoperfusion. Eur Heart J 26(1):91–98
189. Liang F, Takagi S, Himeno R, Liu H (2009) Multi-scale modeling of the human cardiovascular system with applications to aortic valvular and arterial stenoses. Med Biol Eng Comput 47(7): 743–755
190. Schiavazzi DE, Baretta A, Pennati G, Hsia T, Marsden AL (2016) Patient-specific parameter estimation in single-ventricle lumped circulation models under uncertainty. Int J Numer Methods Biomed Eng
191. Sun Y, Beshara M, Lucariello RJ, Chiaramida SA (1997) A comprehensive model for right-left heart interaction under the influence of pericardium and baroreflex. Am J Physiol Heart Circ Physiol 272(3):1499
192. Olsen CO, Tyson GS, Maier GW, Spratt JA, Davis JW, Rankin JS (1983) Dynamic ventricular interaction in the conscious dog. Circ Res 52(1):85–104
193. Santamore WP, Dell'Italia LJ (1998) Ventricular interdependence: significant left ventricular contributions to right ventricular systolic function. Prog Cardiovasc Dis 40(4):289–308

194. Slinker BK, Glantz SA (1986) End-systolic and end-diastolic ventricular interaction. Am J Physiol Heart Circ Physiol 251(5):H1062–H1075
195. Weber KT, Janicki JS, Shroff S, Fishman AP (1981) Contractile mechanics and interaction of the right and left ventricles. Am J Cardiol 47(3):686–695
196. Kerckhoffs RC, Lumens J, Vernooy K, Omens JH, Mulligan LJ, Delhaas T, Arts T, McCulloch AD, Prinzen FW (2008) Cardiac resynchronization: insight from experimental and computational models. Prog Biophys Mol Biol 97(2–3):543–561
197. Trichon BH, O'connor CM (2002) Secondary mitral and tricuspid regurgitation accompanying left ventricular systolic dysfunction: is it important, and how is it treated? Am Heart J 144(3): 373–376
198. Calle S, Delens C, Kamoen V, De Pooter J, Timmermans F (2020) Septal flash: at the heart of cardiac dyssynchrony. Trends Cardiovasc Med 30(2):115–122
199. Leenders GE, Lumens J, Cramer MJ, Boeck BWLD, Doevendans PA, Delhaas T, Prinzen FW (2012) Septal deformation patterns delineate mechanical Dyssynchrony and regional differences in contractility. Circ Heart Fail 5(1):87–96
200. Lumens J, Ploux S, Strik M, Gorcsan J, Cochet H, Derval N, Strom M, Ramanathan C, Ritter P, Haïssaguerre M, Jaïs P, Arts T, Delhaas T, Prinzen FW, Bordachar P (2013) Comparative electromechanical and hemodynamic effects of left ventricular and biventricular pacing in dyssynchronous heart failure: electrical resynchronization versus left-right ventricular interaction. J Am Coll Cardiol 62(25):2395–2403
201. de Roest GJ, Wu L, de Cock CC, Hendriks ML, Delnoy PPH, van Rossum AC, Allaart CP (2014) Scar tissue–guided left ventricular lead placement for cardiac resynchronization therapy in patients with ischemic cardiomyopathy: an acute pressure-volume loop study. Am Heart J 167(4):537–545
202. Lumens J, Tayal B, Walmsley J, Delgado-Montero A, Huntjens PR, Schwartzman D, Althouse AD, Delhaas T, Prinzen FW, Gorcsan J 3rd (2015) Differentiating electromechanical from non-electrical substrates of mechanical discoordination to identify responders to cardiac resynchronization therapy. Circ Cardiovasc Imaging 8(9):e003744
203. Salden F, Huntjens PR, Schreurs R, Willemen E, Kuiper M, Wouters P, Maessen JG, Bordachar P, Delhaas T, Luermans J, Meine M, Allaart CP, van Stipdonk AMW, Prinzen FW, Lumens J, Vernooy K (2021) Pacing therapy for atrioventricular dromotropathy: a combined computational-experimental-clinical study. Europace
204. Huntjens PR, Ploux S, Strik M, Walmsley J, Ritter P, Haissaguerre M, Prinzen FW, Delhaas T, Lumens J, Bordachar P (2018) Electrical substrates driving response to cardiac resynchronization therapy: a combined clinical-computational evaluation. Circ Arrhythm Electrophysiol 11(4):e005647
205. Augustin CM, Gsell MA, Karabelas E, Willemen E, Prinzen F, Lumens J, Vigmond EJ, Plank G (2020) Validation of a 3D-0D closed-loop model of the heart and circulation—modeling the experimental assessment of diastolic and systolic ventricular properties. arXiv preprint arXiv:2009.08802
206. Kuijpers NH, Hermeling E, Lumens J, ten Eikelder HM, Delhaas T, Prinzen FW (2014) Mechano-electrical coupling as framework for understanding functional remodeling during LBBB and CRT. Am J Physiol Heart Circ Physiol 306(12):H1644–H1659
207. Gutenkunst RN, Waterfall JJ, Casey FP, Brown KS, Myers CR, Sethna JP (2007) Universally sloppy parameter sensitivities in systems biology models. PLoS Comput Biol 3(10):e189
208. Bazoukis G, Stavrakis S, Zhou J, Bollepalli SC, Tse G, Zhang Q, Singh JP, Armoundas AA (2021) Machine learning versus conventional clinical methods in guiding management of heart failure patients—a systematic review. Heart Fail Rev 26(1):23–34
209. Donal E, Hubert A, Le Rolle V, Leclercq C, Martins R, Mabo P, Galli E, Hernandez A (2019) New multiparametric analysis of cardiac dyssynchrony: machine learning and prediction of response to CRT. JACC Cardiovasc Imaging 12(9):1887–1888
210. Feeny AK, Rickard J, Patel D, Toro S, Trulock KM, Park CJ, LaBarbera MA, Varma N, Niebauer MJ, Sinha S, Gorodeski EZ, Grimm RA, Ji X, Barnard J, Madabhushi A, Spragg DD,

Chung MK (2019) Machine learning prediction of response to cardiac resynchronization therapy: improvement versus current guidelines. Circ Arrhythm Electrophysiol 12(7):e007316
211. Gallard A, Hubert A, Smiseth O, Voigt J-U, Le Rolle V, Leclercq C, Bidaut A, Galli E, Donal E, Hernandez AI (2020) Prediction of response to cardiac resynchronization therapy using a multi-feature learning method. Int J Cardiovasc Imaging
212. Peressutti D, Sinclair M, Bai W, Jackson T, Ruijsink J, Nordsletten D, Asner L, Hadjicharalambous M, Rinaldi CA, Rueckert D, King AP (2017) A framework for combining a motion atlas with non-motion information to learn clinically useful biomarkers: application to cardiac resynchronisation therapy response prediction. Med Image Anal 35:669–684
213. Cikes M, Sanchez-Martinez S, Claggett B, Duchateau N, Piella G, Butakoff C, Pouleur AC, Knappe D, Biering-Sørensen T, Kutyifa V, Moss A, Stein K, Solomon SD, Bijnens B (2019) Machine learning-based phenogrouping in heart failure to identify responders to cardiac resynchronization therapy. Eur J Heart Fail 21(1):74–85
214. Sánchez J, Luongo G, Nothstein M, Unger LA, Saiz J, Trenor B, Luik A, Dössel O, Loewe A (2021) Using machine learning to characterize atrial fibrotic substrate from intracardiac signals with a hybrid in silico and in vivo dataset. Front Physiol 12:699291
215. Shade JK, Ali RL, Basile D, Popescu D, Akhtar T, Marine JE, Spragg DD, Calkins H, Trayanova NA (2020) Preprocedure application of machine learning and mechanistic simulations predicts likelihood of paroxysmal atrial fibrillation recurrence following pulmonary vein isolation. Circ Arrhythm Electrophysiol 13(7):e008213
216. Shade JK, Prakosa A, Popescu DM, Yu R, Okada DR, Chrispin J, Trayanova NA (2021) Predicting risk of sudden cardiac death in patients with cardiac sarcoidosis using multimodality imaging and personalized heart modeling in a multivariable classifier. Sci Adv 7(31)
217. Trayanova NA, Popescu DM, Shade JK (2021) Machine learning in arrhythmia and electrophysiology. Circ Res 128(4):544–566

Therapeutic Innovations for Heart Failure

Anca Remes, Derk Frank, and Oliver J. Müller

1 Basic Mechanisms of Heart Failure

Pressure overload, triggered by valvular disease or hypertension, is a frequent cause of heart failure [1]. In response to pressure overload, cardiac tissue responds with ventricular remodeling, characterized by cardiomyocyte hypertrophy, extracellular matrix deposition, and intense alterations in metabolic function [2]. Although these processes are initially adaptive in a quest of preserving cardiac output, their chronic activation leads to decompensated disease. Another fundamental mechanism associated with cardiac dysfunction in late stages of the disease is myocyte apoptosis, strongly linked to a reduction in contractile function and development of arrhythmias [3].

On the molecular level, pathological hypertrophy is accompanied by re-activation of fetal genes involved in cardiac contractility and calcium handling and a concomitant downregulation of the expression levels of their adult isoform [4]. Furthermore, various cytokines and growth factors were demonstrated to play key roles in left ventricular growth during remodeling. For example, placental growth factor induces tissue necrosis factor-α (TNF-α) activation by increasing TIMP3 expression [5]. The understanding of neurohormonal activation in conditions of heart failure has opened the path to the major pharmacological treatments that improve the rates of mortality and morbidity in patients [6]. The most well-characterized compensatory homeostatic pathways activated during cardiac hypertrophy are the sympathetic nervous system and the renin–angiotensin–aldosterone system [6].

A. Remes (✉) · D. Frank · O. J. Müller
Department of Internal Medicine III, University of Kiel, Kiel, Germany

German Centre for Cardiovascular Research (DZHK), Partner Site Hamburg/Kiel/Lübeck, Kiel, Germany
e-mail: anca.remes@uksh.de

M. Hecker, D. J. Duncker (eds.), *Cardiac Mechanobiology in Physiology and Disease*, Cardiac and Vascular Biology 9,
https://doi.org/10.1007/978-3-031-23965-6_13

Calcineurin-nuclear factor of activated T cells (NFAT) pathway has been intensively studied in regard to pathological alterations in heart failure and links disturbances in Ca^{2+} signaling to pro-hypertrophic gene program. The first description of its role in the development of the disease has been established in transgenic mice, which present with cardiac hypertrophy that transitioned to decompensated heart failure early in age [7]. Consistently, inhibition of the pathway by administration of inhibitors such as cyclosporin A or FK506 ameliorated myocardial remodeling in various mouse models of heart failure [8, 9]. Calcineurin also regulates the expression of main transcription factors proven to be critical in disease progression and upregulation of fetal genes in pathological conditions.

Recent studies have underlined the main transcription factors involved in cardiac hypertrophy and heart failure and implicated in abnormal gene expression in diseased myocardium. GATA4 is one of the first transcription factors expressed in cardiomyocytes during development, regulating genes such as atrial natriuretic peptide (ANP), brain natriuretic peptide (BNP), cardiac troponin C and I, and α-myosin heavy chain [10]. Moreover, in vitro studies show that GATA4 is required for induction of brain natriuretic peptide (BNP) expression upon adrenergic stimulation [11]. Interestingly, stimulation with isoproterenol led to enhanced NFAT–GATA4 interaction and subsequent activation of fetal genes [12]. Additionally, myocyte enhancer factor 2 (MEF2) family of transcription factors has been identified to be important in calcium signaling in heart failure by activating Ca^{2+}-binding proteins, calmodulin, calmodulin kinases, and calcineurin, thereby inducing myocyte hypertrophy [13]. Moreover, MEF2 binding activity was shown to be enhanced in cardiac dysfunction due to volume or pressure overload [14]. MEF2 also interacts with other transcription factors involved in triggering the hypertrophic gene program, such as GATA4 [15] and NFAT [16]. An extensive characterization of the main transcription factors involved in the development of cardiac hypertrophy and its transition to heart failure is described elsewhere [17]. NFAT has been intensely studied in the context of heart failure. The dephosphorylated form of the protein is translocated into the nucleus and interacts with other transcription factors such as MEF2, GATA4, and activator protein-1 to drive pathological gene expression through multiple pathways [18]. Likewise, NFATc2 was demonstrated to be a necessary inducer of cardiac hypertrophy and heart failure [19].

Inflammation and oxidative stress have been shown to dramatically impair cardiac function during progression to heart failure. Clinical studies point towards a direct correlation between the concentration of circulating pro-inflammatory cytokines and the severity of the disease [20] and demonstrate that their levels can act as reliable prognostic factors in patients [21]. Inflammatory mediators induce cardiac dysfunction by triggering cardiac myocyte apoptosis, impairing intracellular calcium transport, and driving the activation of fibroblasts [22]. For example, TNF-α is a potent inducer of cardiomyocyte hypertrophy and upregulates matrix metalloproteases, leading to ventricular fibrosis [23]. Inflammation of the coronary microvascular endothelial cells causes a drop in nitric oxide (NO) bioavailability [24]. Growing evidence suggests that NO manifests a positive effect on cardiac contractility by decreasing titin-based hyperphosphorylation and hence stiffness

[25]. NO was also demonstrated to blunt pathological left ventricular remodeling after myocardial infarction [26]. The feasibility and efficacy of various anti-inflammatory based therapies in management of heart failure due to pressure overload are described in a separate section of the chapter.

2 Pharmacological Treatment of Heart Failure

The main objectives of the current pharmacological therapy for heart failure are reducing symptom severity and improving cardiac function and contractility, hence reducing the rate of hospitalization and associated mortality.

Chronic pathological activation of the neurohormonal system exerts a negative effect on cardiac function and was shown to be a hallmark of heart failure [6]. Hence, the main strategies to improve the clinical outcome of patients diagnosed with heart failure are the use of diuretics to attenuate edema, and neurohormonal antagonists, proven to reduce the rates of hospitalization and improve the mortality of affected patients [27]. Angiotensin-converting enzyme inhibitors (ACEi) [28] and beta-adrenergic blocking agents [29] are the main class of compounds used for the management of heart failure [30]. The treatment aims to target two major pathways driving disease progression, namely renin–angiotensin–aldosterone system (RAAS) and the sympathetic nervous system.

The beneficial effect of beta-blockers relates to reduction in heart rate, subsequent decrease in cardiac oxygen need, and marked decline in the occurrence of arryhthmias and sudden cardiac arrest. On the other hand, blockage of other adrenergic signaling pathways, such as alpha adrenergic receptor did not exert beneficial effects in patients with heart failure [31].

Additionally, mineralocorticoid receptor antagonists (MRAs) are recommended for patients who remain symptomatic despite being subjected to ACEi, angiotensin AT_1 receptor blockers, or beta-blockers therapy [32]. Inhibition of mineralocorticoid receptor affects pathological myocardial remodeling and decreases the degree of inflammation in diseased myocardium. Combined treatments with the above-mentioned drugs presented with the highest reduction in mortality and rate of hospitalization [33]. Application of the angiotensin receptor neprilysin inhibitor (ARNi) sacubitril/valsartan in rats subjected to experimental pressure overload revealed a superior cardioprotective effect of sacubitril/valsartan over the angiotensin AT_1 receptor antagonist valsartan alone against oxidative stress, mitochondrial dysfunction, and cardiac fibrosis, while there was no reduction in cardiac hypertrophy [34]. A similar study in rats subjected to pressure overload reported no difference in fibrosis but a small difference in antihypertrophic response in the sacubitril/valsartan group compared to valsartan alone which was independent from blood pressure [35]. Further studies are necessary to confirm whether sacubitril/valsartan has an effect on cardiac hypertrophy independent from blood pressure.

Recently, sodium-glucose co-transporter 2 (SGLT2) inhibitors empagliflozin (EMPEROR-Reduced) [36] and dapagliflozin (DAPA-HF) [37] showed a clear improvement in mortality, rate of hospitalization, and improvement of the quality of life when administered in parallel with standard therapies in patients with and without diabetes. However, studies from mice subjected to transverse aortic constriction showed no consistent effect on left ventricular remodeling. While one study suggests that treatment with empagliflozin did not prevent cardiac hypertrophy despite improving left ventricular function [38], another study showed attenuation of the hypertrophic response to pressure overload [39].

3 Anti-inflammatory Therapy

Taking into account the major role of TNF-α in disease development [40], modulating its activity has been intensely studied in large animal models and patients. Although TNF-α signaling was shown to be highly detrimental in experimental heart failure, clinical studies failed to reproduce these observations. The RENEWAL (Randomized Etanercept Worldwide Evaluation) trial, aiming the administration of a TNF-α decoy receptor in patients diagnosed with congestive heart failure, found that the strategy did not improve clinical outcome, but unexpectedly there was a trend toward a harmful effect of the therapy [41]. Consistently, ATTACH (Anti-TNF therapy against congestive heart failure) study demonstrated that high doses of a chimeric monoclonal antibody, infliximab adversely affected the clinical parameters of treated patients and increased the risk of hospitalization [42]. These effects could be explained by infliximab-induced cytotoxicity, observed in other cell types [43].

Dysfunctional IL-1 signaling contributes to cardiac dysfunction through various mechanisms. IL-1 is a main mediator in both acute and chronic stages of inflammation and induces the expression of further secondary cytokines [44]. Similarly, IL-1α induces a deleterious effect on mitochondrial function in cardiomyocytes [45]. Anakinra, an IL-1 receptor antagonist improved coronary flow and contractility in patients with rheumatoid arthritis [46]. Likewise, IL-1 inhibition by subcutaneous administration of anakinra over a period of 12 weeks led to a significant increase in peak V_{O2} in patients with decompensated heart failure [47]. Importantly, the rate of hospitalization was also decreased in patients receiving the antagonist [47]. Similarly, the administration of a human antibody targeting IL-1β (canakinumab) as an adjuvant therapy to statins significantly reduced the risk of cardiovascular death by 15% as compared to placebo (CANTOS clinical trial) [48]. Interestingly, CANTOS study proved that patients who achieved a reduction of C-reactive protein (CRP) below 2 mg/L benefited from the designed treatment, while those who presented with higher values of CRP did not. In addition, exploratory analyses of the CANTOS study pointed out that patients with mutations in *TET2* gene present not only with increased risk of cardiovascular disease, but also with a higher response to canakinumab therapy [49].

Other immunomodulatory-based therapies have been recently described. One of the most important targets for such approaches is C-chemokine receptor type 2–C-chemokine ligand 2 (CCR2-CCL2) axis, specifically affecting monocyte infiltration into the myocardium. Most well-established examples include the use of CCR2 antagonists [50, 51] or antibodies against CCR2 [51]. Despite positive results in preclinical trials, so far no CCR2-modulating approach has been approved for use in patients.

Several other non-specific drugs with anti-inflammatory effects have been tested within Phase III clinical trials for patients with heart failure with reduced ejection fraction, including statins, oxypurinol, and n-1 polyunsaturated fatty acids [52]. Additionally, in case of autoimmune cardiac diseases, administration of peptides binding to autoantibodies against β-adrenergic receptor showed a beneficial effect in several studies; however, Phase III randomized clinical trials are pending [53].

Furthermore, other classes of anti-inflammatory drugs proven effective for other diseases have been repurposed for the treatment of heart failure and tested in various small clinical trials. Low-dose colchicine was demonstrated to reduce the chance of death from cardiovascular complications and decrease the hospitalization rate after myocardial infarction (COLOT trial) [54]. Interestingly, colchicine has been known for decades to improve left ventricular function in a large animal model of pressure overload [55]. Mechanistically, colchicine application targets the increased microtubule network density observed in hypertrophic cardiomyocytes under pressure overload conditions by microtubule depolymerization [55, 56].

4 Gene Therapy

Despite the approval of more than 20 gene therapy-based products on the market for other diseases [57] and development of a multitude of gene therapy clinical trials, at the present moment there is no available option for cardiac complications. However, with a wide variety of targets recently identified, gene therapy can still be a valuable strategy for the treatment of cardiovascular complications and heart failure. A list of gene therapy targets is depicted in Fig. 1.

Adeno-associated viruses (AAVs) represent the gold standard delivery approach for cardiomyocyte gene transfer. In strong contrast to adenovirus-based vectors [58], AAVs offer long-term, non-immunogenic gene expression. Additionally, recent studies focused on the design of highly specific AAV capsids and promoters with effective liver de-targeting [59, 60]. This is of major importance, taking into account that high-dose AAV-mediated liver expression of transgenes can trigger activation of an immune response and hence hepatic toxicity [61–63]. Another approach that led to the development of cardiac-specific AAV variants is the choice of a selective regulatory sequence that provides cardiomyocyte gene transfer with reduced liver transduction [64, 65]. On the other hand, cardiac-specific promoters might exert reduced efficiency during disease progression, as fetal gene expression is reactivated [66]. This challenge could be resolved by using a heart failure-specific B-type natriuretic peptide promoter with low levels of expression under basal conditions but increased under pathological conditions [67]. As no reports are available on

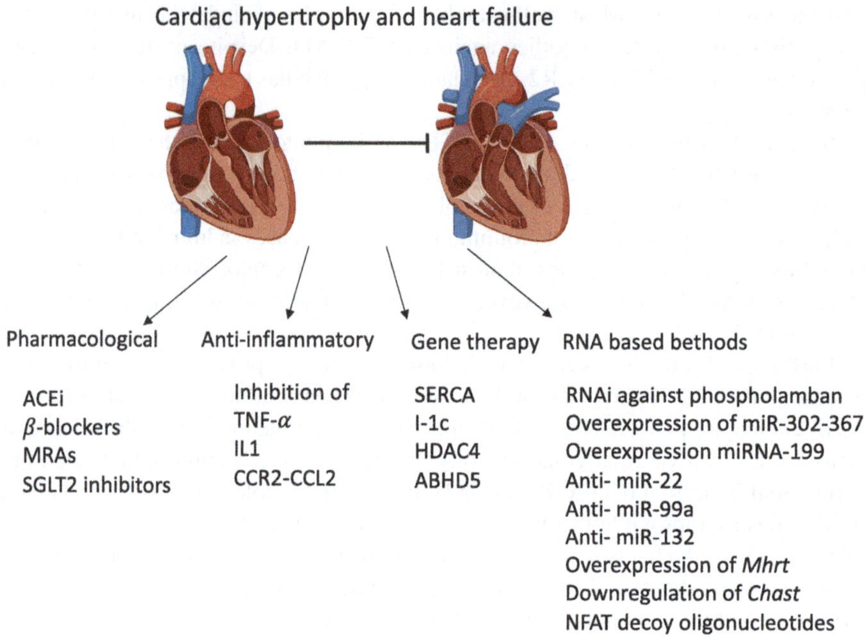

Fig. 1 Cardiac mechanobiology in physiology

lacking or reduced efficiency of promoter sequences currently used for cardiac gene in heart failure models, downregulation of promoters remains rather still a theoretical consideration, which however should be addressed experimentally in future studies. Along these lines, sustained overexpression of distinct proteins by constitutively active promoters in cardiomyocytes may cause stress in the endoplasmic reticulum and unfolded protein response as well as the autophagy lysosomal pathway or other protein degrading pathways resulting in unexpected side effects. An impressive example of the potentially deleterious consequences of sustained therapeutic gene expression was recently reported: persistent overexpression of microRNA-199a in a porcine model of cardiac ischemia resulted in increased mortality due to arrhythmia after initial beneficial effects [68]. Thus, novel regulatory systems are required. For example, drug-inducible promoters may be developed, enabling discontinuation of gene expression in case of unwanted side effects.

The route of administration is another critical factor influencing the degree of cardiomyocyte transduction. The most common and direct strategy is represented by intramyocardial injection or by less invasive percutaneous catheter-based application methods ranging from intracoronary injection to retrograde injection via the cardiac veins with transient block of arterial flow, which are less invasive and hence more easily translated into clinical use [69, 70]. Most preclinical trials performed in animal models focus on prophylactic AAV injection in healthy controls in order to allow efficient transgene expression prior to disease induction. However, transition to heart failure involves deposition of extracellular matrix in perivascular and interstitial areas, which might hinder cardiomyocyte transduction. Therefore, additional experiments assessing the efficiency of gene delivery in diseased myocardium should be performed for translation purposes. In addition, extracellular matrix deposition developed during transition to heart failure might hinder the degree of AAV transduction in target cardiac cells. Hence, gene therapy approaches focusing on reducing fibroblast proliferation is a valuable strategy to improve myocardial contractility and reduce phenotypic changes associated with decompensated heart failure [71, 72].

The first Phase I/II clinical trial for heart failure was based on the upregulation of SERCA2a by adeno-associated virus 1 (AAV1) delivery into cardiac tissue (CUPID study [73]) and showed that this approach is safe. It opened the path to a Phase II, blinded, randomized multicenter trial involving patients with advanced heart failure [74]. The trial proved that the intracoronary administration of a high dose of SERCA2a-overexpressing AAV1 can improve cardiac function and time of hospitalization in the 39 patients enrolled [74], hence setting up the basis for a larger trial (CUPID 2), including 250 patients with heart failure of predominantly ischemic cause, but in 4% also caused by hypertension [75]. Intriguingly, CUPID 2 failed to meet its primary endpoint and there was no evidence of improvement in the clinical course of patients diagnosed with severe heart failure with both ischemic and non-ischemic etiologies [75]. The main reason behind this observation could be the low transduction efficiency within the cardiac tissue following AAV1 transduction [75]. Regardless of the negative results, the CUPID clinical trial points out the crucial issue of efficiency of AAV delivery into the myocardium and underlines the importance of the choice of therapeutic target to be overexpressed [76].

A recently started clinical trial, NAN-CS101 is addressing the effects of overexpression of a constitutively active form of inhibitor-1 c (I-1c) in patients with class II heart failure (NCT04179643). In failing myocardium, I-1 levels are significantly reduced, accordingly leading to diminished SERCA2a activity by increased phospholamban phosphorylation [77]. Consistently, AAV-mediated gene transfer of I-1 in preclinical models of heart failure led to rescue of cardiac phenotype, improved contractility, and reduced cardiomyocyte apoptosis [78–80]. Although mice with either transgenic or adenoviral overexpression of I-1c resulted in improved left ventricular function when subjected to pressure overload [81], AAV-mediated expression of I-1c in mice subjected to transverse aortic constriction (TAC) resulted in even more cardiac hypertrophy and worsening of left ventricular function [82]. However, expression of I-1c using lower doses of AAV vectors in rats with pressure overload induced heart failure, improved left ventricular function, and prevented remodeling [83]. Thus, there remains a challenge of identifying the appropriate dose at least for an application in heart failure due to increased afterload.

Overexpression of the N-terminal fragment of histone deacetylase 4 (HDAC4), derived by proteolytical cleavage by cAMP-dependent protein kinase A has been proven to exert beneficial effects on cardiac function under pathological stress [84]. Mechanistically, the peptide was shown to inhibit the activity of MEF2 without affecting other pro-hypertrophic transcription factors. Interestingly, mice subjected to physiological stress presented with significantly higher levels of proteolytically cleaved HDAC4 [84]. Similar results were obtained through AAV-mediated overexpression of abhydrolase domain containing 5 (ABHD5), a serine protease that cleaves HDAC4 leading to production of the protective N-terminal polypeptide of the histone deacetylase [85]. Specifically, reintroduction of ABHD5 in cardiomyocytes prevented TAC-induced cardiac hypertrophy and heart failure in mice [85].

5 RNA-Based Methods for Heart Failure

Understanding the structure and role of RNAs within a cell has dramatically changed during the last decade, making them powerful targets for the treatment of cardiac complications. RNA-based strategies include RNA interference (RNAi), aiming to silence particular genes, and miRNA modulation, which alters complex gene expression pattern.

Phospholamban downregulation by RNAi was proven to improve cardiac function in the aortic banding mouse model of heart failure [86]. Treated animals presented with reduced left ventricular hypertrophy and decreased extracellular matrix deposition compared to controls [86].

Over the last few years, miRNAs have been proven to be powerful tools for induction of cardiac regeneration and hence recovery of its function after a pathological insult such as myocardial infarction. Indeed, miR-302-367 cluster, physiologically expressed in embryonic stem cells and playing a major role in maintaining their pluripotency, effectively induced cardiomyocyte proliferation in mice [87]. Additionally, other miRNA clusters, first identified to lead to cancer cell proliferation [88], are required and sufficient to trigger adult heart regeneration and blunt cardiac dysfunction in a mouse model of myocardial infarction [89]. An elegant method to deliver therapeutic miRNAs is by means of AAV vector transduction, which shows high efficiency and cardiac tropism. AAV-mediated delivery of miRNA-199 in mice [90] and pigs [68] successfully triggered cardiac regeneration and markedly improved cardiac function in these models. A significant challenge of this approach is the persistent expression of a pro-regenerative molecule, which could lead to abnormal proliferation of cardiac myocytes and hence development of arrhythmias as shown by overexpression of microRNA-199a in a porcine model of cardiac ischemia. As discussed above, AAV-mediated overexpression of microRNA-199a was associated with increased mortality due to arrhythmia after initial beneficial effects on cardiac function [68]. To overcome the problem of sustained gene expression with viral vectors, various lipid nanoformulations were employed to deliver miRNAs and non-coding RNAs, which could be delivered to the myocardial tissue through percutaneous coronary intervention or by direct myocardial injection. While all the aforementioned approaches have been studied in rodent and large animal models, there is still an unmet need to establish the feasibility and efficacy of nanoparticle delivery into the clinical situation.

Other miRNA clusters have been shown to be associated with several cardiovascular complications, including heart failure, and may be used as therapeutic targets. Recently, a wide range of oligonucleotides blocking miRNA function has been established and evaluated in preclinical models. Specifically, these sets of miRNAs are implicated in different cellular mechanisms driving cardiac disease, such as apoptosis, necroptosis, cellular hypertrophy, and autophagy [91]. For instance, treatment with miR-22 anti-miRs triggers autophagy in cardiomyocytes, therefore preventing remodeling in case of myocardial damage after infarction [92]. Similarly, prompting autophagy by miR-99a overexpression led to an increase in mTOR pathway activation and hence improved cardiac function and survival in a large animal model for myocardial infarction [93]. Likewise, inhibition of miR-132 family rescued cardiac function and decreased left ventricular hypertrophy in transverse aortic constriction model in mice [94]. Interestingly, the response to some miRNA therapy has been demonstrated to be gender specific, which needs to be taken into account for further translation into patients [95].

Long non-coding RNAs (lncRNAs) have emerged as valuable molecules involved in cardiomyocyte physiology and pathophysiology [96] (Fig. 1). For example, *Mhrt,* a cluster of lncRNA belonging to Myh7 loci has a cardioprotective role under stress, and restoring its levels in a mouse model for pressure overload-induced hypertrophy blunts pathological myocardial remodeling [97]. Furthermore, cardiomyocyte-specific overexpression of the lncRNA repressor of NFAT (*NRON*) in vitro and in vivo led to exacerbation of the pro-hypertrophic response, while its deletion led to normalization of the phenotype [98]. Additionally, lncRNA profiling enabled the characterization of cardiac hypertrophy-associated transcript (*Chast*) as a key molecular player in pathological cardiac hypertrophy. Chast expression was found to be significantly upregulated upon TAC, whereas its viral-overexpression was sufficient to prompt cardiomyocyte hypertrophy by impairing autophagy [99]. LncRNAs were demonstrated to be deregulated not only in cardiomyocytes, but also in cardiac fibroblasts upon pressure overload-induced hypertrophy, leading to cellular activation, enhanced proliferation, and extracellular matrix deposition [100].

Recent studies point out that RNA-based technology can be successfully used to target abnormal activation of transcription factors involved in induction of cardiac hypertrophy. AAV9-mediated expression of a hairpin RNA containing the consensus binding site of NFAT markedly reduced transcription factor activation, leading to decreased cardiomyocyte hypertrophy in vitro and in vivo [101]. This approach could potentially be translated for targeting other transcription factors involved in pathological remodeling under pressure overload.

6 Conclusion and Future Perspectives

Considering the limited efficiency of current therapies for tackling heart failure due to pressure overload, there is a high need to identify novel target structures and develop appropriate therapeutic strategies. Actual pharmacological approaches indeed decrease the severity of symptoms in affected patients; however, the mortality and hospitalization rates remain high. Thus, it is promising that current preclinical trials have identified novel molecular players that may constitute future therapeutic strategies.

Additionally, gene therapy approaches remain immensely valuable perspectives for the treatment of heart failure. Regardless of the negative outcomes pointed out by CUPID 2 trial, we learned important lessons regarding tissue transduction efficiency, mode of AAV delivery, and choice of the therapeutic target to be investigated. Moreover, preclinical studies aim to establish novel AAV variants with improved cardiomyocyte transduction and immune evasion.

Funding

This work is supported by the EU-Project Cardio-ReGenix to OJM.

Conflict of Interest Authors declare that they have no conflict of interest.

Ethical Approval This article does not contain any studies with human participants or animals performed by any of the authors.

References

1. Miranda-Silva D, Lima T, Rodrigues P, Leite-Moreira A, Falcao-Pires I (2021) Mechanisms underlying the pathophysiology of heart failure with preserved ejection fraction: the tip of the iceberg. Heart Fail Rev 26(3):453–478. https://doi.org/10.1007/s10741-020-10042-0
2. Razeghi P, Young ME, Alcorn JL, Moravec CS, Frazier OH, Taegtmeyer H (2001) Metabolic gene expression in fetal and failing human heart. Circulation 104(24):2923–2931. https://doi.org/10.1161/hc4901.100526
3. Kang PM, Izumo S (2000) Apoptosis and heart failure: a critical review of the literature. Circ Res 86(11):1107–1113. https://doi.org/10.1161/01.res.86.11.1107
4. Dirkx E, da Costa Martins PA (1832) De Windt LJ (2013) Regulation of fetal gene expression in heart failure. Biochim Biophys Acta 12:2414–2424. https://doi.org/10.1016/j.bbadis.2013.07.023
5. Carnevale D, Cifelli G, Mascio G, Madonna M, Sbroggio M, Perrino C, Persico MG, Frati G, Lembo G (2011) Placental growth factor regulates cardiac inflammation through the tissue inhibitor of metalloproteinases-3/tumor necrosis factor-alpha-converting enzyme axis: crucial role for adaptive cardiac remodeling during cardiac pressure overload. Circulation 124(12): 1337–1350. https://doi.org/10.1161/CIRCULATIONAHA.111.050500
6. Hartupee J, Mann DL (2017) Neurohormonal activation in heart failure with reduced ejection fraction. Nat Rev Cardiol 14(1):30–38. https://doi.org/10.1038/nrcardio.2016.163
7. Molkentin JD, Lu JR, Antos CL, Markham B, Richardson J, Robbins J, Grant SR, Olson EN (1998) A calcineurin-dependent transcriptional pathway for cardiac hypertrophy. Cell 93(2): 215–228. https://doi.org/10.1016/s0092-8674(00)81573-1
8. Olson EN, Molkentin JD (1999) Prevention of cardiac hypertrophy by calcineurin inhibition: hope or hype? Circ Res 84(6):623–632. https://doi.org/10.1161/01.res.84.6.623
9. Molkentin JD (2000) Calcineurin and beyond: cardiac hypertrophic signaling. Circ Res 87(9): 731–738. https://doi.org/10.1161/01.res.87.9.731

10. Akazawa H, Komuro I (2003) Roles of cardiac transcription factors in cardiac hypertrophy. Circ Res 92(10):1079–1088. https://doi.org/10.1161/01.RES.0000072977.86706.23
11. He Q, Mendez M, LaPointe MC (2002) Regulation of the human brain natriuretic peptide gene by GATA-4. Am J Physiol Endocrinol Metab 283(1):E50–E57. https://doi.org/10.1152/ajpendo.00274.2001
12. Morimoto T, Hasegawa K, Wada H, Kakita T, Kaburagi S, Yanazume T, Sasayama S (2001) Calcineurin-GATA4 pathway is involved in beta-adrenergic agonist-responsive endothelin-1 transcription in cardiac myocytes. J Biol Chem 276(37):34983–34989. https://doi.org/10.1074/jbc.M005498200
13. van Oort RJ, van Rooij E, Bourajjaj M, Schimmel J, Jansen MA, van der Nagel R, Doevendans PA, Schneider MD, van Echteld CJ, De Windt LJ (2006) MEF2 activates a genetic program promoting chamber dilation and contractile dysfunction in calcineurin-induced heart failure. Circulation 114(4):298–308. https://doi.org/10.1161/CIRCULATIONAHA.105.608968
14. Molkentin JD, Markham BE (1993) Myocyte-specific enhancer-binding factor (MEF-2) regulates alpha-cardiac myosin heavy chain gene expression in vitro and in vivo. J Biol Chem 268(26):19512–19520
15. Morin S, Charron F, Robitaille L, Nemer M (2000) GATA-dependent recruitment of MEF2 proteins to target promoters. EMBO J 19(9):2046–2055. https://doi.org/10.1093/emboj/19.9.2046
16. Blaeser F, Ho N, Prywes R, Chatila TA (2000) Ca(2+)-dependent gene expression mediated by MEF2 transcription factors. J Biol Chem 275(1):197–209. https://doi.org/10.1074/jbc.275.1.197
17. Kohli S, Ahuja S, Rani V (2011) Transcription factors in heart: promising therapeutic targets in cardiac hypertrophy. Curr Cardiol Rev 7(4):262–271. https://doi.org/10.2174/157340311799960618
18. Heineke J, Ritter O (2012) Cardiomyocyte calcineurin signaling in subcellular domains: from the sarcolemma to the nucleus and beyond. J Mol Cell Cardiol 52(1):62–73. https://doi.org/10.1016/j.yjmcc.2011.10.018
19. Bourajjaj M, Armand AS, da Costa Martins PA, Weijts B, van der Nagel R, Heeneman S, Wehrens XH, De Windt LJ (2008) NFATc2 is a necessary mediator of calcineurin-dependent cardiac hypertrophy and heart failure. J Biol Chem 283(32):22295–22303. https://doi.org/10.1074/jbc.M801296200
20. Rauchhaus M, Doehner W, Francis DP, Davos C, Kemp M, Liebenthal C, Niebauer J, Hooper J, Volk HD, Coats AJ, Anker SD (2000) Plasma cytokine parameters and mortality in patients with chronic heart failure. Circulation 102(25):3060–3067. https://doi.org/10.1161/01.cir.102.25.3060
21. Deswal A, Petersen NJ, Feldman AM, Young JB, White BG, Mann DL (2001) Cytokines and cytokine receptors in advanced heart failure: an analysis of the cytokine database from the Vesnarinone trial (VEST). Circulation 103(16):2055–2059. https://doi.org/10.1161/01.cir.103.16.2055
22. Mann DL (2002) Inflammatory mediators and the failing heart: past, present, and the foreseeable future. Circ Res 91(11):988–998. https://doi.org/10.1161/01.res.0000043825.01705.1b
23. Bradham WS, Bozkurt B, Gunasinghe H, Mann D, Spinale FG (2002) Tumor necrosis factor-alpha and myocardial remodeling in progression of heart failure: a current perspective. Cardiovasc Res 53(4):822–830. https://doi.org/10.1016/s0008-6363(01)00503-x
24. Forstermann U, Munzel T (2006) Endothelial nitric oxide synthase in vascular disease: from marvel to menace. Circulation 113(13):1708–1714. https://doi.org/10.1161/CIRCULATIONAHA.105.602532
25. Kruger M, Kotter S, Grutzner A, Lang P, Andresen C, Redfield MM, Butt E, dos Remedios CG, Linke WA (2009) Protein kinase G modulates human myocardial passive stiffness by phosphorylation of the titin springs. Circ Res 104(1):87–94. https://doi.org/10.1161/CIRCRESAHA.108.184408

26. Fraccarollo D, Widder JD, Galuppo P, Thum T, Tsikas D, Hoffmann M, Ruetten H, Ertl G, Bauersachs J (2008) Improvement in left ventricular remodeling by the endothelial nitric oxide synthase enhancer AVE9488 after experimental myocardial infarction. Circulation 118(8): 818–827. https://doi.org/10.1161/CIRCULATIONAHA.107.717702
27. Tromp J, Ouwerkerk W, van Veldhuisen DJ, Hillege HL, Richards AM, van der Meer P, Anand IS, Lam CSP, Voors AA (2021) A systematic review and network-meta-analysis of pharmacological treatment of heart failure with reduced ejection fraction. JACC Heart Fail 10: 73. https://doi.org/10.1016/j.jchf.2021.09.004
28. Investigators S, Yusuf S, Pitt B, Davis CE, Hood WB, Cohn JN (1991) Effect of enalapril on survival in patients with reduced left ventricular ejection fractions and congestive heart failure. N Engl J Med 325(5):293–302. https://doi.org/10.1056/NEJM199108013250501
29. The Cardiac Insufficiency Bisoprolol Study II (CIBIS-II) (1999) A randomised trial. Lancet 353(9146):9–13
30. McDonagh TA, Metra M, Adamo M, Gardner RS, Baumbach A, Bohm M, Burri H, Butler J, Celutkiene J, Chioncel O, Cleland JGF, Coats AJS, Crespo-Leiro MG, Farmakis D, Gilard M, Heymans S, Hoes AW, Jaarsma T, Jankowska EA, Lainscak M, Lam CSP, Lyon AR, McMurray JJV, Mebazaa A, Mindham R, Muneretto C, Francesco Piepoli M, Price S, Rosano GMC, Ruschitzka F, Kathrine Skibelund A (2021) Corrigendum to: 2021 ESC guidelines for the diagnosis and treatment of acute and chronic heart failure: developed by the task force for the diagnosis and treatment of acute and chronic heart failure of the European Society of Cardiology (ESC) with the special contribution of the heart failure association (HFA) of the ESC. Eur Heart J 42(48):4901. https://doi.org/10.1093/eurheartj/ehab670
31. Anon (2000) Major cardiovascular events in hypertensive patients randomized to doxazosin vs chlorthalidone: the antihypertensive and lipid-lowering treatment to prevent heart attack trial (ALLHAT). ALLHAT collaborative research group. JAMA 283(15):1967–1975
32. Durstenfeld MS, Katz SD, Park H, Blecker S (2019) Mineralocorticoid receptor antagonist use after hospitalization of patients with heart failure and post-discharge outcomes: a single-center retrospective cohort study. BMC Cardiovasc Disord 19(1):194. https://doi.org/10.1186/s12872-019-1175-3
33. Komajda M, Bohm M, Borer JS, Ford I, Tavazzi L, Pannaux M, Swedberg K (2018) Incremental benefit of drug therapies for chronic heart failure with reduced ejection fraction: a network meta-analysis. Eur J Heart Fail 20(9):1315–1322. https://doi.org/10.1002/ejhf.1234
34. Li X, Braza J, Mende U, Choudhary G, Zhang P (2021) Cardioprotective effects of early intervention with sacubitril/valsartan on pressure overloaded rat hearts. Sci Rep 11(1):16542. https://doi.org/10.1038/s41598-021-95988-3
35. Norden ES, Bendiksen BA, Andresen H, Bergo KK, Espe EK, Hasic A, Hauge-Iversen IM, Veras I, Hussain RI, Sjaastad I, Christensen G, Cataliotti A (2021) Sacubitril/valsartan ameliorates cardiac hypertrophy and preserves diastolic function in cardiac pressure overload. ESC Heart Fail 8(2):918–927. https://doi.org/10.1002/ehf2.13177
36. Packer M, Anker SD, Butler J, Filippatos G, Pocock SJ, Carson P, Januzzi J, Verma S, Tsutsui H, Brueckmann M, Jamal W, Kimura K, Schnee J, Zeller C, Cotton D, Bocchi E, Bohm M, Choi DJ, Chopra V, Chuquiure E, Giannetti N, Janssens S, Zhang J, Gonzalez Juanatey JR, Kaul S, Brunner-La Rocca HP, Merkely B, Nicholls SJ, Perrone S, Pina I, Ponikowski P, Sattar N, Senni M, Seronde MF, Spinar J, Squire I, Taddei S, Wanner C, Zannad F, Investigators EM-RT (2020) Cardiovascular and renal outcomes with Empagliflozin in heart failure. N Engl J Med 383(15):1413–1424. https://doi.org/10.1056/NEJMoa2022190
37. McMurray JJV, Solomon SD, Inzucchi SE, Kober L, Kosiborod MN, Martinez FA, Ponikowski P, Sabatine MS, Anand IS, Belohlavek J, Bohm M, Chiang CE, Chopra VK, de Boer RA, Desai AS, Diez M, Drozdz J, Dukat A, Ge J, Howlett JG, Katova T, Kitakaze M, Ljungman CEA, Merkely B, Nicolau JC, O'Meara E, Petrie MC, Vinh PN, Schou M, Tereshchenko S, Verma S, Held C, DeMets DL, Docherty KF, Jhund PS, Bengtsson O, Sjostrand M, Langkilde AM, Committees D-HT, Investigators (2019) Dapagliflozin in

patients with heart failure and reduced ejection fraction. N Engl J Med 381(21):1995–2008. https://doi.org/10.1056/NEJMoa1911303
38. Byrne NJ, Parajuli N, Levasseur JL, Boisvenue J, Beker DL, Masson G, Fedak PWM, Verma S, Dyck JRB (2017) Empagliflozin prevents worsening of cardiac function in an experimental model of pressure overload-induced heart failure. JACC Basic Transl Sci 2(4): 347–354. https://doi.org/10.1016/j.jacbts.2017.07.003
39. Li X, Lu Q, Qiu Y, do Carmo JM, Wang Z, da Silva AA, Mouton A, ACM O, Hall ME, Li J, Hall JE (2021) Direct cardiac actions of the sodium glucose co-transporter 2 inhibitor Empagliflozin improve myocardial oxidative phosphorylation and attenuate pressure-overload heart failure. J Am Heart Assoc 10(6):e018298. https://doi.org/10.1161/JAHA.120.018298
40. Schumacher SM, Naga Prasad SV (2018) Tumor necrosis factor-alpha in heart failure: an updated review. Curr Cardiol Rep 20(11):117. https://doi.org/10.1007/s11886-018-1067-7
41. Mann DL, McMurray JJ, Packer M, Swedberg K, Borer JS, Colucci WS, Djian J, Drexler H, Feldman A, Kober L, Krum H, Liu P, Nieminen M, Tavazzi L, van Veldhuisen DJ, Waldenstrom A, Warren M, Westheim A, Zannad F, Fleming T (2004) Targeted anticytokine therapy in patients with chronic heart failure: results of the randomized Etanercept worldwide evaluation (RENEWAL). Circulation 109(13):1594–1602. https://doi.org/10.1161/01.CIR.0000124490.27666.B2
42. Chung ES, Packer M, Lo KH, Fasanmade AA, Willerson JT, Anti TNFTACHFI (2003) Randomized, double-blind, placebo-controlled, pilot trial of infliximab, a chimeric monoclonal antibody to tumor necrosis factor-alpha, in patients with moderate-to-severe heart failure: results of the anti-TNF therapy against congestive heart failure (ATTACH) trial. Circulation 107(25):3133–3140. https://doi.org/10.1161/01.CIR.0000077913.60364.D2
43. Mitoma H, Horiuchi T, Tsukamoto H, Tamimoto Y, Kimoto Y, Uchino A, Kentaro T, Harashima S, Hatta N, Harada M (2008) Mechanisms for cytotoxic effects of anti-tumor necrosis factor agents on transmembrane tumor necrosis factor alpha-expressing cells: comparison among infliximab, etanercept, and adalimumab. Arthritis Rheum 58(5):1248–1257. https://doi.org/10.1002/art.23447
44. Van Tassell BW, Raleigh JM, Abbate A (2015) Targeting interleukin-1 in heart failure and inflammatory heart disease. Curr Heart Fail Rep 12(1):33–41. https://doi.org/10.1007/s11897-014-0231-7
45. Zell R, Geck P, Werdan K, Boekstegers P (1997) TNF-alpha and IL-1 alpha inhibit both pyruvate dehydrogenase activity and mitochondrial function in cardiomyocytes: evidence for primary impairment of mitochondrial function. Mol Cell Biochem 177(1–2):61–67. https://doi.org/10.1023/a:1006896832582
46. Ikonomidis I, Lekakis JP, Nikolaou M, Paraskevaidis I, Andreadou I, Kaplanoglou T, Katsimbri P, Skarantavos G, Soucacos PN, Kremastinos DT (2008) Inhibition of interleukin-1 by anakinra improves vascular and left ventricular function in patients with rheumatoid arthritis. Circulation 117(20):2662–2669. https://doi.org/10.1161/CIRCULATIONAHA.107.731877
47. Van Tassell BW, Canada J, Carbone S, Trankle C, Buckley L, Oddi Erdle C, Abouzaki NA, Dixon D, Kadariya D, Christopher S, Schatz A, Regan J, Viscusi M, Del Buono M, Melchior R, Mankad P, Lu J, Sculthorpe R, Biondi-Zoccai G, Lesnefsky E, Arena R, Abbate A (2017) Interleukin-1 blockade in recently decompensated systolic heart failure: results from REDHART (recently decompensated heart failure Anakinra response trial). Circ Heart Fail 10(11):e004373. https://doi.org/10.1161/CIRCHEARTFAILURE.117.004373
48. Everett BM, Cornel JH, Lainscak M, Anker SD, Abbate A, Thuren T, Libby P, Glynn RJ, Ridker PM (2019) Anti-inflammatory therapy with Canakinumab for the prevention of hospitalization for heart failure. Circulation 139(10):1289–1299. https://doi.org/10.1161/CIRCULATIONAHA.118.038010
49. Yura Y, Sano S, Walsh K (2020) Clonal hematopoiesis: a new step linking inflammation to heart failure. JACC Basic Transl Sci 5(2):196–207. https://doi.org/10.1016/j.jacbts.2019.08.006

50. Liao X, Shen Y, Zhang R, Sugi K, Vasudevan NT, Alaiti MA, Sweet DR, Zhou L, Qing Y, Gerson SL, Fu C, Wynshaw-Boris A, Hu R, Schwartz MA, Fujioka H, Richardson B, Cameron MJ, Hayashi H, Stamler JS, Jain MK (2018) Distinct roles of resident and nonresident macrophages in nonischemic cardiomyopathy. Proc Natl Acad Sci USA 115(20):E4661–E4669. https://doi.org/10.1073/pnas.1720065115
51. Patel B, Bansal SS, Ismahil MA, Hamid T, Rokosh G, Mack M, Prabhu SD (2018) CCR2(+) monocyte-derived infiltrating macrophages are required for adverse cardiac remodeling during pressure overload. JACC Basic Transl Sci 3(2):230–244. https://doi.org/10.1016/j.jacbts.2017.12.006
52. Adamo L, Rocha-Resende C, Prabhu SD, Mann DL (2020) Reappraising the role of inflammation in heart failure. Nat Rev Cardiol 17(5):269–285. https://doi.org/10.1038/s41569-019-0315-x
53. Mann DL (2001) Autoimmunity, immunoglobulin adsorption and dilated cardiomyopathy: has the time come for randomized clinical trials? J Am Coll Cardiol 38(1):184–186. https://doi.org/10.1016/s0735-1097(01)01310-9
54. Tardif JC, Kouz S, Waters DD, Bertrand O, Diaz R, Maggioni AP, Pinto FJ, Ibrahim R, Gamra H, Kiwan GS, Berry C, Lopez-Sendon J, Ostadal P, Koenig W, Angoulvant D, Gregoire JC, Lavoie MA, Dube MP, Rhainds D, Provencher M, Blondeau L, Orfanos A, L'Allier PL, Guertin MC, Roubille F (2019) Efficacy and safety of low-dose colchicine after myocardial infarction. N Engl J Med 381(26):2497–2505. https://doi.org/10.1056/NEJMoa1912388
55. Koide M, Hamawaki M, Narishige T, Sato H, Nemoto S, DeFreyte G, Zile MR, Cooper GI, Carabello BA (2000) Microtubule depolymerization normalizes in vivo myocardial contractile function in dogs with pressure-overload left ventricular hypertrophy. Circulation 102(9): 1045–1052. https://doi.org/10.1161/01.cir.102.9.1045
56. Zile MR, Koide M, Sato H, Ishiguro Y, Conrad CH, Buckley JM, Morgan JP, Gt C (1999) Role of microtubules in the contractile dysfunction of hypertrophied myocardium. J Am Coll Cardiol 33(1):250–260. https://doi.org/10.1016/s0735-1097(98)00550-6
57. Shahryari A, Saghaeian Jazi M, Mohammadi S, Razavi Nikoo H, Nazari Z, Hosseini ES, Burtscher I, Mowla SJ, Lickert H (2019) Development and clinical translation of approved gene therapy products for genetic disorders. Front Genet 10:868. https://doi.org/10.3389/fgene.2019.00868
58. St George JA (2003) Gene therapy progress and prospects: adenoviral vectors. Gene Ther 10(14):1135–1141. https://doi.org/10.1038/sj.gt.3302071
59. Grimm D, Buning H (2017) Small but increasingly mighty: latest advances in AAV vector research, design, and evolution. Hum Gene Ther 28(11):1075–1086. https://doi.org/10.1089/hum.2017.172
60. Geisler A, Jungmann A, Kurreck J, Poller W, Katus HA, Vetter R, Fechner H, Muller OJ (2011) microRNA122-regulated transgene expression increases specificity of cardiac gene transfer upon intravenous delivery of AAV9 vectors. Gene Ther 18(2):199–209. https://doi.org/10.1038/gt.2010.141
61. Buning H, Schmidt M (2015) Adeno-associated vector toxicity-to be or not to be? Mol Ther 23(11):1673–1675. https://doi.org/10.1038/mt.2015.182
62. Sun CP, Wu TH, Chen CC, Wu PY, Shih YM, Tsuneyama K, Tao MH (2013) Studies of efficacy and liver toxicity related to adeno-associated virus-mediated RNA interference. Hum Gene Ther 24(8):739–750. https://doi.org/10.1089/hum.2012.239
63. Anon (2020) High-dose AAV gene therapy deaths. Nat Biotechnol 38(8):910. https://doi.org/10.1038/s41587-020-0642-9
64. Domenger C, Grimm D (2019) Next-generation AAV vectors-do not judge a virus (only) by its cover. Hum Mol Genet 28(R1):R3–R14. https://doi.org/10.1093/hmg/ddz148
65. Werfel S, Jungmann A, Lehmann L, Ksienzyk J, Bekeredjian R, Kaya Z, Leuchs B, Nordheim A, Backs J, Engelhardt S, Katus HA, Muller OJ (2014) Rapid and highly efficient

inducible cardiac gene knockout in adult mice using AAV-mediated expression of Cre recombinase. Cardiovasc Res 104(1):15–23. https://doi.org/10.1093/cvr/cvu174
66. van der Pol A, Hoes MF, de Boer RA, van der Meer P (2020) Cardiac foetal reprogramming: a tool to exploit novel treatment targets for the failing heart. J Intern Med 288(5):491–506. https://doi.org/10.1111/joim.13094
67. Miyazaki Y, Ikeda Y, Shiraishi K, Fujimoto SN, Aoyama H, Yoshimura K, Inui M, Hoshijima M, Kasahara H, Aoki H, Matsuzaki M (2012) Heart failure-inducible gene therapy targeting protein phosphatase 1 prevents progressive left ventricular remodeling. PLoS One 7(4):e35875. https://doi.org/10.1371/journal.pone.0035875
68. Gabisonia K, Prosdocimo G, Aquaro GD, Carlucci L, Zentilin L, Secco I, Ali H, Braga L, Gorgodze N, Bernini F, Burchielli S, Collesi C, Zandona L, Sinagra G, Piacenti M, Zacchigna S, Bussani R, Recchia FA, Giacca M (2019) MicroRNA therapy stimulates uncontrolled cardiac repair after myocardial infarction in pigs. Nature 569(7756):418–422. https://doi.org/10.1038/s41586-019-1191-6
69. Sherman W, Martens TP, Viles-Gonzalez JF, Siminiak T (2006) Catheter-based delivery of cells to the heart. Nat Clin Pract Cardiovasc Med 3(Suppl 1):S57–S64. https://doi.org/10.1038/ncpcardio0446
70. Tilemann L, Ishikawa K, Weber T, Hajjar RJ (2012) Gene therapy for heart failure. Circ Res 110(5):777–793. https://doi.org/10.1161/CIRCRESAHA.111.252981
71. Francisco J, Zhang Y, Jeong JI, Mizushima W, Ikeda S, Ivessa A, Oka S, Zhai P, Tallquist MD, Del Re DP (2020) Blockade of fibroblast YAP attenuates cardiac fibrosis and dysfunction through MRTF-A inhibition. JACC Basic Transl Sci 5(9):931–945. https://doi.org/10.1016/j.jacbts.2020.07.009
72. Yoo SY, Jeong SN, Kang JI, Lee SW (2018) Chimeric adeno-associated virus-mediated cardiovascular reprogramming for ischemic heart disease. ACS Omega 3(5):5918–5925. https://doi.org/10.1021/acsomega.8b00904
73. Jaski BE, Jessup ML, Mancini DM, Cappola TP, Pauly DF, Greenberg B, Borow K, Dittrich H, Zsebo KM, Hajjar RJ, Calcium Up-Regulation by Percutaneous Administration of Gene Therapy In Cardiac Disease Trial I (2009) Calcium upregulation by percutaneous administration of gene therapy in cardiac disease (CUPID trial), a first-in-human phase 1/2 clinical trial. J Card Fail 15(3):171–181. https://doi.org/10.1016/j.cardfail.2009.01.013
74. Jessup M, Greenberg B, Mancini D, Cappola T, Pauly DF, Jaski B, Yaroshinsky A, Zsebo KM, Dittrich H, Hajjar RJ, Calcium Upregulation by Percutaneous Administration of Gene Therapy in Cardiac Disease I (2011) Calcium Upregulation by percutaneous Administration of Gene Therapy in cardiac disease (CUPID): a phase 2 trial of intracoronary gene therapy of sarcoplasmic reticulum Ca2+-ATPase in patients with advanced heart failure. Circulation 124(3):304–313. https://doi.org/10.1161/CIRCULATIONAHA.111.022889
75. Greenberg B, Butler J, Felker GM, Ponikowski P, Voors AA, Desai AS, Barnard D, Bouchard A, Jaski B, Lyon AR, Pogoda JM, Rudy JJ, Zsebo KM (2016) Calcium upregulation by percutaneous administration of gene therapy in patients with cardiac disease (CUPID 2): a randomised, multinational, double-blind, placebo-controlled, phase 2b trial. Lancet 387(10024):1178–1186. https://doi.org/10.1016/S0140-6736(16)00082-9
76. Kieserman JM, Myers VD, Dubey P, Cheung JY, Feldman AM (2019) Current landscape of heart failure gene therapy. J Am Heart Assoc 8(10):e012239. https://doi.org/10.1161/JAHA.119.012239
77. Wittkopper K, Dobrev D, Eschenhagen T, El-Armouche A (2011) Phosphatase-1 inhibitor-1 in physiological and pathological beta-adrenoceptor signalling. Cardiovasc Res 91(3):392–401. https://doi.org/10.1093/cvr/cvr058
78. Chen G, Zhou X, Florea S, Qian J, Cai W, Zhang Z, Fan GC, Lorenz J, Hajjar RJ, Kranias EG (2010) Expression of active protein phosphatase 1 inhibitor-1 attenuates chronic beta-agonist-induced cardiac apoptosis. Basic Res Cardiol 105(5):573–581. https://doi.org/10.1007/s00395-010-0106-3

79. Nicolaou P, Rodriguez P, Ren X, Zhou X, Qian J, Sadayappan S, Mitton B, Pathak A, Robbins J, Hajjar RJ, Jones K, Kranias EG (2009) Inducible expression of active protein phosphatase-1 inhibitor-1 enhances basal cardiac function and protects against ischemia/reperfusion injury. Circ Res 104(8):1012–1020. https://doi.org/10.1161/CIRCRESAHA.108.189811
80. Watanabe S, Ishikawa K, Fish K, Oh JG, Motloch LJ, Kohlbrenner E, Lee P, Xie C, Lee A, Liang L, Kho C, Leonardson L, McIntyre M, Wilson S, Samulski RJ, Kranias EG, Weber T, Akar FG, Hajjar RJ (2017) Protein phosphatase Inhibitor-1 gene therapy in a swine model of nonischemic heart failure. J Am Coll Cardiol 70(14):1744–1756. https://doi.org/10.1016/j.jacc.2017.08.013
81. Pathak A, del Monte F, Zhao W, Schultz JE, Lorenz JN, Bodi I, Weiser D, Hahn H, Carr AN, Syed F, Mavila N, Jha L, Qian J, Marreez Y, Chen G, McGraw DW, Heist EK, Guerrero JL, DePaoli-Roach AA, Hajjar RJ, Kranias EG (2005) Enhancement of cardiac function and suppression of heart failure progression by inhibition of protein phosphatase 1. Circ Res 96(7):756–766. https://doi.org/10.1161/01.RES.0000161256.85833.fa
82. Schwab DM, Tilemann L, Bauer R, Heckmann M, Jungmann A, Wagner M, Burgis J, Vettel C, Katus HA, El-Armouche A, Muller OJ (2018) AAV-9 mediated phosphatase-1 inhibitor-1 overexpression improves cardiac contractility in unchallenged mice but is deleterious in pressure-overload. Gene Ther 25(1):13–19. https://doi.org/10.1038/gt.2017.97
83. Pritchard TJ, Kawase Y, Haghighi K, Anjak A, Cai W, Jiang M, Nicolaou P, Pylar G, Karakikes I, Rapti K, Rubinstein J, Hajjar RJ, Kranias EG (2013) Active inhibitor-1 maintains protein hyper-phosphorylation in aging hearts and halts remodeling in failing hearts. PLoS One 8(12):e80717. https://doi.org/10.1371/journal.pone.0080717
84. Lehmann LH, Jebessa ZH, Kreusser MM, Horsch A, He T, Kronlage M, Dewenter M, Sramek V, Oehl U, Krebs-Haupenthal J, von der Lieth AH, Schmidt A, Sun Q, Ritterhoff J, Finke D, Volkers M, Jungmann A, Sauer SW, Thiel C, Nickel A, Kohlhaas M, Schafer M, Sticht C, Maack C, Gretz N, Wagner M, El-Armouche A, Maier LS, Londono JEC, Meder B, Freichel M, Grone HJ, Most P, Muller OJ, Herzig S, Furlong EEM, Katus HA, Backs J (2018) A proteolytic fragment of histone deacetylase 4 protects the heart from failure by regulating the hexosamine biosynthetic pathway. Nat Med 24(1):62–72. https://doi.org/10.1038/nm.4452
85. Jebessa ZH, Shanmukha KD, Dewenter M, Lehmann LH, Xu C, Schreiter F, Siede D, Gong XM, Worst BC, Federico G, Sauer SW, Fischer T, Wechselberger L, Muller OJ, Sossalla S, Dieterich C, Most P, Grone HJ, Moro C, Oberer M, Haemmerle G, Katus HA, Tyedmers J, Backs J (2019) The lipid droplet-associated protein ABHD5 protects the heart through proteolysis of HDAC4. Nat Metab 1(11):1157–1167. https://doi.org/10.1038/s42255-019-0138-4
86. Suckau L, Fechner H, Chemaly E, Krohn S, Hadri L, Kockskamper J, Westermann D, Bisping E, Ly H, Wang X, Kawase Y, Chen J, Liang L, Sipo I, Vetter R, Weger S, Kurreck J, Erdmann V, Tschope C, Pieske B, Lebeche D, Schultheiss HP, Hajjar RJ, Poller WC (2009) Long-term cardiac-targeted RNA interference for the treatment of heart failure restores cardiac function and reduces pathological hypertrophy. Circulation 119(9):1241–1252. https://doi.org/10.1161/CIRCULATIONAHA.108.783852
87. Tian Y, Liu Y, Wang T, Zhou N, Kong J, Chen L, Snitow M, Morley M, Li D, Petrenko N, Zhou S, Lu M, Gao E, Koch WJ, Stewart KM, Morrisey EE (2015) A microRNA-hippo pathway that promotes cardiomyocyte proliferation and cardiac regeneration in mice. Sci Transl Med 7(279):279ra238. https://doi.org/10.1126/scitranslmed.3010841
88. Volinia S, Calin GA, Liu CG, Ambs S, Cimmino A, Petrocca F, Visone R, Iorio M, Roldo C, Ferracin M, Prueitt RL, Yanaihara N, Lanza G, Scarpa A, Vecchione A, Negrini M, Harris CC, Croce CM (2006) A microRNA expression signature of human solid tumors defines cancer gene targets. Proc Natl Acad Sci USA 103(7):2257–2261. https://doi.org/10.1073/pnas.0510565103

89. Chen J, Huang ZP, Seok HY, Ding J, Kataoka M, Zhang Z, Hu X, Wang G, Lin Z, Wang S, Pu WT (2013) Mir-17-92 cluster is required for and sufficient to induce cardiomyocyte proliferation in postnatal and adult hearts. Circ Res 7:12
90. Eulalio A, Mano M, Dal Ferro M, Zentilin L, Sinagra G, Zacchigna S, Giacca M (2012) Functional screening identifies miRNAs inducing cardiac regeneration. Nature 492(7429): 376–381. https://doi.org/10.1038/nature11739
91. Zhou SS, Jin JP, Wang JQ, Zhang ZG, Freedman JH, Zheng Y, Cai L (2018) miRNAS in cardiovascular diseases: potential biomarkers, therapeutic targets and challenges. Acta Pharmacol Sin 39(7):1073–1084. https://doi.org/10.1038/aps.2018.30
92. Gupta SK, Foinquinos A, Thum S, Remke J, Zimmer K, Bauters C, de Groote P, Boon RA, de Windt LJ, Preissl S, Hein L, Batkai S, Pinet F, Thum T (2016) Preclinical development of a MicroRNA-based therapy for elderly patients with myocardial infarction. J Am Coll Cardiol 68(14):1557–1571. https://doi.org/10.1016/j.jacc.2016.07.739
93. Li Q, Xie J, Li R, Shi J, Sun J, Gu R, Ding L, Wang L, Xu B (2014) Overexpression of microRNA-99a attenuates heart remodelling and improves cardiac performance after myocardial infarction. J Cell Mol Med 18(5):919–928. https://doi.org/10.1111/jcmm.12242
94. Ucar A, Gupta SK, Fiedler J, Erikci E, Kardasinski M, Batkai S, Dangwal S, Kumarswamy R, Bang C, Holzmann A, Remke J, Caprio M, Jentzsch C, Engelhardt S, Geisendorf S, Glas C, Hofmann TG, Nessling M, Richter K, Schiffer M, Carrier L, Napp LC, Bauersachs J, Chowdhury K, Thum T (2012) The miRNA-212/132 family regulates both cardiac hypertrophy and cardiomyocyte autophagy. Nat Commun 3:1078. https://doi.org/10.1038/ncomms2090
95. Bernardo BC, Ooi JY, Matsumoto A, Tham YK, Singla S, Kiriazis H, Patterson NL, Sadoshima J, Obad S, Lin RC, McMullen JR (2016) Sex differences in response to miRNA-34a therapy in mouse models of cardiac disease: identification of sex-, disease- and treatment-regulated miRNAs. J Physiol 594(20):5959–5974. https://doi.org/10.1113/JP272512
96. Hobuss L, Bar C, Thum T (2019) Long non-coding RNAs: at the heart of cardiac dysfunction? Front Physiol 10:30. https://doi.org/10.3389/fphys.2019.00030
97. Han P, Li W, Lin CH, Yang J, Shang C, Nuernberg ST, Jin KK, Xu W, Lin CY, Lin CJ, Xiong Y, Chien H, Zhou B, Ashley E, Bernstein D, Chen PS, Chen HV, Quertermous T, Chang CP (2014) A long noncoding RNA protects the heart from pathological hypertrophy. Nature 514(7520):102–106. https://doi.org/10.1038/nature13596
98. Hoepfner J, Leonardy J, Lu D, Schmidt K, Hunkler HJ, Biss S, Foinquinos A, Xiao K, Regalla K, Ramanujam D, Engelhardt S, Bar C, Thum T (2021) The long non-coding RNA NRON promotes the development of cardiac hypertrophy in the murine heart. Mol Ther 30: 1265. https://doi.org/10.1016/j.ymthe.2021.11.018
99. Viereck J, Kumarswamy R, Foinquinos A, Xiao K, Avramopoulos P, Kunz M, Dittrich M, Maetzig T, Zimmer K, Remke J, Just A, Fendrich J, Scherf K, Bolesani E, Schambach A, Weidemann F, Zweigerdt R, de Windt LJ, Engelhardt S, Dandekar T, Batkai S, Thum T (2016) Long noncoding RNA Chast promotes cardiac remodeling. Sci Transl Med 8(326):326ra322. https://doi.org/10.1126/scitranslmed.aaf1475
100. Piccoli MT, Gupta SK, Viereck J, Foinquinos A, Samolovac S, Kramer FL, Garg A, Remke J, Zimmer K, Batkai S, Thum T (2017) Inhibition of the cardiac fibroblast-enriched lncRNA Meg3 prevents cardiac fibrosis and diastolic dysfunction. Circ Res 121(5):575–583. https://doi.org/10.1161/CIRCRESAHA.117.310624
101. Remes A, Wagner AH, Schmiedel N, Heckmann M, Ruf T, Ding L, Jungmann A, Senger F, Katus HA, Ullrich ND, Frey N, Hecker M, Muller OJ (2021) AAV-mediated expression of NFAT decoy oligonucleotides protects from cardiac hypertrophy and heart failure. Basic Res Cardiol 116(1):38. https://doi.org/10.1007/s00395-021-00880-w

The manufacturer's authorised representative in the EU is Springer Nature Customer Service Centre GmbH, Europaplatz 3, 69115 Heidelberg, Germany. If you have any concerns regarding our products, please contact ProductSafety@springernature.com

Printed and bound by CPI Group (UK) Ltd, Croydon, CR0 4YY

28/09/2026

02224071-0017